37th Congress of the European Brewery Convention (EBC 2019)

Antwerp, Belgium
2-6 June 2019

ISBN: 978-1-5108-9758-8

Printed from e-media with permission by:

Curran Associates, Inc.
57 Morehouse Lane
Red Hook, NY 12571

CURRAN ASSOCIATES INC. proceedings.com

Some format issues inherent in the e-media version may also appear in this print version.

European Brewery Convention
c/o The Brewers of Europe
Rue Caroly 23-25
1050 Brussels
Belgium

Phone: 32 2 551 1810
Fax: 32 2 660 9402

info@brewersofeurope.org

Additional copies of this publication are available from:

Curran Associates, Inc.
57 Morehouse Lane
Red Hook, NY 12571 USA
Phone: 845-758-0400
Fax: 845-758-2633
Email: curran@proceedings.com
Web: www.proceedings.com

TABLE OF CONTENTS

ORAL PRESENTATIONS

POSTER SESSION

SCIENTIFIC AND TECHNICAL PROGRAMME – EBC 2019

MONDAY, 3 JUNE

PLENARY SESSION Queen Elisabeth Hall (Level 1)

Sustainability and Climate Change

Chair: Tiago Brandao (Project Director, The Browers Company, Porto, Portugal)

8h30 – 9h00 **L1:** Sustainable brewing of the future

Zoran Gojkovic[1]
1 - Carlsberg, Denmark

9h00 – 9h30 **L2:** Impact of the climate change on hops – exemplified by the Hallertau Growing Region.

Florian Schüll[1]; Adrian Forster[1]; Andreas Gahr[2]
1 - HVG Hopfenverwertungsgenossenschaft; 2 - Hopfenveredlung St. Johann

9h30 – 10h00 **L3:** Processing Brewer's Spent Grains & Trub for Food Ingredients.

Timothy Hobley[1]; Preben Hansen[1]; Matias Bjerregaard[1]; Radhakrishna Shetty[1,2]; Helena Pastell[3]; Shiwen Zhuang[1]; Angelos Charalampidis[1]; Rasmus Froding[1]; Katrine Rorby[1]; Charlotte Jacobsen[1]; Manuel Pinelo[1]; Nina Gringer[1]
1 - Technical University Denmark; 2 - Carlsberg Research Laboratory; 3 - Finnish Food Safety Authority Evira

10h00 – 10h30 **L4:** Comparison of Saaz growing in climatically different and challenging years.

Andreas Gahr[1]; Adrian Forster[2]
1 - Hopfenveredlung St. Johann; 2 - HVG Hopfenverwertungsgenossenschaft Eg

10h30 – 11h00 Coffee / Tea, followed by joint BF / EBC plenary 11h00 – 12h30

12h30 – 14h00 Lunch

AFTERNOON PARALLEL SESSION 1 Darwin Room (Level 1)

Yeast & Fermentation

Chair: Sandra Stelma (Head of Science, Diageo, Dublin, Ireland)

14h00 – 14h30 **L5:** Use of wine non-Saccharomyces yeast for brewing.

Fotini Drosou[1]; Panagiotis Tataridis[2]; Vasso Oreopoulou[1]; Vassilis Dourtoglou[2]
1 - Department of Chemical Engineering, National Technical Univesity of Athens; 2 - Department of Wine, Vine & Beverage Sciences, University of West Attica

14h30 – 15h00 **L6:** Yeast proteomics during beer fermentation: challenges and prospects.

Magda Costa[1]; David Laureys[1]; Soren Planckaert[2]; Bart Devreese[2]; Anita Van Landschoot[1]; Jessika De Clippeleer[1,3]
1 - Ghent University, Faculty of Bioscience Engineering, Department of Biotechnology, Research Group Brewing Science & Technology, Campus Schoonmeersen, Valentin Vaerwyckweg 1, B-9000 Ghent, Belgium; 2 - Ghent University, Faculty of Science, Department of Biochemistry and Microbiology, Laboratory for Microbiology, Campus Ledeganck, KL Ledeganckstraat 35, B-9000 Ghent, Belgium; 3 - University College Ghent, Faculty of Science and Technology, Department of Natural and Food Sciences, Research Group Brewing Science & Technology, Campus Schoonmeersen, Valentin Vaerwyckweg 1, B-9000 Ghent, Belgium

15h00 – 15h30 **L7:** Saccharomyces cerevisiae var. diastaticus- friend or foe?

Tim Meier-Dornberg[1]; Fritz Jacob[1]; Mathias Hutzler[1]
1 - TU Munchen - Research Center Weihenstephan for Brewing and Food Quality

15h30 – 16h00 **L8:** The Guinness Yeast: 250 Years in the Making.

Daniel Kerruish[1]; Sandra Stelma[2]

1 - Group Microbiologist, Diageo Global Supply, Science and Technology, St James's Gate Brewery, Dublin 8, Ireland; **2** - Head of Science, Diageo Global Supply, Science and Technology, St James's Gate Brewery, Dublin 8, Ireland

AFTERNOON PARALLEL SESSION 2 Gorilla 4-5 Room (Level 1)

Technical reviews and insights

Chair: Stefan Lustig (President DBMB: German Brew- and Malt-Masters Association, Munich)

14h00 – 14h30 **L9:** Teff (Eragrostis tef (zucc.) Trotter) and quinoa (Chenopodium quinoa W.) as novel raw materials for brewing.

Ombretta Marconi[1]; Valeria Sileoni[1]; Giovanni De Francesco[1]; Lidia Di Ghionno[2]; Giuseppe Perretti[1]

1 - University of Perugia-Italian Brewing Research Centre; **2** - University of Perugia, Department of Agricultural, Food and Environomental Sciences

14h30 – 15h00 **L10:** Process innovation at Asahi Breweries – The pursuit of ultimate consistency of quality in Asahi Super Dry.

Yoshinori Ito[1]; Shigekuni Noba[1]; Minoru Kobayashi[1]; Kazuhiko Uemura[1]

1 - Asahi breweries, Ltd, Japan

15h00 – 15h30 **L11:** Identification of factors at the origin of the phenomenon of beer gushing in order to develop a premature detection technique of gushing risk.

Julien Billard[1]; Sophie Schwebel[1]; Romain Kapel[2]; Marc Schmitt[1]

1 - IFBM; **2** - University of Lorraine, Nancy, France

15h30 – 16h00 **L12:** Sourdough cultures as a source of speciality yeast for the brewing industry.

Brian Gibson[2]; Jarkko Nikulin[2]; Linnea Johansson[1]; Frederico Magalhaes[2]; Kristoffer Krogerus[2]; Elina Sohlberg[2]; Paula Jouhten[2]; Linnea Laitila[2]

1 - Metropolia University of Applied Sciences; **2** - VTT Technical Research Centre of Finland

16h00 – 16h30 Coffee / Tea / Beer

LATE - AFTERNOON PARALLEL SESSION 1 Darwin Room (Level 1)

Yeast & Fermentation

Chair: Sandra Stelma (Head of Science, Diageo, Dublin, Ireland)

16h30 – 17h00 **L13:** Exploiting Yeast Diversity For Product and Process Optimisation.

Katherine Smart[1]

1 - University of Cambridge, UK

17h00 – 17h30 **L14:** Impact of Fermentation Parameters and Hops on the Flavour Expression of Beer Yeasts.

Gabriela Montandon[1]; Philippe Janssens[1]; Yves Gosselin[1]; Stephane Meulemans[1]

1 - Fermentis, Lille, France

17h30 – 18h00 **L15:** Rapid Assessment of Yeast Phenotypic Diversity through the Application of Raman Spectroscopy.

Yang He[1]; Hua Yin[1]; Jianjun Dong[1]; Junhong Yin[1]; Lu Chen[1]

1 - Tsingtao Brewery Co. Ltd., China

LATE - AFTERNOON PARALLEL SESSION 2 Gorilla 4-5 Room (Level 1)

Technical reviews and insights

Chair: Stefan Lustig (President DBMB: German Brew- and Malt-Masters Association, Munich)

16h30 – 17h00 **L16:** The copper binding affinity of wort during the brewing process.

Marcus Pagenstecher[1]; Morten J. Bjerrum[2]; Mogens L. Andersen[1]

1 - Department of Food Science, University of Copenhagen, Denmark; 2 - Department of Chemistry, University of Copenhagen, Denmark

17h00 – 17h30 **L17:** Unlocking the potential of lactic acid bacteria and their application in malting and brewing.

Elke Arendt[1,2]; Kieran Lynch[3]; Emanuele Zannini[3]

1 - School of Food and Nutritional Sciences and APC Microbiome Institute, University College Cork, Ireland; 2 - Cork; 3 - School of Food and Nutritional Sciences, University College Cork, Ireland

17h30 – 18h00 **L18:** Challenging the assumptions around the pasteurisation requirements of beer spoilage bacteria.

Grzegorz Rachon[1]; Christopher Rice[1]; Karin Pawlowsky[1]; Christopher Raleigh[1]

1 - CampdenBRI, Nutfield (Surrey) UK

TUESDAY, 4 JUNE

MORNING PARALLEL SESSION 1 Darwin Room (Level 1)

Hops in Beer

Chair: Paul Lefebvre (Owner, Brasserie Lefebvre, Belgium)

8h30 – 9h00 **L19:** Taking hop aroma apart: Supercritical CO2 fractionated oils offers new flavour possibilities for brewers.

Christina Dietz[1]; David Cook[1]; Ray Marriott[2]; Colin Wilson[2]; Rebecca Ford[1]

1 - University of Nottingham, School of Biosciences, Division of Food Sciences, Sutton Bonington Campus, Loughborough, UK; 2 - Totally Natural Solutions Ltd, Unit 3a Arnold Business Park, Branbridges Road, East Peckham, Kent, UK

9h00 – 9h30 **L20:** Exploring hop derived contributors to beer bitterness using data-independent acquisition (DIA).

Nils Rettberg[1]; Laura Knoke[1]; Julia Hildebrandt[1]; Sarah Thorner[1]; Jorg Maxminer[1]

1 - VLB Berlin - Research Institute for Beer and Beverage Analysis, Berlin, Germany

9h30 – 10h00 **L21:** Dimethyl sulfide in beer: the potential role of hops.

Elise Salanouve[1]; Stephane Delpech[1]; Clement Viel[1]; Scott R. Lafontaine[1]; Thomas H. Shellhammer[2]; Laurent Dagan[1]

1 - NYSEOS, ZA Parc 2000, 53 rue Claude Francois, 34080 Montpellier, France; 2 - Department of Food Science and Technology, Oregon State University, 100 Wiegand Hall, Corvallis, Oregon 97331, United States

MORNING PARALLEL SESSION 2 Gorilla 4-5 Room (Level 1)

Wood & Beer

Chair: Lene Bech (Senior Laboratory and Regulatory Manager, Carlsberg A/S, Copenhagen, Denmark)

8h30 – 9h00 **L22:** Beech wood sawdust as a sustainable filter aid for pre-coat filtration.

Max Panglisch[1]; Thomas Kunz[1]; Frank-Jurgen Methner[1]

1 - Technische Universitat Berlin

9h00 – 9h30 **L23:** Wooden barrels are an additional source of microorganisms for lambic beer production.

Jonas De Roos[1]; Luc De Vuyst[1]

1 - Research Group of Industrial Microbiology and Food Biotechnology, Vrije Universiteit Brussel, Pleinlaan 2, B-1050 Brussels, Belgium

9h30 – 10h00 **L24:** How aging in tropical wood barrels may create an identity for Brazilian beer.

Giovanni Casagrande Silvello[1]; Aline Marques Bortoletto[1]; Ana Carolina Correa[1]; Mariana Costa De Castro[1]; Andre Ricardo Alcarde[1]

1 - University of Sao Paulo, USP-ESALQ, Piracicaba, Sao Paulo, Brazil

10h00 – 10h30 Coffee / Tea

MID - MORNING PARALLEL SESSION 1 Darwin Room (Level 1)

Malting Processes

Chair: Frank-Jurgen Methner (Former Head of Dept. of Brewing Science, TU Berlin)

10h30 – 11h00 **L25:** Malting barley agronomy & quality: evolution of the raw material qualities other 20 years in France.

Marc Schmitt[1]

1 - IFBM-Qualtech, Nancy, France

11h00 – 11h30 **L26:** Germination Kilning Combo Drum – an old method combined with new technology.

Christoph Remmelberger[1]; Johannes Lauer[2]

1 - Kaspar Schulz Brauereimaschinenfabrik& Apparatebauanstalt GmbH; **2 -** Kaspar Schulz Brauereimaschinenfabrik & Apparatebauanstalt GmbH

11h30 – 12h00 **L27:** First time quantification of cysteinylated aldehydes in malt and brewery samples.

Paula Bustillo Trueba[1]; Barbara Jaskula-Goiris[1]; Mark Sanders[3]; Jean-Paul Vincken[3]; Jessika De Clippeleer[2]; Erik Van Der Eycken[1]; Joshep De Brabanter[1]; Luc De Cooman[1]; Gert De Rouck[1]; Guido Aerts[1]

1 - KU Leuven; **2 -** Ghent University, Belgium; **3 -** Wageningen University & Research, The Netherlands

12h00 – 12h30 **L28:** Quantitative monitoring of beer staling aldehydes during the malting process.

Weronika Filipowska[1]; Maciej Ditrych[1]; Barbara Jaskula-Goiris[1]; Guido Aerts[1]; Gert De Rouck[1]; Luc De Cooman[1]

1 - KU Leuven, Bioengineering Technology TC, Ghent Technology Campus, Belgium

MID - MORNING PARALLEL SESSION 2 Gorilla 4-5 Room (Level 1)

Beer Styles, including non-alc. / low-alc. Beer

Chair: David De Schutter (Innovation & Technology Development Director Europe, AB-Inbev, Leuven, Belgium

10h30 – 11h00 **L29:** Non-Starch Carbohydrates from unmalted cereals as mouthfeel contributors in NABLAB and Light beers.

Niels A. Langenaeken[1]; Pieter Ieven[1]; Christian Clasen[3]; David De Schutter[2]; Christophe M. Courtin[1]

1 - Laboratory of Food Chemistry and Biochemistry and Leuven Food Science and Nutrition Research Centre (LFoRCe), KU Leuven, Kasteelpark Arenberg 20, 3001 Heverlee, Belgium; **2 -** Fund Baillet-Latour, Brouwerijplein 1, 3000 Leuven, Belgium; **3 -** Soft Matter, Rheology and Technology, KU Leuven, Celestijnenlaan 200f, 3001 Heverlee, Belgium

11h00 – 11h30 **L30:** Chance and Challenge: Non-Saccharomyces yeasts in non-alcoholic and low alcohol beer brewing.

Konstantin Bellut[1]; Elke Arendt[1,2]

1 - University College Cork, Cork, Ireland; **2 -** APC Microbiome Institute

11h30 – 12h00 **L31:** An innovative production of low alcohol beer using a Mrakia gelida.

Valeria Sileoni[1]; Giovanni De Francesco[1]; Ombretta Marconi[1]; Ciro Sannino[2]; Giuseppe Perretti[3]

1 - Italian Brewing Research Centre, University of Perugia; **2 -** Department of Agriculture, Food and Environmental Science, Industrial Yeasts
Collection DBVPG, University of Perugia; **3 -** Department of Agriculture, Food and Environmental Science, University of Perugia

12h00 – 12h30 **L32:** **Temporal dominance of sensations (TDS) paired with dynamic wanting: A new method of sensory evaluation with consumers for a better understanding of beer drinkability.**

Takahiro Wakihira[1]; Seiko Miyashita[1]; Minoru Kobayashi[1]; Kazuhiko Uemura[1]; Pascal Schlich[2]

1 - Asahi Breweries, Ltd., Research Laboratories for Alcohol Beverages, Moriya, Japan; **2 -** Centre des Sciences du Gout et de l'Alimentation, AgroSup Dijon, CNRS, INRA, Univ. Bourgogne Franche-Comte, F-21000 Dijon, France

12h30 – 14h00 Lunch

AFTERNOON PARALLEL SESSION 1 Darwin Room (Level 1)

Malt, Hops & Flavour

Chair: Monica Mandrutiu (Brewing & Innovation Lead, Asahi Breweries Europe Group, Czech Republic)

14h00 – 14h30 **L33:** **Understanding aromatic stability in dry-hopped beer.**

Margaux Huismann[1]; Fraser Gormley[2]; Dzeti Dzait[2]; Alex Speers[3]; Dawn Maskell[1]

1 - International Centre for Brewing & Distilling, Edinburgh, UK; **2 -** BrewDog; **3 -** Canadian Institute of Fermentation Technology

14h30 – 15h00 **L34:** **Influence of whirlpool temperature on hop aroma intensity in ale beers.**

Marius Hartmann[1]; Christina Schoenberger[1]; Mark Zunkel[1]

1 - Barth-Haas Group, Nuremberg, Germany

15h00 – 15h30 **L35:** **How to optimize the utilization of hop cysteine and glutathione S-conjugates in late and dry hopping: Focus on dual hops and Saaz.**

Cécile Chenot[1]; Sonia Collin[1]

1 - UCL, Louvain-la-Neuve, Belgium

15h30 – 16h00 **L36:** **Functionality of hop proanthocyanidins in brewing: an alternative to tannic acid?**

Arthur Gadon[1]; David Cook[1]

1 - University of Nottingham, UK

AFTERNOON PARALLEL SESSION 2 Gorilla 4-5 Room (Level 1)

Filtration, Stabilisation & Packaging

Chair: Carsten Zufall (Corporate Manager Technology and Innovation, Polar Group, Venezuela)

14h00 – 14h30 **L37:** **Microplastic in beer.**

Ursula Brendel-Thimmel[1], Reiner Gaub[1]

1 - Pall Filtersystems GmbH, Germany

14h30 – 15h00 **L38:** **Functional prediction of beer cross-flow membrane filtration efficiency from malt quality.**

Evan Evans[1]; Claudio Cornaggia[2]; Claudio Anna[2]; David Mangan[2]

1 - TBD Consulting and School of Natural Science, University of Tasmania; **2 -** Megazyme, Ireland

15h00 – 15h30 **L39:** **New insights into the cleanliness of washed returnable bottles.**

Roland Pahl[1]; Mick Holewa[1]; Alfons Ahrens[1]

1 - VLB Berlin, Germany

15h30 – 16h00 **L40:** **Status and Trends in Beer Filtration – A Brewery Managers' Perspective.**

Stefan Lustig[1]

1 - DBMB (German Brew- and Malt-Masters Association, Munich)

16h30 – 18h00 Coffee / Tea / Beer & Poster Session. Joint sundowner beers (BF & EBC)

WEDNESDAY, 5 JUNE

MORNING PARALLEL SESSION 1 Okapi 1-2-3 Room (Level 0)

Flavour Development in Fermentation

Chair: Martina Gastl (Head of Raw Materials and Beverage Technology, TU Munich, Germany)

9h00 – 9h30 **L41:** Sulphite content of beer : impact of raw material & fermentation parameters.

Sophie Schwebel[1]; Marc Schmitt[1]

1 - IFBM-Qualtech, Nancy, France

9h30 – 10h00 **L42:** Yeast Strain Specific Release of Cystein-Conjugated Thiols like 4-MSP by ß-Lyase Activity.

Maximilian Michel[1]; Korbinian Haslbeck[1]; Aofei Cheng[1]; Mathias Hutzler[1]; Mehmet Coelhan[1]; Martin Zarnkow[1]; Aofei Jacob[1]

1 - Research Center Weihenstephan for Brewing and Food quality, TU Munchen, Germany

10h00 – 10h30 **L43:** Metabolization of glycated amino acids from wort by different S. cerevisiae strains.

Michael Hellwig[1]; Anna-Lena Kertsch[1]; Thomas Henle[1]

1 - Technische Universitat Dresden, Chair of Food Chemistry

MORNING PARALLEL SESSION 2 Gorilla 1 Room (Level 1)

Brewing Technology

Chair: Jeff Potter (R&D Director, Lion Supply Group, Sydney, Australia)

9h00 – 9h30 **L44:** A rapid and objective method for in situ determination of a degree of sensory deterioration of beer caused by aging process.

Jana Olšovská[1]; Tomaš Vrzal[1,2]; Karel Štěrba[1]

1 - Research Institute of Brewing and Malting; 2 - Charles University, Faculty of Science, Prague, Czech Republic

9h30 – 10h00 **L45:** More sustainable brewing and increased flavour stability due to the AB InBev Simmer & Strip technology.

David De Schutter[1]; Mario Rottiers[2]; Mario Thiry[3]

1 - R&D Director Europe at AB InBev; 2 - Process Development Specialist; 3 - Global Process Development Specialist, Leuven, Belgium

10h00 – 10h30 **L46:** Influence of mashing parameters on filter cake structure and filterability in the lauter tun.

Martin Hennemann[1], Martina Gastl[1], Thomas Becker[1], Florian Lehnhardt[1]

1 - TU Munchen-Weihenstephan, Germany

10h30 – 11h00 Coffee / Tea

MID - MORNING PARALLEL SESSION 1 Okapi 1-2-3 Room (Level 0)

The concept of terroir in brewing materials

Chair: Chuck Skypeck (Technical Brewing Projects Manager, Brewers Association, Boulder / CO, USA)

11h00-11h30 **L47:** Development of the methodology for determination of the geographical origin of hops.

Iztok Jože Košir[1]; Marijan Nečemer[2]; Miha Ocvirk[1]

1 - Slovenian Institute of Hop Research and Brewing; 2 - Institute Jozef Stefan

11h30 – 12h00 **L48:** **Characterising the diversity of wild hops (Humulus lupulus) from Portugal.**

Julio Cesar Machado Junior[1]; Miguel A. Faria[1]; Zita E. Martins[1]; Armindo Melo[1]; Sara C. Cunha[1]; Florian Lehnhardt[2]; Hubert Kollmannsberger[2]; Martina Gastl[2]; Thomas Becker[2]; Isabel M. P. L. V. O. Ferreira[1]

1 - LAQV/ REQUIMTE, Departamento de Ciencias Quimicas, Faculdade de Farmacia, Universidade do Porto; **2** - School of Life Sciences Weihenstephan, Chair of Brewing and Beverage Technology, Technische Universitat Munchen

12h00 – 12h30 **L49:** **Native yeasts from Ecuadorian chichas: insights into their potential application in beer production.**

Nubia Jimena Grijalva Vallejos[1]; Kristoffer Krogerus[2,3]; Jarkko Nikulin[2,4]; Agustin Aranda[1]; Emilia Matallana[1]; Brian Gibson[2]

1 - Institute for Integrative Systems Biology (I2SysBio), Universitat de Valencia-CSIC, 46980 Paterna, Valencia, Spain.; **2** - VTT Technical Research Centre of Finland Ltd, Tietotie 2, P.O. Box 1000, FI-02044 VTT, Espoo, Finland; **3** - Department of Biotechnology and Chemical Technology, Aalto University, School of Chemical Technology, Kemistintie 1, Aalto, P.O. Box 16100, FI-00076 Espoo, Finland; **4** – Chemical Process Engineering, Faculty of Technology, University of Oulu, P.O. Box 8000, FI-90014 Oulun Yliopisto, Finland

MID - MORNING PARALLEL SESSION 2 Gorilla 1 Room (Level 1)

Beer & Brewing Materials Analysis

Chair: Esko Pajunen (Former President EBC)

11h00 – 11h30 **L50:** **Critical assessment of calibration strategies for effective beer flavor analysis by solid-phase microextraction (SPME).**

Sarah Thorner[1]; Johanna Dennenlohr[1]; Nils Rettberg[1]

1 - VLB Berlin - Research Institute for Beer and Beverage Analysis

11h30 – 12h00 **L51:** **International Calibration Standards for hop and beer analysis – Overview and News.**

Biendl Martin[1]; Foster Robert, T.[2]; Maye John Paul[3]

1 - Hopsteiner HHV GmbH Mainburg, Germany; **2** - Molson Coors; **3** - S.S. Steiner, Inc.

12h00 – 12h30 **L52:** **The influence of dry-hopping on selected chemical-physical characteristics of beer.**

Martin Zarnkow[1]; A. Stallforth[1]; K. Haselbeck[1]; S. Cocuzza[2]; F. Jacob[1]

1 - TU Munchen Research Center for Brewing and Food Quality; **2** - Hopsteiner, Simon H. Steiner Hopfen GmbH, Mainburg, Germany

12h30 – 14h00 Lunch

AFTERNOON PARALLEL SESSION 1 Okapi 1-2-3 Room (Level 0)

The concept of terroir in brewing materials

Chair: Chuck Skypeck (Technical Brewing Projects Manager, Brewers Association, Boulder / CO, USA)

14h00 – 14h30 **L53:** **Genetic and biochemical approach to study the impact of terroir on the brewing value of hops.**

Ann Van Holle[1,2]; Hilde Muylle[3]; Tom Ruttink[3]; Anita Van Landschoot[2]; Geert Haesaert[2]; Dirk Naudts[1]; Denis De Keukeleire[2]; Isabel Roldan-Ruiz[2,3]

1 - De Proefbrouwerij; **2** - Ghent University; **3** - Institute for Agricultural and Fisheries Research (ILVO), Ghent, Belgium

14h30 – 15h00 **L54:** **Effect of harvest time on polyphenols profiles of Saaz hops and beer taste.**

Takeshi Kumagai[1]; Takako Inui[1]; Takamasa Hasegawa[1]; Taichi Maruhashi[1]; Akira Kogin[1]; Seisuke Takaoka[1]

1 - Suntory Beer Ltd., Beer Development Department, Tokyo, Japan

AFTERNOON PARALLEL SESSION 2 Gorilla 1 Room (Level 1)

Beer & Brewing Materials Analysis

Chair: Esko Pajunen (Former President EBC)

14h00 – 14h30 **L55: Insights on the lager beer volatile terpenic compounds through an advanced chromatographic tool.**

Cátia Martins[1]; Tiago Brandao[2]; Adelaide Almeida[3]; Silvia M. Rocha[1]

1 - Departamento de Quimica & QOPNA, Universidade de Aveiro, Campus Universitario Santiago, 3810-193 Aveiro, Portugal; **2** - Super Bock Group, Via Norte, 4465-764 Leca do Balio, Portugal; **3** - Departamento de Biologia & CESAM, Universidade de Aveiro, Campus Universitario Santiago, 3810-193 Aveiro, Portugal

14h30 – 15h00 **L56: Arabinoxylans in the brewing process – an underestimated substance class?**

Michael Kupetz[1]; Martina Gastl[1]; Thomas Becker[1]

1 - Lehrstuhl fur Brau- und Getranketechnologie, TU Munchen, Germany

PLENARY SESSION Okapi 1-2-3 Room (Level 0)

15h00 – 16h30 CLOSING PANEL DISCUSSION: **Brewing Science: past milestones, future endeavours**

Moderated by Dr. Ina Verstl

Participants:

Prof. Dr. Guido Aerts, Prof. Dr. Charlie Bamforth, Ms. Kamini Dickie, Prof. em. Dr. Ludwig Narziss, Dr. Sandra Stelma

19h30 – 23h15 FAREWELL PARTY

Antwerpse Brouw Compagnie (Indiestr. 21, 2000 Antwerp).

SESSION CHAIRS

MONDAY, 3 JUNE

SUSTAINABILITY AND CLIMATE CHANGE

TIAGO BRANDÃO

President of the European Brewery Convention (EBC), nominated since 2016. Currently, accumulates roles as New Business Director for Beer & Brewing Projects at Super Bock Group.

Has more than 17 years of professional experience at the main brewing and beverage company in Portugal. Among other, experienced the following roles and responsibilities:

- Human Resources Director (between 2014 and 2017).
- New Product Development & Innovation Director, including the responsibility to manage the Research & Development team (from 2006 to 2014).
- Executive Board member at Maltiberica, the Malting company of Super Bock Group (from 2006 to 2013, in accumulation).
- Since 2001, performed several roles within Beer production and quality departments.

Owns a Biochemistry degree from Porto Science Faculty, and two Master of Science (MSc): in Biotechnology – by Leicester University; and in Brewing Engineering by the Scandinavian School of Brewing. Married with 2 daughters, he lives in Porto.

YEAST AND FERMENTATION

SANDRA STELMA

Sandra has a PhD in Biochemistry and brewing qualifications plus a 23 year career in Brewing R&D and brewing operations. She started her career with SABMiller in South Africa, while working in R&D she obtained first her MSc. and later her PhD on brewing related topics.

After finishing her brewing qualifications she moved across into production and held various roles in South Africa, Switzerland and Netherlands before joining Diageo, based in Dublin, 4 years ago. Here she is currently Head Of Science on the beer side of the business with a global remit heading up the R&D agenda, extensively supporting the African markets and the innovation agenda. Sandra is member of the EBC Executive team and the EBC Brewing Science Group and is active in the Brewers of Europe.

TECHNICAL REVIEWS AND INSIGHTS

STEFAN LUSTIG

Stefan Lustig is a brewing engineer graduated fromTU Munchen-Weihenstephan. After completion of his PhD in research on flavor stability under guidance of Prof. Ludwig Narziss, he joined Beck's brewery as head of development & technology.

After several leadership roles in quality, production, packaging and plant management, he joined the board of Anheuser-Busch InBev Germany as Director of Brewery Operations. Next he moved to Brau-Holding International, Germanys 6th largest brewing company, a joint venture with Heineken involvement. As Chief Operating Officer and Managing Director he was responsible for Supply, covering a network of 12 breweries and the construction of a 3.5 Mhl greenfield brewery. Currently he is president of "Deutscher Braumeister und Malzmeister Bund", Germanys largest brewmasters association.

Stefan's experience combines know how on brewing science/technology with the operational world of a brewery manager. His expertise covers the supply chain of the brewing industry from raw materials to logistics.

TUESDAY, 4 JUNE

HOPS IN BEER

PAUL LEFEBVRE

Paul Lefebvre was graduated as Industrial Engineer in Biochemistry (Master), Fermentation Section at Institut Meurice (Brussels) in 2002. He started to work in the family brewery as production, R&D and environnemental manger.
During 10 years, he completed the range of beer proposed by the brewery with some new creations such fruity and hoppy beers while the brewery was raised at a capacity of 120 000hl/year. Since 2012, he is the CEO and 6th generation brewer of Brasserie Lefebvre. Active in the world of beer, Paul is member of the EBC Executive Team since 2014 and the Steering Committee of the Belgian Brewers since 2018.

WOOD & BEER

LENE BECH

Lene Mølskov Bech is Senior Laboratory and Regulatory Manager at Carlsberg Breweries A/S, responsible for Analytical and Regulatory in Carlsberg Group. Lene is MSc and Ph.D. from University of Copenhagen. For many years Lene's primary tasks have been Regulatory and Analytical support as well as development of new advanced analytical methods. She has previously worked with many aspects of beer quality, e.g. foam and foam proteins, flavour characterization, flavour stability and sensory analyses. Lene has a long tradition working with EBC as member of EBC Brewing Science Group, several of the Subgroups and of the expert team supporting the Analysis Group. Since January 2018 she is the Chair of the EBC Analysis Group. Lene is also a Carlsberg representative in Brewers of Europe's Product IMT dealing with regulatory activities such as labelling, additives, enzymes, ingredients and beer safety and member of the expert groups on Consumer Information, Food Improvement Agents, and Technical Information Sheets.

MALTING PROCESSES

FRANK-JURGEN METHNER

Prof. Dr. Frank-Jürgen Methner was graduating with a Diploma in Engineering at Technical University of Berlin in 1881. From 1982 to 1986 he was PhD-Student at TU Berlin graduating 1987. From 1987 til 2004 he was working as a Director at Bitburger Brauerei, Bitburg, Germany, with responsibilities in Brewing Technology, Development and Quality Management. Since winter-semester 2004/2005 he is leading the Department of Brewing Science at the TU Berlin.

BEER STYLES, INCLUDING NON-ALC./ LOW ALC. BEER

DAVID DE SCHUTTER

David De Schutter obtained his PhD degree at the Center for Malting and Brewing Science of the KULeuven under supervision of Prof. Delvaux and Prof. Derdelinckx. Before his career at AB InBev he made a passage outside the brewing industry at the R&D Center of Procter & Gamble in Brussels. He joined AB InBev as Product & Process Development Specialist where he was involved in the development of the first new generation non-alcoholic beer Then he became Global Director for Process Development and R&D Transformation in the Global Innovation & Technology Center in Leuven (GITEC). His current position is R&D Director for Europe at AB InBev. David is also representing AB InBev in the EBC Executive Committee, member of the BIR&D (Belgian Industrial Research and Development organization) and coordinator of the Brewing program of Fund Baillet Latour.

MALT, HOPS & FLAVOUR

MONICA MANDRUTIU

Monica Mandrutiu is Quality Lead at Asahi Breweries Europe Group.
She ensures the alignment with the regional and global quality strategy & standards, acting as quality improvement catalyst for eleven breweries across five countries. She joined the brewing industry in 1990 at Ursus Breweries in Romania, where she had various roles, such as QC Manager, Chief Brewer and Quality & Brewing Director. Since 2013, she held regional roles as Brewing Lead-Europe at SABMiller European Hub and Brewing & Innovation Lead at Asahi Breweries Europe Group. She has an MSc in Food technology (Galati University) and an MBA degree from Edinburgh Business School, Heriot-Watt University.
Monica is member of EBC Executive Team and EBC Brewing Science Group, Associate member of IBD, member of ASBC and MBAA. She is accredited Beer sommelier (Beer Academy, UK) and member of the tasting panels of many key beer competitions in the world.

FILTRATION, STABILISATION & PACKAGING

CARSTEN ZUFALL

Carsten Zufall is Head of Technology and Innovation at Polar Breweries (Venezuela), where he is responsible for brewing processes and technology, brewing materials, applied research and new product design. He graduated in Brewing Science from Berlin University of Technology in 1990 and subsequently completed a Ph. D. (Dr.-Ing.). Following postdoctoral lecture qualification, he was awarded Associate Professorship (Priv.-Doz.) in Brewing Science in 2001. Carsten is chairman of the EBC Brewing Science Group and a member of ASBC, MBAA, the German Brewmasters' and Maltmasters' Association (DBMB), the VLB Alumni Association, the Berlin Brewers' Guild and the Cerveceros Latinoamericanos association.
He serves with the Editorial Boards of the Journal of the Institute of Brewing, Brewing Science and the ASBC Journal and he is a past president of MBAA's District Venezuela. His current research activities include beer flavour stability, flavour chemistry, sensory analysis, microbiology and environmental topics.

WEDNESDAY, 5 JUNE

FLAVOUR DEVELOPMENT IN FERMENTATION

MARTINA GASTL

Martina Gastl apprenticed as a brewer and maltster from 1994 to 1996. She studied brewing and beverage technology at the Technische Universitat Munchen-Weihenstephan, Germany. She graduated as an engineer at 2002. From 2002 until 2006 she completed her phD concerning 'Technological influence on lipid degradation in terms of improvement of beer flavor stability'. She is currently assistant professor and head of the research group Raw Materials based Brewing and Beverage Technology at the 'Lehrstuhl fur Brau- und Getranketechnologie' in Weihenstephan.

BREWING TECHNOLOGY

JEFF POTTER

Jeff began his brewing career with Lion at Toohey's Brewery in Sydney in 1990. He has since held a wide range of roles including Brewer, QA Manager, NPD Manager, Head Brewer at both Tooheys and later Malt Shovel Breweries and several Group Technical roles providing brewing, technical and NPD support and leadership across Lions operations in Australia and New Zealand.

He is currently the Research & Development Director for Lion Supply Chain – responsible for brewing and beverage product and process innovation, licensed brewing and technical regulatory compliance. Jeff is a Fellow of the Institute of Brewing & Distilling, having served for 6 years on the Asia Pacific section Board of Management with 4 years as Chairman. He represents Lion on industry associations including Brewers Association of Australia and New Zealand, a Board Member of Barley Australia and is a member of the EBC Brewing Science Group.

THE CONCEPT OF TERROIR IN BREWING MATERIALS

CHUCK SKYPECK

A pioneer of craft brewing in the southeastern U.S., Chuck Skypeck opened Tennessee's first brewpub, Boscos, in 1992. As a founding owner and director of brewery operations for Boscos Brewing Company, he opened additional brewpubs in Memphis, Nashville and Little Rock, AR. In 2007, Skypeck opened and operated Ghost River Brewing in Memphis. He has served as chair of the Association of Brewers Board of Advisors, chair of the Association of Brewers Board of Directors and was a member of the first Board of Directors for the Brewers Association. Winner of the Brewers Association Recognition Award in 2006, Skypeck joined the Brewers Association staff in 2013 and serves as Technical Brewing Projects Manager.
In this capacity at the Brewers Association, Chuck works with the Technical Committee and Subcommittees to develop member resources and encourage and facilitate best practices in all phases of member's brewery operations.

BEER & BREWING MATERIALS ANALYSIS

ESKO PAJUNEN

Born 1945, graduated 1970 from Helsinki University of Technology as Chemical Engineer on biochemistry and food technology. Joined 1973 VTT Biotechnical Laboratory with focus on brewing and malting. Short visit to Finnish Food Industry Federaration as R&D Director, before joining Sinebrychoff Brewery as R&D Manager in 1979. In 1987 started as Production Director with responsibility on Kerava new brewery and continued as Senior Vice President R&D&I from 1992.Joined Carlsberg as Vice President R&D in 2006 heading Carlsberg Research Center until retirement in 2009. Esko joined the EBC Biochemistry Group in 1978, became EBC Council Member in 1987, Vice President in 1997 and acted as EBC President 1999 -2003 and Board Member until 2010. Acted also Director and Chairman of PBL, the ResearchLaboratory of the Finnish Brewers and Maltsters.

SENIOR'S PANEL

INA VERSTL

Dr Ina Verstl is Editor of Brauwelt International. Together with Ernst Faltermeier she wrote the book *The Beer Monopoly* (2016).

POSTER LIST / POSTER SESSION

POSTER SESSION on Tuesday, 4 June 2019 at 4.30 pm

TOPIC: ANALYSIS AND METHODOLOGY (raw materials, product, research)

P001	TURBIDITY IDENTIFICATION VIA RAMAN MICRO SPECTROSCOPY	Eva-Maria Kahle	Research Center Weihenstephan for Brewing and Food Quality
P002	NON-DESTRUCTIVE DETERMINATION OF HOP CONSTITUENTS BY NEAR INFRARED SPECTROSCOPY (NIRS)	Florian Schull	RMI Analytics
P003	SENSOMICS AS A PROMISING TOOL IN BREWING SCIENCE	Tomáš Vrzal	Research Institute of Brewing and Malting, PLC, Prague, Czech Republic
P004	THE CHARACTERIZATION OF A HOP OIL PROFILE OF HOP BY FLUIDIZED BED EXTRACTION AND GC-MS-QQQ WITH A HEART-CUT TECHNIQUE	Karel Štěrba	Research Institute of Brewing and Malting, Prague
P005	ANALYSIS OF HOP AROMA COMPONENTS AFTER FERMENTATION BASED ON CLOSE GENETIC BACKGROUND	Tetsu Sugimura	Kirin Company, Limited
P006	NON-DISCRIMINANT ANALYSIS OF BEER AGEING CHEMISTRY	Laura Knoke	VLB Berlin
P007	SENSORIAL AND ANALYTICAL CHARACTERIZATION OF THE NEW HÜLLER "SPECIAL FLAVOUR HOPS"	Klaus Kammhuber	Bavarian Hop research center
P008	COMPREHENSIVE ANALYSIS OF POLYFUNCTIONAL THIOLS IN A NEW FLAVOR HOP MOSAIC BY THIOL-SPECIFIC EXTRACTION WITH SELECTABLE 1D/2D/3D GC-SCD/Q-TOF-MS	Koji Takazumi	Frontier Laboratories for Value Creation, SAPPORO HOLDINGS LTD.
P009	UNIQUE VARIETAL AROMA OF THE 'SORACHI ACE' HOP ~EFFECTS OF THE COEXISTENCE OF GERANIC ACID AND OTHER HOP-DERIVED COMPOUNDS ON THE VARIETAL AROMAS	Ayako Sanekata	Frontier Laboratories of Value Creation, SAPPORO BREWERIES LTD.;
P010	WHICH HOP VARIETY FOR A BETTER FLAVOUR AND COLLOIDAL STABILITY OF DRY-HOPPED BEERS: EVIDENCE OF THE KEY-ROLE OF MINOR FLAVAN-3-OLS	Carlos Silva Ferreira	Université catholique de Louvain
P011	MYCOBIOME IN MODERN BREWERIES	Riikka Juvonen	VTT Technical Research Centre of Finland Ltd;
P012	EFFECT OF HOP VARIETIES ON BEER AROMA DURING DRY HOPPING	Damien Steyer	TWISTAROMA
P013	CANCELLED	---------------------	----------------------------
P014	EVALUATION OF WATER AND ETHANOL EXTRACTS OF LEMON VERBENA, SAGE, CINNAMON AND HIBISCUS IN THE AROMATIC AND ORGANOLEPTIC PROFILE OF BEER	Avgoustinos Evripiotis	Department of Wine, Vine & Beverage Sciences, University of West Attica, Ag. Spyridonos str., 12243 Egaleo, Greece
P015	BENEFITS OF BLEND-HOPPING WITH CASCADE, CENTENNIAL, AND CHINOOK DURING DRY-HOPPING	Thomas Shellhammer	Oregon State University, Food Science and Technology
P016	DRY HOPPING POTENTIAL OF EUREKA!, A NEW HOP VARIETY.	Christina Schmidt	Hopsteiner
P017	A NOVEL TECHNIQUE FOR ASSESSING ENERGY AND UTILITY CONSUMPTION IN BREWING	Sebastian Schmid	Technical University of Munich
P018	THE RAPID VISCO ANALYSER AS A RHEOLOGICAL TOOL TO DETERMINE THE POTENTIAL BREWING QUALITY OF BARLEY AND MALT	Bruno Godin	Walloon Agricultural Research Center
P019	INFLUENCE OF PASTEURIZATION ON DRY HOPPED BEERS	Frank Peifer	Hopsteiner

P020	FACTORS AFFECTING HUMULINONE FORMATION IN HOPS	Philip Wietstock	Technische Universität Berlin
P021	ARABINOXYLAN AND Β-GLUCAN IN BARLEY ENDOSPERM CELL WALLS: MICROSTRUCTURE ANALYSIS USING CLSM AND CRYO-SEM	Niels A. Langenaeken	Laboratory of Food Chemistry and Biochemistry and Leuven Food Science and Nutrition Research Centre (LFoRCe), KU Leuven, Kasteelpark Arenberg 20, 3001 Leuven, Belgium;
P022	BREWING YEAST STRAIN DIFFERENTIATION: PRACTICAL RESULTS AND COMPARISON OF GENOTYPE AND PHENOTYPE BASED TECHNIQUES	Marta Orive Camprubí	Mahou-San Miguel
P023	UNEXPECTED RESULTS IN ALE YEAST DNA FINGERPRINTING	Nuria Feliu-Besora	Mahou S.A
P024	BEEROMICS FROM MALT AND MASHING TO BEER	Ida Kallehauge Nielsen	Technical University of Denmark
P025	NORWEGIAN PRODUCTION OF MALTING BARLEY AND HOPS	Mette Goul Thomsen	NIBIO
P026	ION MOBILITY SPECTROMETRY BASED HEADSPACE DETECTION OF VOLATILE ORGANIC COMPOUNDS PRODUCED BY MOLD FUNGI	Thomas Kunz	Technical University Berlin
P027	NON-INVASIVE ON-LINE MONITORING OF THE SECONDARY BOTTLE FERMENTATION PROCESS USING NEAR INFRARED SPECTROSCOPY	Manuel Zimmer	Institute of Food Technology NRW
P028	PORTUGUESE WILD HOPS GENOTYPING BY HRMA OF A MINIMAL SNP SET	Miguel Faria	LAQV/REQUIMTE, Laboratório de Bromatologia e Hidrologia, Departamento de Ciências Químicas, Faculdade de Farmácia, Universidade do Porto
P029	METABOLOMICS STRATEGY FOR MAPPING OF VOLATILE EXOMETABOLOME FROM SACCHAROMYCES SPP. WIDELY USED IN BREWING BASED ON COMPREHENSIVE TWO-DIMENSIONAL GAS CHROMATOGRAPHY	Silvia Rocha	Departamento de Química & QOPNA, Universidade de Aveiro, Campus Universitário Santiago, 3810-193 Aveiro, Portugal
P030	IN SITU IDENTIFICATION OF ANTIOXIDANT MOLECULES IN BEER BY HPTLC/BIOASSAY/MS	Pablo Murath	Universidad de Concepcion
P031	WINEBEER: A BEER WITH IMPROVED FUNCTIONAL PROPERTIES THROUGH GRAPE-SKIN EXTRACT ADDITION	Stephanie Hoffmann	Universidad de Concepcion
P032	THE USE OF HIBISCUS SABDARIFFA FLOWERS IN BREWING	Panagiotis Tataridis	Department of Wine, Vine & Beverage Sciences, University of West Attica, Athens, Greece
P033	THE EFFECT OF MAGNESIUM AND ZINC IONS ON THE RATE OF LACTIC ACID FERMENTATION FOR SOUR BEER PRODUCTION	Aleksander Poreda	University of Agriculture in Krakow, Faculty of Food Technology
P034	OPTIMIZATION OF MASHING PROCESSES USING INLINE CARBOHYDRATE PROFILING	Andreas Kunov-Kruse	Specshell ApS
P035	CANCELLED	------------------	--------------------------------
P036	A SIMPLE AND ACCURATE METHOD FOR THE DETERMINATION OF VERY LOW ALCOHOL CONCENTRATIONS IN 0.0 BEERS AND 0.0 BEER MIXED BEVERAGES USING NEAR INFRARED SPECTROSCOPY	Helmut Klein	Brau Union Österreich AG

TOPIC: BARLEY BREEDING AND AGRONOMY

P037	SCREENING OF DROUGHT TOLERANCE IN SPRING BARLEY	Bram Marynissen	University College Ghent
P038	THE IMPACT OF A REDUCED INPUT GROWING REGIME ON MALTING CAPABILITY OF BARLEY AND WHEAT	Calum Holmes	Heriot-Watt University
P039	TR15245: A NEW TWO-ROW MALTING BARLEY COMBINES DESIRABLE AGRONOMICS, DISEASE RESISTANCE AND QUALITY INCLUDING REDUCED PHYTATE	Ana Badea	Agriculture and Agri-Food Canada, Brandon Research and Development Centre
P040	EFFECTS OF PLANT GROWTH REGULATOR APPLICATION ON THE MALTING QUALITY OF BARLEY	Marta Izydorczyk	Grain Research Laboratory, Canadian Grain Commission, Winnipeg MB Canada

TOPIC: BEER QUALITY - FOAM

P041	RELATIONSHIP BETWEEN BEER FOAM AND SURFACE MOLECULES INVESTIGATED USING SUM-FREQUENCY GENERATION SPECTROSCOPY	Takayuki Miyamae	National Institute of Advanced Industrial Science and Technology
P042	TOWARDS A BIOMIMETIC SENSOR FOR HYDROPHOBIN DETECTION IN BEER	Fatemeh Bajoul Kakahi	Soft-Matter Physics and Biophysics Section, Department of Physics and Astronomy, KULeuven
P043	INVESTIGATION OF FACTORS AFFECTING BEER FOAM COLOR APPLYING A NOVEL TECHNIQUE USING TWO-DIMENSIONAL COLORIMETER	Azusa Asai	SAPPORO BREWERIES LTD., Research and Development Institute for Alcoholic Beverages, Yaizu, Japan

TOPIC: BEER QUALITY - SENSORY

P044	INFLUENCE OF POST-FERMENTATION PROCESSES ON HOP AROMA IN BEER	Ray Marriott	Totally Natural Solutions Ltd
P045	HIDDEN SECRETES OF THE NEW ENGLAND IPA	John Paul Maye	Hopsteiner
P046	THE CONTRIBUTION OF STALING ALDEHYDES TO THE FLAVOR (IN)STABILITY OF TOP FERMENTED, HOPPY BEERS	Sarah Thörner	VLB Berlin - Research Institute for Beer and Beverage
P047	DISENTANGLING SENSORY EXPERIENCES OF BEER WITH ARTIFICIAL INTELLIGENCE	Supinya Piampongsant	VIB – KU Leuven Center for Microbiology, Gaston Geenslaan 1, B-3001 Leuven, Belgium
P048	A NEW METHODOLOGY FOR DYNAMIC "EASINESS TO DRINK" WITH TEMPORAL DOMINANCE OF SENSATIONS (TDS) - GOOD KIRE (CRISPNESS) OF BEER LEADING TO EASINESS OF DRINKING	Takahiro Wakihira	Asahi Breweries, Ltd., Research Laboratories for Alcohol Beverages, Moriya, Japan
P049	THE IMPACT OF BREWING WITH UNMALTED CEREAL ADJUNCTS ON THE SENSORY AND ANALYTICAL PROFILE OF BEER	Joanna Yorke	University of Nottingham
P050	INFLUENCE ON BEER BITTER QUALITY BY USING DIFFERENT HOP PRODUCTS	Dominique Dixius	Bitburger Braugruppe GmbH
P051	WOOD AGED IMPACTS: COMPARISON BETWEEN BRAZILIAN SPECIES AND AMERICAN OAK CHIPS	Ana Carolina Corrêa	University of São Paulo - College of Agriculture "Luiz de Queiroz"
P052	EUROPEAN HOPS VARIETIES ANALYSED WITH SENSORY AND ANALYTICAL BEST PRACTICES	Boris Gadzov	FLAVORACTIV LIMITED

P053	DETERMINATION OF POTENTIAL FLAVOUR DEFICIENCIES OF COMMERCIAL NON-ALCOHOLIC BEERS	Jeroen Bauwens	KU Leuven, Department of Microbial and Molecular Systems (M^2S), Cluster for Bioengineering Technology (CBeT), Laboratory of Enzyme, Fermentation and Brewing Technology (EFBT), Technology Campus Ghent, GebroedersDe Smetstraat1, 9000 Ghent, Belgium
P054	FLAVOR COMPOSITION OF BEER AGED IN BARRELS OF OAK AND BRAZILIAN WOODS - A PROPOSED SENSORY WHEEL	Aline Marques Bortoletto	University of São Paulo
P055	ASSESSMENT OF QUALITY CONTROL METRICES FOR MEASURING BEER FRESHNESS IN THE LAB USING THE MICROESR TECHNOLOG	Mangethe Zwane	FlavorActiV
P056	INTRODUCTION OF A STANDARD BEER FOR QUALITY CONTROL ASSESSMENT OF BEER FRESHNESS IN THE LABORATORY USING THE MICROESR TECHNOLOGY	Mangethe Zwane	FlavorActiV
P057	SENSORY IMPACT OF ESTER-RICH SPECIAL BEER VIA RETRONASAL AROMA ANALYSIS	Kazuki Maruyama	SAPPORO HOLDINGS LTD

TOPIC: BEER, SOCIETY, HEALTH AND NUTRITION

P058	DEVELOPMENT OF NOVEL, MATURED, HOP-DERIVED MILD BITTER ACIDS THAT IMPROVE COGNITIVE FUNCTION	Takafumi Fukuda	Kirin Company, Limited, Research Laboratories for Health Science and Food Technologies, Yokohama, Japan
P059	SHOULD WE WORRY ABOUT NICKEL CONTENT IN BEERS ON THE BELGIAN MARKET?	Anneleen Decloedt	Ghent University, Department of Veterinary Public Health and Food Safety, Laboratory of Chemical Analysis, Salisburylaan 133, B-9820 Merelbeke, Belgium
P060	CAN BEER CONSUMPTION CONTRIBUTE TO THE DAILY INTAKE OF ESSENTIAL MINERALS ELEMENTS?	Edgar Pinto	University of Porto -Faculty of Pharmacy

TOPIC: CONSUMER STUDIES

P061	CANCELLED	------------------------	--------------------------------
P062	REVEALING PATTERNS OF CONSUMER INTEREST IN GLUTEN-FREE BEERS	Edyta Kordialik-Bogacka	Institute of Fermentation Technology and Microbiology, Lodz University of Technology, Wólczańska 171/173, 90-530, Łódź, Poland
P063	PATHS FOR SMART TOURISM OPPORTUNITIES TOWARDS BEER CONSUMERS	Jan Lichota	UNED - Universidad Nacional de Educación a Distancia

TOPIC: END OF BEER PROCESSING INCLUDING AUXILIARY AND PROCESS AIDS

P064	YEAST PROTEIN EXTRACT (YPE), A NATURAL INNOVATIVE AND SUSTAINABLE SOLUTION FOR WHITE BEER HAZE STABILITY	Olivier Caille	FERMENTIS Lesaffre for beverage
P065	STUDY OF BEER-SPOILAGE LACTOBACILLUS NAGELII HARBORING HOP RESISTANCE GENE HORA	Minami Umegatani	Asahi Breweries, Ltd., Research Laboratories for Alcohol Beverages, Moriya, Japan
P066	IMPACT OF HIGH PRESSURE PROCESSING ON THE FOAM STABILITY OF UNFILTERED BEER	Katerina Stulikova	University of Chemistry and Technology, Prague

TOPIC: ENVIRONMENTAL ISSUES AND SUSTAINABILITY

P067	RENEWABLE ENERGIES IN BREWERY MICRO GRIDS	Holger Schmidt	ABB Automation Products GmBH
P068	UNLOCKING PRODUCTION EFFICIENCY	Holger Schmidt	ABB Automation Products GmBH
P069	THE RE-USE OF PVPP-REGENERATION-CAUSTIC FOR CIP-CLEANING	Deniz Bilge	VLB Berlin
P070	HEATING WITH BEER - DISTRICT HEATING BY HEAT RECOVERY FROM THE COOLING PROCESS OF THE BREWERY PUNTIGAM	Gerald Zanker	Brau Union Österreich - Heineken
P071	BREWING WITH GREEN MALT FOR AN ENERGY AND WATER EFFICIENT FUTURE	Celina Dugulin	International Centre for Brewing Science, Bioenergy and Brewing Science Building, University of Nottingham, School of Biosciences, Sutton Bonington Campus, Loughborough, LE12 5RD, UK
P072	BULIDING A PILOT REACTOR - VALORIZING CO2 THROUGH POWER-TO-GAS AND BIOCATALYTIC METHANISATION	Timo Broeker	Institute of Food Technology ILT.NRW

TOPIC: FILTRATION

P073	REDUCED IRON ENTRY DURING KIESELGUHR FILTRATION VIA PRE-TREATED FUNCTIONAL VISCOSE FIBERS TO IMPROVE OXIDATIVE AND COLLOIDAL BEER STABILITY	Frank Jürgen Methner	Technical University Berlin, Institute of Food Technology and Food Chemistry, Chair of Brewing Science, Seestraße 13, 13353 Berlin, Germany
P074	BEECH WOOD SAWDUST AND GALLOTANNIN COMBINATION AS A SUSTAINABLE FILTER AID	Thomas Kunz	Technische Universität Berlin, Institute of Food Technology and Food Chemistry, Chair of Brewing Science, Seestraße 13, 13353 Berlin, Germany
P075	FIRST EXPERIENCE OF THE 'BREWERY 4.0' – THE INTERNET OF THINGS CONCEPT APPLIED IN AN OPERATIONAL BEER FILTRATION UNIT	Roland Folz	Roland Folz
P076	A NEW APPROACH TO ASSESS THE CROSSFLOW MEMBRANE FILTRATION OF BEER	Michael Kupetz	Lehrstuhl für Brau- und Getränketechnologie, TU München
P077	PRACTICAL EXPERIENCE WITH THE KRONES MEMBRANE FILTER PHOEBUS	Joerg Zacharias	Krones AG
P078	NOVEL CERAMIC MEMBRANE FOR CROSSFLOW FILTRATION OF BEER	Kohei Watanabe	Asahi Breweries, Ltd., Production Technology Center, Moriya, Japan

TOPIC: HOP BREEDING, AGRONOMY AND PROCESSING

P079	HOP OIL PROFILING TO MITIGATE CLIMATE CHANGE IMPACT ON TRADITIONAL HOP VARIETIES	Ray Marriott	Totally Natural Solutions Ltd
P080	BREWING QUALITY ATTRIBUTES OF UMBRIAN WILD HOPS	Valeria Sileoni	Italian Brewing Research Centre, University of Perugia;
P081	THE APPLICATION OF METABOLOMICS AND GENOMICS IN HOP BREEDING	Alexander Feiner	Simon H. Steiner, Hopfen, GmbH;
P082	IMPACT OF KILN TEMPERATURES ON THE AROMA AND ENZYMATIC POTENTIAL OF HOPS	Scott Lafontaine	Oregon State University, Department of Food Science and Technology
P083	"VAKUPACK" – WHOLE HOPS PACKAGED UNDER AN INERT ATMOSPHERE	Adrian Forster	HVG Hopfenverwertungsgenossenschaft e.G
P084	EVALUATION OF CASCADE HOPS QUALITY CULTIVATED IN NORTH OF ITALY	Paolo Passaghe	University of Udine, Italy, Dept. of Agricultural, Food, Animal and Environmental Sciences
P085	WHAT CAN CRYO HOPS® AND AMERICAN NOBLE HOPS™ DO FOR YOUR BEER?	Cynthia Almaguer	Yakima Chief, Louvain-la-Neuve

TOPIC: MALTING: MALT PRODUCTION AND QUALITY

P086	CHARACTERIZATION OF MALTS BY SENSORY ANALYSIS AND ANALYSIS OF VOLATILES	Michael Féchir	Trier University of Applied Sciences;
P087	AN EXAMINATION OF CHANGES IN PASTING BEHAVIOR OF IN-PROCESS MALT USING RAPID VISCO ANALYZER (RVA)	Xiang S Yin	Rahr Corporation, USA
P088	UNDERSTANDING THE COMPONENTS OF SPECIFIC WEIGHT IN MALTING BARLEY	Aaron Hoyle	Scotland's Rural College
P089	ECONOMIC CONSIDERATIONS ON A SMALL SIZE DRUM MALTING PLANT	Stefano Buiatti	University of Udine, Italy, Dept. of Agricultural, Food, Animal and Environmental Sciences;
P090	EVALUATION OF GERMINATION PARAMETERS DURING MALTING REGARDING PHENOLIC ACID RELEASE DURING WORT PRODUCTION USING RESPONSE SURFACE METHODOLOGY	Torsten Seewald	Technische Universität Berlin, Institute of Food Technology and Food Chemistry, Chair of Brewing Science

TOPIC: NEW PROCESS DEVELOPMENT

P091	LABORATORY PLANT FOR A CONTINUOUS CLOSED LOOP CONTROLLED MASHING AIDED BY DIGITAL TECHNOLOGIES	Patrick Wefing	Institute of Food Technology ILT.NRW;
P092	PRE-TREATMENT OF BARREL: CONTROL TOOL FOR AGING OF BEERS	Mariana Costa De Castro	University of São Paulo, USP-ESALQ, Piracicaba, São Paulo, Brazil
P093	CANCELLED	--------------	---------------------
P094	ADJUNCT CUM MASH TUN FIRST OF ITS KIND IN COMMERCIAL SCALE BREWING IN ASIA.	Gopal Krishnan	Carlsberg India
P095	MICROBIAL EXOPOLYSACCHARIDES AS AN ALTERNATIVE FOR DECLARABLE STABILIZERS AND OPACIFIERS	Julian Huchtmann	ILT.NRW;

P096	SIMULATION OF DRY HOPPING IN LABORATORY-SCALE HOP EXTRACTOR	Lukáš Jelínek	University of Chemistry and Technology Prague, Department of Biotechnology

TOPIC: NEW PRODUCT DEVELOPMENT

P097	FASTER AND SAFE TANK CLEANING USING CONTROLLED JET MOTION	Thomas Weyrauch	Hohe Tanne GmbH
P098	USE OF WOOD CHIPS IN BREWING : IMPACT OF PROCESS AND MATERIALS	Brabant Pierre	IFBM
P099	TRANSFER OF CHARACTERISTIC MARKERS DURING AGING OF BEER WITH DIFFERENT OAK CHIPS	Marcel Karabín	University of Chemistry and Technology Prague, Department of Biotechnology, Prague, Czech Republic
P100	HIGH PROLYL-ENDOPEPTIDASE ENZYME DOSAGES FOR THE PRODUCTION OF GLUTEN-FREE WHEAT BEER: A LAB-SCALE CASE STUDY	Hellen Watson	Ghent University, Faculty of Bioscience Engineering, Department of Biotechnology, Brewing Science and Technology Research group, Ghent, Belgium
P101	WHY BREWER'S YEASTS ARE THE BEST "HEALTH BENEFIT" PARTNERS OF KOMBUCHA MICROORGANISMS CONSORTIUM?	Van Nedervelde Laurence	Labiris

TOPIC: NOVEL RAW MATERIALS FOR BEER PRODUCTION

P102	BREWING WITH 10/20 % LENTIL MALT ADDITION IN A MICROBREW SCALE	Jonas Trummer	University of Agriculture in Krakow
P103	HOW RESEARCH ON ALTERNATIVE GRAINS CAN BOOST THE BREWING INDUSTRY	Jessika De Clippeleer	Ghent University
P104	RAW MATERIALS INCLUDING ANTHOCYANINS COULD CHANGE THE CONCENTRATION OF DIACETYL IN BREWING.	Atsushi Tanigawa	SAPPORO BREWERIES LTD
P105	TASTE PROPERTIES AFFECTING BEER OF POLYPHENOLS DERIVED FROM HERBS AND SPICES	Takako Inui	Suntory Beer Ltd., Beer Development Department, Osaka, Japan
P106	TRITORDEUM A NOVEL INGREDIENT FOR BREWING	Jose Luis Olmedo	Hijos de Rivera Inversiones Corporativas, S.L.
P107	OLIVE LEAVES AS NUTRACEUTICAL INGREDIENT IN CRAFT BEER PRODUCTION	Manfredi Guglielmotti	University of Udine, Italy, Dept. of Agricultural, Food, Animal and Environmental Sciences

TOPIC: WORT PRODUCTION

P108	SPECIAL MALT ADAPTED HOP DOSAGE TO IMPROVE OXIDATIVE BEER STABILITY	Thomas Kunz	Technische Universität Berlin, Institute of Food Technology and Food Chemistry, Chair of Brewing Science, Seestraße 13, 13353 Berlin, Germany
P109	AND IN THE END, IT'S ALWAYS BEER - A COMPARISON OF THREE LAUTERING SYSTEMS	Nele Bastgen	Ziemann Holvrieka GmbH

P110	INVESTIGATING THE EVOLUTION OF STALING ALDEHYDES DURING THE BREWING PROCESS AND BEER AGING	Maciej Ditrych	KU Leuven, Faculty of Engineering Technology, Technology Campus Ghent, Laboratory of Enzyme, Fermentation and Brewing Technology, Ghent, Belgium
P111	A FRESH LOOK AT SMALL AND LARGE STARCH GRANULES IN BARLEY MALT	Charlotte De Schepper	Laboratory of Food Chemistry and Biochemistry (KU Leuven)
P112	INFLUENCE OF COMPOSITION OF WORT ON THE QUALITY OF LAGERS	Pavel Dostálek	Department of Biotechnology, University of Chemistry and Technology, Prague
P113	ASSESSMENT OF METAL CHELATION DURING THE MASHING STAGE OF BREWING	Tuur Mertens	Technical University of Berlin, Institute of Food Technology and Food Chemistry, Department of Brewing Science, Berlin (Germany)
P114	PREDICTION MODEL FOR EXTRACT AND FERMENTABLE SUGARS	Héctor Hugo Rivera Yerena	Tecnológico Nacional de México en Celaya
P115	NEAR INFRARED SPECTROSCOPY AND MASHING – A PROMISING APPROACH FOR REAL TIME INLINE QUALITY CONTROL?	Florian Conradi	ILT.NRW - Institute of Food Technology.NRW
P116	HYDROGEN PEROXIDE FORMATION AND -INACTIVATION BY MALT-DERIVED THIOLS DURING MASHING	Matthias Baldus	Technische Universität Berlin, Institute of Food Technology and Food Chemistry, Chair of Brewing Science
P117	VARIETAL DEPENDENCE OF FE, MN AND CU CONTENT AND CONTRIBUTION INTO BREWERS WORT IN POLISH AND AMERICAN HOPS	Olga Szczepanik	University of Agriculture in Krakow
P118	MASH HOPPING - A PERFECT SOLUTION FOR A HOPPY SOUR BEER	Aleksander Poreda	University of Agriculture in Krakow, Faculty of Food Technology

TOPIC: YEAST AND FERMENTATION

P119	A NEW METHOD FOR SCREENING INDUSTRIAL YEASTS WITH FERULIC ACID DECARBOXYLATION ACTIVITY BASED ON SNP MARKERS	Chen Lu	State Key Laboratory of Biological Fermentation Engineering of Beer
P120	INFLUENCE OF LATE HOPPING ON FERMENTATIONPPERFORMANCE OF YEAST	Stefan Hanke	Bitburger Braugruppe GmbH
P121	IDENTIFICATION AND EXPLOITATION OF MANNITOL-PRODUCING LEUCONOSTOC CITREUM STRAIN TO PRODUCE NATURALLY-SWEETENED, REDUCED-SUGAR, FERMENTED OAT AND BARLEY BASED FUNCTIONAL BEVERAGES	Kieran Lynch	School of Food and Nutritional Sciences, University College Cork, Cork, Ireland;
P122	ISOLATION, IDENTIFICATION AND CHARACTERIZATION OF WILD YEAST, SACCHAROMYCES CEREVISIAE AFY-5 FROM SOUTH KOREA FOR THE BEER BREWING	Ha Yeon Lee	Agro-food Research Institute, Gangwon Agricultural Research and Extension Services (GARES)
P123	REDUCTION OF DIACETYL FORMATION DURING BREWING USING CRISPR/CAS9 GENETICALLY MODIFIED YEAST	Ida Kallehauge Nielsen	Technical University of Denmark

P124	VOLATILE ORGANIC COMPOUNDS MONITORING DURING LAGER BEER FERMENTATION: EFFECTS OF YEAST RE-PITCHING AND PROCESS VARIABILITY	Ana C. Vieira	Faculdade de Ciências Exatas e da Engenharia, Universidade da Madeira, Campus Universitário da Penteada, 9020-105 Funchal, Portugal
P125	SYNERGIES BETWEEN LACTIC ACID BACTERIA AND BREWING YEASTS ON FLAVOUR PROFILE OF SOUR BEERS	Van Nedervelde Laurence	Labiris
P126	THE IMPACT OF TEMPERATURE ON KVEIK FERMENTATION KINETICS AND FLAVOUR COMPOUND PRODUCTION	Barret Foster	University of Guelph

New Barley Solutions for all brewers

Carlsberg Research Laboratory
Zoran Gojkovic, PhD & Brew Master
Brewing Science & Technology

Carlsberg Group

BREWING FOR A BETTER TODAY & TOMORROW

Carlsberg Group

Global Scientific Collaboration Network

Carlsberg Group

Barley breeding capabilities

Hop breeding capabilities

Barley field trials

Hop field trials

Fermentation expertise

Yeast breeding know-how

Brewing capabilities

The barley as taste driver

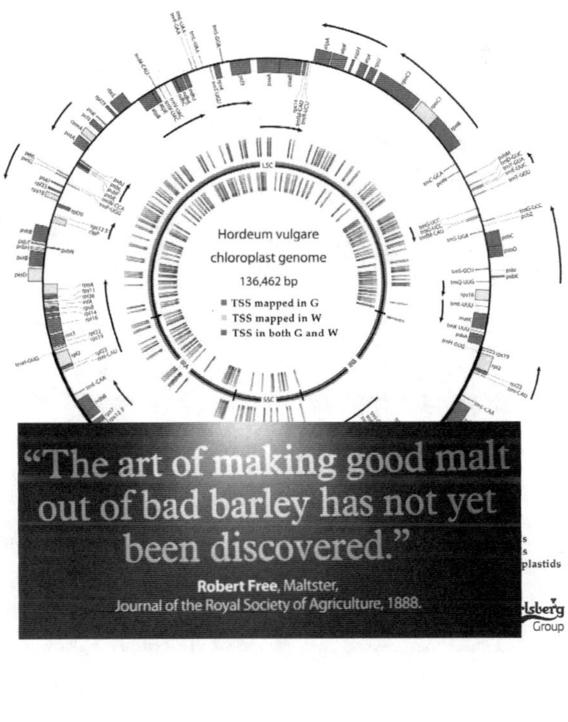

New barley types

Different malt stages

Day 1 Day 3 Day 5

Carlsberg Group

Barley malt – different color, flavour & body

Pilsner malt / Pale malt
Bready / grainy aroma
Lots of carbohydrates for fermenting into alcohol

Black malt
Coffee/chocolate aroma
No sugar molecules left for the yeast to eat

Crystal malt
Tastes like caramel, dark bread, nut and more. Sugar molecules are caramelized.

Carlsberg Group

Other cereal malts

Wheat

Rye

Rice (koji)

Adjuncts

Adjuncts are defined as a source of fermentable sugar NOT originating from malt.

Adjuncts can be sugar/cane sugar, unmalted grains such as corn, rice, rye, oats, barley, wheat and different fruit juices or other sources of sugar.

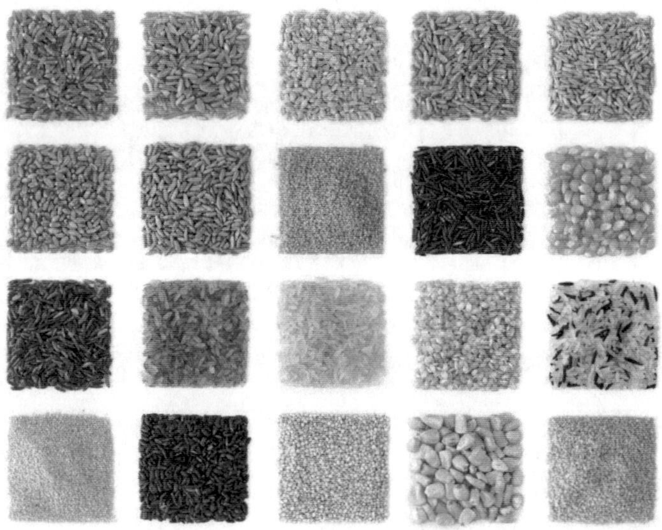

New scientific breakthrough in barley for improved beer quality and higher efficiency

Null-LOX 1G barley
Freshness
No papery off flavor
Flavour stable
Foam stability

Null-LOX 3G barley
Freshness
Foam stability
Energy savings,
Less water loss Improved beer quality

Null-LOX 4G barley
Same as for Null-LOX 3G
+ Improved beer clarity

Climate tolerant barley
Heat and drought tolerance

CARLSBERG
RESEARCH
LABORATORY

Carlsberg Group

Barley lipoxygenase pathway

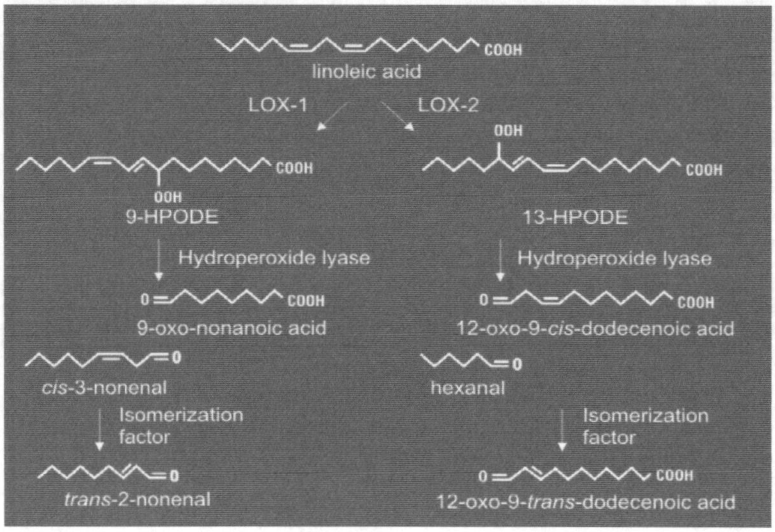

Beer freshness & T2N

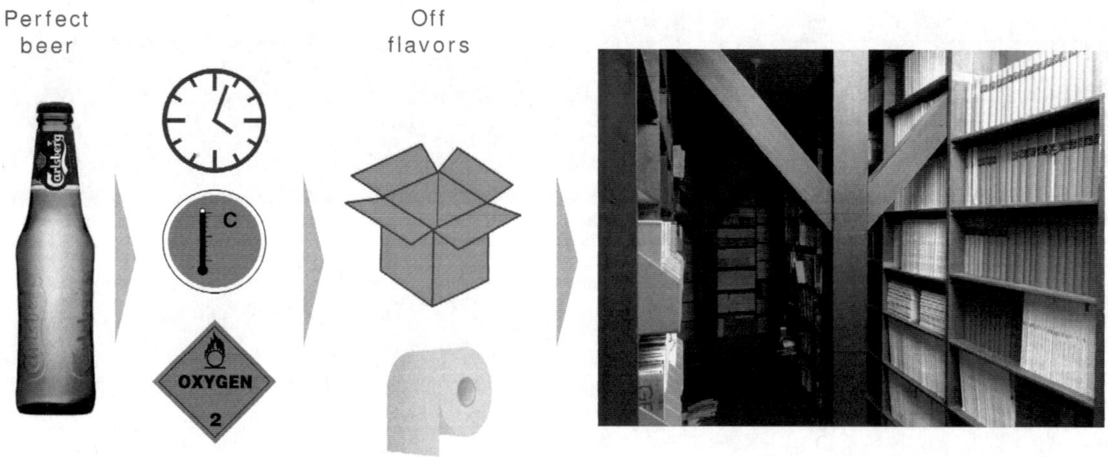

Double Null-LOX barley - **unpreceded beer freshness**

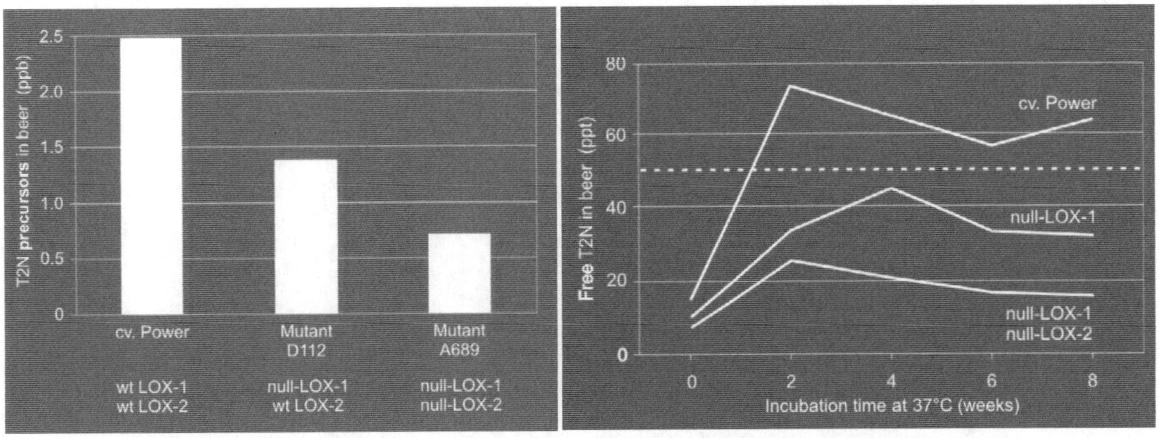

External Null-LOX breeders: Secobra, Sejet, Ackermann, Uni. Adelaide, KWS
Null-LOX1G and 3G available to all malsters, brewers and breeders

Carlsberg Group

DMS in beer

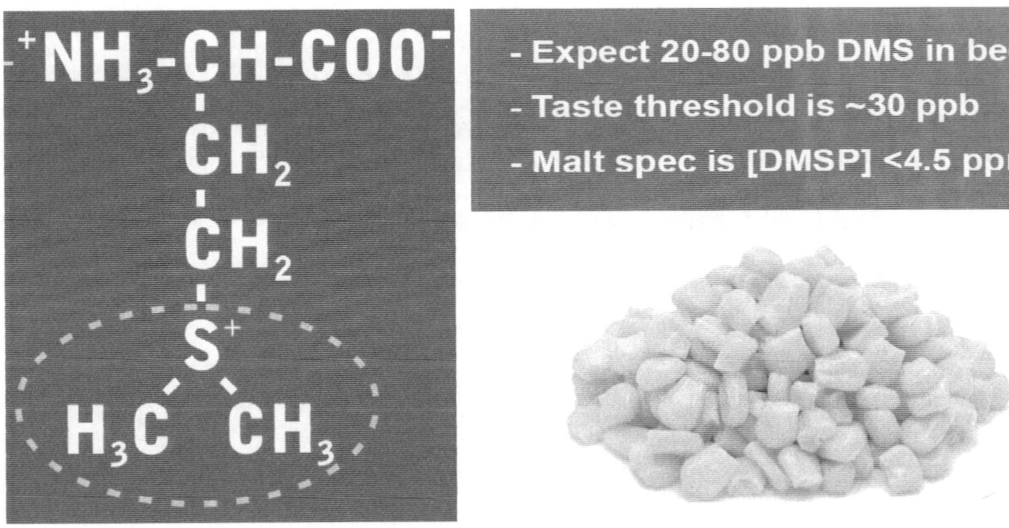

- Expect 20-80 ppb DMS in beer
- Taste threshold is ~30 ppb
- Malt spec is [DMSP] <4.5 ppm

Carlsberg Group

Brewing with null-MTT malt gives null-DMS beer

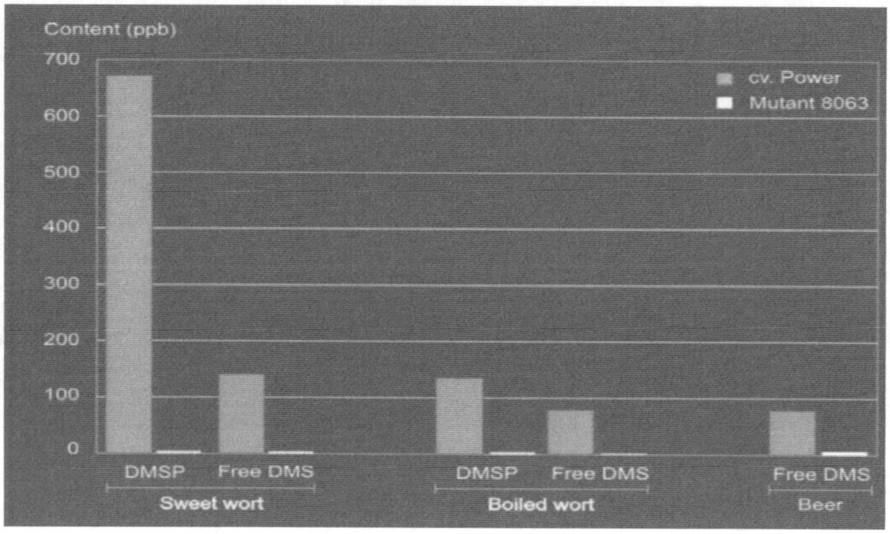

Carlsberg Group

Carlsberg barley key traits

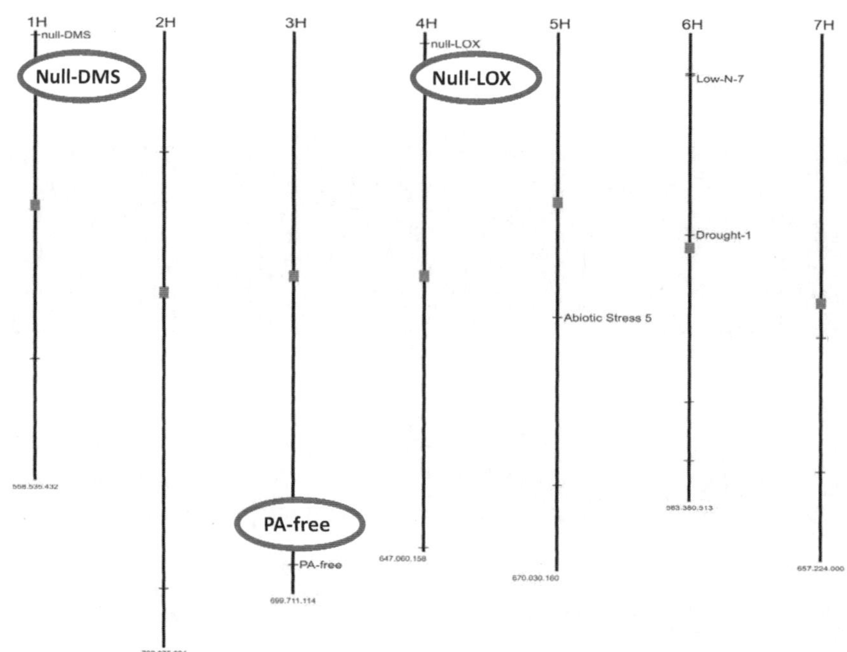

Success rate ~ 1.5%

28

Barley up-scaling from few grains to 10 K tones

50 grains January 2019

1 kg April 2019

40 kg **4 tons** **160 tons**

5.000-10.000 tons August 2021

29

Carlsberg Group

Benefits of Null-LOX3G barley varieties

Malting

- Lower kilning temperature / shorter time
- Energy savings (approx. 5%)
- Higher malting capacity (up to 10%)
- Climate – greenhouse gas reduction

Brewing and Processing

- Higher enzyme activity (approx.25%)
- Less evaporation
  - Energy savings (approx. 5%)
  - 16-20,000 t CO_2 for CB group
- <u>Potential</u> higher RDF (up to 3%)

Beer Quality Improvements

- Freshness, flavour and foam stability

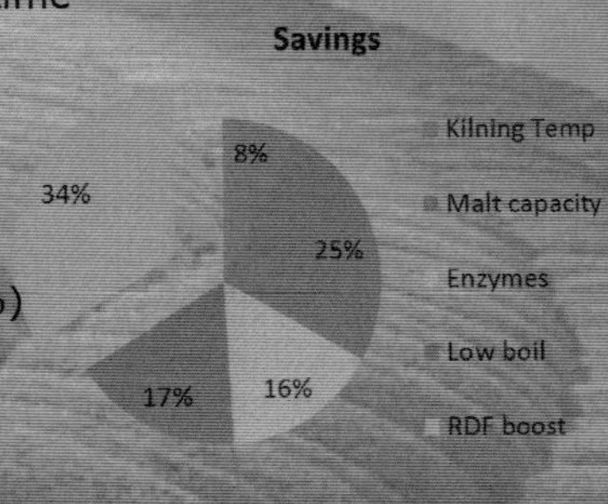

Null-LOX3G malting and brewing:
Beer taste stability during 12 month of storage

Reduced kiln time: No taste effect of short malting regime

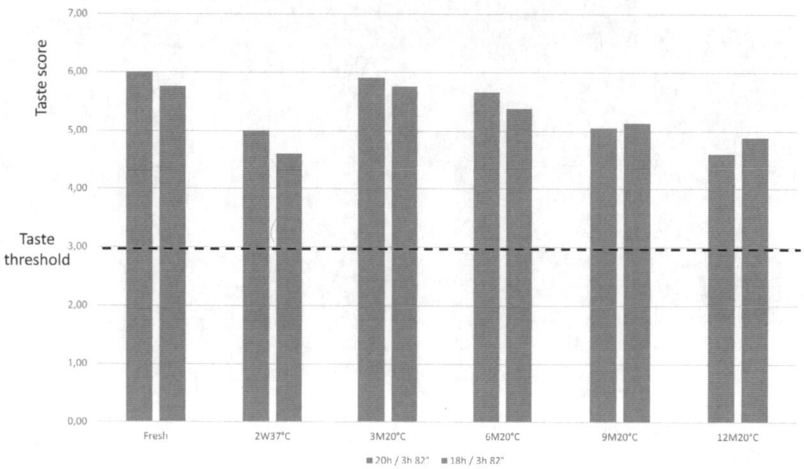

Malting Barley
CB Canut

Sample set **4 x 320 t** batches

New Carlsberg varieties with Null-LOX3G offer **uncompromised taste** when reducing kiln time

31

Carlsberg Group

What not to do

Impact of the climate change on hops – exemplified by the Hallertau Growing Region

Florian Schüll
Adrian Forster

02. Juni 2019

Climate during the main growth period

Daily average temperature in **Bavaria** June-August **1881 to 2017**

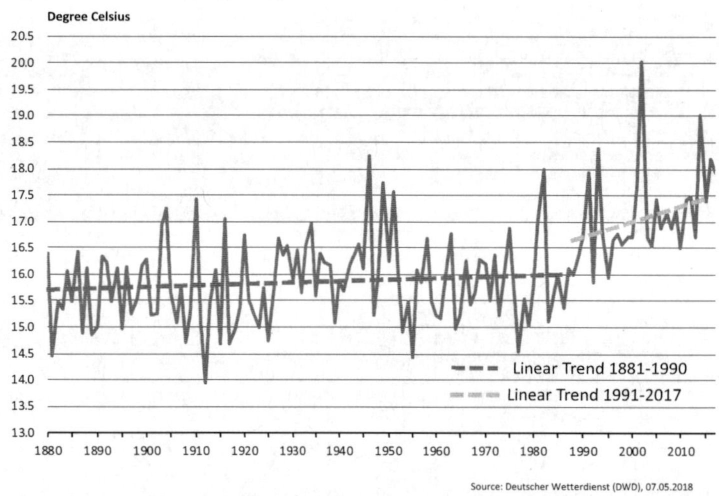

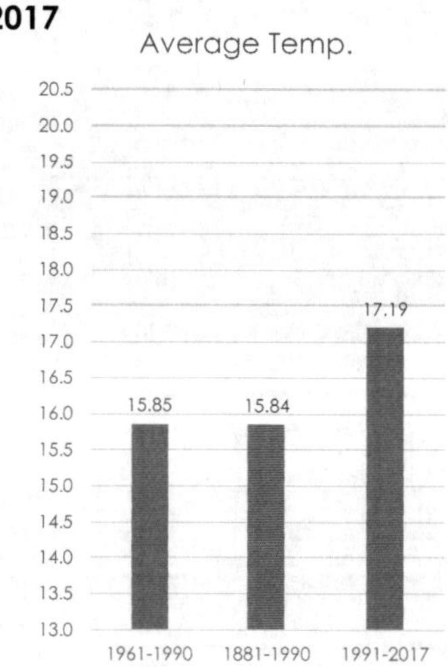

Climate during the main growth period

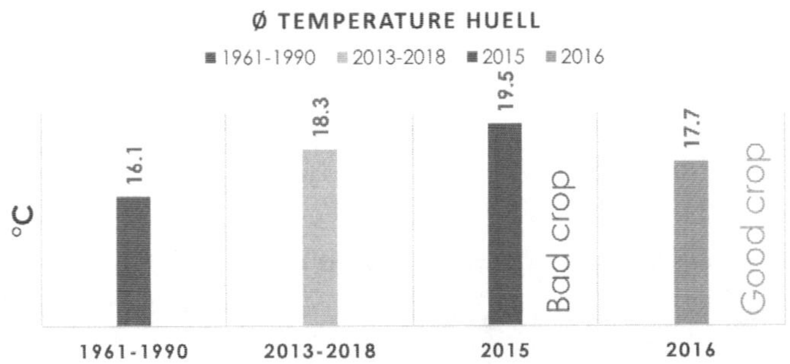

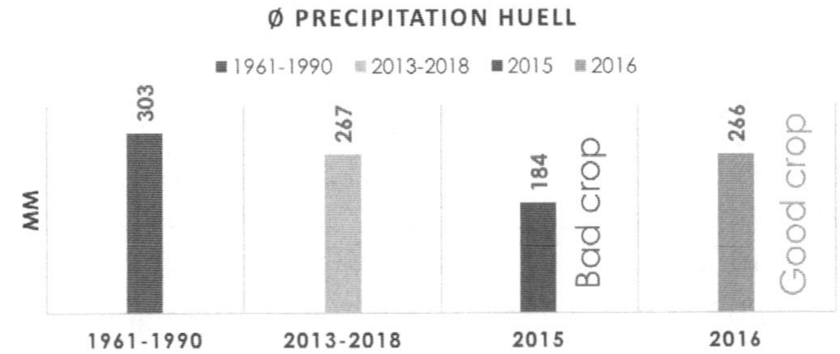

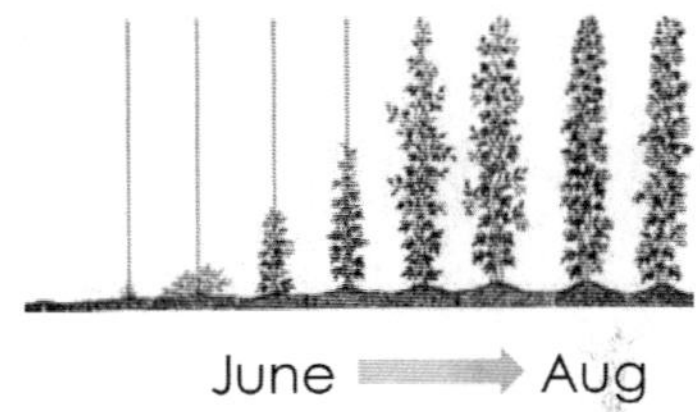

Impact of weather conditions on hop compounds

Our research shows for the relevant hop compound groups :

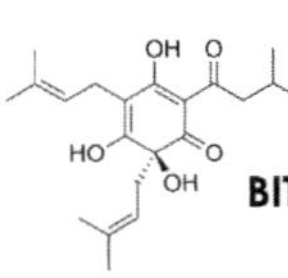

BITTERNESS

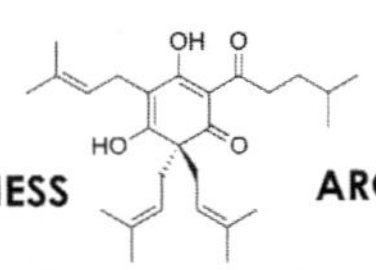

AROMA

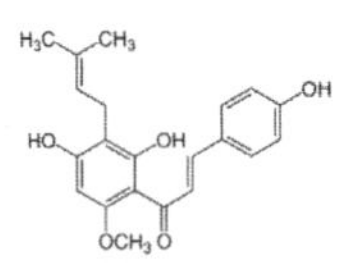

Alpha Acids ⟶Beta Acids⟶Total oil content ⟶ Polyphenols

Bitter compounds suffer more than aroma compounds and esp. polyphenols

⟶ we focus in this presentation on the alpha yield of different crop years

Impact of the crop year on yield shown on 2 varieties

Method 1: Comparison of successive harvests with completely different weather conditions.

2015: bad crop (dry and hot)
2016: good crop ("normal" weather conditions)

	Perle			Magnum		
	2015	2016	2015/2016	2015	2016	2015/2016
Yield [to/ha]	1,4	2,3	60 %	1,73	2,14	81 %
alpha acids [%-w/w]	4,5	8,2	55 %	12,6	14,3	88 %
Yield [kg alpha acids /ha]	**63**	**191**	**34 %**	**218**	**306**	**71 %**

⇨ **Perle: Differences of 2/3 possible** **Magnum: less than 1/3**

5 HVG

Impact of the crop year on alpha yield per hectare

Comparison of a poor (2015) and a good (2016) hop harvest in Germany

$$Factor = \frac{\text{yield alpha kg/ha in 2015}}{\text{yield alpha kg/ha in 2016}}$$

Aroma	Spalter	28%
	Perle	34%
	Nothern Brewer	36%
	Spalter Select	39%
	Hallertauer mfr.	40%
	Saphir	42%
	Hall. Tradition	45%
	Opal*	45%
	Tettnang Tettnanger	46%
	Hersbrucker	56%
	Smaragd*	66%
	Average	43%

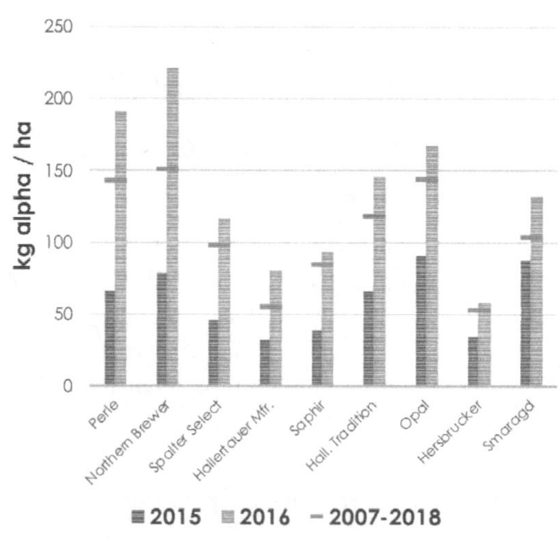

6 HVG

Impact of the crop year on Yield

Comparison of a poor (2015) and a good (2016) hop harvest in Germany

$$Factor = \frac{\text{yield alpha kg/ha in 2015}}{\text{yield alpha kg/ha in 2016}}$$

Bitter	Taurus	52%
	Nugget	56%
	Herkules*	64%
	Magnum	71%
	Polaris*	77%
	Average	64%

Flavor	Cascade	52%
	Hall. Blanc*	56%
	Hüll Melon*	65%
	Mandarina Bavaria*	71%
	Average	61%

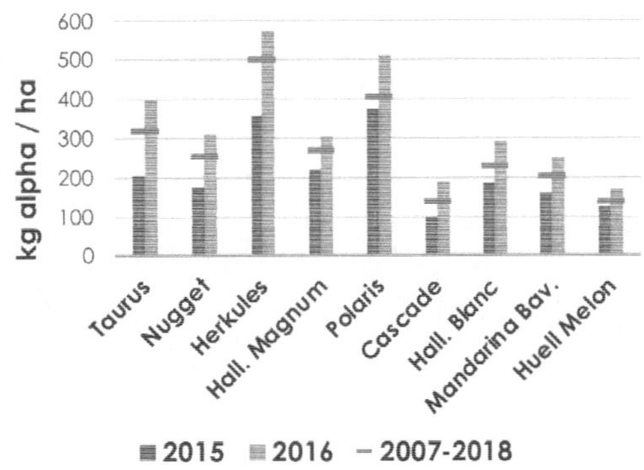

⇨ **Traditional aroma hops react more sensitive** to negative weather conditions **than bitter and recent aroma hops (flavor).**

7

Background information for the calculation of the yield kg alpha / ha for german varieties

All needed data used for the calculations is publicly available.

Acreage [ha]	https://www.deutscher-hopfen.de/de/Hopfen-Info/Anbauflaechen
Harvest yields [tons]	https://www.hallertauerhopfen.de/de/Hopfen-Infos/Marktbericht-und-EU-Bericht
Alpha acid content (AHA)	www.brauwelt.de

What we claculate is: yield alpha $[\frac{\text{kg alpha}}{\text{ha}}] = \frac{\text{harvest yield [kg]} \times \text{alpha acids[\%]}}{\text{acreage [ha]}}$

What has to be considered for all varieties marked with a * in the tables on page 6 & 7:

The **harvest yield** of **new acreage** is **much lower** than of **production acreage!**

Depeneding on the **variety** there is a **wide range for the harvest yield of new acreage**. It differs from 0 up to 50 %.

Especially for new varieties with a huge percentage of new acreage it is necessary to consider an empiric correction factor.

8

Background information for the calculation of the yield kg alpha / ha for german varieties

Yield evaluation of new acreage exemplified on **Mandarina Bavaria**

$$\textbf{yield alpha}\left[\frac{\text{kg alpha}}{\text{ha}}\right]=\frac{\textbf{harvest yield}\,[\text{kg}]\times\textbf{alpha acids}[\%]}{(\textbf{production acreage}+(\text{new acreage}\times\textbf{correction factor})\,[\text{ha}]}$$

$$\frac{\textbf{harvest yield new acreage}\,[\text{kg}]}{\textbf{harvest yield production acreage}\,[\text{kg}]}=\text{correction factor}$$

Mandarina Bavaria	2014	2015	2016
Production acreage [ha]	44	118	230
New acreage	55	89	116
Total *acreage*	**99**	**207**	**346**
Total acreage **corrected**	44+(55x0,3)= **61**	118+(89x0,3)= **145**	230+(116x0,3)= **265**

Used empiric **correction factors** for this study :
Herkules: 0,45; **Opal**: 0,25; **Smaragd**:0,25; **Mandarina**: 0,3; **Polaris:**0,3; **Melon**:0,25; **Blanc**: 0,3

HVG

Approach to estimate the relationship between hop yield and climate for the Hallertau region - Method 2

The billing of **all available weather data against the hop yield** is **Sisyphus work.** It creates more confusion than clarity.

Calculation model for hop yield

A (simple) approach to estimate the relationship between hop yield and climate:

$$\textbf{Climate Factor}=\frac{\sum \textbf{Rainfall}\text{ June to August [mm]}}{\text{Average }\textbf{Temperature}\text{ June to August [°C]}}$$

IF Climate Factor is "high": enough rain and/or moderate Temp

IF Climate Factor is "low": too less rain and/or higher Temp

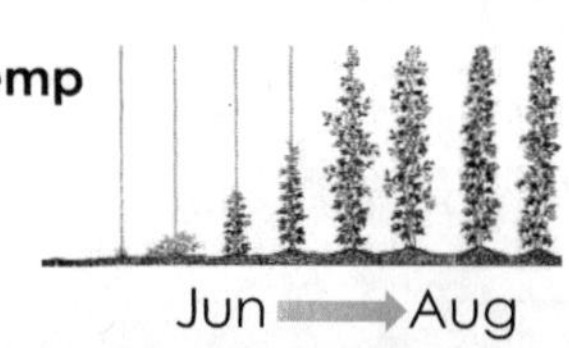

10 HVG

Approach to estimate the relationship between hop yield and climate for the Hallertau region – Method 2

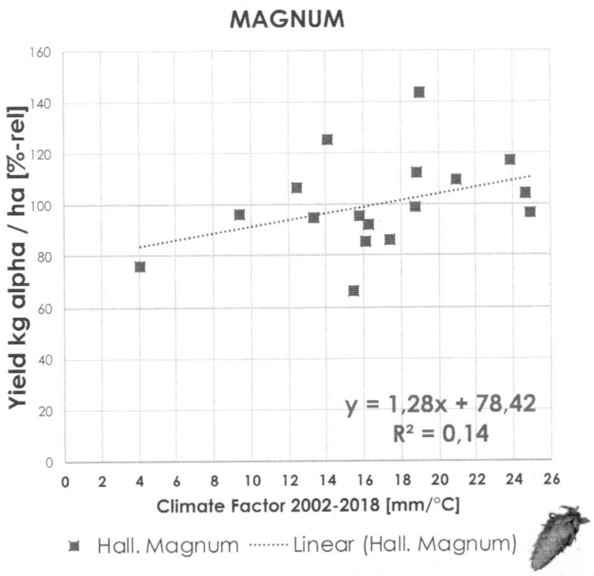

11

Approach to estimate the relationship between hop yield and climate for the Hallertau region – Method 2

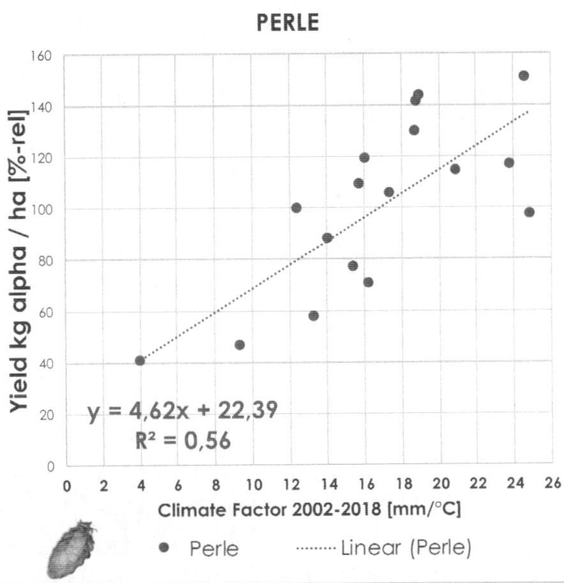

12

Approach to estimate the relationship between hop yield and climate for the Hallertau region – Method 2

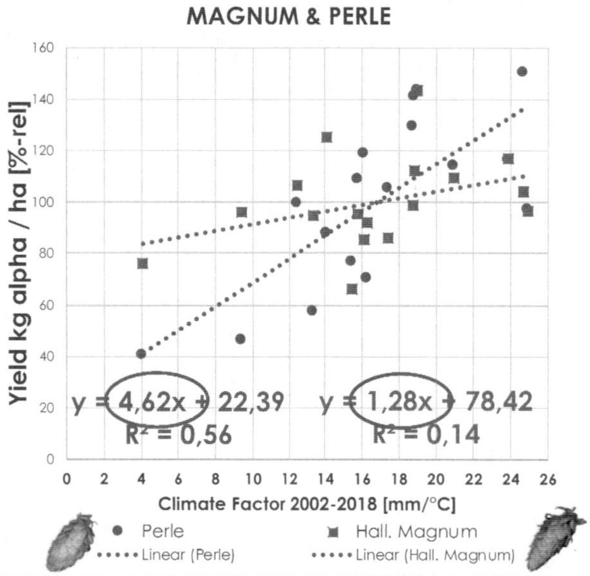

In this kind of plot, the **slope of the regression line** is an **indication of how sensitive a variety reacts on weather conditions.**

- **Flat:** low sensitivity
- **Steep:** high sensitivity

13

Approach to estimate the relationship between hop yield and climate for the Hallertau region – Method 2

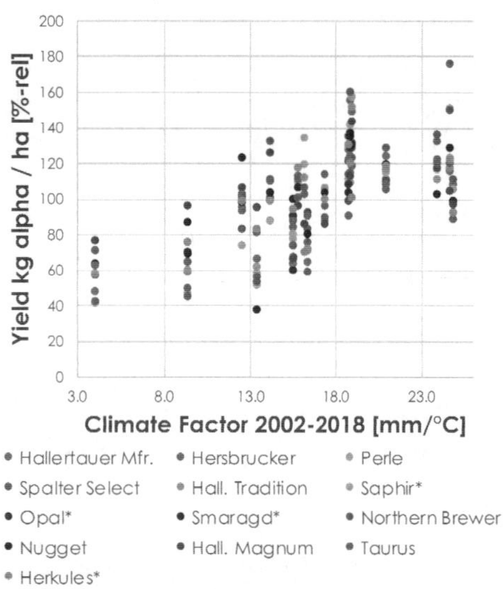

Slope of the regression line

Hall. Magnum	1,3
Herkules*	2,3
Taurus	2,5
Nugget	2,7
Hall. Tradition	3,1
Hersbrucker	3,9
Spalter Select	4,1
Smaragd*	4,2
Hallertauer Mfr.	4,2
Opal*	4,3
Perle	4,6
Saphir*	4,8
Northern Brewer	4,9

*new varieties - less datapoints

14

Approach to proof the relationship between hop yield and climate for the Hallertau region with Method 2

	Slope of the regression line	2003	2006	2013	2015
Hall. Magnum	1,3				
Herkules	2,3	-----			
Taurus	2,5				
Nugget	2,7				
Hall. Tradition	3,1				
Hersbrucker	3,9				
Spalter Select	4,1				
Hallertauer Mfr.	4,2				
Perle	4,6				
Northern Brewer	4,9				

Yield in BAD crop years
kg alpha / ha [%-rel]

Color sheme: dark green (99%) to red (37 %) = ratio of the long term average alpha yield/ha of a variety and the respective crop year

- Varieties with **steeper regression lines** show more often **lower yields in critical crop years**
- There are exceptions which can not be explained with our approach
  (as a result of e.g. of hail, late frost...)

15 HVG

Climate factor used for different growing regions

$$\textbf{Climate Factor} = \frac{\sum \textbf{Rainfall} \text{ June to August [mm]}}{\text{Average } \textbf{Temperature} \text{ June to August [°C]}}$$

	2002	2003	2004	2005	2006	2007	2008	2009	2010	2011	2012	2013	2014	2015	2016	2017	2018	Average
Hallertau	15,8	4,1	14,1	23,9	16,3	16,1	19,0	17,4	24,9	24,7	21,0	13,4	18,8	9,4	18,9	12,5	15,5	**16,8**
Tettnang	22,1	10,1	16,9	25,3	23,5	28,7	22,4	23,3	35,0	30,5	20,5	15,4	21,5	14,4	29,9	26,6	13,7	**22,3**
Spalt	14,4	7,3	12,1	17,3	11,4	19,4	11,6	11,6	17,5	16,4	9,0	14,8	15,7	6,8	10,0	16,4	3,9	**12,7**

(Nothing new) but **the fact**, that there are **different climate conditions** in the same year in **different growing regions** can be clearly shown with this kind of calculation.

16 HVG

Summary

We propose two ways for evaluating variety dependency on weather conditions

Method 1. Comparison of a good and a poor harvest

With 2015 and 2016 we had two consecutive harvests with **very different weather conditions** in the recent past.

The comparison of the alpha yield per hectare 2015/16 documents the differences in traditional varieties as well as younger ones such as Herkules, Mandarina, Melon, Blanc, Opal, Polaris and Smaragd. (For the younger ones the area correction is necessary.)

Advantage of this method:

Also new varieties, which were already in cultivation during these years, can be described.

Disadvantage of the method:

No long-term consideration. In general there is always a risk to draw wrong conclusions when considering in statistic only two points.

Summary

Method 2. Billing of long term information

Basis of this investigation is the **correlation of a climate factor** (which takes into account precipitation and temperature during the main growth period) and the **alpha yield per hectare** of several crop years (**2002 to 2018**).

The **slope of the regression line** is a statistical value to describe the **climate dependence of a variety**.

Advantage of the method:

Long-term observation with a more reliable informative value than 1st method

Disadvantage:

In this kind of consideration new varieties can not be included.

Summary

3. Correlation of method 1 and 2

for

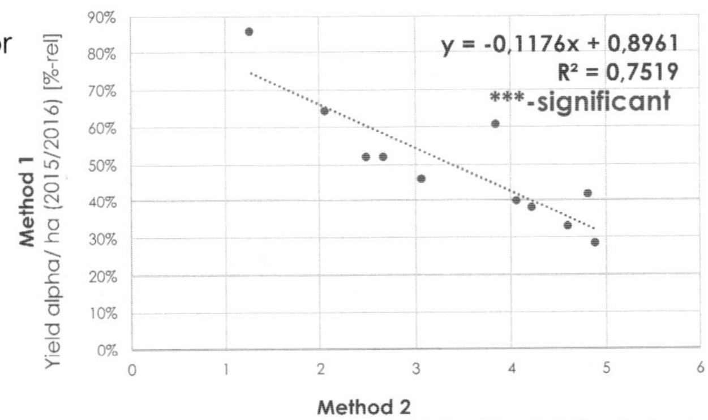

Slope of the regression line alpha yield /ha (%-rel) /climate factor

(11 Varieties; 2002-2018)

The results on climate dependency of german hop varieties in a **2-point analysis** (2015/16) and the **associated long-term** consideration **are well together**.

We conclude:
considering only two climatically different years, is acceptable for a first assessment of the climate stability of a new variety.

Conclusion

- **Traditional aroma hops** are regarding their alpha yield likely **stronger affected by extreme weather conditions** than **bitter and new aroma hops (flavor).**
- A **substitution** of such varieties **is recommended**! It is open when, to what extent and with which "better" varieties.
- **Breeders have a big job to do**!
- **Irrigation helps**, but is only part of the solution.
- **Climate change is a global issue and affects all growing areas, but in different ways.**
  (what happens if the water supply for irrigating agricultural products gets more strictly regulated, cause there is too less for the public?)

Climate Change
is no longer some far-off problem;
it is happening here,
it is happening now

Barack Obama

DTU

Processing Brewer's Spent Grains & Trub for Food Ingredients.

Tim Hobley
tjho@food.dtu.dk

100% DTU employed

(oh no, not again... :-)

DTU

Why has the brewing industry lagged behind in turning by-products into ingredients?

- Conservatism?
- Been tried before and failed?
- No/poorly defined business case?
- Lack of technical solutions?
- Perishes too fast?
- Not on company radar?

Compare to milk: Arla (4th largest dairy company in world) earns <u>much</u> more money from by-products than from milk (https://www.dr.dk/nyheder/regionale/oestjylland/fra-spildevand-til-stoerste-indtaegtskilde-vallen-er-arlas-svar-paa)

DTU

Putting numbers on the waste (DTU)

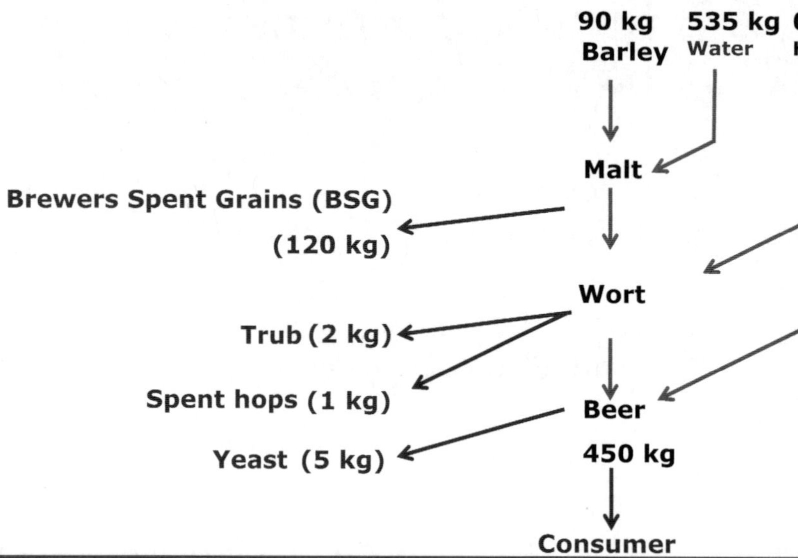

DTU

The potential in Europe

- Ca. 400 million hecto litres of beer produced in 2017*
  - Equates to ca. 10 million tonne of spent grains/year in EU
- Assume 4% protein in the BSG
  - 0.4 million tonne of protein...
  - Assume 270 euro/tonne protein: 108 million euro
- Protein from BSG is only a small fraction of the by-products

- What are by-products current value?
- +ve, 0, -ve
- Why???

*https://brewersofeurope.org/uploads/mycms-files/documents/publications/2018/EU-beer-statistics-2018-web.pdf

A New Technical Solution

The rotary drum filter press

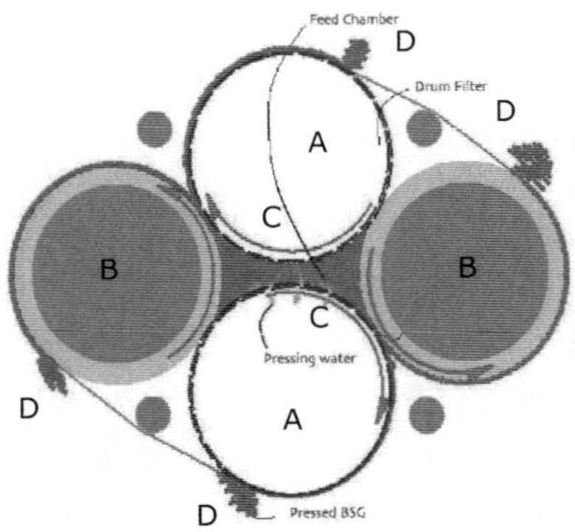

Fig. 1. Top view schematic of rotary drum filter.

Fig. 2. Picture of prototype rotary drum filter

What is the Rotary drum filter?

- A continuous rotary (no vacuum) filter press for dewatering solids
- DTU invention, patented

- Well developed via 3 innovation grants
- 1 tonne/h development and research machine at DTU
- Food approved contact materials
- Low energy consumption
- 80 tonne/h prototype tested by Larsson Starch Technology A/B
- Process directly from mash filter/lauter tun while hot

DTU

The processing concept we examined:

De-water the spent grains, keep the food chain intact

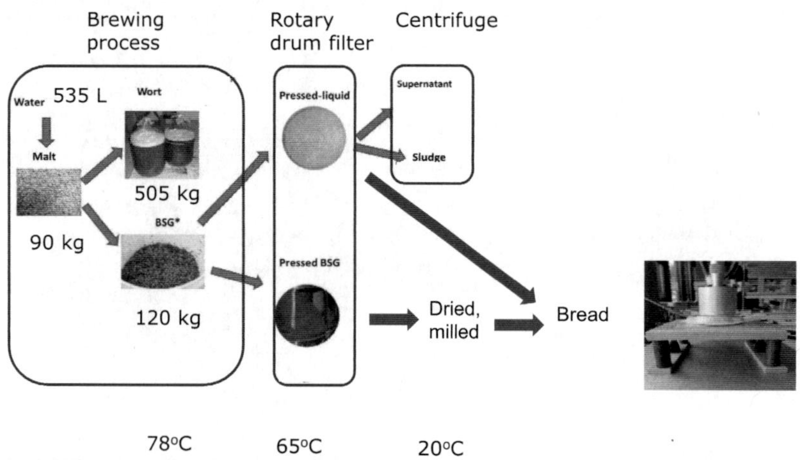

DTU

Rate of BSG filtration for liquid fraction (100 micron filter)

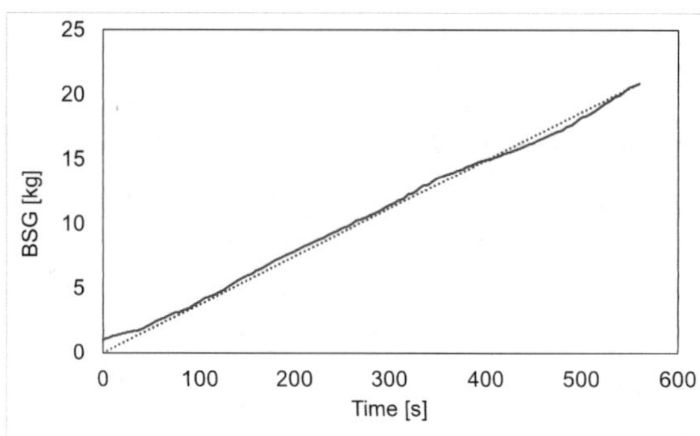

- Filter runs continuously, at between 0.5 – 1 tonne/h
- Rate is essentially constant over the 10 minute period examined
- No change in flux over extended periods of time.

DTU

Mass ballances from BSG fractionation

Liquid fraction

100 micron filter

Solids fraction

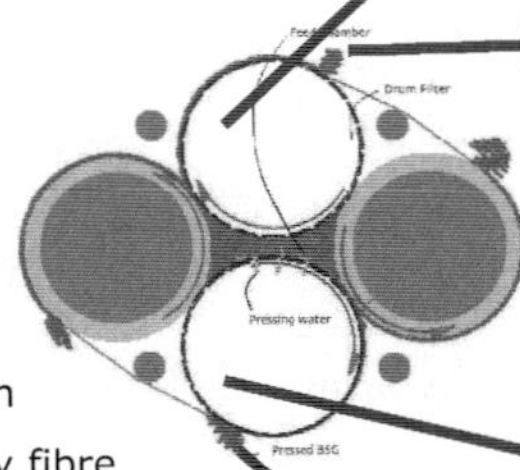

BSG

120 kg

23% DW of which

16.3% of DW is protein

42.3% of DW is dietary fibre

0.029 kg antioxidants

Liquid fraction

300 micron filter

Solids fraction

DTU

Key figures so far

- Half of the spent grains is recovered in a liquid fraction, half in a solid fraction
- 300 micron filter delivers more of both liquid and solid fraction than 100 micron filter
- Spent grains is dewatered from 23% to 35% dry matter, we have seen up to 42% DW.

After Pressing:

- Liquid fraction contains 0.03 % protein
- Liquid fraction contains 1-1.5% fibre

- Solid fraction contains ca. 7 % protein
- Solid fraction is enriched in fibre from ca. 10% to 20%

DTU

Centrifugation of liquid from filter (all in kg)

100 micron

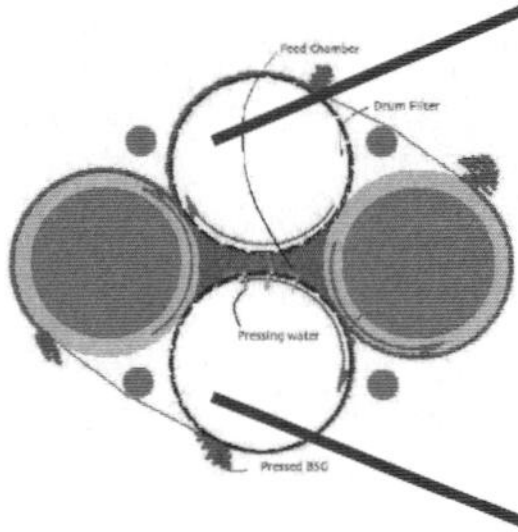

Liquid fraction

kg
25 wet
2.4 DW
0.4 protein
0.23 fibre
0.9 sugars
0.002 antioxidants

Centrifuge

Supernatant

- 18 wet
- 1.3 DW
- 0.055 protein
- 0.1 fibre
- 0.75 sugars
- 0.001 antioxidants

Sludge

- 6.7 wet
- 1.2 DW
- 0.38 protein
- 0.23 fibre
- 0.27 sugars
- 0.001 antioxidants

300 micron

Liquid fraction

kg
35 wet
3.6 DW
0.7 protein
0.5 fibre
1.4 sugars
0.002 antioxidants

Centrifuge

Supernatant

- 22 wet
- 1.7 DW
- 0.07 protein
- 0.1 fibre
- 1 sugars
- 0.001 antioxidants

Sludge

- 12.7 wet
- 2.4 DW
- 0.8 protein
- 0.55 fibre
- 0.54 sugars

June 3, 2019 DTU

DTU

Summary of data

Properties measured after filtration			BSG	Liquid fraction, filtrate		Solid fraction, filter cake	
				100μm	300μm	100μm	300μm
Spent grains from Unmalted Oat	Wet matter	Wet mass (kg)	85.1	19.26	29.53	10.42	20.17
		Percentage of wet mass	100	22.6	34.7	12.24	23.7
	Dry matter	Dry mass (kg)	27.15	4.29	6.50	4.56	8.41
		Dry weight-% [a]	31.9	22.26	22.0	43.8	41.7
		Percentage of dry mass	100	15.8	23.9	16.8	31
	Total Protein (% of the dry weight) by Dumas		8.53	6.32	6.45	9.03	8.57
	Soluble protein (mg/ml; by Bradford)		Na	0.374	0.417	Na	Na
	FAN (mg/ml)		Na	58	112	Na	Na
	Soluble sugars (g/L)	Maltose	Na	32.33	60.27	Na	Na
		Glucose	Na	2.68	3.49	Na	Na
	Specific gravity (g/ml)		Na	1.0564	1.0521	Na	Na
	Ash (kg)		0.55	0.015	0.024	0.20	0.34
Spent grans from Unmalted Rye	Wet matter	Wet mass (kg)	149.3	15.86	49.26	29.4	42.79
		Percentage of the wet mass	100	10.6	33.0	19.7	28.7
	Dry matter	Dry mass (kg)	47.0	3.43	9.95	12.17	17.59
		Dry weight-%	31.5	21.6	20.2	41.4	41.1
		Percentage of the dry mass	100	7.3	21.2	25.9	37.4
	Total Protein (% of the dry weight) by Dumas		3.77+-0.34	13.01+-2.31	4.73+-1.73	5.85+-0.39	5.55+-0.36
	FAN (mg/ml)		Na	72.85+-8.01	70.89+-1.52	Na	Na
	Soluble sugars (g/L)	Maltose [d]	Na	70.65+-5.31	71.18+-5.71	Na	Na
		Glucose [d]	Na	0.26	0.91+-0.22	Na	Na
	Ash (kg)		0.7	0.1	0.2	1.2	1.2
Trub	Wet mass (kg)		27.31	16.25	7.36	0.61	1.52
	% Dry weight		Na	13.7	13.7	26.6	27.1
	Total protein by Dumas % of the dry weight		Na	0.60±0.052	0.61±0.064	6.51±0.074	6.69±0.288

June 3, 2019 DTU

DTU

Studies at DTU of BSG fractions as ingredients

Bread

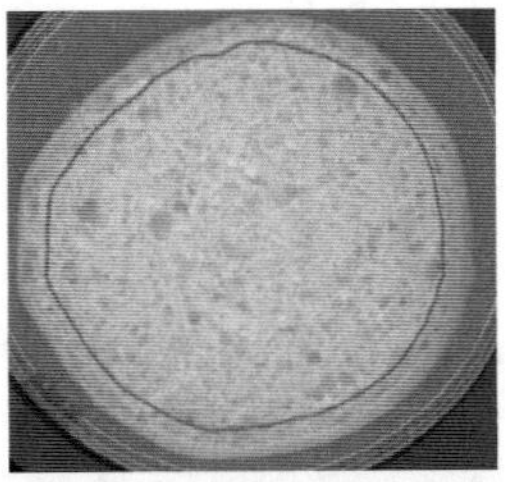

Reference

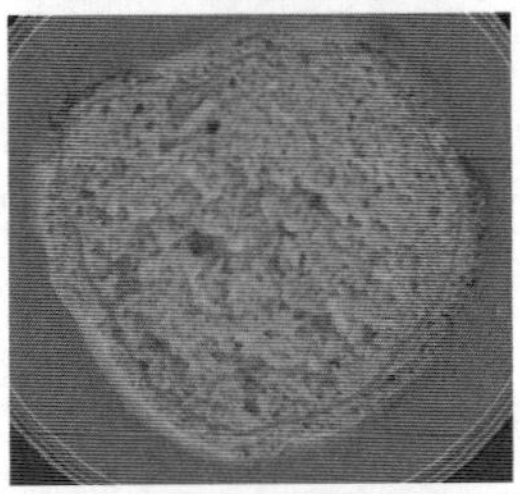

80% of water replaced

Sausages

RGB image

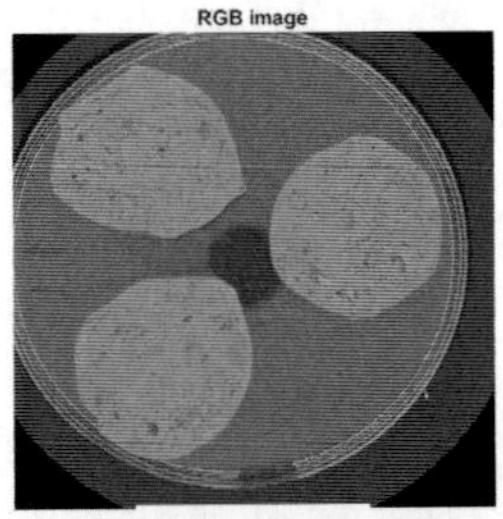

Reference

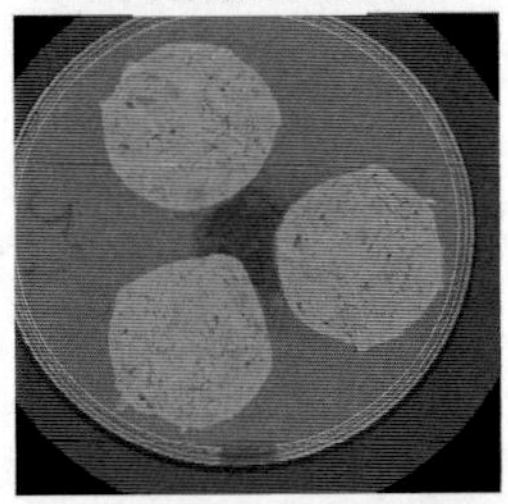

80% of flour and water replaced

Drinks

DTU

Texture analysis of water/flour substituted dough & bread

Sample	Hardness (N)	Resilience (%)	Cohesiveness	Springiness (%)	Chewiness
Dough					
Control	9.2	6.5	0.26	10.7	25.8
50%/10%	11.0	7.5	0.31	13.9	47.4
80%/10%	10.1	7.3	0.30	13.6	41.0
Bread					
Control	6.3	52.3	0.84	94.3	483.4
50%/10%	5.9 b	53.1 b	0.83 b	87.3	433.8 b
80%/10%	6.1 b	51.1 b	0.80 b	83.3	414.7 b

b = not statistically different to control (ANOVA)

- Wheat flour 550g, yeast 34.4g, tap water 232.1g, fine salt 13.8g.
- The straight-dough method, yeast and 'water' mixed for one minute before adding other ingredients and mixing.
- 50g and 30g dough pieces were raised at 30°*C* and a relative humidity of 80% for an hour.
- 50g pieces were baked in a preheated oven at 230°*C* for 20 minutes
- 30g pieces were tested after rising

DTU

Comparison of bread crusts

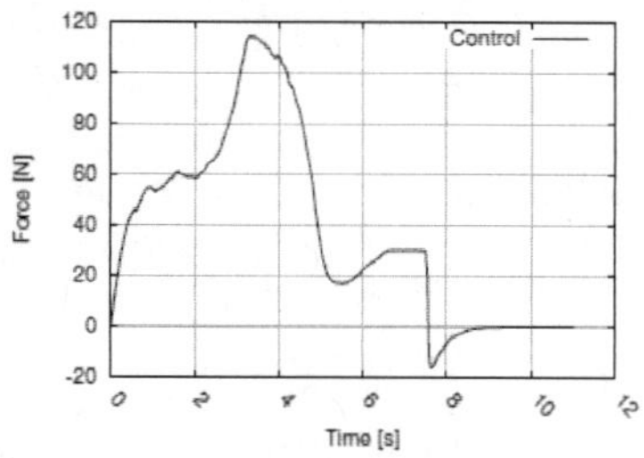

Control bread

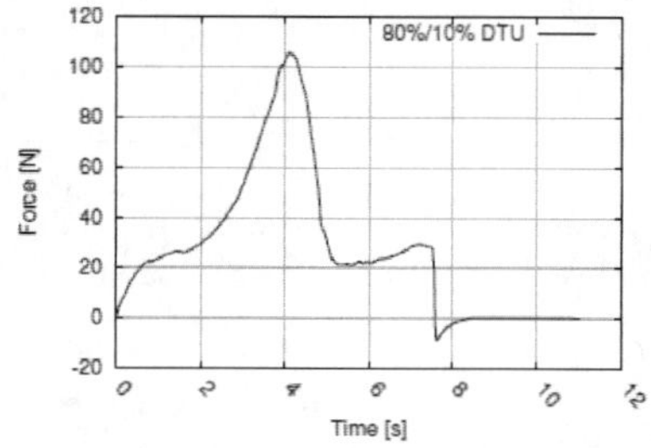

Bread with 10% flour and 80% water substituted with milled filtercake and filtrate

DTU

A potential process:

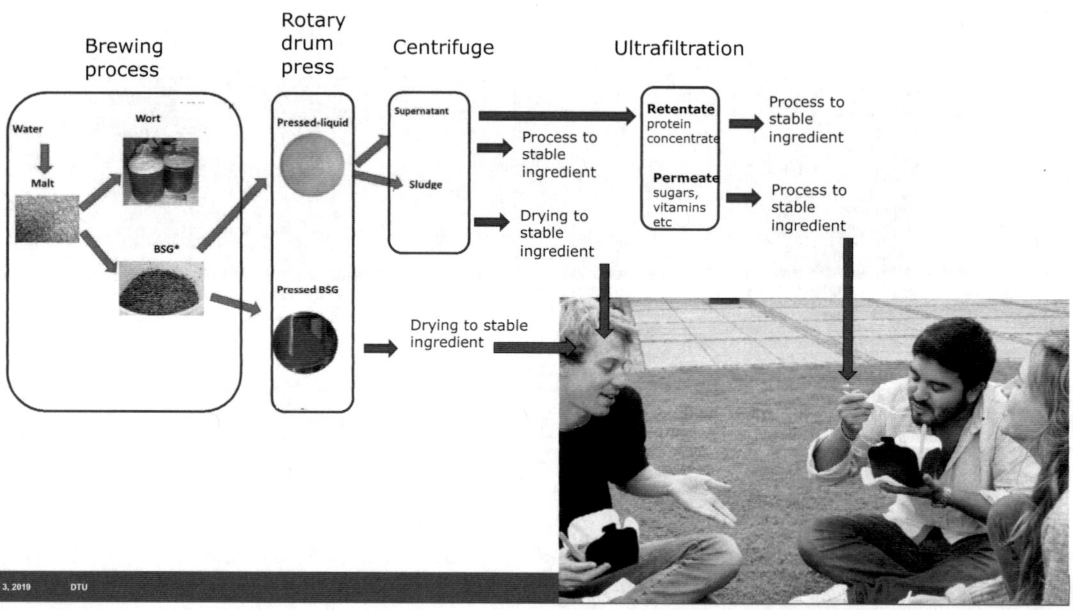

DTU

Conclusions

- The rotary drum press can be applied for dewatering and partial fractionation of BSG and trub.
- The fractions contain a variety of nutritional ingredients that can be used in bread baking without detrimental functional effects.
- Substitution of up to 80% of the water and 10% of the flour in bread can be an attractive way to improve the sustainability of beer brewing and bread production.

June 3, 2019 DTU

DTU

Thanks ☺

Matias Bjerregaard,
Katrine Kiib Rørby,
Daniel Stender Nøgaard,
Rasmus Frøding,
Radhakrishna Shetty,

Charlotte Jacobsen,
Jakob Lerce Skytte,
Preben Bøje Hansen,
Manuel Pinelo,
Nina Gringer.

Some of this work is published in: Bjerregaard, et al. (2019) Processing of brewing by-products to give food ingredient streams. *European Food Research and Technology*, 245: 545-558

Comparison of Saaz Hops Grown in Two Regions in Climatically Different Years

Andreas Gahr and
Adrian Forster

37TH CONGRESS
EBC
ANTWERP
EUROPEAN BREWERY CONVENTION 02–06 June 2019

Research Brewery St. Johann

HVG

Czech Hop Growing Areas – Hop Production (%)

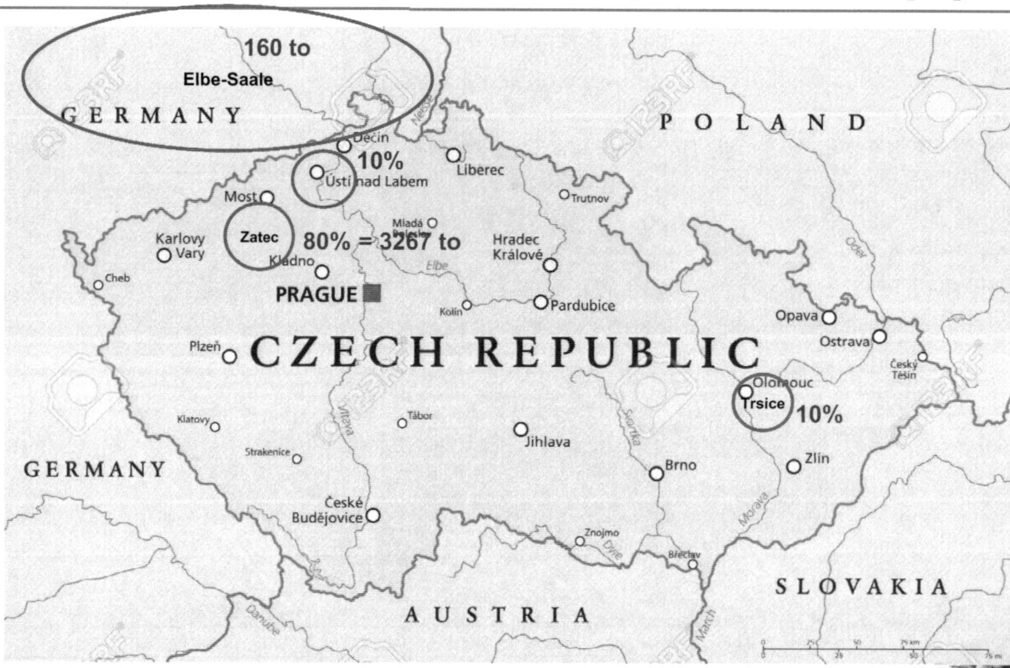

Climate Change – Interklim Study (DE/CZ 2014)

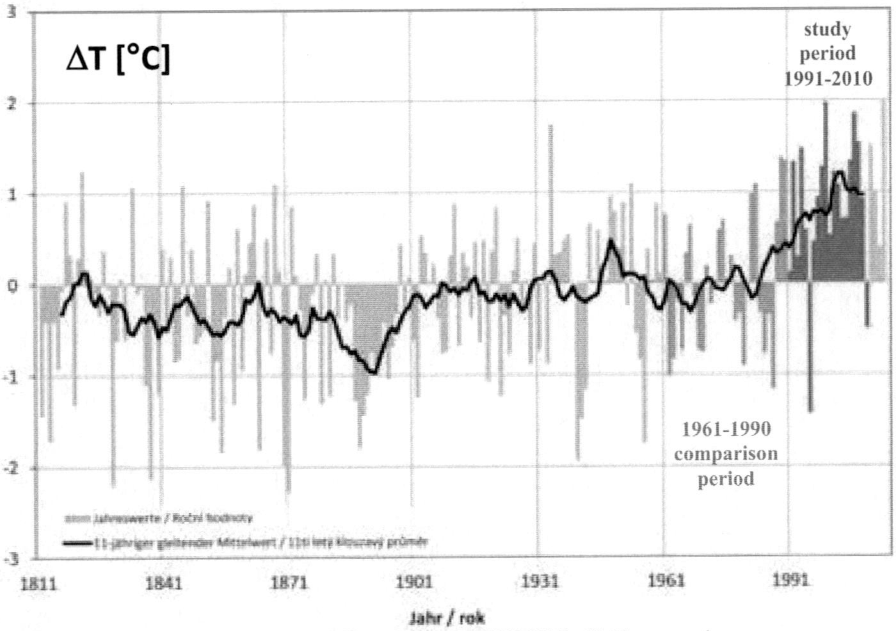

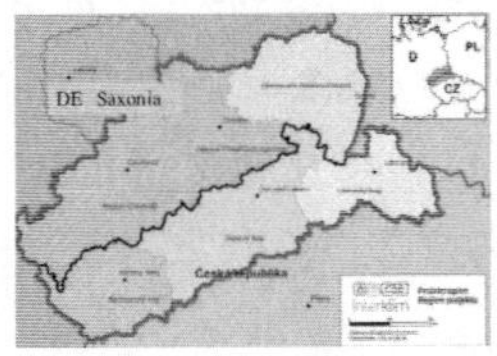

Yearly Average of the Air Temperature Difference (Dresden, 1812-2014) with 11-year running mean (Bernhofer et. al., 2009)

Warming in the Project Area

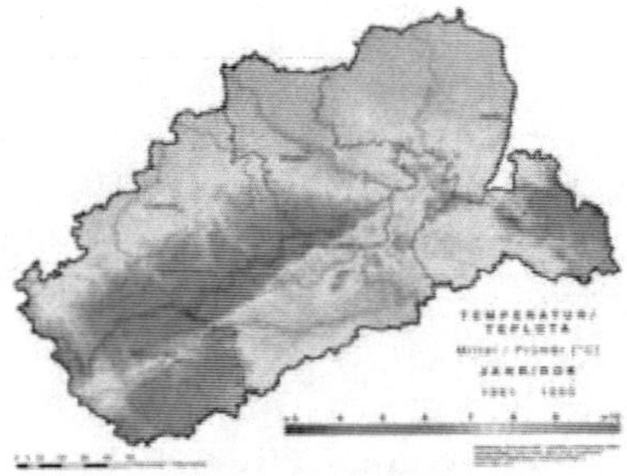

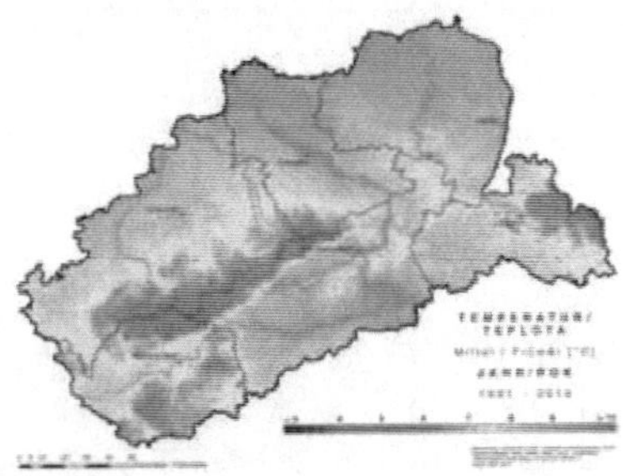

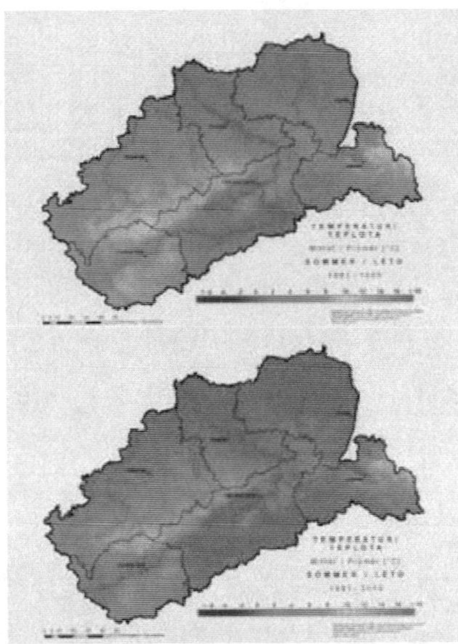

Average yearly temperature 1991-2010 compared to 1961-1990 (above)

Average summer temperature 1991-2010 compared to 1961-1990 (above)

Av. α-acid Content of CZ-Saaz Hops (1994-2018)

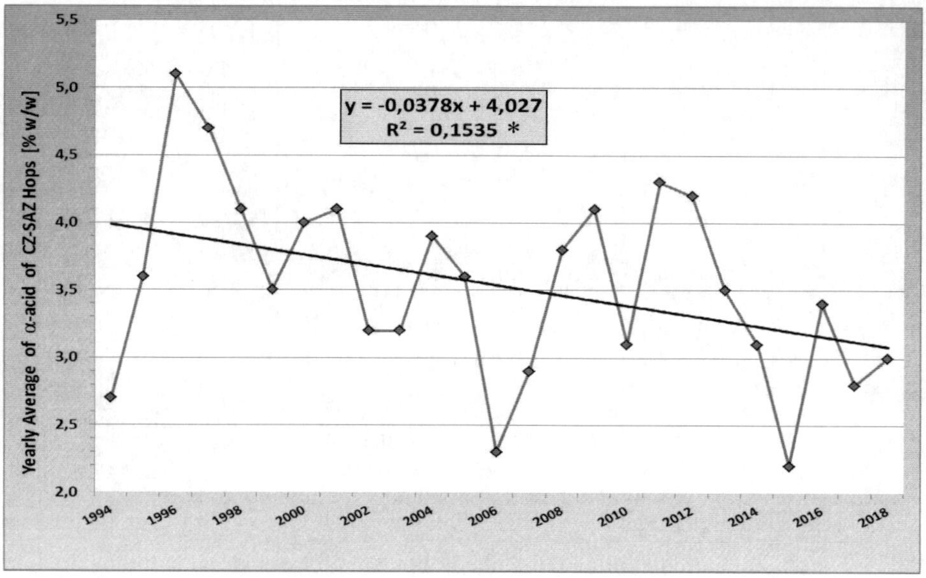

Data from: Mikyška, A. and Jurková, M. (2019) "Analysis and prognosis of bitter acids content in Czech hop varieties - year 2018 and long-term comparisons and trends", KVASNY PRUMYSL, 65 (2019), doi: 10.18832/kp2019.65.23

Andreas Gahr– EBC 2019 – Antwerp - 5

Introducing the „Climate Factor" (CF)

$$CF = \frac{Sum\ of\ Precipitation\ (June - August)}{Av.\ daily\ temperature\ (June - August)}\ [\frac{mm}{°C}]$$

CF = the higher the better in a range between 5 and 25

S P = the lower, the worse

Ø T = the higher the worse

Correlation between CF and av. alpha acid content (% w/w) and yield (kg alpha/ha).

Andreas Gahr– EBC 2019 – Antwerp - 6

CF and Rel. Hop Yield (%) for CZ-SAZ (2011-2018)

$$\text{Rel. Hop Yield } [\%] = \frac{\text{Av. Yield in the Year X } [\text{kg/ha}]}{\text{Av. Yield in the Years } 2011-2018 \; [\text{kg/ha}]}$$

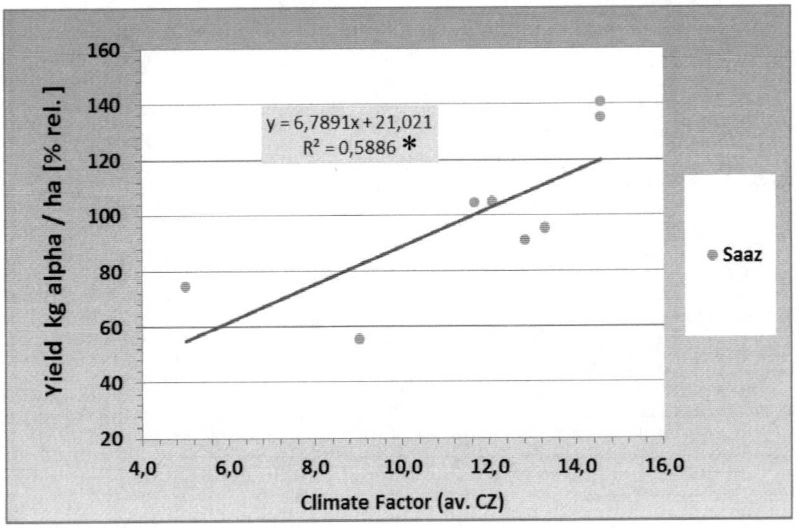

CZ-SAZ **relative hop yield (%) is signifcantly correlated with the „climate factor" CF in the observation period.**

Data from the Barth Report 2011-2017, Johann Barth & Sohn, Nuremberg, Bohemia Hop, Crop Report 2018 and Czech Hydrometeorological Institute, fully available since 2011, http://portal.chmi.cz/?l=en

CF and Rel. Yield (%) for Other CZ Hops (2011-18)

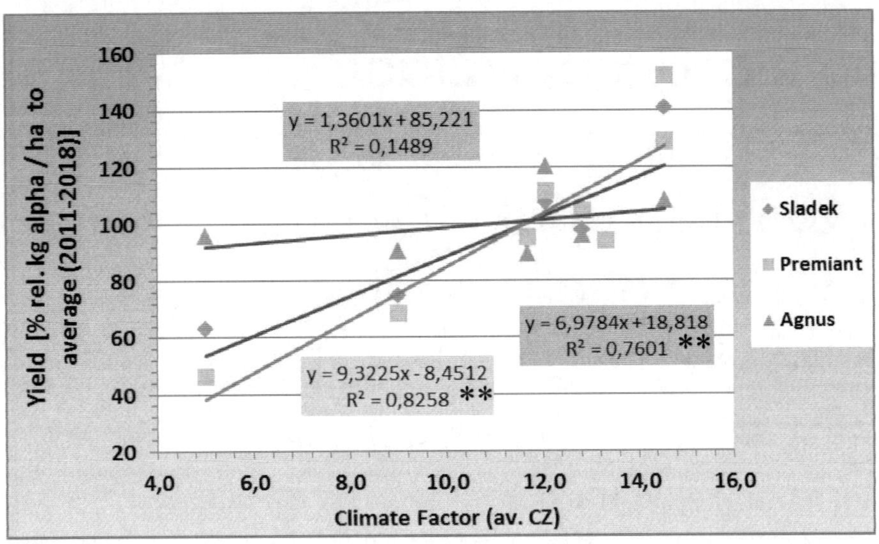

Yield of *Sladek* and *Premiant* also highly correlated with the „climate factor". *Agnus* yield shows low weather impact. The slope is a measure for the climate tolerance.

Data from the Barth Report 2011-2017, Johann Barth & Sohn, Nuremberg, and Bohemia Hop, Crop Report 2018

Comparison 2018 / 2016

	Variety			
	Saaz		Sladek	
Year	to/ha	kg α/ha	to/ha	kg α/ha
2016	1,52	51,7	2,39	174,5
2018	0,94	27,3	1,56	78,0
Difference	-38%	-47%	-35%	-55%

	Variety			
	Premiant		Agnus	
Year	to/ha	kg α/ha	to/ha	kg α/ha
2016	2,42	217,8	2,37	229,9
2018	1,41	66,3	2,16	235,4
Difference	-42%	-70%	-9%	+2%

Summary Climate Influence on CZ Hops

- Climate change is a fact.
- Hops react to the changing conditions.
- Hop varieties react differently.
- Introduction of climate factor (CF)
- Relative yield (kg α / ha) shows correlation to climate factor
- Security of supply ?

Future Strategy for CZ-SAZ Hops

Besides the development of **new varieties** that are less susceptible to drought:

- Conservation breeding program
- Renewal of the planted hop gardens
- More irrigation
- Growing the variety in other areas ?
- ...

Andreas Gahr– EBC 2019 – Antwerp - 11

Brewing Trials Crop 2016/2018

- Saaz hops have been grown already in the GDR in a considerable volume and in Germany since 2008 again due to a shortage in Saaz hops and to improve the reliability of supply.
- Brewing trials with Saaz variety hops from Elbe-Saale/Germany (ESAZ) and Žatec/Czech Republic (CZ-SAZ) cropped in 2016 and 2018.

Research Brewery St. Johann

Andreas Gahr– EBC 2019 – Antwerp - 12

Hops for the Brewing Trials

- Hops from 4 commercial pellet productions between 2 and 33 to stored and processed in a similar way
- 1 each from 2016 and 2018
- 1 each from CZ-Saaz and DE-Elbe-Saale
- Hops were comparable e.g. the HSI

Hop Analyses SAZ Hops 2016/2018

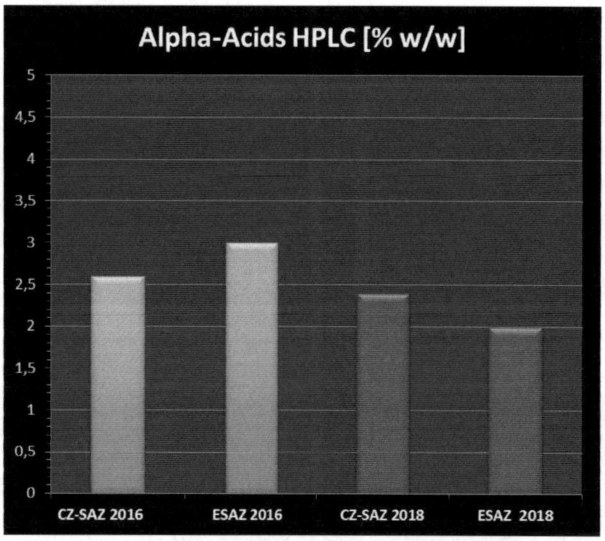

α-acids (R=0.3):

- ESAZ seems to be bit more susceptible to hot and dry weather conditions
- 2018 a bit lower in α as 2016

Hop Analyses SAZ Hops 2016/2018

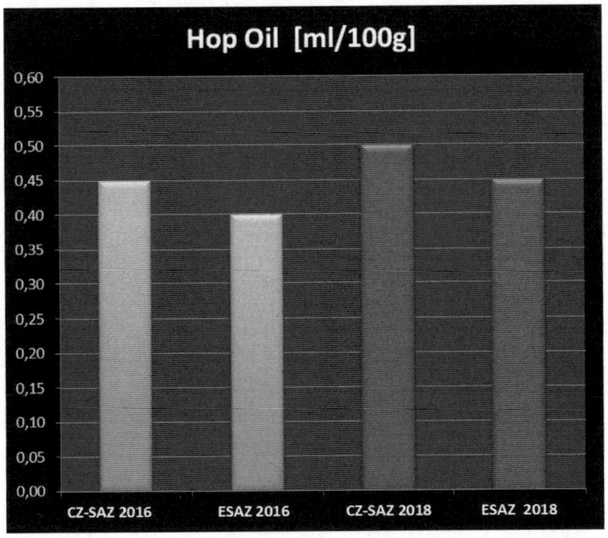

Hop Oil (EBC 7.10):

- Differences in crop year and growing area within the variance of the analysis (R=0.15)

0.17 0.13 0.21 0.22 Ratio oil/alpha [ml/g]

Hop Analyses SAZ Hops 2016/2018

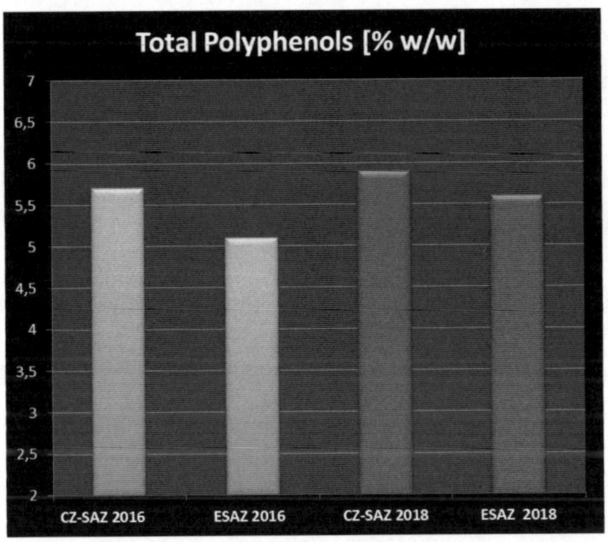

Total Polyphenols (R=1):

- ESAZ a bit less (ns) in comparison to CZ-SAZ in both years
- No influence of weather conditions

2.2 1.7 2.5 2.8 Ratio PP/alpha [g/g]

Brewing Trials (Crop 2016/2018)

- 100% Pilsener Malt
- Infusion mashing 52/64/72/78 °C
- Single variety hopping:
  - begin of boil: 4 gα/hl
  - end of boil: 150g/hl
  - whirlpool: 150g/hl
- Bottom fermented with a W34/70 yeast (@ 9°C)
- Kieselguhr filtration

Beer Analyses (Crop 2016/2018)

- General beer analyses like OG, alcohol, pH etc. comparable

		CZ-SAZ 16	ESAZ 16	CZ-SAZ 18	ESAZ 18
Bitterness	*IBU*	19	17	17	16
Isohumulones	*mg/l*	13,8	13,2	11,8	11,7
R-Linalool	*µg/l*	56	37	45	41
Transfer Rate Linalool	*%*	63	41	58	67
Total Polyphenols	*mg/l*	299	277	270	293
Transfer Rate Hop PP	*%*	58	59	48	53

- Linalool and TPP may seem different, but when having a closer look to dosage, transfer rates and analytical variation have to be regarded as similar.

Tasting Results: ESAZ and CZ-SAZ beers (2016/2018)

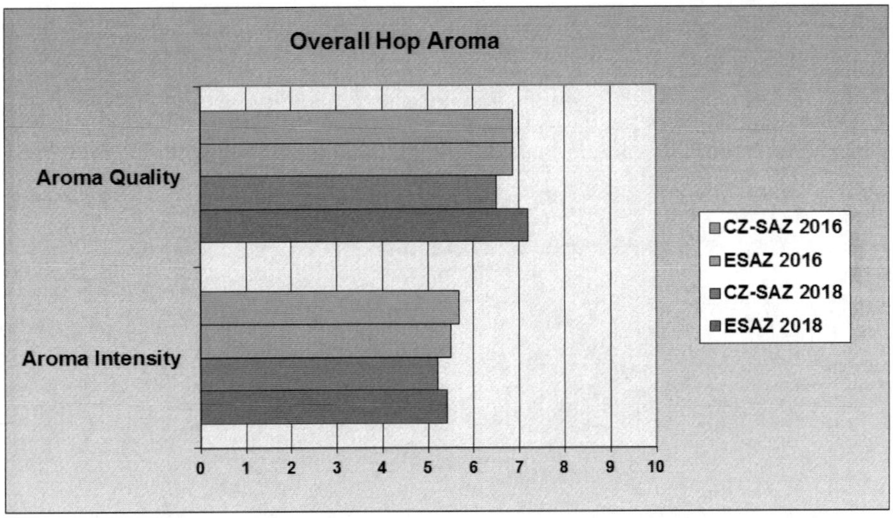

- Slight differences in hop aroma quality and intensity in different crop years and growing areas

Tasting Results: ESAZ and CZ-SAZ beers (2016/2018)

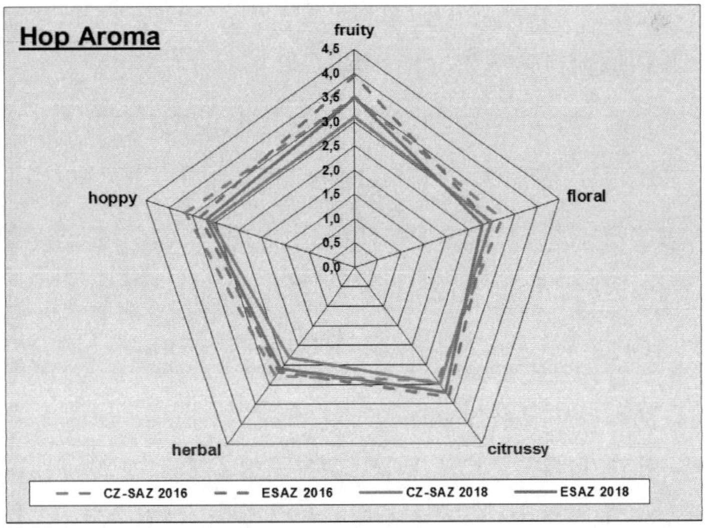

- Intensity of aroma descriptors similar in different crop years and growing areas
- DLG results similar (not shown)

Tasting Results: ESAZ and CZ-SAZ beers (2016/2018)

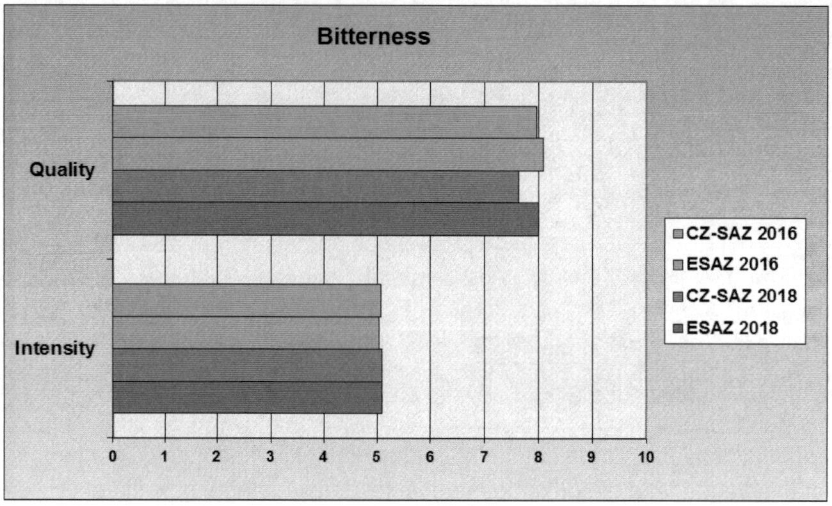

- Bitterness quality and intensity similar in different crop years and growing areas

CONCLUSION

- Saaz hops react sensitively to the changes of the weather conditions in CZ.
- The transfer of the cultivation to Elbe-Saale area had brought hops of similar quality and yield.
- The climate in the Elbe-Saale area does not differ substantially from that of the Saaz area.
- So is this the solution for the climate problem?
  Certainly not.
- The Czech authorities meanwhile promote a program to intensify irrigation in hop growing.

CONCLUSION

- The obvious climate change is more decisive to a hop varieties´ survive than the growing area.
- All partners in the hop supply chain (breeders, growers, merchants and brewers) have to work together on a strategy to overcome the upcoming problems in the field of tension between sustainable hop growing and security of supply.

Andreas Gahr– EBC 2019 – Antwerp - 23

USE OF WINE NON-SACCHAROMYCES YEASTS FOR BREWING

Fotini Drosou[1,2]; P. Tataridis[2]; V. Oreopoulou[1]; V. Dourtoglou[2]
[1] Department of Chemical Engineering, National Technical University of Athens;
[2] Department of Wine, Vine & Beverage Sciences, University of West Attica

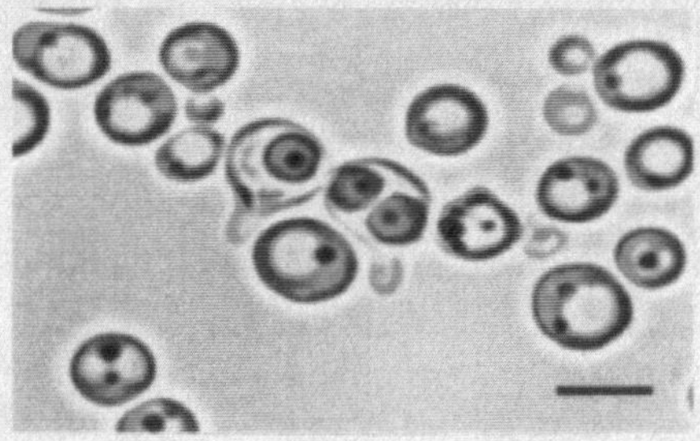

Beer Ingredients

MALT → Enzymes + fermentable sugars

WATER → 90-95 % of beer

HOP → Bitterness +foam +aromas

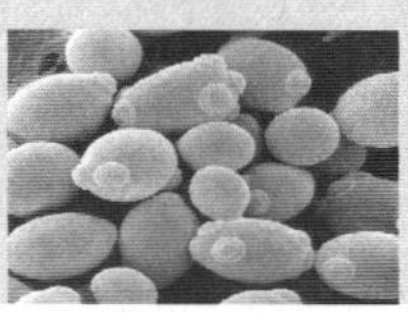

YEASTS → Metabolize fermentable sugars→ production of ethanol, CO_2, secondary metabolites

Non-*Saccharomyces* Yeasts

Past

- Known from the wine industry
- Spoilage microorganisms
- Off-flavors to the product
- Low fermentability

Present

- Enhance the aroma complexity
- Production of glycerine→ mouthfeel
- Production of more monoterpenes
- New products with low alcohol content or free-alcohol

Non-*Saccharomyces* yeasts already used in brewing

- *Brettanomyces species* (in specific styles such as lambic and gueuze beers)
- *Torulaspora delbrueckii* – some strains metabolize maltose

Saccharomycodes ludwigii
Kluyveromyces lactis
Zygosaccharomyces rouxii
Pichia kluyveri

} Do not metabolize maltose but they used in production of low alcohol (<1% v/v) and non-alcohol beer

Non-*Saccharomyces* Yeasts

Torulaspora delbrueckii

- Small yeast cells
- Enhance the products with rose-like aroma
- Low production of off-flavors (i.e. acetaldehyde and acetic acid)
- High tolerenance in ethanol and low temperatures
- It is already been used in wine industry as the main yeast
- Different yeast strains

Metschinikowia pulcherrima

- High production of higher alcohols and esters
- Production of Phenolic Off-flavor (POF character)
- Tolerance in high acidity, ethanol and sugar concentration (>100 g/l)
- Production of pulcherrimin reddish pigment
- Most of the strains cannot metabolize maltose

EXPERIMENTS

A. SYNTHETIC SUBSTRATES

- 200 ml of synthetic substrates rich in nutrients
  + 100 g/L glucose/ fructose/ maltose
- Mix Substrate (86% maltose, 9% glucose, 5% fructose)
- Inoculum pure cultures of each yeast with $6*10^6$ cfu/ml

B. PALE ALE BEER

- 100 L Pale Ale wort
- Inoculum with pure culture of each yeast (11.5 gr dry yeast / 20 L)
- Inoculum with mixed culture *S.cerevisiae : T. delbrueckii 1:10*

Fermentation at 20° C and 13° C

ANALYSIS

A. SYNTHETIC SUBSTRATES

- DURING FERMENTATION
  - Yeast growth and viability
- AFTER FERMENTATION
  - Final pH
  - Final FAN (mg/L)

B. PALE ALE BEER

- DURING FERMENTATION
  - Sugar consumption (°P)
  - pH
  - FAN (mg/L)
- AFTER FERMENTATION
  - Alcohol
  - Final Color (SRM)
  - Final Bitterness (IBU)
  - Sensory analysis (GC-MS)

RESULTS

A. SYNTHETIC SUBSTRATES

Yeast growth 13° C

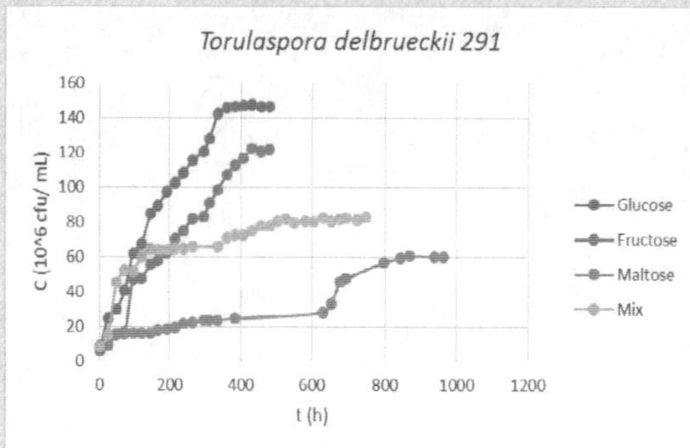

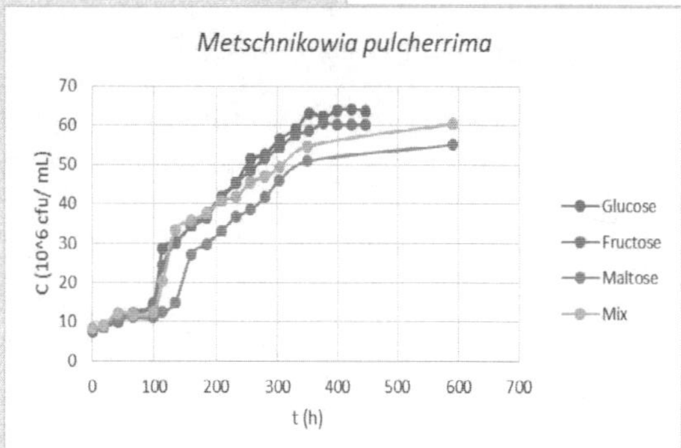

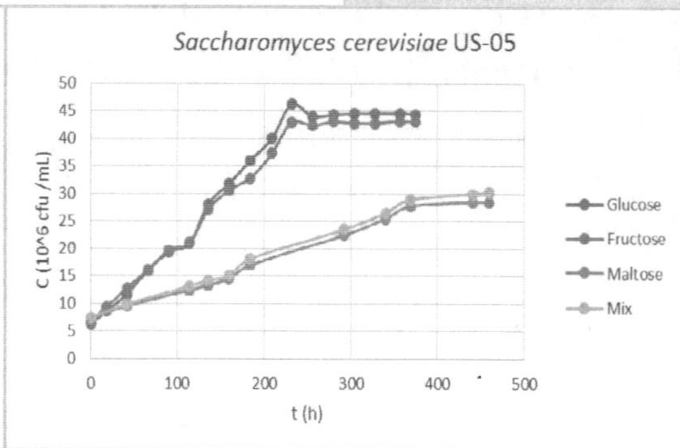

Yeast growth 20° C

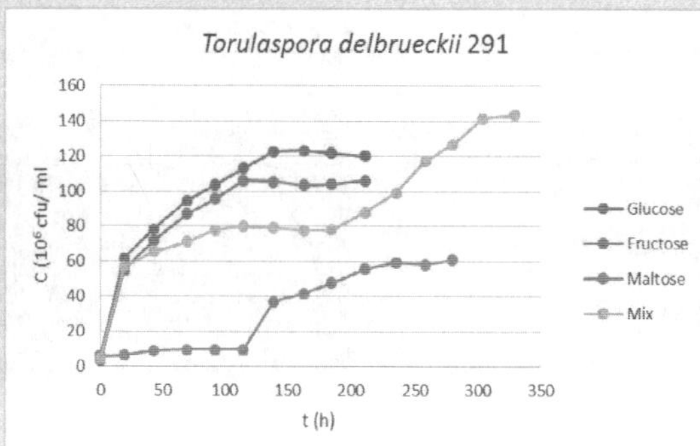

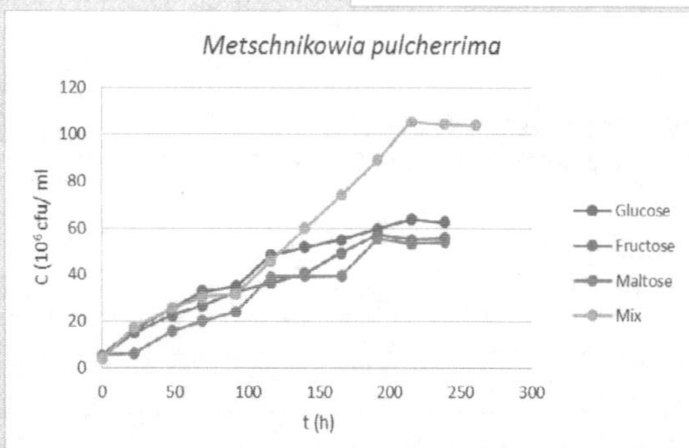

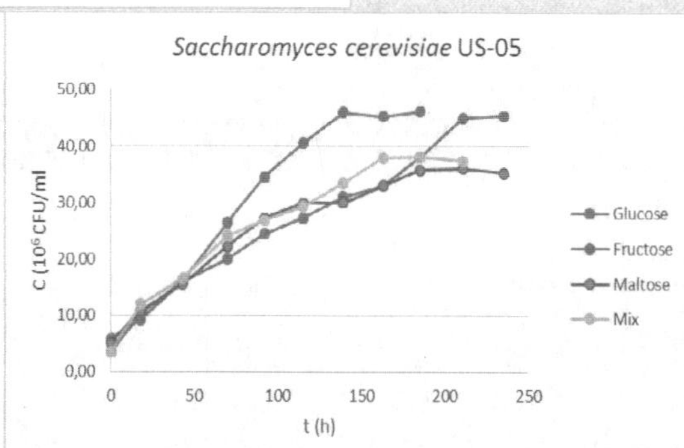

Results of final Ph and FAN

20° C					
Yeast strains	Initial pH	Final pH			
		Glucose	Fructose	Maltose	Mix
T.delbrueckii 291	4,5	3,17	3,12	3,19	2,85
M. pulcherrima	4,5	2,91	3,12	3,03	2,87
S.cerevisiae US-05	4,5	3,27	3,1	3,17	3,2

13° C					
Yeast strains	Initial pH	Final pH			
		Glucose	Fructose	Maltose	Mix
T.delbrueckii 291	4,5	3,01	3,04	3,28	2,78
M. pulcherrima	4,5	3,01	2,93	3,09	3,02
S.cerevisiae US-05	4,5	3,1	3,04	3,22	3,15

20° C					
Yeast strains	Initial FAN (mg/L)	Final FAN (mg/L)			
		Glucose	Fructose	Maltose	Mix
T.delbrueckii 291	283,27	164,82	146,9	212,25	142,49
M. pulcherrima	283,27	180,24	201,19	168,38	147,23
S.cerevisiae US-05	189,25	138,82	130,01	146,2	140,65

13° C					
Yeast strains	Initial FAN (mg/L)	Final FAN (mg/L)			
		Glucose	Fructose	Maltose	Mix
T.delbrueckii 291	189,25	134,46	124,73	110,96	129,35
M. pulcherrima	189,25	119,82	117,6	112,12	102,22
S.cerevisiae US-05	189,25	127,86	125,76	118,69	109,4

RESULTS

B. PALE ALE BEER

°Plato

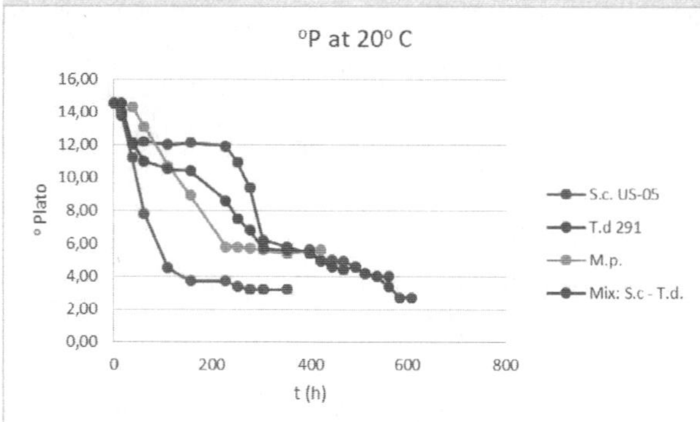

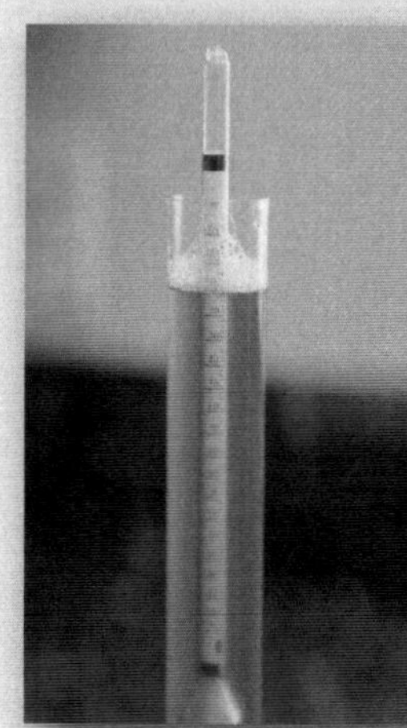

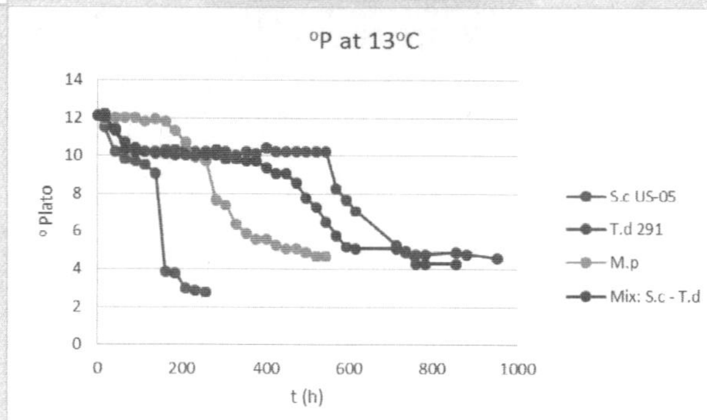

Free AminoNitrogen (mg/L)

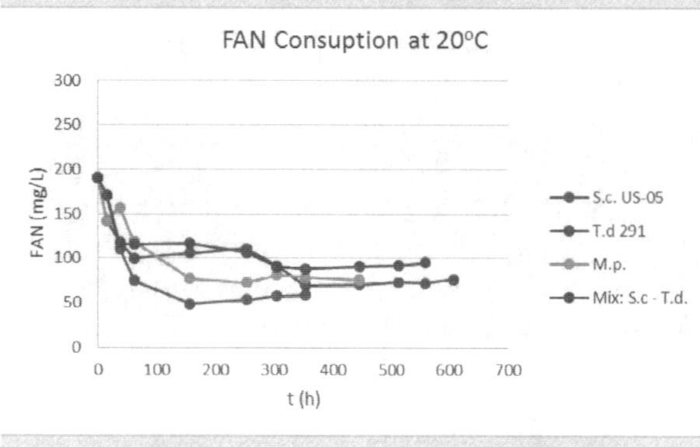

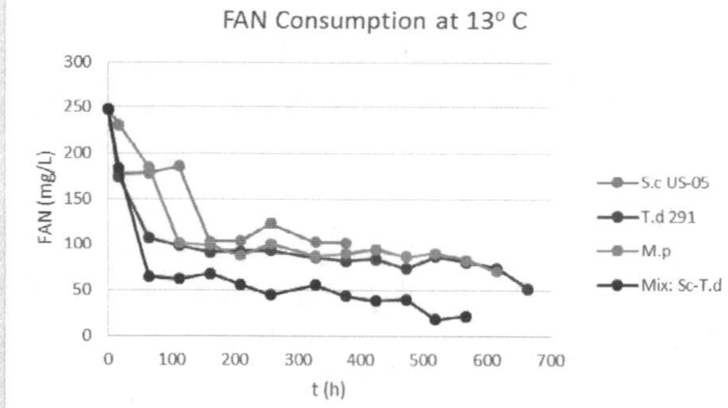

ALCOHOL, BITTERNESS, COLOR

20° C			
Yeasts	% ABV	IBU	SRM
S.cerevisiae US-05	6.6	32.65	9.21
T.delbrueckii 291	6.2	34.54	9.08
M.pulcherrima	5.2	36.52	8.85
Mix: S.c-T.d	6.5	31.56	9.08

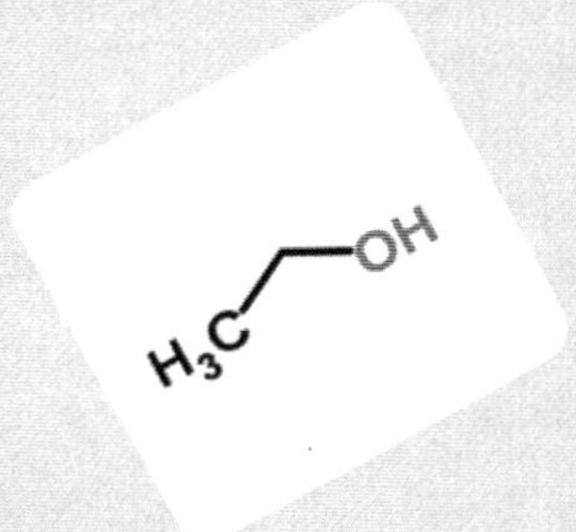

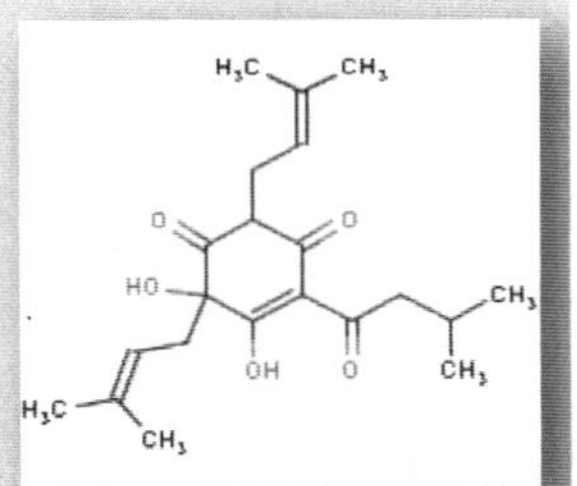

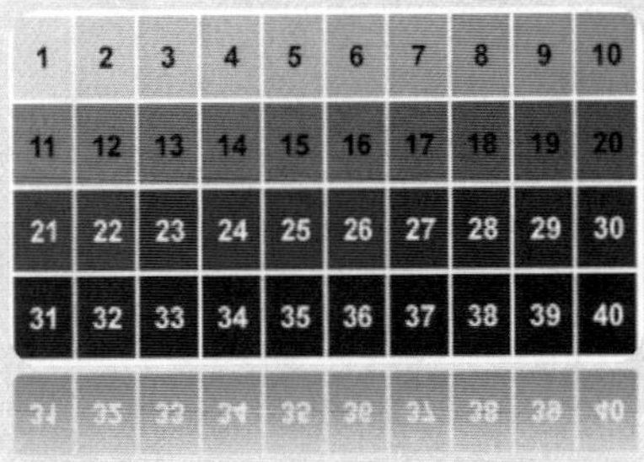

13° C			
Yeasts	% ABV	IBU	SRM
S.cerevisiae US-05	4,8	22,1	6,37
T.delbrueckii 291	3,9	23,45	8,12
M.pulcherrima	4	22,8	8,59
Mix: S.c-T.d	4,7	31,15	7,61

GC-MS ANALYSIS

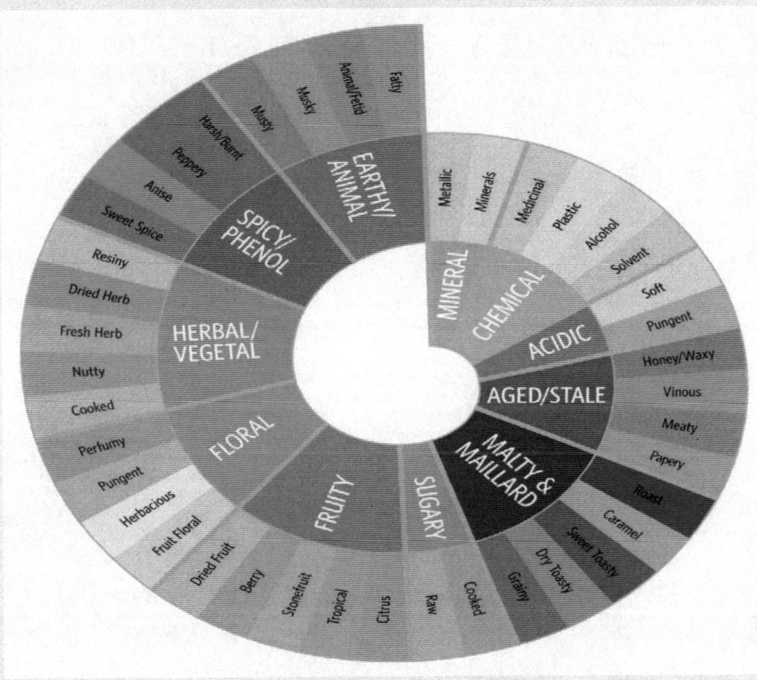

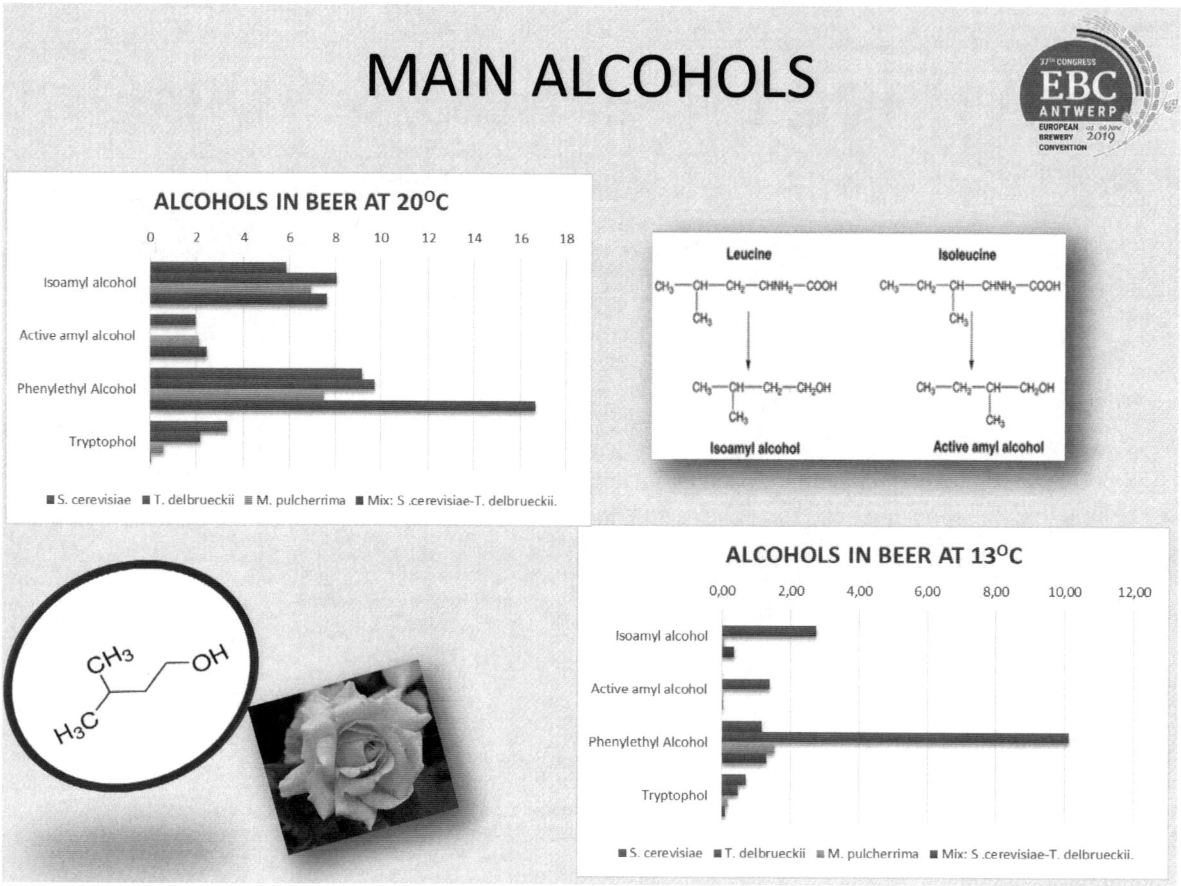
MAIN ALCOHOLS
37th CONGRESS EBC ANTWERP EUROPEAN BREWERY CONVENTION 2019
ALCOHOLS IN BEER AT 20ºC
0 2 4 6 8 10 12 14 16 18
Isoamyl alcohol
Active amyl alcohol
Phenylethyl Alcohol
Tryptophol
S. cerevisiae
T. delbrueckii
M. pulcherrima
Mix: S .cerevisiae-T. delbrueckii.
Leucine
Isoleucine
Isoamyl alcohol
Active amyl alcohol
ALCOHOLS IN BEER AT 13ºC
0,00 2,00 4,00 6,00 8,00 10,00 12,00
Isoamyl alcohol
Active amyl alcohol
Phenylethyl Alcohol
Tryptophol
S. cerevisiae
T. delbrueckii
M. pulcherrima
Mix: S .cerevisiae-T. delbrueckii.

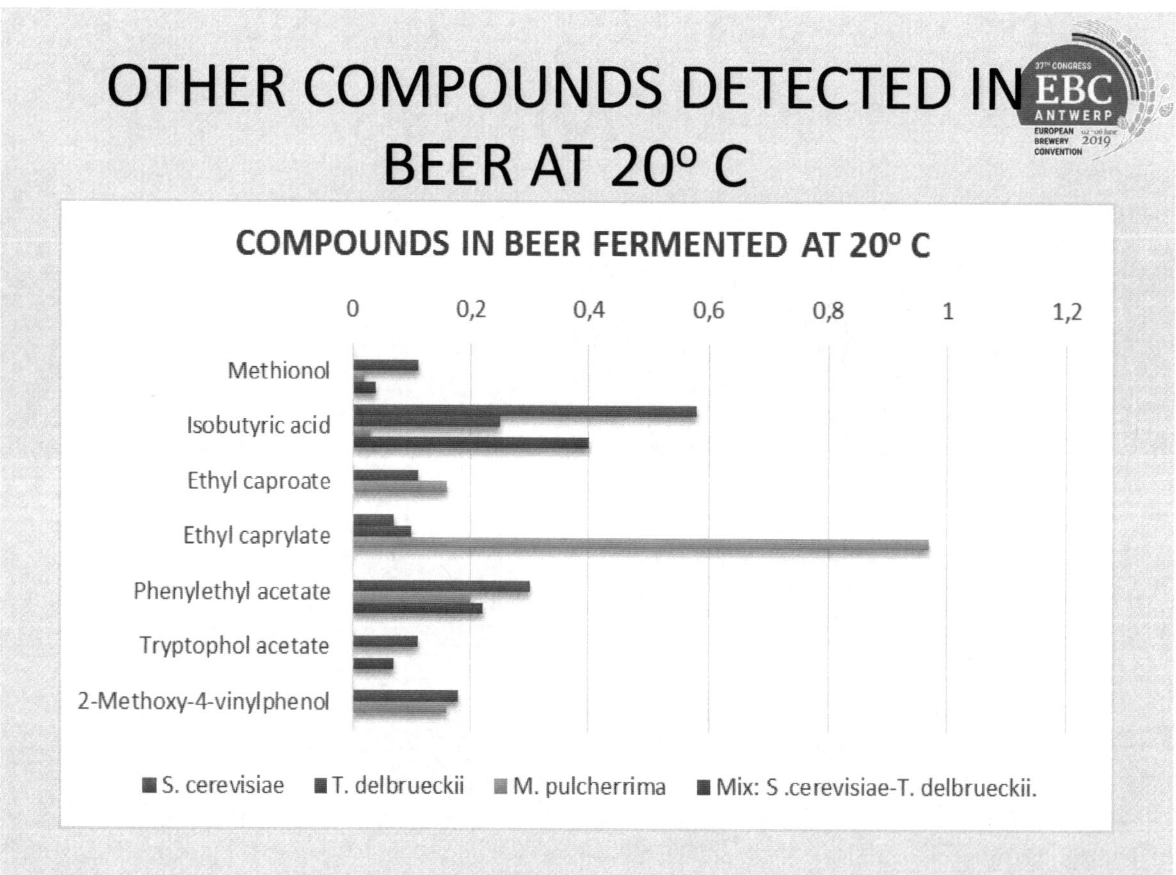
OTHER COMPOUNDS DETECTED IN BEER AT 20º C
37th CONGRESS EBC ANTWERP EUROPEAN BREWERY CONVENTION 2019
COMPOUNDS IN BEER FERMENTED AT 20º C
0 0,2 0,4 0,6 0,8 1 1,2
Methionol
Isobutyric acid
Ethyl caproate
Ethyl caprylate
Phenylethyl acetate
Tryptophol acetate
2-Methoxy-4-vinylphenol
S. cerevisiae
T. delbrueckii
M. pulcherrima
Mix: S .cerevisiae-T. delbrueckii.

OTHER COMPOUNDS DETECTED IN BEER AT 13° C

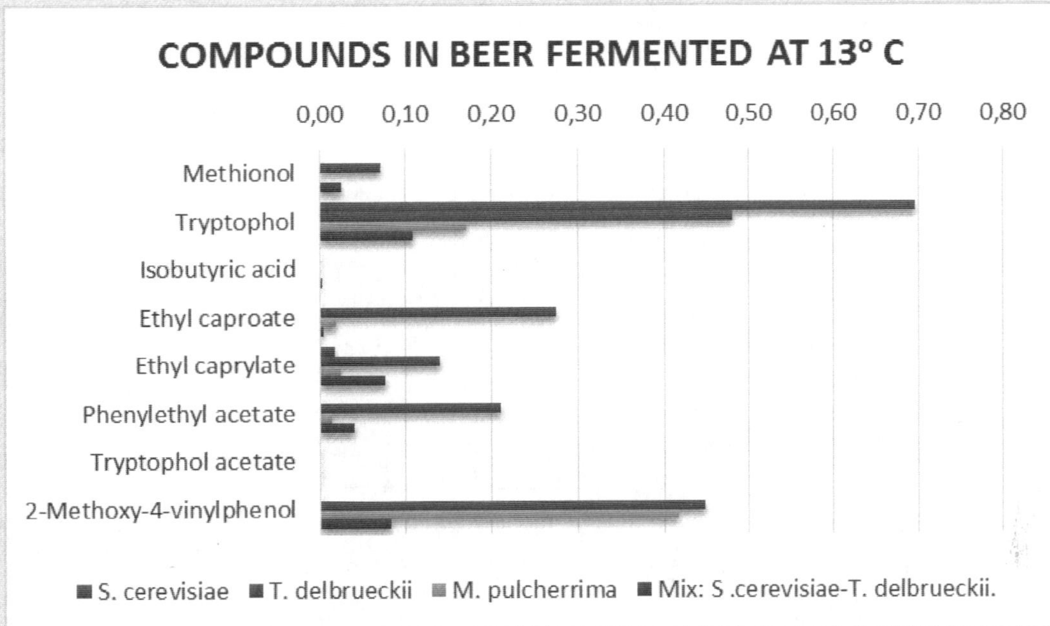

CONCLUSION

- Concerning the fermentation ability:
  - ✓ The studied non-*Saccharomyces* yeasts are capable of metabolizing maltose
  - ✓ The fermentation rate of *M. pulcherimma* was found to be slightly slower than that of *S. cerevisiae*
  - ✓ *T. delbrueckii* requires more time to activate maltase
  - ✓ Depending on the *T. delbrueckii* strain and temperature of fermentation the activation time differs
- Concerning the final product:
  - ✓ *M. pulcherimma* has higher percentages of residual sugar and gives products with more bitterness
  - ✓ The studied non-*Saccharomyces* yeasts provide the aromatic complexity needed for a product to be different than the usual
  - ✓ 2-phenylethanol produced by T.*delbrueckii* → same concentration despite temperature
  - ✓ Lower temperature fermentation → more elegant aroma and flavor
  - ✓ 2-Methoxy-4-vinylphenol → higher concentration produced by non-*Saccharomyces* in lower temperature

37TH CONGRESS
EBC
ANTWERP
EUROPEAN BREWERY CONVENTION
02–06 June 2019
Questions?
Thank You

L6 - YEAST PROTEOMICS DURING BEER FERMENTATION: CHALLENGES AND PROSPECTS

Magda Costa[1]; David Laureys[1]; Sören Planckaert[2]; Bart Devreese[2]; Anita Van Landschoot[1]; Jessika De Clippeleer[1,3]

1 - Ghent University, Faculty of Bioscience Engineering, Department of Biotechnology, Research Group Brewing Science & Technology, Campus Schoonmeersen, Valentin Vaerwyckweg 1, B-9000 Ghent, Belgium; 2 - Ghent University, Faculty of Science, Department of Biochemistry and Microbiology, Laboratory for Microbiology, Campus Ledeganck, KL Ledeganckstraat 35, B-9000 Ghent, Belgium; 3 - University College Ghent, Faculty of Science and Technology, Department of Natural and Food Sciences, Research Group Brewing Science & Technology, Campus Schoonmeersen, Valentin Vaerwyckweg 1, B-9000 Ghent, Belgium

Abstract

The alcoholic fermentation of wort by yeast is a crucial step during the production of beer. The overall conditions during this process have an impact on the behaviour of the yeast cells, which is itself determined by the yeast proteome. Therefore, monitoring the yeast proteome via targeted proteomics might provide more insight into the response of the yeast cells to the fermentation conditions. Mass spectrometry proteomics has already been used to describe the dynamics of the wort untargeted proteome during nanoscale beer production but has not yet been applied to follow and quantify the yeast proteome during beer production. In this study, a framework is being developed to monitor the yeast proteome during the fermentation process of beer via liquid chromatography electrospray ionization mass spectrometry (LC-ESI-MS). Hereto, a selection of around 25 enzymes responsible for the production of desirable (e.g. esters and higher alcohols) and undesirable flavours (e.g. DMS and diacetyl), are quantified during the fermentation process, and the results are compared with the substrates assimilation profiles and the metabolites production profiles. A variety of yeast enzymes is already monitored with this technique, but the reliable quantification remains a challenge and is currently being investigated. The established framework will be used to shed new light on the reaction of different yeast strains to innovative raw materials and adjuncts, and differences in gravity and temperature. These insights may help the development of novel beer flavours and more sustainable production processes for beer.

Short biography

Magda Costa (1993, Portugal) is currently a PhD student of the Research Group Brewing Science & Technology in the Department of Biotechnology at Ghent University, Belgium, under the project 'European Joint Doctorate in Food Science', funded by the European Union's Horizon 2020 research and innovation program with the Marie Skłodowska-Curie grant agreement No 722166. Magda has a MSc degree in Biological Engineering from Técnico Lisboa, Portugal (2016), and an Erasmus experience at the University of Amsterdam, The Netherlands, for her MSc thesis in biocatalysis (2016). She moved to Belgium (2017) to start her doctoral thesis research work related to the fermentation in the brewing process, more specifically the study of the yeast cell proteome.

37th Congress of the European Brewery Convention — Antwerp, Belgium, May 2019

„*Saccharomyces cerevisiae* var. *diastaticus* - friend or foe "

Meier-Dörnberg, T., Jacob, F., Hutzler, M.

37th CONGRESS EBC ANTWERP 02–06 June 2019 EUROPEAN BREWERY CONVENTION

Forschungszentrum Weihenstephan für Brau- und Lebensmittelqualität

TUM Technische Universität München

S. cer. var. diastaticus -a brewing yeast?-

OXFORD FEMS JOURNALS investing in science

FEMS Yeast Research, 18, 2018, foy023

doi: 10.1093/femsyr/foy023
Advance Access Publication Date: 6 March 2018
Research Article

RESEARCH ARTICLE

Saccharomyces cerevisiae variety diastaticus friend or foe?—spoilage potential and brewing ability of different *Saccharomyces cerevisiae* variety diastaticus yeast isolates by genetic, phenotypic and physiological characterization

Tim Meier-Dörnberg[1], Oliver Ingo Kory[2], Fritz Jacob[1], Maximilian Michel[1] and Mathias Hutzler[1,*]

[1]Technical University of Munich, Research Center Weihenstephan for Brewing and Food Quality, Alte Akademie 3, 85354 Freising, Germany and [2]Miho Inspektionssysteme GmbH, Obervellmarsche Str. 12, 334292 Ahnatal, Germany

*Corresponding author: Technical University of Munich, Research Center Weihenstephan for Brewing and Food Quality, Alte Akademie 3, 85354 Freising, Germany. Tel: +49(0)8161713100; Fax: +49(0)8161714181; E-mail: m.hutzler@tum.de

One sentence summary: This paper presents novel findings on the spoilage potential and the use of *S. cerevisiae* var. *diastaticus* yeast strains as beer fermentation starter cultures.

Editor: Isak Pretorius

S. cer. var. *diastaticus* -strain selection-

Saccharomyces cerevisiae var. *diastaticus* strain selection

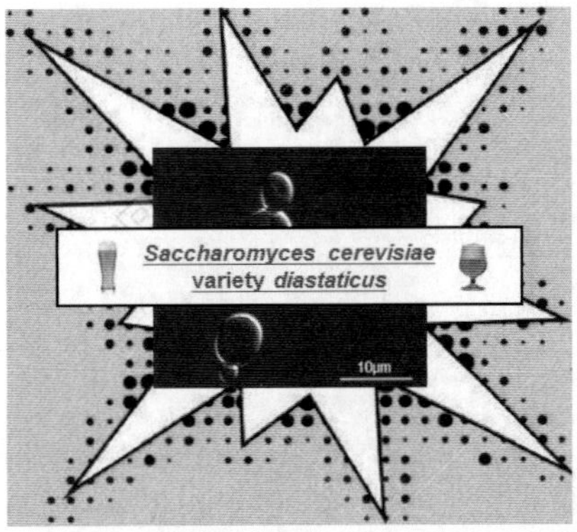

S. cer. var. *diastaticus* -species identificatin and strain determination-

DNA-extraction → PCR → Real time PCR

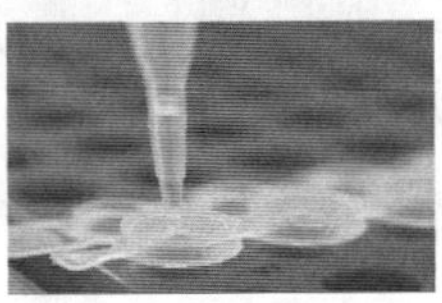

e. g. Real Time PCR screening on
Saccharomyces cerevisiae
Saccharomyces pastorianus
Saccharomyces diastaticus
Saccharomyces bayanus
***Saccharomyces* eubayanus**

DNA sequencing and fingerprinting by capillary gel electrophoresis

D1/D2 26S
as well as
ITS1-5.8S-ITS2 rDNA

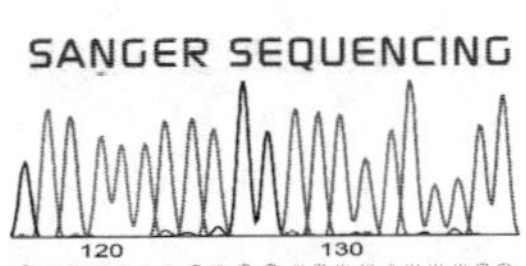

IGS2-314 rDNA

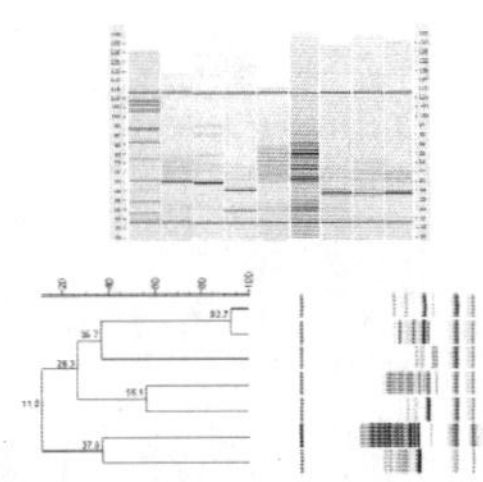

Method -*S. cer. var. diastaticus test strains*-

S. cer. var. *diastaticus* -strain selection-

Table 1. *Saccharomyces cerevisiae* var. diastaticus yeast isolates with TUM identifier and the as reference strains used brewing culture yeast strains.

TUM yeast isolates/strains			
TUM yeast isolate/strain	Yeast species	Industrial application	Obtained from
DSM 70487	*Saccharomyces cerevisiae* var. *diastaticus*	spoilage yeast (control strain)	Leibniz Institute DSMZ—German Collection of Microorganisms and Cell Cultures GmbH
TUM PI BA 31	*Saccharomyces cerevisiae* var. *diastaticus*	spoilage yeast	Isolated from brewery
TUM PI BA 45	*Saccharomyces cerevisiae* var. *diastaticus*	spoilage yeast	Isolated from brewery
TUM PI BA 109	*Saccharomyces cerevisiae* var. *diastaticus*	spoilage yeast	Isolated from brewery
TUM PI BA 124	*Saccharomyces cerevisiae* var. *diastaticus*	spoilage yeast	Isolated from brewery
TUM PI BB 105	*Saccharomyces cerevisiae* var. *diastaticus*	spoilage yeast	Isolated from beer-mixed beverage
TUM PI BB 121	*Saccharomyces cerevisiae* var. *diastaticus*	spoilage yeast	BLQ
TUM PI BB 124	*Saccharomyces cerevisiae* var. *diastaticus*	spoilage yeast	BLQ
TUM PI BB 125	*Saccharomyces cerevisiae* var. *diastaticus*	spoilage yeast	BLQ
TUM PI BB 133	*Saccharomyces cerevisiae* var. *diastaticus*	spoilage yeast	BLQ
TUM PI BB 159	*Saccharomyces cerevisiae* var. *diastaticus*	spoilage yeast	BLQ
TUM 1-B-8	*Saccharomyces cerevisiae* var. *diastaticus*	spoilage yeast	BLQ
TUM 3-D-2	*Saccharomyces cerevisiae* var. *diastaticus*	spoilage yeast	Isolated from brewery
TUM PI BB 182 (17-E-7)	*Saccharomyces cerevisiae* var. *diastaticus*	spoilage yeast	Isolated from lemonade
TUM 2-F-1	*Saccharomyces cerevisiae* var. *diastaticus*	spoilage yeast	BLQ
TUM 1-G-7	*Saccharomyces cerevisiae* var. *diastaticus*	spoilage yeast	BLQ
TUM 1-H-7	*Saccharomyces cerevisiae* var. *diastaticus*	spoilage yeast	BLQ
TUM 3-H-2	*Saccharomyces cerevisiae* var. *diastaticus*	spoilage yeast	Isolated from brewery
TUM 71	*Saccharomyces cerevisiae* var. *diastaticus*	spoilage yeast	BLQ
LeoBavaricus—TUM 68®	*Saccharomyces cerevisiae*	wheat beer production	BLQ commercial yeast culture
LunaBavaria—TUM 127®	*Saccharomyces cerevisiae*	wheat beer production	BLQ commercial yeast culture
Colonia—TUM 177®	*Saccharomyces cerevisiae*	kölsch and alt beer production	BLQ commercial yeast culture
Vetus—TUM 184®	*Saccharomyces cerevisiae*	alt beer production	BLQ commercial yeast culture
Pensum—TUM 210®	*Saccharomyces cerevisiae*	ale beer production	BLQ commercial yeast culture
Mel—TUM 211®	*Saccharomyces cerevisiae*	ale beer production	BLQ commercial yeast culture
TUM 213	*Saccharomyces cerevisiae*	wheat beer production	BLQ commercial yeast culture
Monacus—TUM 381®	*Saccharomyces cerevisiae*	wheat beer production	BLQ commercial yeast culture
TUM 503	*Saccharomyces cerevisiae*	ale beer production	BLQ commercial yeast culture
Tropicus—TUM 506®	*Saccharomyces cerevisiae*	ale beer production	BLQ commercial yeast culture
Harmonia—TUM 511®	*Saccharomyces cerevisiae*	ale and wheat beer production	BLQ commercial yeast culture
Frisinga—TUM 34/70®	*Saccharomyces pastorianus*	lager beer production	BLQ commercial yeast culture
Securitas—TUM 193®	*Saccharomyces pastorianus*	lager beer production	BLQ commercial yeast culture

1 reference strain

18 isolates

13 culture yeast strains for comparison

S. cer. var. *diastaticus* -friend or foe-

S. cer. var. *diastaticus* - fermentation

Saccharomyces cerevisiae var. *diastaticus* as brewing yeast

Method *-Brewing trials-*

Yeast characterization at standard conditions

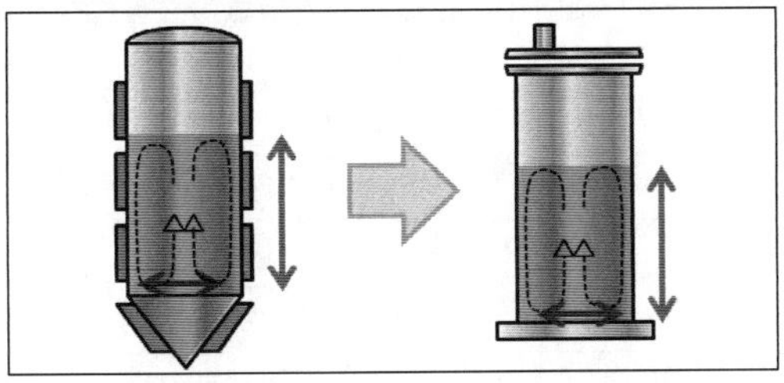

Stainless steel vessels
Ratio diameter to high: 1:3
20 % headspace
Nominal-/Filling quantity: 2,5 / 2 Liter
Pressure proofed

- TÜV SV 15-2055 safety bung valve, spring loaded, 0.5 bar, to ensure practical fermentation conditions
- Oxygen and contamination-free filling
- Evaluation in triplicate

Method *-Brewing trials-*

Tempered cooling chamber for constant fermentation temperature

1 additional fermentation vessel for daily measurement of extract, pH-value und yeast cells in suspension

➔ The system will not be influenced by taking the samples

Method *-Brewing trials-*

Experimental beer production parameters

Parameter	Set value
pressure	0.5 bar
medium	stand. wort 12.4 °P (base malt extract)
propagation	standardised and in parallel (20 °C)
pitching cell concentration	15 x 10^6 cells/ml
temperature (pit./ ferm./mat.)	TF 20 °C
volume	2 l (mass adjustment 2000 g)
system number	4 vessels per strain
brief mixing	N_2
reference strain	(e. g. DSMZ 70487)
number of simultaneously tested strains	7
enforced fermentation and maturation	Fermentation until final attenuation (2 d constant) /maturation: 5 d lager conditions: 7 d; 0 °C

37th EBC Congress Antwerp, Belgium, May 2019

Method *-Brewing trials-*

Analysis

Parameter	Wort	Pitching wort	Daily analysis	Finished beer
Original gravity	×		×	×
pH	×		×	×
Alcohol				×
Degree of attenuation				×
Nitrogen (N_2) soluble/ coagulable	×			
Thiobarbituric acid (TBZ)	×			
Iodine test spectrophotometric	×			
Beta glucans	×			
Zinc	×			
Dimethyl sulfide (DMS) total and free/ DMS-P	×			×
Alpha-amino nitrogen (FAN)	×	×		×
Total amino acids	×	×		×
Sugar composition	×	×		×
EBC-bittering units	×			
Yeast cells in suspension		×	×	×
Total yeast cell count		×		×
Diacetyl			×	
Sulfur dioxide (SO_2)				×
Fermentation by-products				×
Vicinal diketones total / free				×
4-Vinylguajacol				×
Sensory analysis				×

Results *-Brewing trials-*

S. cer. var. *diastaticus* - fermentation

Drop in specific gravity measured in a single reference vessel compared with the average in final gravity (marked with box) measured in triplicate for the tested yeast strains DSM 70487, TUM 1-H-7, TUM PI BB 121, TUM PI BA 124, TUM 71, TUM 3-H-2 and TUM 3-D-2; confidence level 95 %

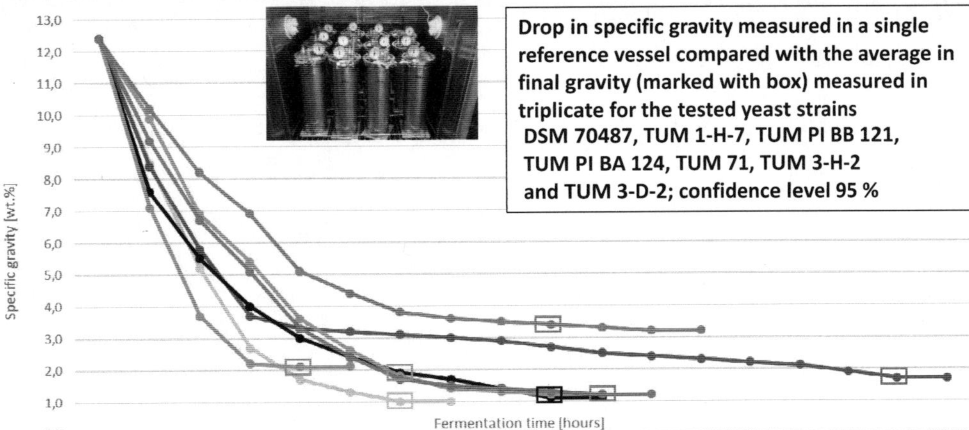

	0	24	48	72	96	120	144	168	192	216	240	264	288	312	336	360	384	408
DSM 70487	12,4	8,5	5,2	2,7	1,7	1,3	1,0	1,0										
TUM 1-H-7	12,4	8,4	5,8	3,7	3,3	3,2	3,1	3,0	2,9	2,7	2,5	2,4	2,3	2,2	2,1	1,9	1,7	1,7
TUM PI BB 121	12,4	7,6	5,5	4,0	3,0	2,4	1,9	1,7	1,4	1,1	1,1							
TUM PI BA 124	12,4	7,1	3,7	2,2	2,1	2,1												
TUM 71	12,4	9,9	6,9	5,4	3,6	2,6	1,8	1,4	1,3	1,2	1,2							
TUM 3-H-2	12,4	10,2	8,2	6,9	5,1	4,4	3,8	3,6	3,5	3,4	3,3	3,2	3,2					
TUM 3-D-2	12,4	9,2	6,7	5,1	3,3	2,4	1,7	1,5	1,4	1,3	1,2	1,2						

Results *-Brewing trials-*

S. cer. var. *diastaticus* -sugar consumption-

Pay attention:
Diastaticus ≠ Diastaticus
→ Spoilage potential

TUM 1-H-7

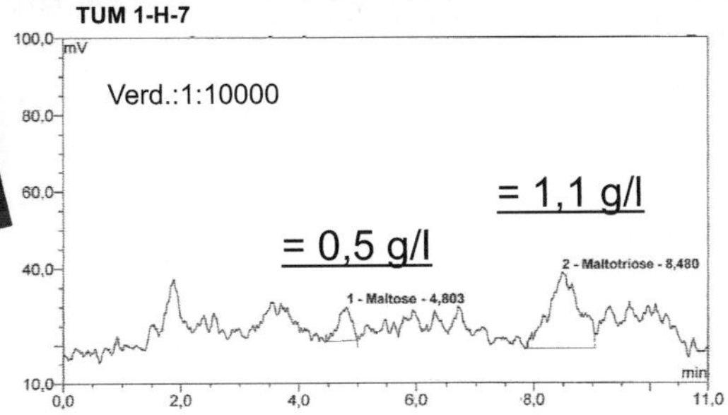

„Normal" Sugar fermentation at Vs= ~ 82,7%

All sugars fermented

Maltose gap

DSMZ 70487

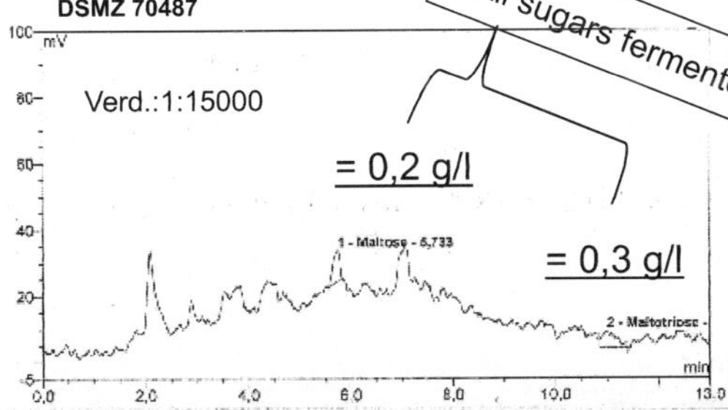

TUM 3-H-2

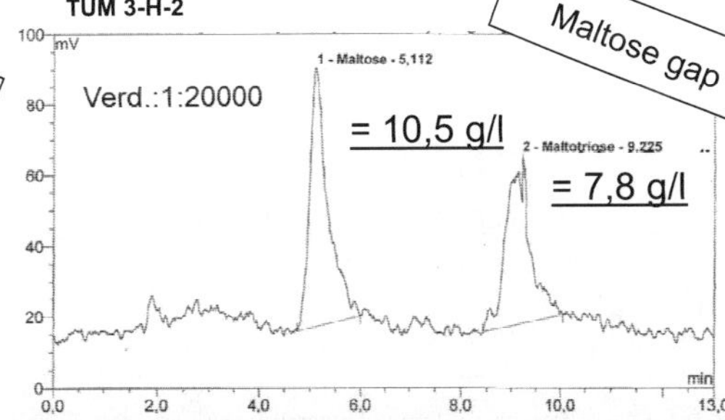

Method and Results *-Sensory analysis-*

Technische Universität München

Sensory analysis

- Expected beer type
- DLG*-scheme for beer
- Triangle test
- Phenolic off-flavor
- Descriptive sensory analysis

Sensory analysis			
S. cer. var. diastaticus strain	DSM 70487	TUM 3-H-2	TUM 1-H-7
Expected beer type	Wheatbeer	"Bavarian ale"	Ale
DLG (taste)	4.6	4.7	4.5

101 821 948 666 *

411 007 918 326 310 376 587 455

* DLG = Deutsche Landwirtschaft Gesellschaft

* Triangle test = symbolic illustration

37th EBC Congress Antwerp, Belgium, May 2019

Method and Results *-Sensory analysis-*

Technische Universität München

- Expected beer type
- DLG-scheme for beer
- Triangle test
- Phenolic off-flavor
- Descriptive sensory analysis

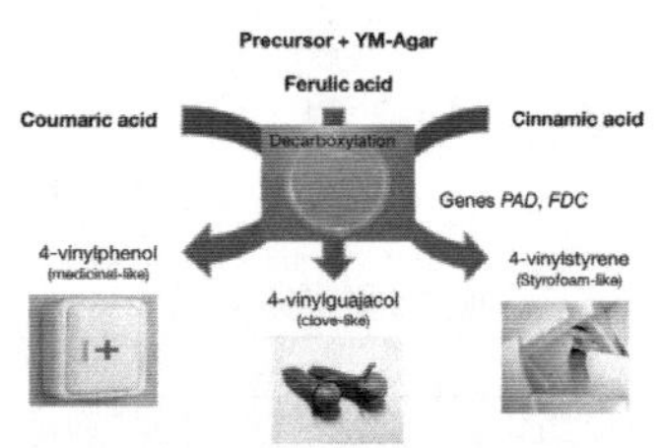

Precursor 100ppm coumaric acid, cinnamic acid, ferulic acid in YM-Agar

Evaluation by sniffing after 3 days of incubation at 24° C

Table 7. POF potential of the investigated *Saccharomyces cerevisiae* var. *diastaticus* and the reference strains Frisinga—TUM 34/70® and LeoBavaricus—TUM 68®.

Yeast strain (TUM identifier)	POF test/sniffing perception of Product/precursor 4-VG/ ferulic acid	4-VP/ coumaric acid	4-VS/ cinnamic acid
DSM 70487	+	+	+
TUM PI BA 31	+	+	−
TUM PI BA 45	+	+	+
TUM PI BA 109	+	+	+
TUM PI BB 105	+	−	+
TUM PI BB 121	+	+	+
TUM PI BB 124	+	+	+
TUM PI BB 125	+	+	+
TUM PI BB 133	+	+	+
TUM PI BB 159	+	+	+
TUM 1-B-8	+	−	+
TUM 3-D-2	+	−	+
TUM 17-E-7	+	+	+
TUM 2-F-1	+	−	+
TUM 1-G-7	+	−	+
TUM 1-H-7	+	+	+
TUM 3-H-2	+	−	+
TUM 71	+	+	+
LeoBavaricus—TUM 68®	+	+	+
Frisinga—TUM 34/70®	−	−	−
TUM PI BA 124	−	−	−

Method and Results *-Sensory analysis-*

Sensory analysis

- Expected beer type
- DLG-scheme for beer
- Triangle test
- Phenolic off-flavor
- Descriptive sensory analysis

4.) Descriptive sensory analysis (aroma)

Please mark the appropriate aroma impression according to their intensity
(1 ≙ slightly noticeable, 5 ≙ extremely noticeable)
Pleaser write perceived non-listed flavor impressions in the category called „Other".

	BEER 056					BEER 932					BEER 155					BEER 264				
	1	2	3	4	5	1	2	3	4	5	1	2	3	4	5	1	2	3	4	5
Tropical fruity																				
Banana (green)																				
Banana																				
Pineapple																				
Passionfruit																				
Peach																				
Mango																				
Melon																				
Tangerine																				
Lemon																				
Grapefruit																				
Lime																				
Spicy																				
Clove																				
Juniper																				
Pepper																				
Cinnamon																				
Sweet																				
Malty																				
Wortlike, sweet																				
Honey																				
Toffee																				
Bitter almond, marzipan																				
Vanilla																				
Gummy bear																				
Chocolatey																				
Floral																				
Floral																				
Green gras																				
Herbs																				
Resinous																				
Other																				
Yeasty																				
Sulfurous																				
Phenolic off																				
Rotten egg																				
Butty (Diacetyl)																				
Sweaty, cheesy																				
Fungoid, forrest floor																				
Solvent like																				
Alcoholic																				
Vinous																				
Smoky																				

Method and Results *-Sensory analysis-*

S. cer. var. *diastaticus* -descriptive sensory analysis-

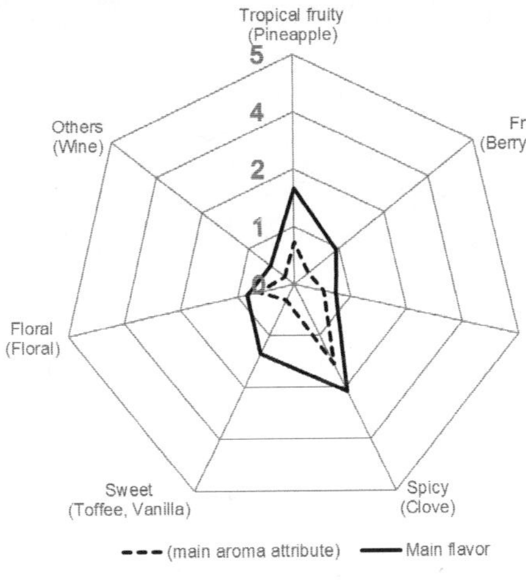

TUM PIBB 121

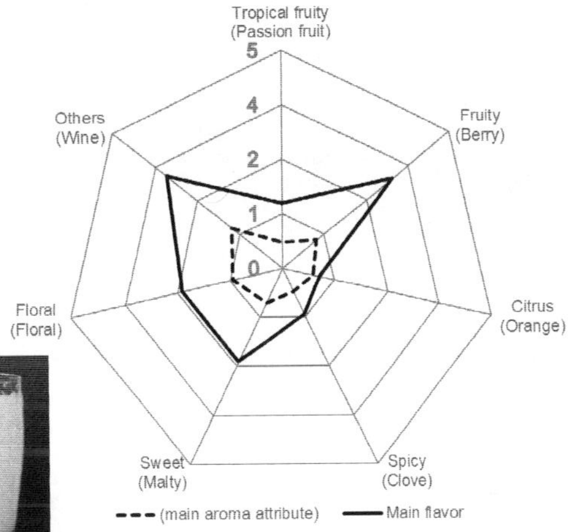

TUM 71

Comparison of the flavors grouped according to the main categories and the respective main aroma attributes for *S. cerevisiae* var. *diastaticus* yeast strains TUM PIBB (left) and TUM 71 (right).

S. cer. var. *diastaticus* - fermentation

Saccharomyces cerevisiae var. *diastaticus* as spoilage yeast

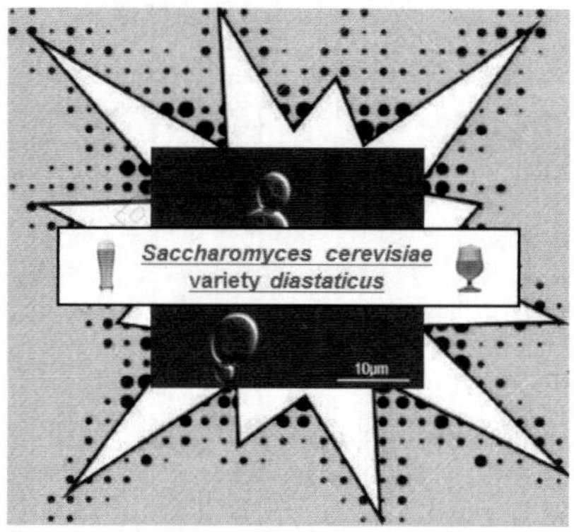

Saccharomyces cerevisae var. *diastaticus*

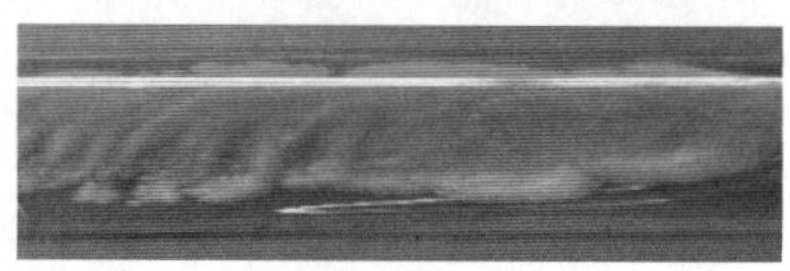

	S. cerevisiae	*S. cerevisiae* var. *diastaticus*
Fermentation:	Normal fermenting, fermentation of the major wort sugars → Residual extract up to 1,8 ° P	Super attenuation, fermentation of the major wort sugars, dextrins and starch → Residual extract up to 0,9 ° P
Occurence:	Desired culture yeast in brewing	Beer and beer mixed bevearges, mostly secondary contamination
Consequences/ Hazards:	None	Bottle bursting (↑CO_2), cloudiness, dry taste, tapping impossible (Gushing) → surprise effect

Results *-Incidences-*

S. cer. var. diastaticus -most dangerous beer spoiler- Incidence from 1997 - *2017*

126 investigated cases of 52 companys from 15 european countries
(Breweries, Yeast- and Hopsuppliers, Laboratories)

→ 62 confirmed primary contaminations
→ 71% secondary contaminations

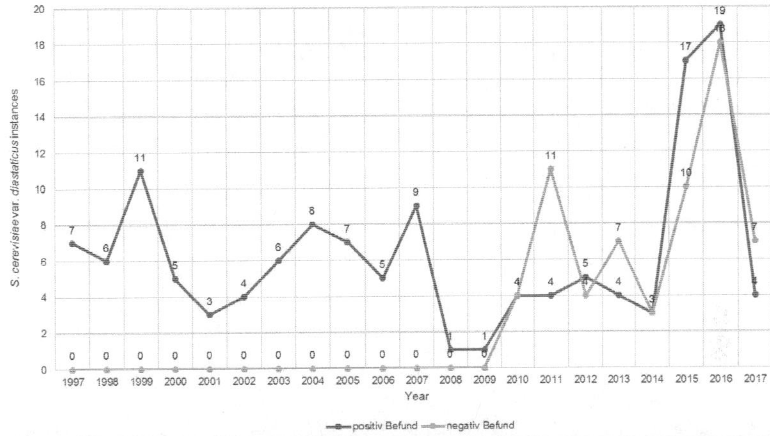

37th EBC Congress Antwerp, Belgium, May 2019

Results *-Incidences-*

Technische Universität München

Incidence of *Saccharomyces cer.* var. *diastaticus* from *2008 - 2017*

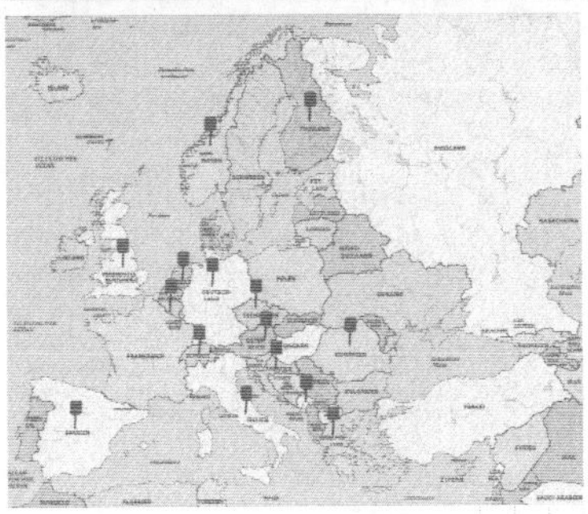

126 investigated cases of 52 companys from 15 european countries
(Breweries, Yeast- and Hopsuppliers, Laboratories)

→ 62 confirmed primary contaminations
→ 71% secondary contaminations

Positive contamination especially in the ¾ quarter of the year (summer months)

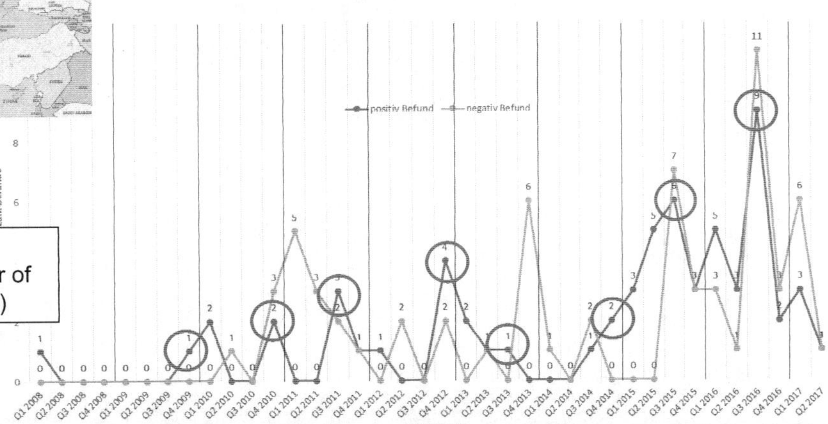

37th EBC Congress Antwerp, Belgium, May 2019

Results *-Incidences-*

Technische Universität München

Incidence of *S. cer.* var. *diastaticus* from *2008 - 2017*

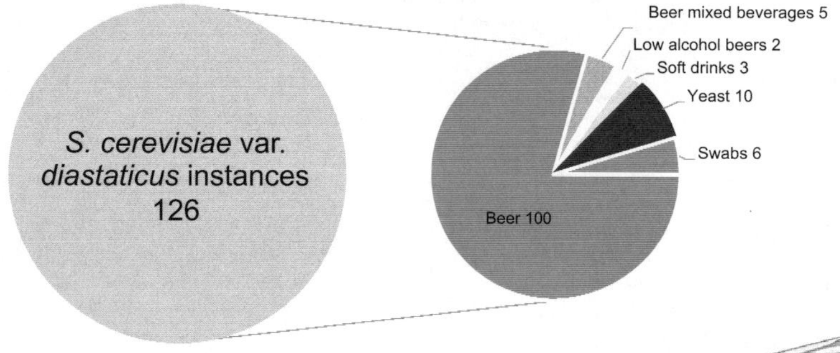

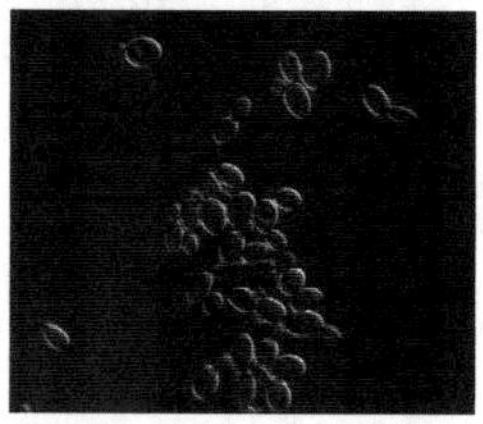

Forschungszentrum Weihenstephan
für Brau- und Lebensmittelqualität

37th EBC Congress Antwerp, Belgium, May 2019

Results *-Incidences-*

Technische Universität München

Incidence of *S. cer.* var. *diastaticus* from *2008 - 2017*

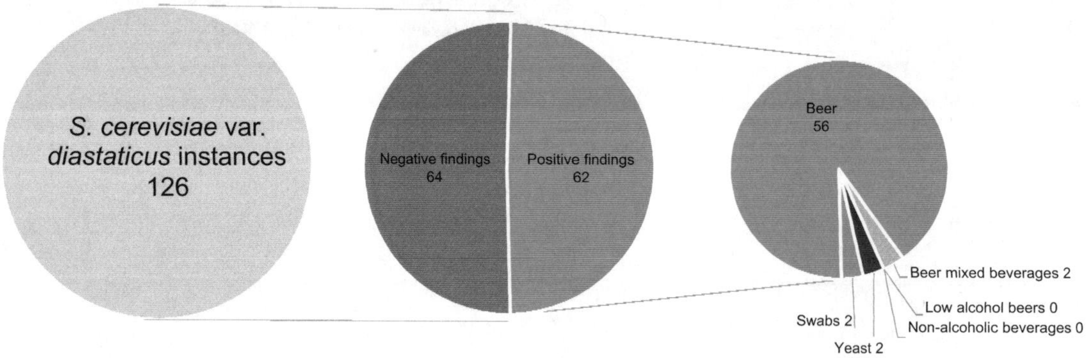

Forschungszentrum Weihenstephan
für Brau- und Lebensmittelqualität

Results *-Spoilage potential-*

S. cer. var. *diastaticus* -spoilage potential-

Pay attention:
Diastaticus ≠ Diastaticus
→ Spoilage potential

S. cer. var. diastaticus stain	fermentation degree %	starch agar	Durham	quantity ranking
DSMZ 70487	92.77	strong	1	1
TUM 3-D-2	90.67	strong	0.75	2
TUM 71	86.47	strong	0.75	
PIBB 121	91.73	strong	0.75	
PIBB 105	n.v	middle	0.37	3
TUM 1-H-7	87.47	weak	0.25	4
TUM 3-H-2	73.80	negative	negative	5

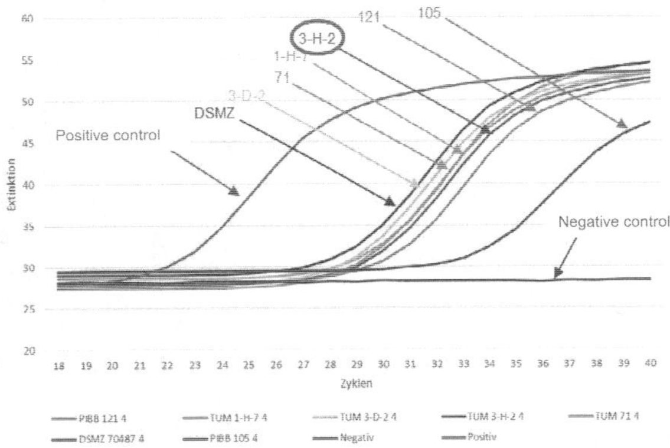

. IAC = Internal amplification control / approaches carried out in triplicate

Mismatching:

PIBB 105 — Physiological –middle- Genetically –weak-

TUM 3-H-2 — Physiological –negative- Genetically –weak-

TUM 1-H-7 — Physiological –weak- Genetically –middle-

Results *-Spoilage potential-*

S. cer. var. *diastaticus* spoilage potential - Tests

Dextrin Agar

Starch Agar + Bromphenol blue

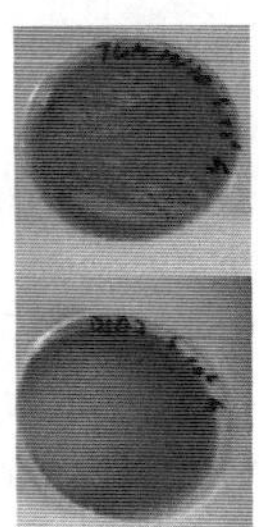

Modified Durham tube

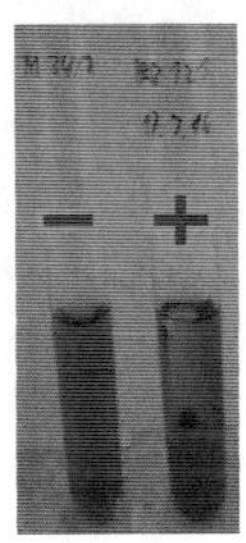

Agar + Carbohydrate source + Yeast nitrogen base + Bidistilled water and pH adjustement

- Growth control
- Fast and easy to handle

Pasteurized fermented beer medium

- Gas formation
- Time consuming test

Results -*Physiological tests*-

S. cer. var. *diastaticus* spoilage potential (physiological tests)

		Physiological test (anaerob)		
Yeast species	Yeast strain	starch agar	starch agar (+ indicator)	durham test
S. cerevisiae (TF)	LeoBavaricus - TUM 68	negative	negative	negative
	LunaBavaria - TUM 127	negative	negative	negative
	Mel- TUM 211	negative	negative	negative
	Tropicus - TUM 506	negative	negative	negative
	Harmonia - TUM 511	negative	negative	negative
	Vetus - TUM 184	negative	negative	negative
	Monacus - TUM 381	negative	negative	negative
	Colonia - TUM 177	negative	negative	negative
S. pastorianus (BF)	Frisinga - TUM 34/70	negative	negative	negative
	Securitas - TUM 193	negative	negative	negative
S. cer. var. diastaticus (TF)	DSMZ 70487	strong	strong	1
	TUM 1-H-7	weak	weak	0.25
	PIBB 121	strong	strong	0.75
	TUM 3-H-2	negative	neagtive	negative
	TUM 71	strong	strong	0.75
	TUM 3-D-2	strong	strong	0.75
	PIBB 105	middle	middle	0.38

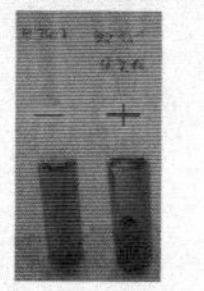

<u>Growth control</u>

→ Qualitive test

<u>Spoilage potential</u>

→ Quantitive test
(CO_2 gas formation)

Notification -*S. cer. var. diastaticus*-

Notification in case of an contamination with *S. diastaticus*

<u>§ 44a 1. LFGB:</u>

Food business operator (brewery, beverage manufacturer) is obliged„[…]
in the presence of positive and confirmed test results on levels of pathogenic substances (mycotoxins, etc.) u. Microorganisms (diastaticus) in food (beer) […]"

Laboratory is obliged
„[…] Laboratory analysis positive and reason to believe that the food (beer) would be subject to a traffic ban under Article 14 (1) of Regulation (EG) No 178/2002 […]"

→ **Notification** to notify the competent authority

Notification -*S. cer. var. diastaticus*-

Contaminated goods with potential hazard (bottle bursting) already retailed?

§ 44a 1. LFGB:
„German laws for food and feed"

Notification to the competent authority

↓ Silent callback

Already handed over by end user to retailer

↱ **Public callback**

Goods must be taken back by the manufacturer (brewery).

↓

End user must be informed about potential danger
(notices from trading partners, press, etc.)

↓

If the obligation to notify is not, not correctly, not on time or not fully investigated
→ will be issued depending on the case an administrative offense
(fine up to € 20,000)

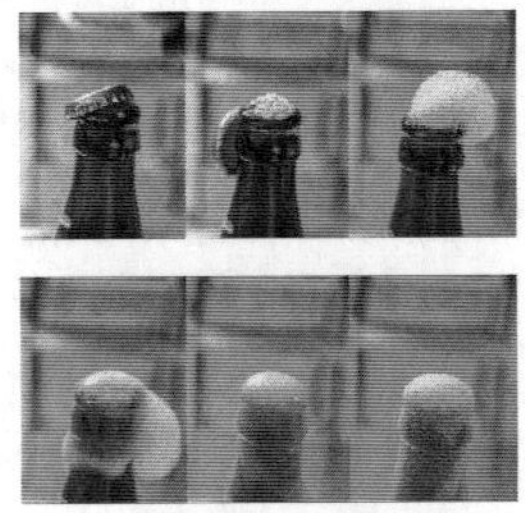

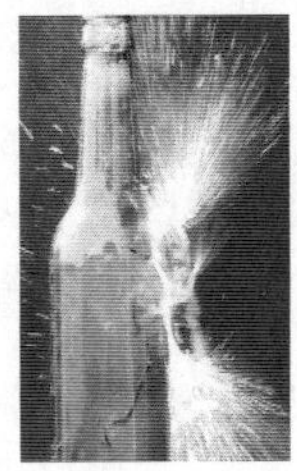

Results -Morphologie-

***S. cer.* var. *diastaticus* -morphological and cultural characteristics-**

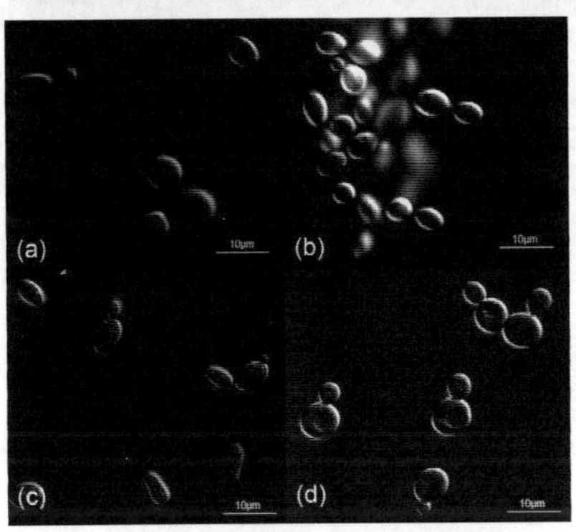

a)
c)
d)

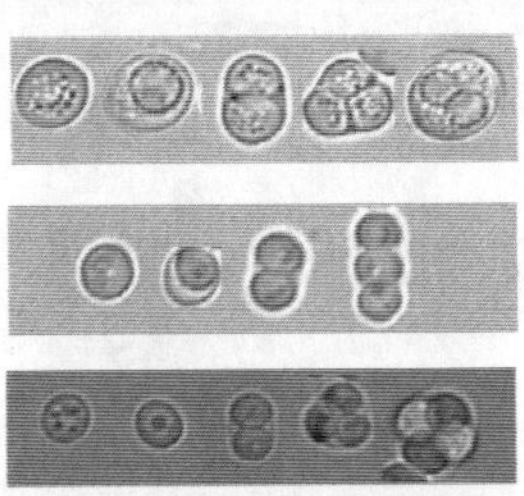

Culture yeast *S. past.* (a) u. *S. cer.* (b)		*S. cer. var. diastaticus* *(c) u. (d)*
oval (egg)	Cell shape:	round- spheric
6-10 µm diameter 22-30 µm² square	Cellsize:	3-4 µm diameter 13-14 µm² square
1-2 (BF) or 6-8 cells (TF)	Cellagglomeration:	1-4 Cells
middle	Spore formation:	high

37th EBC Congress Antwerp, Belgium, May 2019

Summary -*S. cer. var. diastaticus*-

Summary of the investigated *S. cer.* var. *diastaticus* brewing strains

TUM identifier	Species identification by qPCR	Sugar metabolism		Fermentation degree	Spoilage potential	Phenolic off flavor	Flocculation behaviour	Sporulation behaviour	Sensory acceptance	Recommended beer style
		wort sugars	higher dextrins							
DSM 70487	*S. cerevisiae* var. *diastaticus*	+	+	SF	↑	+	flocculent	low	✔	Wheat beer
TUM PI BB 121	*S. cerevisiae* var. *diastaticus*	+	+	SF	↑	+	powdery	medium	✔	Wheat beer
TUM 3-D-2	*S. cerevisiae* var. *diastaticus*	+	+	SF	↑	+	powdery	medium	✔	Wheat beer
TUM 1-H-7	*S. cerevisiae* var. *diastaticus*	+	+	SF	→	+	flocculent	high	✔	Ale
TUM 71	*S. cerevisiae* var. *diastaticus*	+	+	SF	↑	+	powdery	medium	✔	Wheat beer / Ale ("Bavarian Ale")
TUM 3-H-2	*S. cerevisiae* var. *diastaticus*	-	-	WF	↓	+	flocculent	high	✔	Wheat beer / Ale ("Bavarian Ale")
TUM PI BA 124	*S. pastorianus*	+	-	HF	↓	-	flocculent	low	✔	Dark beer / Ale

? Calorie reduced beers ?

Forschungszentrum Weihenstephan für Brau- und Lebensmittelqualität

37th EBC Congress Antwerp, Belgium, May 2019

Conclusion -*S. cer. var. diastaticus*-

A brewing yeast –conclusion-

- *S. cer.* var. *diastaticus* can be used as brewing yeast
- Variable aroma profiles (all POF +) and fermentation performance

diastaticus friend

- Some strains produce very acceptable specific beers (potential for special beers → fruity, citric, wine-like)

- Some „saison" strains, some „Belgian strains" and some top fermenting mixed strains are/contain *S. cer.* var. *diastaticus*
- Some produce super-stable foam

- Cross contamination risk

or foe?

Forschungszentrum Weihenstephan für Brau- und Lebensmittelqualität

Thanks for your attention…

and thanks to:

Dr. Tim Meier-Dörnberg
(Research Center Weihenstephan)
Dr. Maximilian Michel
(Research Center Weihenstephan)
Prof. Dr. Fritz Jacob
(Research Center Weihenstephan)
Oliver Ingo Kory
(Miho Inspektionssysteme GmbH)

Dr.-Ing. Mathias Hutzler
Alte Akademie 3
85354 Freising-Weihenstephan

Telefon: +49 (0) 8161 / 71-3100
Telefax: +49 (0) 8161 / 71-4181

E-Mail: m.hutzler@tum.de
www.blq-weihenstephan.de

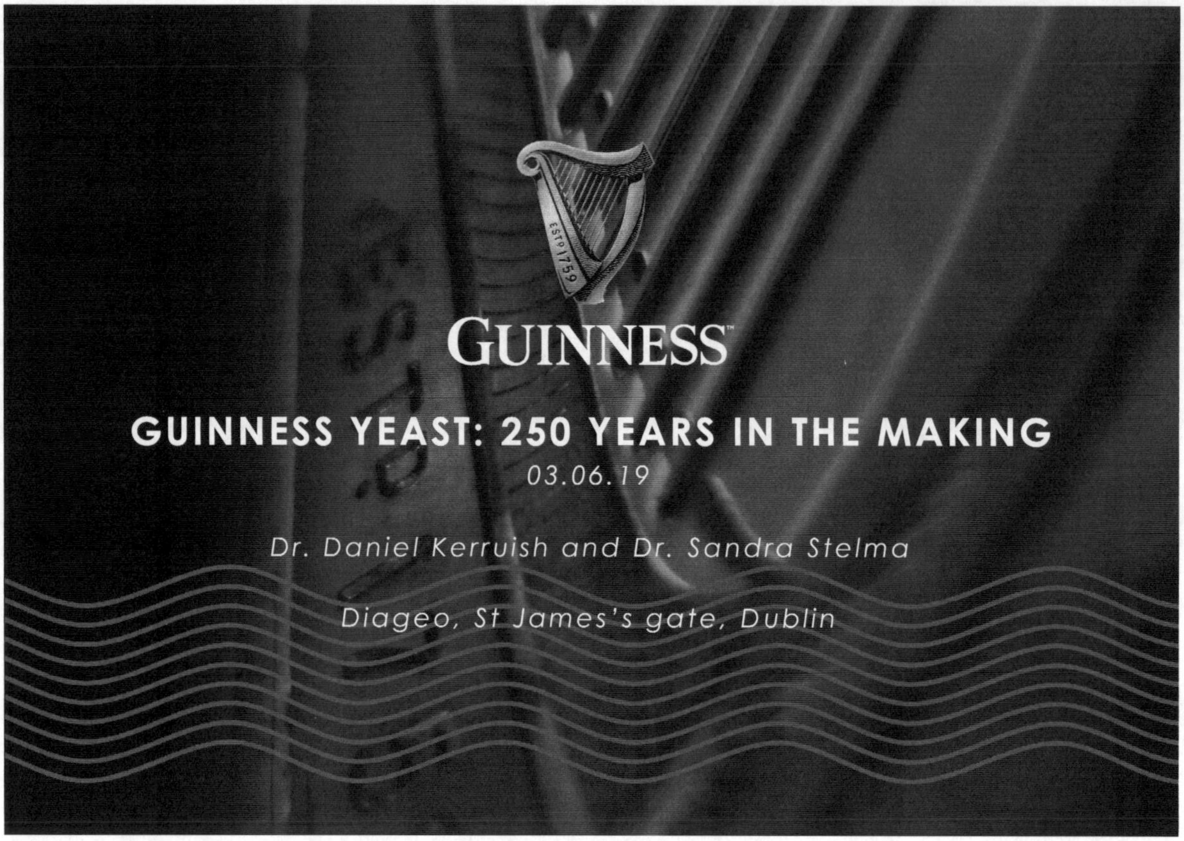

BACKGROUND

- The St James's Gate Yeast Library contains over 300+ yeast and 200+ bacterial species.
- The Yeast Library was established in 1903.
- Meticulous records of each yeast are kept within the Guinness Archive.
- Today the Yeast Library is stored in liquid Nitrogen at the St James's Gate Brewery site Dublin.
- Within the Yeast Library there are number of yeast that are termed 'Guinness Stout Yeast'.
- These *Saccharomyces cerevisiae* yeast have been used to brew Guinness stout and its variants.

Confidential 2

OVERVIEW

- The history of Guinness Yeast
- Genotype and Phenotype analysis of the Guinness yeast
- Pilot Plant Brewing of Guinness Stout
- Full Genome Sequence analysis of the Guinness yeast

Confidential 3

THE HISTORY OF GUINNESS' YEAST

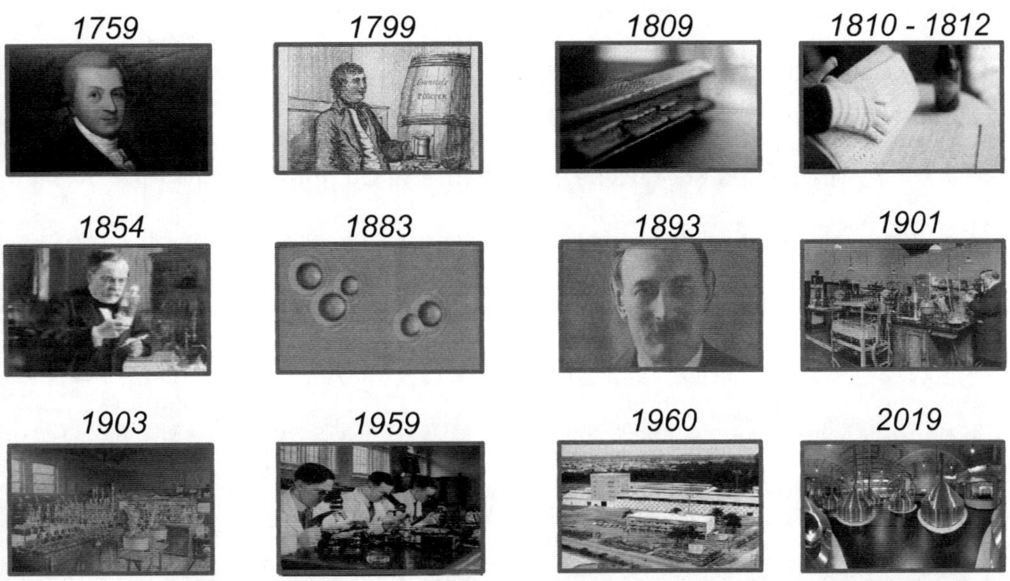

Confidential 4

THE HISTORY OF GUINNESS' YEAST

- In 1903 the Guinness yeast has been 'stored down' allowing Brewers the opportunity to avail of fresh yeast
- In 1959 following extensive work by Dr B. Gilliland and colleagues the IDS (Irish Draught Stout) yeast was selected from the original 1903 yeast
- In 1960 a variant of the IDS yeast was chosen to brew Guinness in Africa; the FES (Foreign Extra Stout) Yeast

Yeast Number	Guinness Stout Yeast
1	Foreign Extra Stout
2	Irish Draft Stout
3	47' Pitching Yeast
4	50' Pitching Yeast
5	55' Pitching Yeast
6	59' Exp Yeast
7	59' Brewing School

Yeast Number	Guinness Stout Yeast
8	59' Ikeja Nigeria Brewery
9	63' Park Royal
10	63' Research Yeast
11	79' Park Royal
12	81' Pitching Yeast
13	86' Park Royal

OVERVIEW

- The history of Guinness Yeast
- Genotype and Phenotype analysis of the Guinness yeast
- Pilot Plant Brewing of Guinness Stout
- Full Genome Sequence analysis of the Guinness yeast

EXPERIMENTAL PROCEDURE

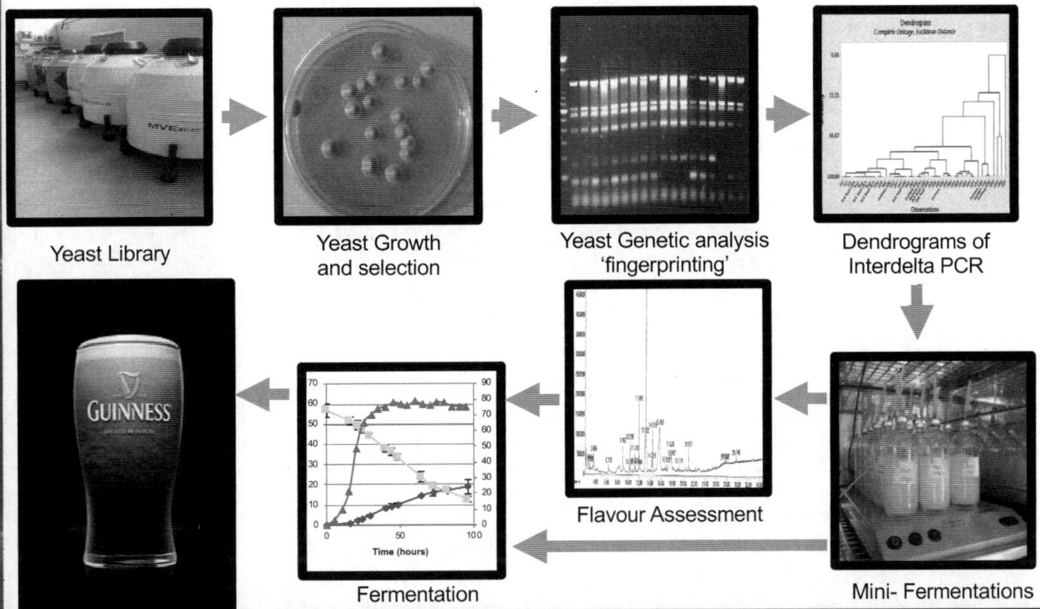

INTERDELTA PCR ASSESSMENT

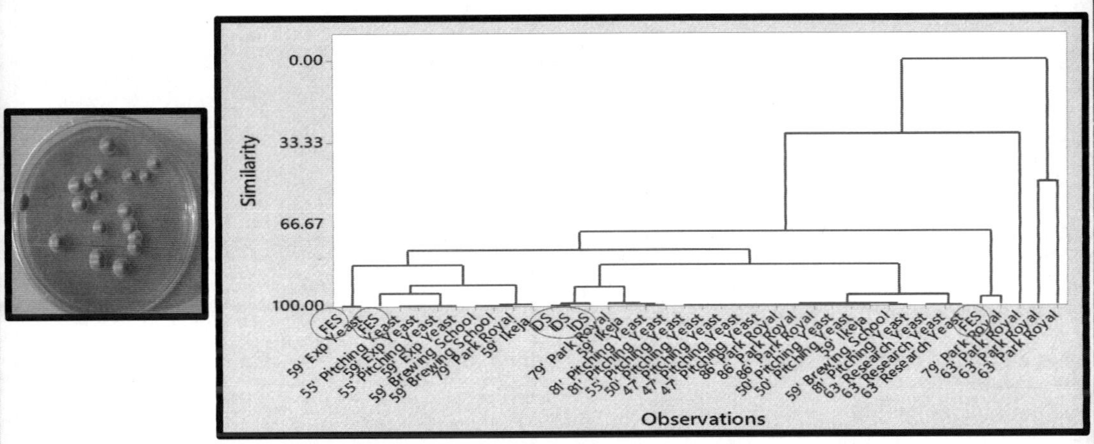

Figure 1: Hierarchical clustering analysis of PCR products, determined using Bioanalzer 2.0 DNA chips, of the interdelta specific primers Δ2/ Δ12 (Legras and Kurst 2003). PCRs were conducted on typical and atypical Guinness yeast morphology as observed through Giant Colony Morphology.

GUINNESS YEAST PHENOTYPE

The Guinness Yeast have a number of phenotypic characteristics, these include:

- Strong Alcohol production
- Phenolic off flavour (POF) positive
- Guinness yeast lineage
- Rapid fermentation
- Organoleptic profile
- Poor flocculation

Yeast Name	Phenotype Assesment
Foreign Extra Stout (FES)	dg_15,17 & 24
Irish Draught Stout (IDS)	dg_9 & 23
47' Pitching Yeast	dg_16
50' Pitching Yeast	dg_3
55' Pitching Yeast	dg_5
59' Brewing School	dg_21
59' Exp Yeast	dg_18
59' Ikeja Nigeria Brewery	dg_22
63' Park Royal	dg_20
63' Research Yeast	dg_11
79' Park Royal	dg_12
81' Pitching Yeast	dg_11
86' Park Royal	dg_14

Confidential

9

FERMENTATIONS

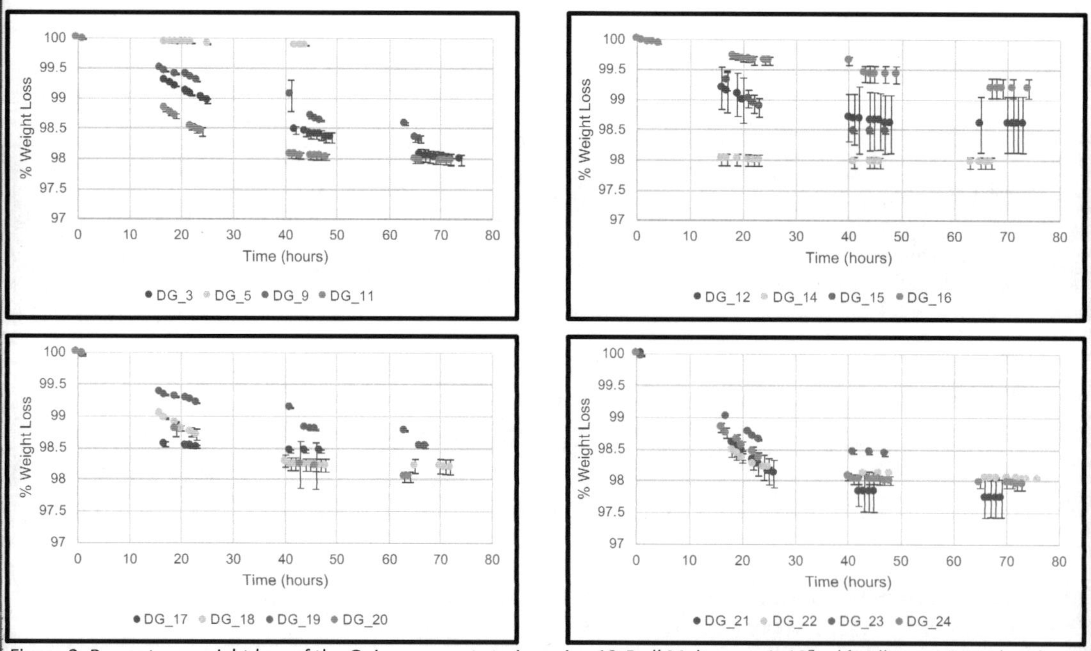

Figure 2: Percentage weight loss of the Guinness yeast strains using 12°P all Malt wort. 1x10^{7} ml^{-1} cells were inoculated into 100ml of wort, incubated at 25°C and stirred at 250rpm. Each point represents the mean of 3 independent replicates and standard errors are shown as SD ± error bars

10

ABV AND DIACETYL PRODUCTION

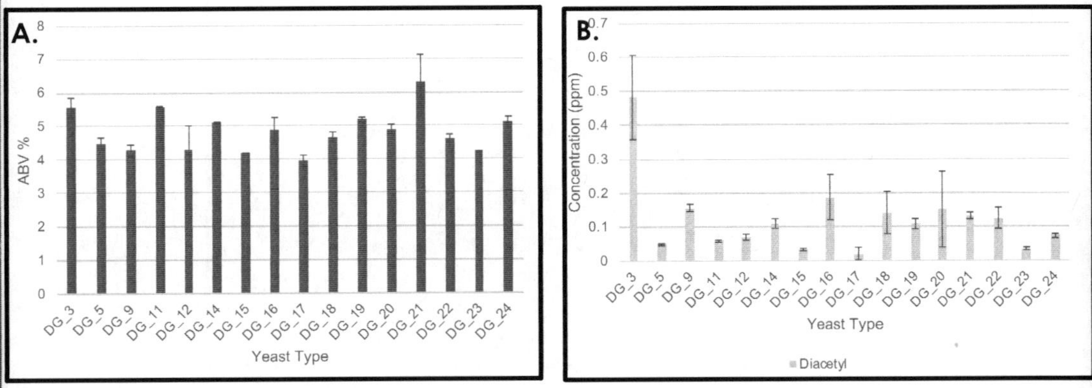

Figure 3: ABV (A.) and Diacetyl (B.) production of the Guinness yeast strains at the cessation of fermentation as determined by Anton Paar and GC respectively. $1x10^7$ ml^{-1} cells were inoculated into 100ml of 12°P all Malt wort, incubated at 25°C and stirred at 250rpm. The fermentation profile did not include a VDK rest. Each point represents the mean of 3 independent replicates and standard errors are shown as SD ± error bars

ETHYL ESTER & HIGHER ALCOHOL PRODUCTION

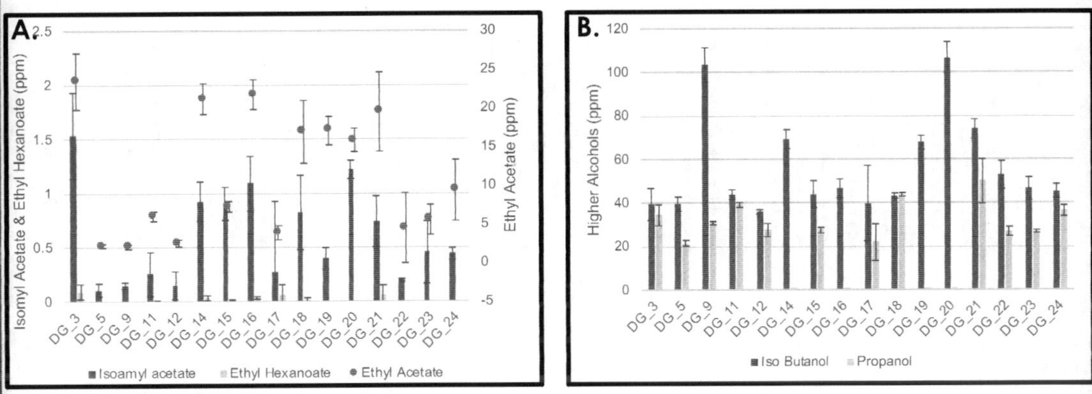

Figure 4: assessment of Ethyl Ester (A.) and Higher Alcohol (B.) production using GC at the cessation of fermentation with the Guinness yeast strains using 12°P all Malt wort. $1x10^7$ ml^{-1} cells were inoculated into 100ml of wort, incubated at 25°C and stirred at 250rpm. Each point represents the mean of 3 independent replicates and standard errors are shown as SD ± error bars

POF PRODUCTION

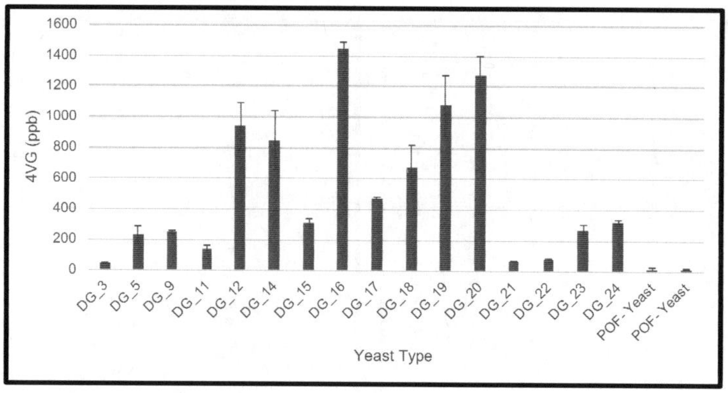

Figure 5: Phenolic Off Flavours (POF) production by different Guinness Yeast strains. The Guinness yeast at a concentration of $1x10^7$ ml^{-1} cells were inoculated into 80mls of YM Broth containing ferulic acid and incubated for 48 hours at 25°C (EBC method 2.3.9.5). The concentration of 4 Vinyl Guaiacol (4 VG) was determined using GC-MS. Each measurement is taken from 3 independent replicate samples; standard errors are shown as SD $\pm$ error bars.

FLOCCULENCE

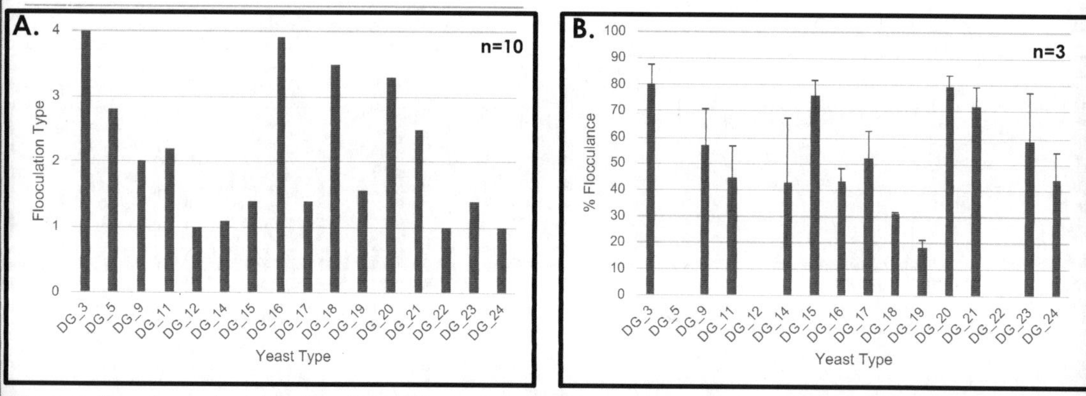

Figure 6: Flocculence characteristics of the different Guinness Yeast strains using EBC 3.5.3.1. Gilliland method (A.) and the ASBC Flocculence Absorbance Method (B.). The EBC Gilliland method class yeast flocculence in terms of non-flocculant (1), slight flocculant (2), moderately flocculant (3) and highly flocculant (4). The ASBC Absorbance Flocculence Method class yeast flocculence in terms of relative percentage with <20% indicating non-flocculant yeast, 20-80% moderately flocculant and >85% highly flocculant. The EBC 3.5.3.1. Gilliland method measurements are the average of 10 independent replicates. The ASBC Flocculence Absorbance Method measurements are taken from 3 independent replicate samples; where there is no recorded sample the experimental procedure produced a negative results and is recorded as zero as per the ASBC method; standard errors are shown as SD $\pm$ error bars.

GUINNESS YEAST PHENOTYPE CONCLUSION

- Interdelta PCR assessment of the individual Guinness yeast with typical and atypical morphology indicate that many of the Guinness yeast are pure or contain closely related variants.
- The Interdelta sequence analysis between the different Guinness yeast show high degree of similarity, suggesting that the Guinness yeast are closely related.
- Phenotypic assessment of the Guinness yeast show many different behavioural traits these include:
  - ABV production and subsequent carbon flux
  - Speed of fermentation
  - Levels of Diacetyl Production
  - Ethyl ester production
  - Flocculant Characteristics – the 'Guinness yeast' is both highly and non flocculant
- All Guinness yeast are POF+, but the level of 4VG production is Guinness yeast strain specific.

Confidential 15

OVERVIEW

- The history of Guinness Yeast
- Genotype and Phenotype analysis of the Guinness yeast
- Pilot Plant Brewing of Guinness Stout
- Full Genome Sequence analysis of the Guinness yeast

Confidential 16

BREWING GUINNESS: OPEN GATE BREWERY DUBLIN

From the phenotype data the yeasts dg_9 and dg_14 were used to ferment Guinness wort at 12°P in the Guinness Open Gate Brewery 1hl pilot plant in Dublin.

Irish Draught Stout (IDS) yeast dg_9

- Genetically identical to the IDS control
- Slower fermentation profile compared to the IDS control
- Similar POF production to the IDS control
- Similar ethyl ester profile to the IDS control
- Greater Diacetyl and Isobutanol production compared to the IDS control

86' Park Royal Yeast dg_14

- Genetically different from the IDS yeast
- Rapid fermentation capability compared to IDS
- Significant POF production
- Less flocculence than the IDS control
- Significant IAA production and only 1 of 7 yeast to produce Ethyl Hexanoate

The Guinness Yeast dg_9 and dg_14 as well as a 3rd generation Guinness yeast from Dublin St James's Gate Brewery were fermented in the Guinness Open Gate brewery 1hl pilot plant using standard Guinness fermentation profile. Yeast were inoculated at a concentration of $1x10^7$ cells ml^{-1} into 12°P Guinness Wort.

The resulting 'Guinness' was tasted by the Diageo Beer taste panel.

SENSORY ANALYSIS AND FERMENTATION PROFILES

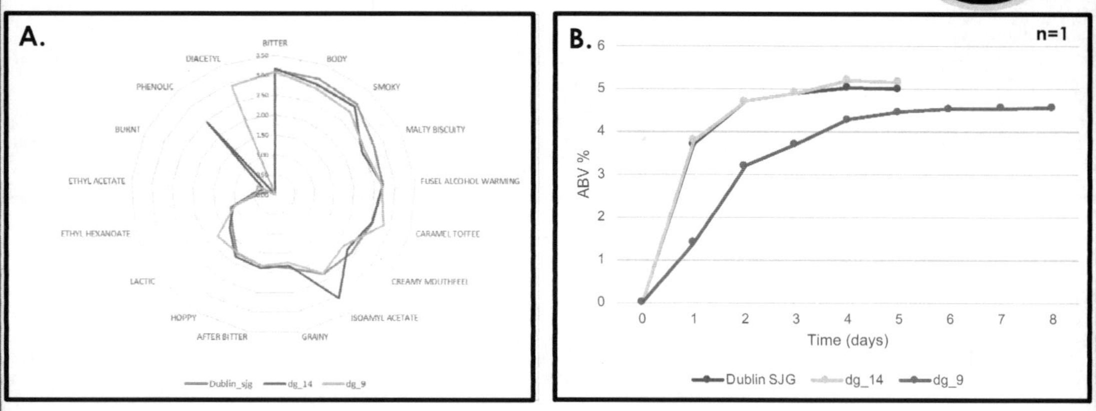

Figure 7: Guinness Stout brewed using the control yeast from Dublin St James's Gate (SJG) and the Guinness variant yeast dg_9 and dg_14. The Spider chart details the different flavour attributes of the Guinness stouts (A.). The different fermentation profiles of the brewed beers are presented in the line graph (B.). The tasting samples were assessed, in duplicate, using the Guinness Draught attribute list by an expert panel using Quantitative Descriptive methodology.

SENSORY ANALYSIS

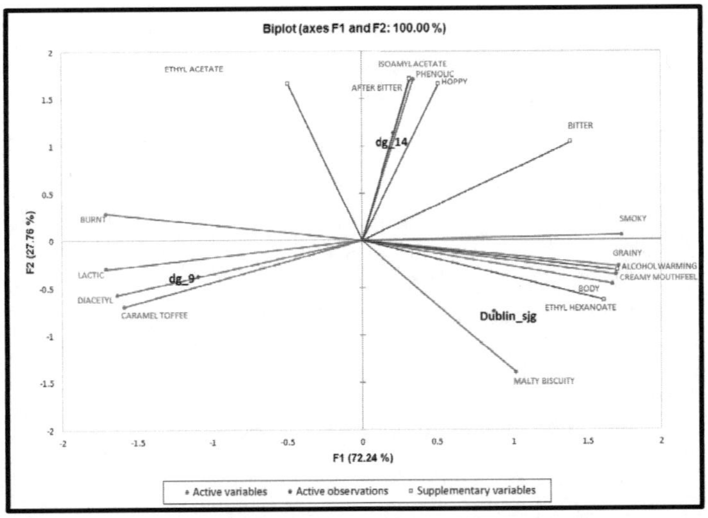

Figure 8: Sensory Principal Component Analysis (PCA) of Guinness Stout brewed using the control yeast from Dublin St James's Gate and the Guinness variant yeast dg_9 and dg_14.

ANALYTICAL ASSESSMENT OF FLAVOUR ATTRIBUTES

Figure 9: Ester (A.), Diacetyl (B.), Higher Alcohol (C.) and 4VG (D.) production of final pack Guinness stout product brewed with the control yeast Dublin SJG and the Guinness variant yeast dg_9 and dg_14. Each measurements are taken from 3 replicate samples; standard errors are shown as SD ± error bars.

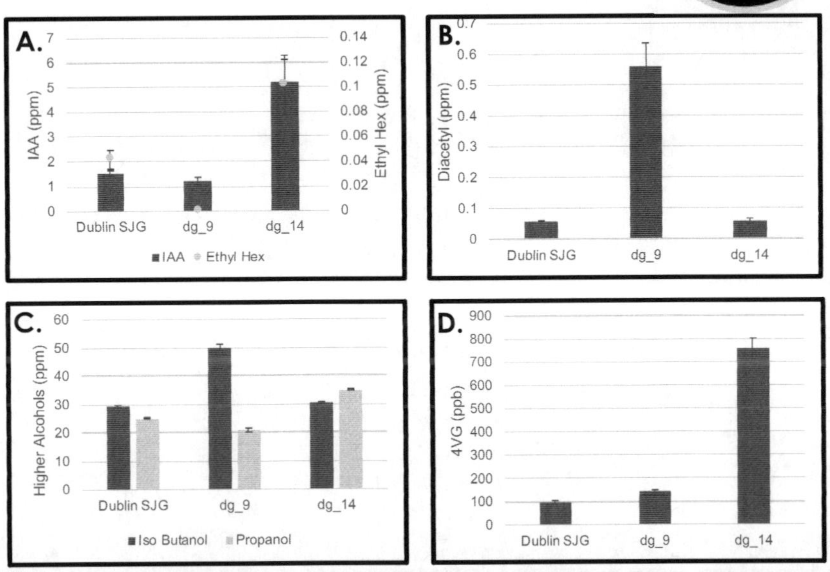

BREWING GUINNESS: CONCLUSIONS

- Guinness brewed with different Guinness' yeast strains produce different Guinness Products
- The PCA analysis confirmed that for key attributes such as POF, Diacetyl, IAA, mouth feel and malty- the three tested yeast have significant variance.
- The brewed Guinness confirmed the phenotype observations of the mini fermentations.
- Yeast dg_9 IDS Variant:
  - Poor fermentation ability compared to SJG control
  - Similar POF production to the IDS/SJG control
  - Greater Diacetyl and Isobutanol production compared to the IDS/SJG control
- Yeast dg_14 86' Park Royal:
  - Rapid fermentation capability compared to IDS
  - Significant POF production
  - Significant IAA and Ethyl Hexanoate production

Confidential 21

OVERVIEW

- The history of Guinness Yeast
- Genotype and Phenotype analysis of the Guinness yeast
- Pilot Plant Brewing of Guinness Stout
- Full Genome Sequence analysis of the Guinness yeast

Confidential 22

GENOME ANALYSIS OF THE GUINNESS YEAST

To gain a better understanding of Guinness Yeast all 16 yeast of interest have been sequenced.

Gallone *and* Steensels *et al.,* 2016 sequenced 157 *S . cerevisiae* yeasts. They placed these yeasts within 5 clades.

The beer 1 clade of yeast attributed geographical location to domestication.

The sequence data from Gallone *and* Steensels *et al.,* 2016 is available to download. This has allowed scientists to place their yeast within yeast clades i.e Norwegian Kveik yeast (Preiss *et al.,* 2018).

The Guinness yeast were placed within the context of the 157 *S . cerevisiae* yeasts sequenced.

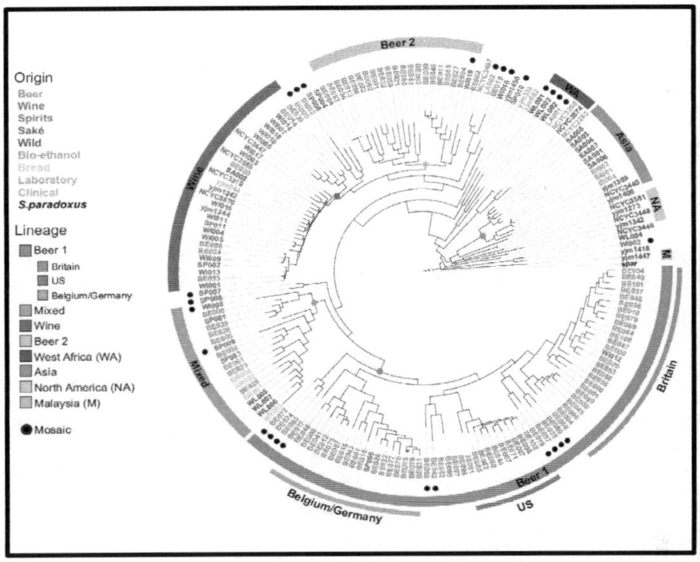

Figure 9: Phylogenetic tree of *S . cerevisiae* (Gallone *and* Steensels *et al.,* 2016)

Confidential

23

GENOME ANALYSIS OF THE GUINNESS YEAST

- All of the Guinness yeast are intimately related forming their own monophyletic group.
- The Guinness yeast share lineages with yeast within Beer 1 clade.
- The Guinness Yeast share an ancestor common to the US and UK yeast.
- The Guinness yeast shares a more recent ancestor with the US yeasts.
- The Guinness yeasts are the only yeast within the US and UK clade to have POF+ characteristics.

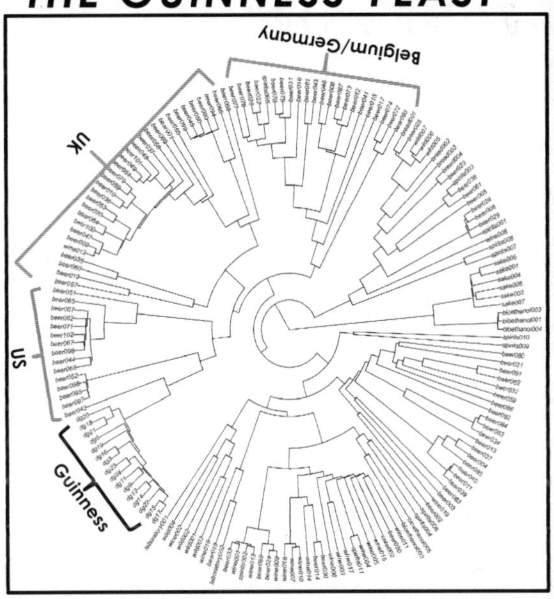

Figure 10: Guinness yeast within the phylogenetic tree of S . cerevisiae as determined by Gallone and Steensels et al., 2016. The Guinness yeast were sequenced using an Illumina MiSeq machine reading with a minimum depth of 30X coverage using 2 x 250bp paired reads. Sequencing data for 157 samples (Gallone and Steensels et al., 2016) were downloaded from the SRA and combined with the Guinness sequence data. Cleaned reads from each sample were aligned to S. cerevisiae S288c reference genome using bwa mem (version 0.7.17). Variants were filtered to remove non bi-allelic sites , sites with Minor allele frequency < 0.05 and sites with calls in < 90% of samples in the study using vcftools (version 0.1.16). This resulted in 154,284 segregating variants for population structure analysis. SNPs were LD-PRUNED using strict ld.threshold of 0.2 and a sliding window of 5000 base pairs. This resulted in 10,399 markers remaining for population structure analysis using SNPRELATE (version 1.18.0)

24

GUINNESS YEAST PROJECT CONCLUSIONS

- Historically Guinness have sought to understand the brewing process through scientific understanding.
- Guinness Yeast are are intimately related forming their own monophyletic group.
- The group of Guinness yeast share a common ancestor with the UK and US yeast Beer 1 clade
- The Guinness yeast are phenotypically diverse
- All Guinness yeast are Phenolic Off Flavour (POF+) positive
- The Guinness yeast produce different Guinness Stout.

Confidential 25

FUTURE WORK

- Further assessment of the Guinness yeast using the full genome sequence analysis i.e. understanding genes related to phenotype especially flavour production and process.
- Using the full genome analysis establish further the relationships between the different Guinness yeast.
- Continue brewing trials with the Guinness yeast in the Open Gate Brewery pilot plant; establishing flavour attributes of the different Guinness yeast.
- Assess the other Irish brewing yeast within the Guinness library to establish whether there is an Irish yeast clade.

Confidential 26

THANK YOU

Micro Technicians:

Jessica Kearns and Edel Stone

Brewers:

Patrick Kerr and Marcus Bengelstorff

Archivists:

Eibhlin Colgan and Fergus Brady

Full Genome sequencing and Bioinformatics:

ELDA Bioscience Dr Elaine Kenny

DIAGEO

Confidential

27

THANK YOU

Teff (Eragrostis tef (Zucc.) Trotter) and quinoa (Chenopodium quinoa W.) as novel raw materials for brewing

Ombretta Marconi*, Valeria Sileoni, Giovanni De Francesco, Lidia Di Ghionno, Giuseppe Perretti

ombretta.marconi@unipg.it

Italian Brewing Research Centre

UNIVERSITÀ DEGLI STUDI DI PERUGIA

37th EBC Congress 2-6 June 2019, Antwerp

Innovative raw materials…

- Innovative food ingredients for innovative food beverage.
- Some cereals and pseudocereals possess antioxidative potential, anticarcinogenic activities and health-promoting effects that were also partly observed from barley, wheat, and rye.
- High nutritional beer: antioxidant , soluble fibre, vitamins, microelements.

Antwerp 04-06-19

Innovative raw materials...

Gluten content < 20 ppm

(CE Regulation 41/2009)

GLUTEN FREE BEER

1 By gluten-free cereals and pseudo cereals or non-grain sources (sugar, syrups)

2 By gluten minimization treatments of barley malt wort

Antwerp
04-06-19

37th EBC Congress 2-6 June 2019, Antwerp — Ombretta Marconi

Innovative raw materials...

Ceccaroni, D. , Sileoni, V., Marconi, O., De Francesco, G., Lee, E.G., Perretti, G. LWT - *Food Science and Technology* 99, 2019, 299-305

Ceccaroni, D., Marconi, O., Sileoni, V., Wray, E., Perretti, G. *Journal of the Science of Food and Agriculture* 99, (6) 2019, 2726-2734

Di Ghionno, L., Marconi, O., Lee, E.G., Rice, C.J., Sileoni, V., Perretti, G. *Journal of Agricultural and Food Chemistry*, 65, 23, 2017, 4777-4785

Di Ghionno, L., Sileoni, V., Marconi, O., De Francesco, G., Perretti, G. LWT - *Food Science and Technology* 84, 2017, 746-752

De Meo B., Freeman G., Marconi O., Booer C, Perretti G., Fantozzi P. *Journal of the Institute of Brewing*, 117 (4), (2011)

Sileoni V., De Francesco G., Ceccaroni D., Marconi O., Perretti G. *Proceedings Trends in Brewing* (2018)

Antwerp
04-06-19

37th EBC Congress 2-6 June 2019, Antwerp — Ombretta Marconi

TEFF (*Eragrostis tef*)

Teff cultivated at the experimental fields of the Department of Agricultural, Food and Environmental Science, University of Perugia

Antwerp
04-06-19

37th EBC Congress 2-6 June 2019, Antwerp — Ombretta Marconi

TEFF (*Eragrostis tef*)

- Teff is a gluten-free cereal native to Ethiopia
- Used to make flatbread (*Injera*), porridge or home brewed alcoholic drinks (*tella*)
- Rich in starch (73 %) and good content of protein (11-12%)
- Grain is less than 1 mm in diameter which make it a challenging grain to produce malt

Antwerp
04-06-19

37th EBC Congress 2-6 June 2019, Antwerp — Ombretta Marconi

White teff variety called *Witkop*

Parameters	
Moisture (%)	13.3 ± 0.3
Germinative Energy 4ml (%)	98 ± 1
Germinative Energy 8ml (%)	93 ± 1
total nitrogen (% d.m.)	1.96 ± 0.09
protein (% d.m.)	12.25 ± 0.7
Lipids (% d.m.)	1.63 ± 0.2

data shown as mean ± standard deviation; n= 2 technological replicate; d.m.= dry matter.

Antwerp
04-06-19

37th EBC Congress 2-6 June 2019, Antwerp | Ombretta Marconi

Micromalting

400 g *Witkop* in nylon bags a at two different temperatures (22°C and 24°C) and five different times of germination (2, 3, 4, 5 and 6 days)

Steeping
T: 24°C
I step: 3 h + 2 h aria
II step: 2 h

Germination
6 giorni

Drying
20h a 30°C
4h a 60°C
3h a 65°C

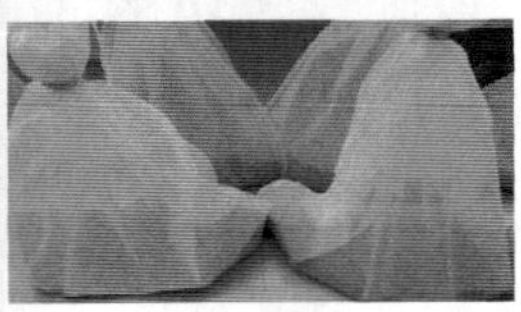

Witkop in nylon bags

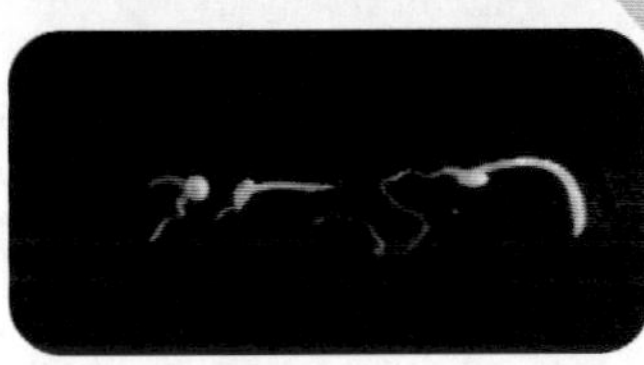

Micromalting pilot plant
(Campden BRI (UK))

Antwerp
04-06-19

37th EBC Congress 2-6 June 2019, Antwerp | Ombretta Marconi

Quality attributes of teff malts

	Germination temperature (22 °C)					Germination temperature (24 °C)				
Germination time (days)	*2*	*3*	*4*	*5*	*6*	*2*	*3*	*4*	*5*	*6*
Extract (% dm)	80.4^{b}	80.6^{bc}	80.6^{b}	81.0^{bc}	80.9^{bc}	74.2^{a}	80.8^{bc}	82.3^{d}	81.3^{c}	**80.4^{b}**
Fermentability (%)	75.9^{b}	76.2^{b}	78.4^{d}	81.2^{f}	80.3^{ef}	54.3^{a}	77.1^{c}	79.7^{e}	78.3^{d}	**80.9^{f}**
Viscosity (mPas)	1.48^{a}	1.47^{a}	1.54^{c}	1.58^{e}	1.66^{g}	1.60^{f}	1.53^{c}	1.51^{b}	1.56^{d}	**1.53^{c}**
Colour (EBC-U)	5.6^{a}	6.3^{b}	6.7^{c}	6.9^{cd}	6.8^{cd}	5.9^{a}	6.3^{b}	7.0^{d}	7.5^{e}	**7.4^{e}**
Soluble nitrogen (% dm)	0.38^{b}	0.45^{cd}	0.46^{cde}	0.47^{ef}	0.49^{fg}	0.31^{a}	0.44^{c}	0.47^{de}	0.50^{g}	**0.51^{g}**
Kolbach Index (%)	26.5^{b}	29.0^{cd}	30.7^{de}	31.3^{ef}	31.5^{ef}	21.0^{a}	28.8^{c}	30.8^{e}	32.8^{fg}	**33.4^{g}**
FAN (mg/L at 8.6°P)	100^{b}	111^{c}	118^{d}	120^{de}	123^{ef}	94^{a}	113^{c}	121^{def}	125^{fg}	**129^{g}**
Diastatic power (WK)	29^{a}	41^{c}	62^{e}	75^{g}	85^{h}	37^{b}	54^{d}	67^{f}	76^{g}	**87^{h}**
Filtration speed (min)	> 60	> 60	> 60	> 60	> 60	> 60	> 60	> 60	> 60	**> 60**
Alpha-amylase (CU/g)	10^{a}	11^{a}	22^{cd}	19^{bc}	18^{b}	17^{b}	22^{cd}	24^{de}	24^{de}	**26^{e}**
Beta-amylase (CU/g)	2.2^{a}	3.1^{b}	6.4^{c}	8.1^{g}	7.9^{f}	2.1^{a}	6.4^{c}	7.1^{d}	7.5^{e}	**8.2^{g}**
Limit dextrinase (U/Kg)	408^{a}	553^{b}	584^{c}	597^{d}	716^{h}	559^{b}	639^{e}	677^{f}	691^{g}	**776^{i}**

n= 2 technological replications; values on the same row with different superscript letters are statistically different (p<0.05) FAN: Free Amino Nitrogen

Antwerp
04-06-19

37th EBC Congress 2-6 June 2019, Antwerp

Ombretta Marconi

1

Brewing trials with raw teff and malted teff

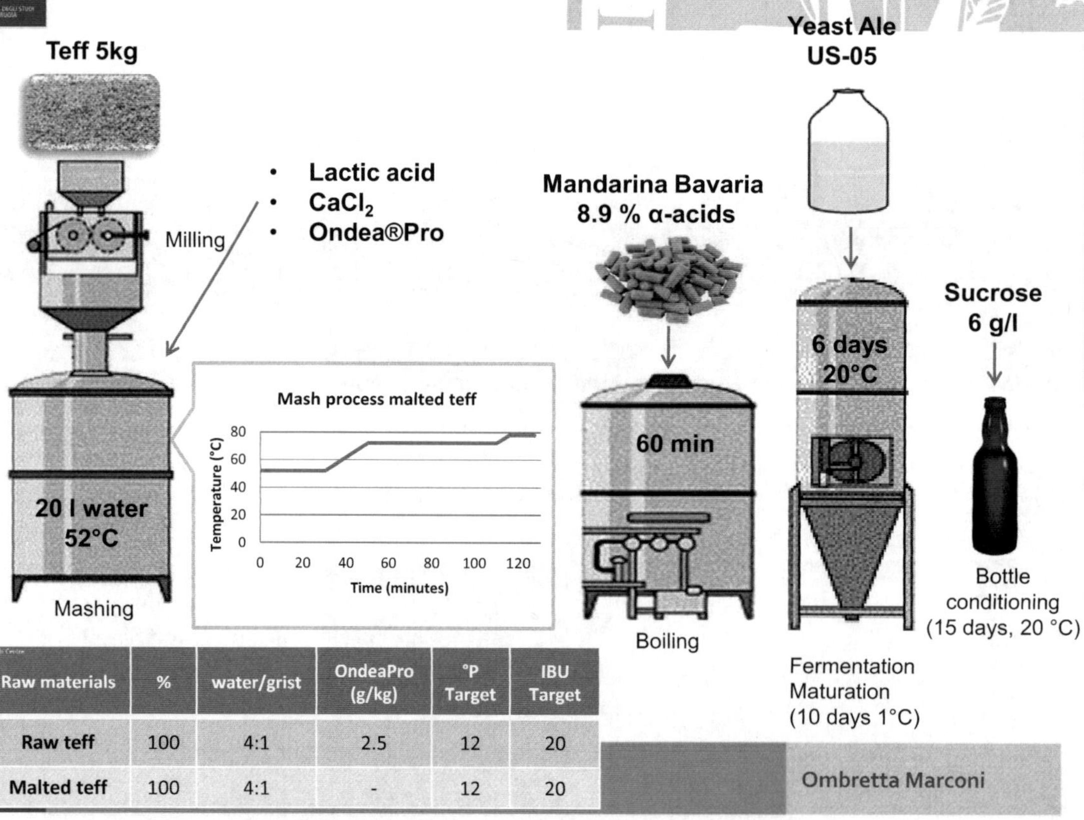

Raw materials	%	water/grist	OndeaPro (g/kg)	°P Target	IBU Target
Raw teff	100	4:1	2.5	12	20
Malted teff	100	4:1	-	12	20

Ombretta Marconi

Teff (raw and malted) wort qulity attributes

Teff Wort	Raw teff	Malted teff
Extract (°P)	11.73 ± 0.09[a]	12.18 ± 0.44[a]
Fermentability (%)	71.0 ± 1.4[a]	75.0 ± 1.4[a]
pH	5.27 ± 0.04[a]	5.12 ± 0.04[a]
Color (EBC-Unit)	7.9 ± 0.4[a]	11.4 ± 0.8[b]
Soluble nitrogen (mg/L at 12°P)	590 ± 2[a]	751 ± 11[b]
FAN (mg/L at 12°P)	106 ± 3[a]	176 ± 6[b]
Viscosity (mPa s at 12 °P)	1.75 ± 0.01[a]	1.78 ± 0.04[a]
Turbidity (U-EBC)	83.2 ± 2.6[a]	104.0 ± 2.8[b]

data shown as mean ± standard deviation; values with different superscript letters are statistically different ($p<0.05$).

- Good values of fermentability for both raw and malted teff were obtained
- Soluble nitrogen and free amino nitrogen were higher in teff malt

- However, satisfactory fermentations were achieved for both raw and malted teff

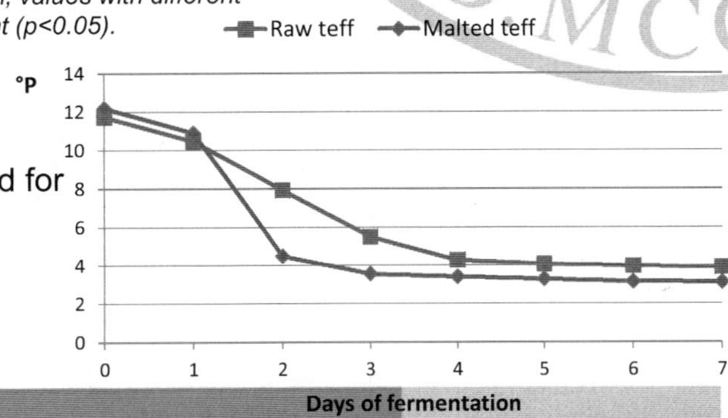

Antwerp 04-06-19

37th EBC Congress 2-6 June 2019, Antwerp

Ombretta Marconi

Teff beer	Raw teff	Malted teff
Original extract (% P)	11.65 ± 0.35[a]	11.70 ± 0.42[a]
Alcohol (%v/v)	4.37 ± 0.06[a]	4.68 ± 0.52[a]
Apparent degree of fermentation (%)	69.8 ± 1.1[a]	74.7 ± 1.9[a]
pH	4.29 ± 0.04[a]	4.23 ± 0.19[a]
Color (EBC-Unit)	5.6 ± 0.28[a]	9.8 ± 0.85[b]
Total nitrogen (at 12 °P) (mg/L)	333 ± 8.9[a]	360 ± 3.5[a]
FAN (at 12 °P) (mg/L)	41 ± 10[a]	72 ± 4[a]
Viscosity at 12°P	1.52 ± 0.02[a]	1.57 ± 0.05[a]
Nibem 30 (sec)	154 ± 7.02[a]	181. ± 9.73[a]

n = 2 technological repetitions. Mean ± standard deviation. Values in the same row followed by a different letter are statistically different ($p \leq 0.05$).

Antwerp 04-06-19

37th EBC Congress 2-6 June 2019, Antwerp

Ombretta Marconi

Sugar profiles of teff worts and beers

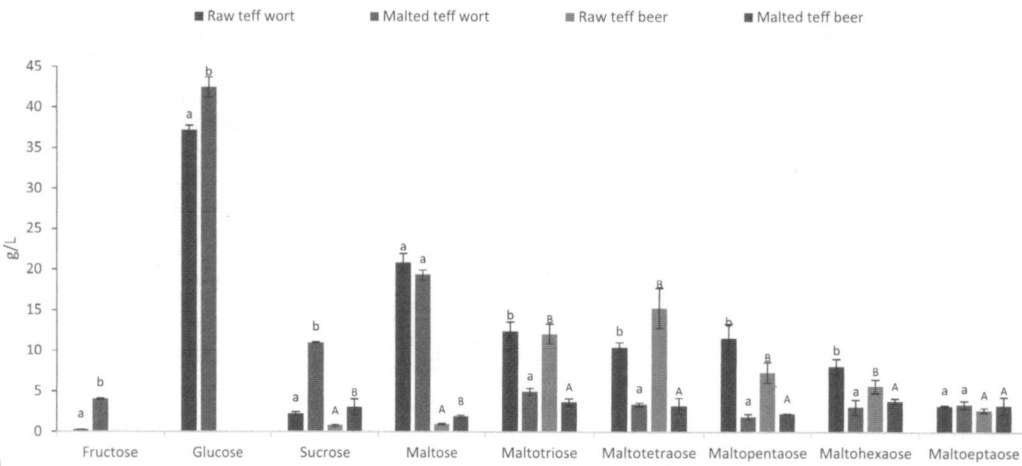

Antwerp
04-06-19

37th EBC Congress 2-6 June 2019, Antwerp

Ombretta Marconi

Amino acid profiles of teff worts and beers

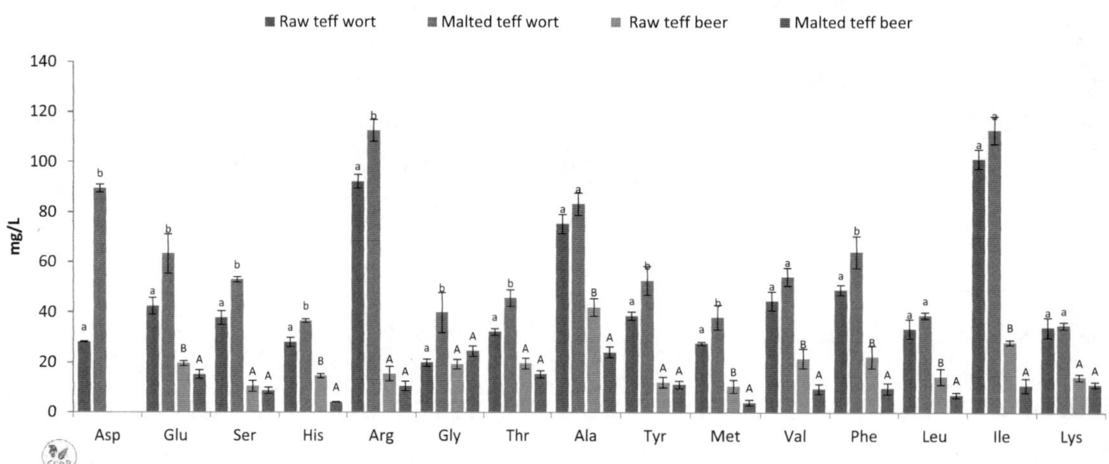

Antwerp
04-06-19

37th EBC Congress 2-6 June 2019, Antwerp

Ombretta Marconi

Volatile compounds in raw and malted teff beer

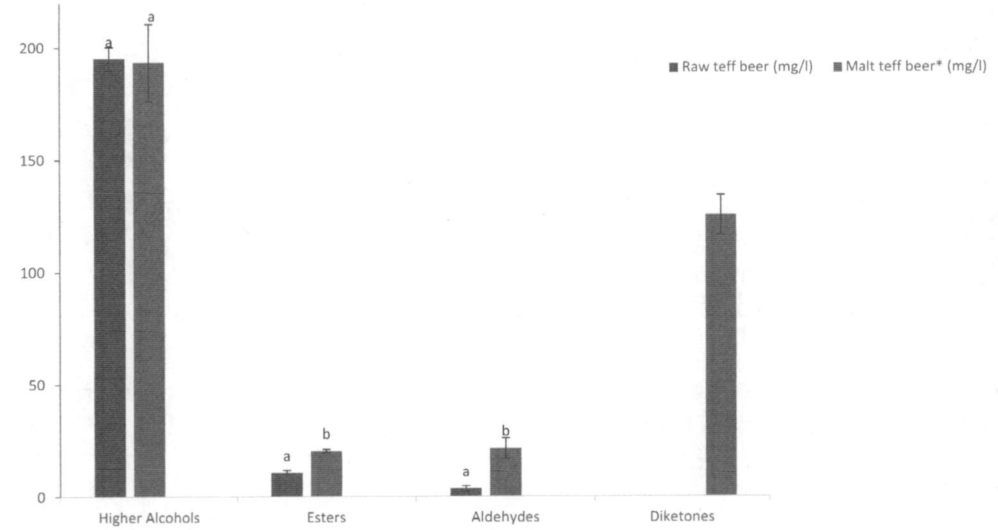

Antwerp
04-06-19

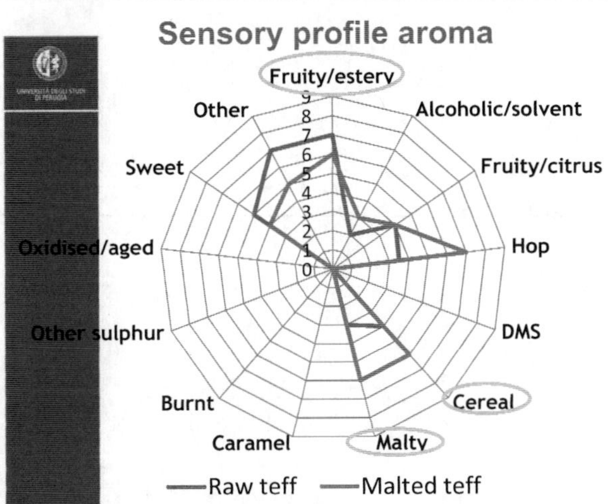

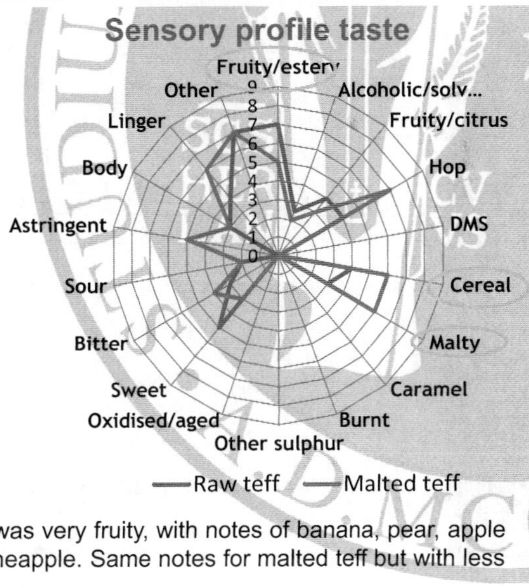

- The aroma of raw teff was very fruity, with notes of banana, pear, apple and tropical fruits as pineapple. Same notes for malted teff but with less intensity
- Malted teff beer was also characterized by a strong malty and cereal aroma/taste, with peanuts, biscuit, vanilla, hay and straw notes. Hop notes were also high and were perceived as grassy and spicy. Aldehydes were also picked up with notes of green apple
- In both beers were perceived a strong aroma/taste of **rhubarb and earthy (others)**
- Both beers had a light body with a good sweet/bitter balance for malted teff, and more high sweetness for raw teff.

1

Teff conclusions and future perspectives

- This work showed that teff can be malted with acceptable quality attributes especially for malt germinated at 24°C for 6 days
- However it was concluded that brewing with raw teff and malted teff result in acceptable gluten-free beer in terms of quality attributes and sensory profile
- The small grain size may lead to difficulties in handling in malting plants and wort separation issues at the industrial scale. Thus, a specific malting and brewing plant needs to be developed and further studies are required to optimise the separation technology of teff wort.
- Teff could also be used as an adjunct with other gluten free cereal or pseudoceral. Distinctive flavours could be achieved with teff as only part of the source of extract

Antwerp
04-06-19

37th EBC Congress 2-6 June 2019, Antwerp — Ombretta Marconi 1

QUINOA *(CHENOPODIUM QUINOA WILLD.)*

Antwerp
04-06-19

37th EBC Congress 2-6 June 2019, Antwerp — Ombretta Marconi

Quinoa

Quinoa (Chenopodium quinoa) is a pseudocereal crop of the Andean region.

The worldwide interest on quinoa is mainly due to:

- Wide genetic variability and strong agronomic potential;
- Nutritional characteristics of the seeds;
- Absence the protein fraction gluten-forming of the seeds.

Ombretta Marconi

Malting

The experimentation was carried out using two varieties of quinoa:

Real

Black Collana

Micromalting system Custom Products Laboratory, Keith, Scozia, UK.

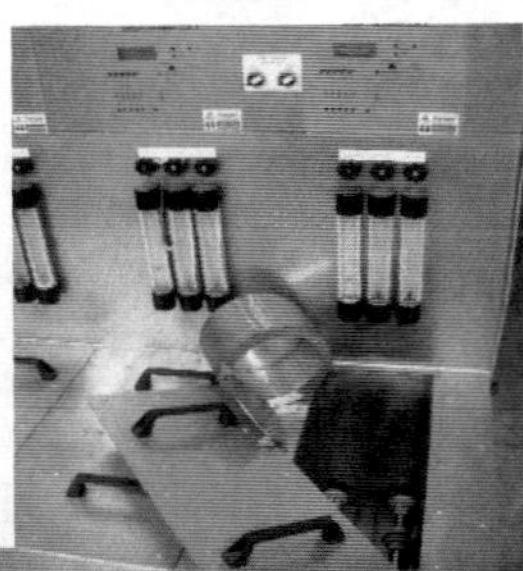

Antwerp
04-06-19

37th EBC Congress 2-6 June 2019, Antwerp

Ombretta Marconi

Malting quality attitude of quinoa

Parameters	Real Mean ± Ue	*Black Collana* Mean ± Ue
Moisture (%)	13.0 ± 0.5	12.3 ± 0.5
Germinative energy 4ml (%)	62 ± 13	92 ± 6
Germinative energy 8ml (%)	61 ± 18	91± 9
Total nitrogen (% d.m.)	2.64 ± 0.09	2.25 ± 0.09
Protein (% d.m.)	15 ± 0.6	12.8 ± 0.6

n= 2 repliche analitiche; Ue= incertezza estesa;
s.s.= sostanza secca.

Antwerp
04-06-19

37th EBC Congress 2-6 June 2019, Antwerp

Ombretta Marconi

Micromalting

- steeping/germination temperature(10°C; 15°C; 20°C);
- germination time (3, 5 and 7 days).

steeping
5 h wet + dry rest

Germination

Kilning
20h at 30°C
2h at 60°C
3h at 65°C

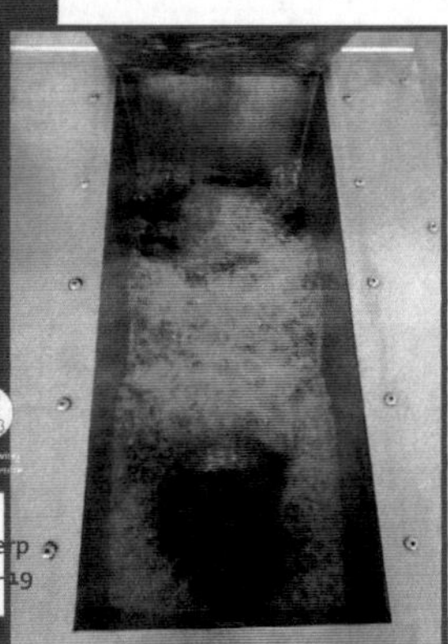

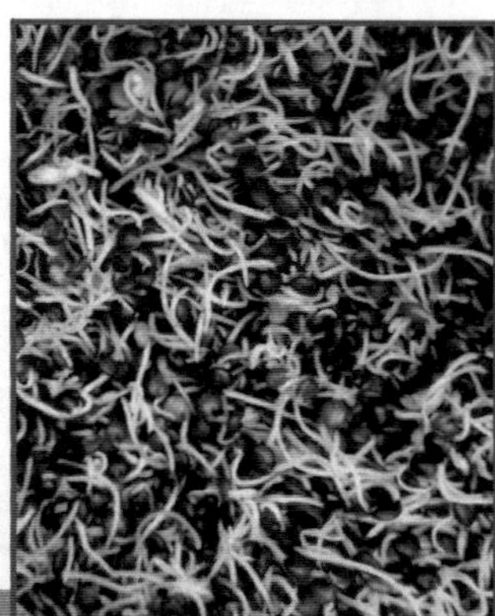

Antwerp
04-06-19

37th EBC Congress 2-6 June 2019, Antwerp

Ombretta Marconi

Results

3 days of germination at 10°C ⟶ extract value of 64.15 (%d.m.)

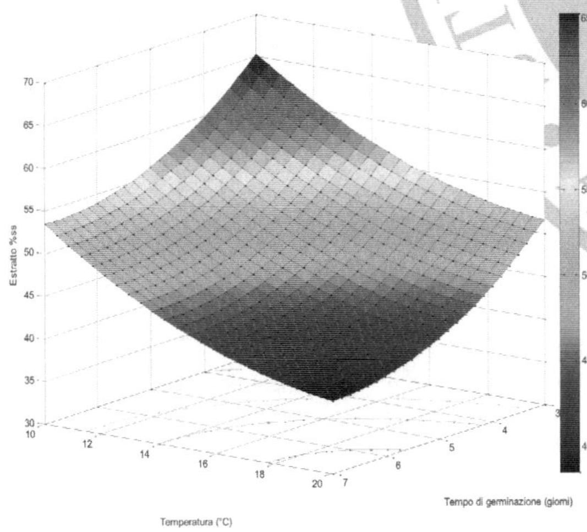

Antwerp
04-06-19

37th EBC Congress 2-6 June 2019, Antwerp

Ombretta Marconi

Effect of temperature on Black Collana germination

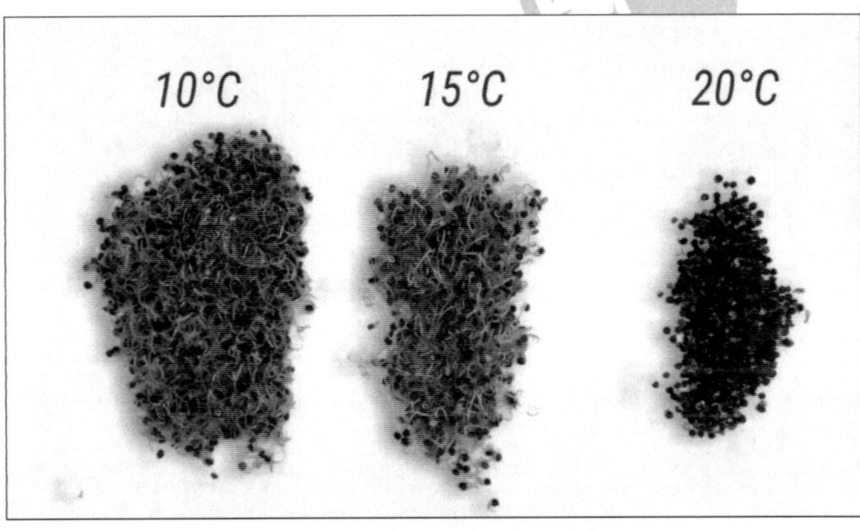

-days of germination: 7

Antwerp
04-06-19

37th EBC Congress 2-6 June 2019, Antwerp

Ombretta Marconi

Quinoa malt

	mean ± Ue
Extract (% d.m.)	63.1 ± 0.8
Filtration	slow
Saccharification time (min)	**> 60**
Fermentability (%)	11.6 ± 0.6
Diastatic power (WK/ s.s.)	n.d.
Kolbach index	31.9 ± 1.4
FAN (mg/l)	154.0 ± 14
α-amylase (CU/g)	n.r.
β-amylase (CU/g)	1.2 ± 0.1
Limit destrinase (U/Kg malt)	3079 ± 206

n=2 analytical replication; d.m.=dry matter; Ue= uncertainty;

Antwerp
04-06-19

37th EBC Congress 2-6 June 2019, Antwerp

Ombretta Marconi

Brewing trials with raw quinoa

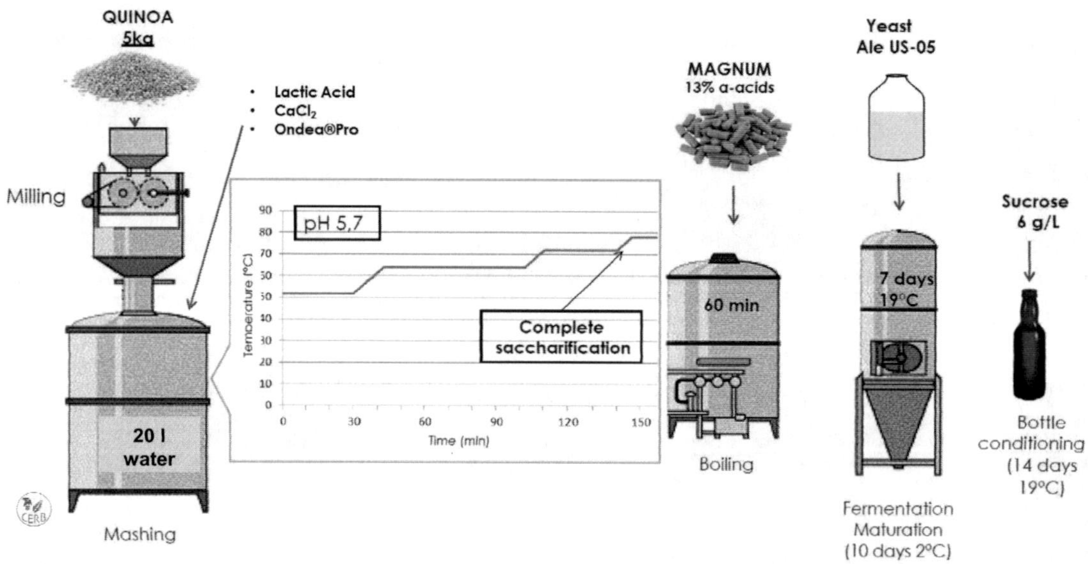

Antwerp
04-06-19

37th EBC Congress 2-6 June 2019, Antwerp

Ombretta Marconi

Antwerp
04-06-19

Wort standard quality attributes

	WHITE QUINOA WORT	BLACK QUINOA WORT
Extract (°P)	12.47[a]	11.88[a]
pH	5.30[a]	5.3[a]
Colour (U-EBC)	8.98[b]	5.7[a]
Total Nitrogen (mg/L at 12°P)	1645[b]	1108[a]
FAN (mg/L at 12°P)	170[b]	115[a]
Viscosity (mPa*s at 12°P)	1.55[a]	1.53[a]
Fermentability (%)	84[a]	84[a]

n = 2 technological replicates. Values on the same row with different letters show a significant difference $p<0.05$. FAN = Free Amino Nitrogen. U-EBC = European Brewery Convention Units.

Antwerp
04-06-19

Fermented beverages standard quality attributes

	WHITE QUINOA beer	BLACK QUINOA beer
Original Extract (°P)	12.70[a]	12.00[a]
Real Extract (°P)	3.49[a]	3.12[a]
Alcohol (% v/v)	6.07[a]	5.80[a]
Real Attenuation (%)	73.90[a]	75.20[a]
Colour (U-EBC)	6.95[b]	4.70[a]
pH	4.59[a]	4.26[a]
Total Nitrogen (mg/L)	1464[b]	837[a]
FAN (mg/L)	82[b]	58[a]
Viscosity (mPa*s)	1.42[a]	1.39[a]

n = 2 technological replicates. Values on the same row with different letters show a significant difference $p<0.05$. FAN = Free Amino Nitrogen. U-EBC = European Brewery Convention Units.

Sugar profiles of quinoa worts and beers

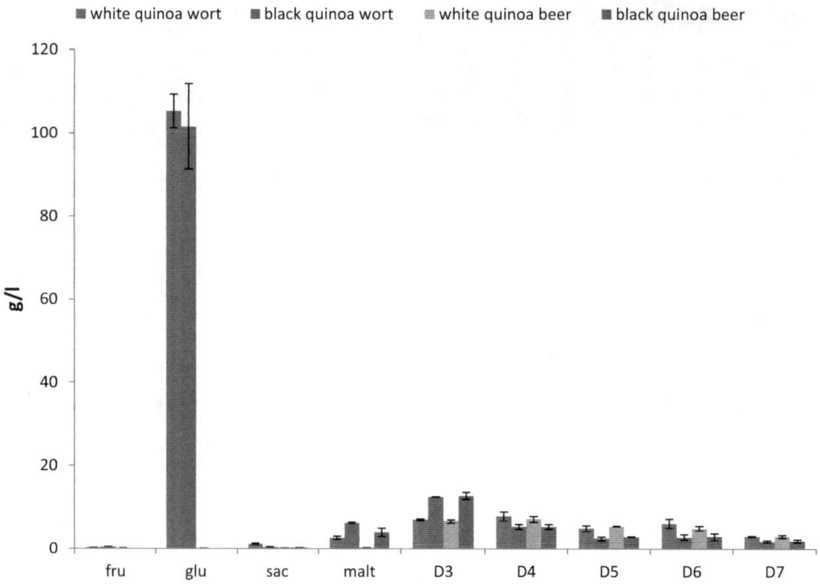

Antwerp
04-06-19

37th EBC Congress 2-6 June 2019, Antwerp

Ombretta Marconi

Amino acid profiles of quinoa worts and beers

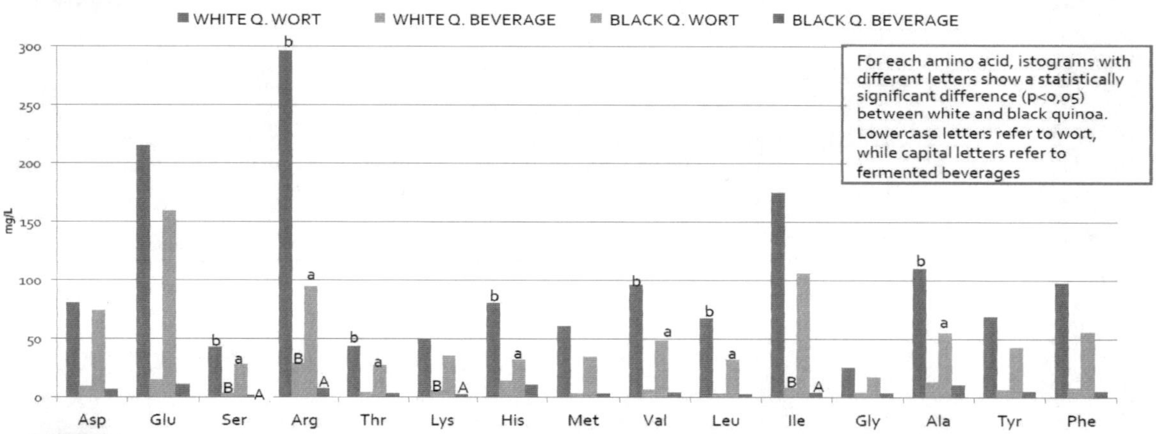

Antwerp
04-06-19

37th EBC Congress 2-6 June 2019, Antwerp

Ombretta Marconi

Antioxidant properties of quinoa worts and beers

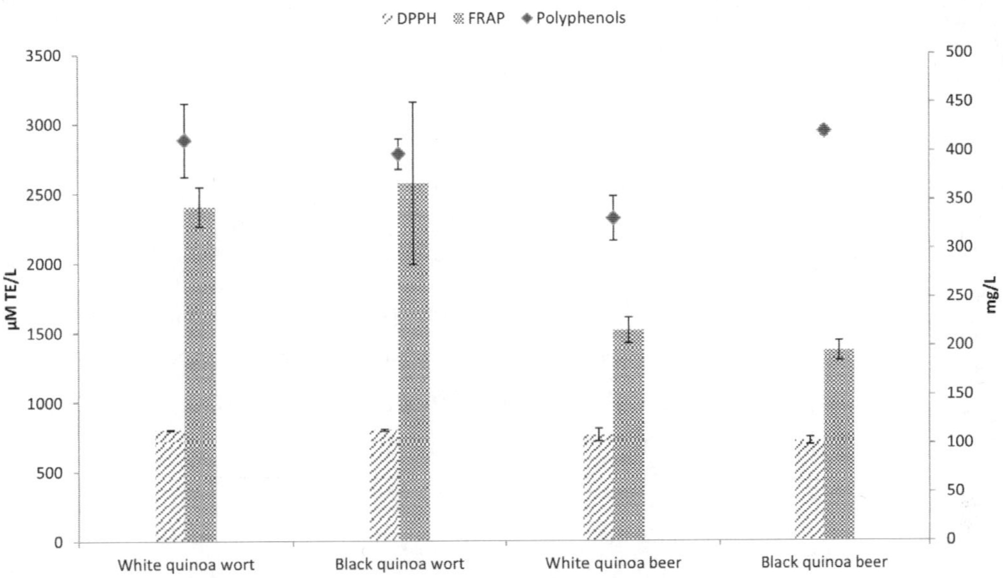

Antwerp
04-06-19

37th EBC Congress 2-6 June 2019, Antwerp

Ombretta Marconi

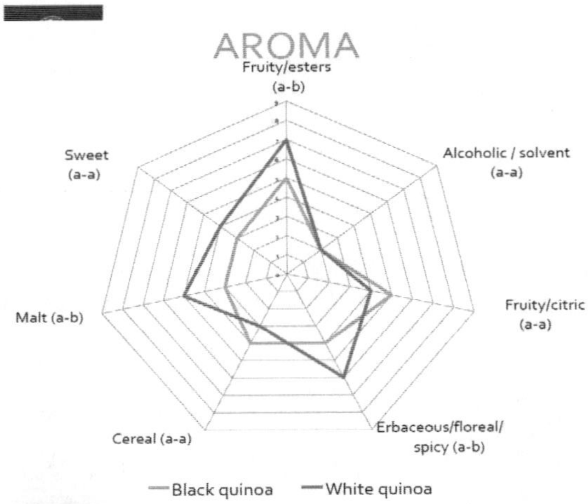

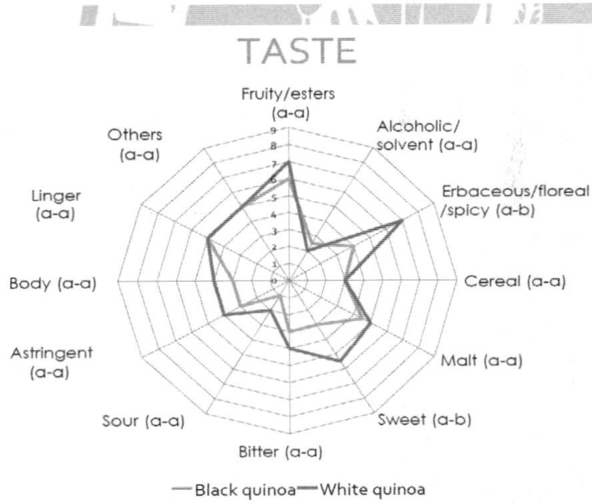

Fermented beverages based on white (A) and black (B) quinoa

Antwerp
04-06-19

37th EBC Congress 2-6 June 2019, Antwerp

Ombretta Marconi

Quinoa - Conclusion

- The *quinoa* malting conditions are low steeping/germination temperature and short germination time obtaining good protein modification but low saccharification and fermentability, slow filtration time;
- the use of quinoa and exogenous enzymes have provided a fermented beverage with characteristics similar to other cereal-based drinks currently on the market;
- the obtained innovative product shows high nutritional value, due to the high content of essential amino acids and the gluten-free protein, indicating a valid alternative for the valorization of this pseudocereal.

University of Perugia

Italian Brewing Research Centre

Prof. Giuseppe Perretti

Dr. Elio Moretti

Dr. Simona Floridi

Dr. Giovanni De Francesco

Dr. Valeria Sileoni

Dr. Lidia Di Ghionno, Dr. Eng. Heidi Mayer, Dr. Francesco Scipioni,
Dr. Raffaele Stupazzini

Antwerp
04-06-19

37th EBC Congress 2-6 June 2019, Antwerp   Ombretta Marconi

Process Innovation at Asahi Breweries -The pursuit of ultimate quality consistency for Asahi Super Dry –

Yoshinori Ito

Asahi breweries, Ltd

"Quality is the achievement of consistency and the elimination of unwanted surprises."

Maintaining consistency is very difficult.

Vintage? Lot-to-Lot Variations?

Vintage? Lot-to-Lot Variations?

We place the utmost importance on producing beer with an extremely high levels of consistency.

Our Flagship Brand: *SUPER DRY*

"KARAKUCHI" TASTE

Karakuchi means "dry," and this word best describes the sophisticated yet congenial character of ASAHI SUPER DRY.

Its **refreshingly crisp, clear taste** makes it an excellent match for ar

SUPER DRY is a Japanese-style pilsner, brewed with some adjuncts (rice, corn, and sta
to control overall body and taste

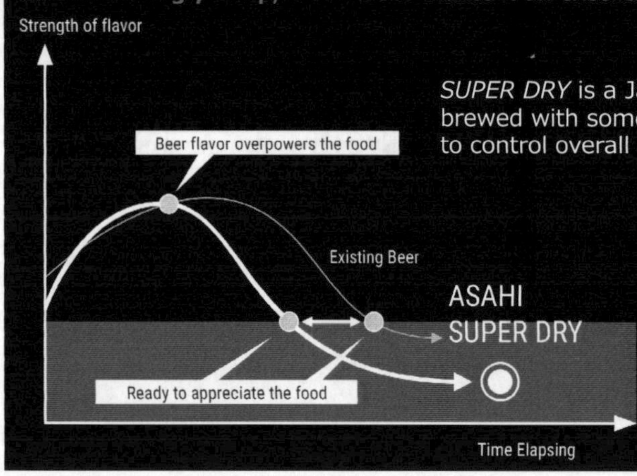

Our goal

Weitertrinken !
= "One more, please!"

・KIRE (a balance of clear taste and body)
・no residual or unpleasant flavors after drinking

It requires delicate control for consistent taste.

Past 10 year's presentations for consistency of beer

Homogeneity of flavor
Kobayashi, WBC (2012)
Nakamura, WBC (2012)
Koizumi, WBC (2012)
Noba, WBC (2016)
Noba, EBC (2017)
Noba, Brewing summit (2018)
Oomuro, EBC (2017)
Okada, Brewing summit (2018)
Irie, EBC(2013)
Irie, EBC (2017)
Irie, ASBC (2017)
Irie, Brewing summit (2018)

sensory evaluation
Miyashita, WBC (2016)
Miyashita, ASBC (2017)
Miyashita, EBC (2019) poster presentation
Wakihira, EBC (2019) oral June/4 12:00

Freshness (flavor stability)
Koyano, ASBC (2017)

Consistent taste of SUPER DRY in all breweries

different climates/facilities/materials

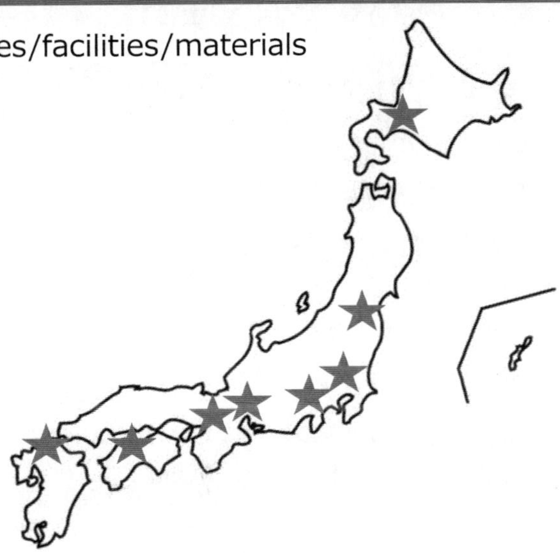

To achieve consistency

- **Technological-innovation capability**
  - Identification and quantification of trace compounds
  - Pursuit of fundamental principles/understandings
- **Cross-divisional coordination**
  - Transnational Management
  - Quality Conference
- **Improvement capability on the manufacturing site**
  - "Thinking" operators: self-modification of the working process
  - Process-capability-index (Cp)-based approach

To achieve consistency

- **Technological-innovation capability**
  - Identification and quantification of trace compounds
  - Pursuit of fundamental principles/understandings
- **Cross-divisional coordination**
  - Transnational Management
  - Quality Conference
- **Improvement capability on the manufacturing site**
  - "Thinking" operators: self-modification of the working process
  - Process-capability-index (Cp)-based approach

R&D center

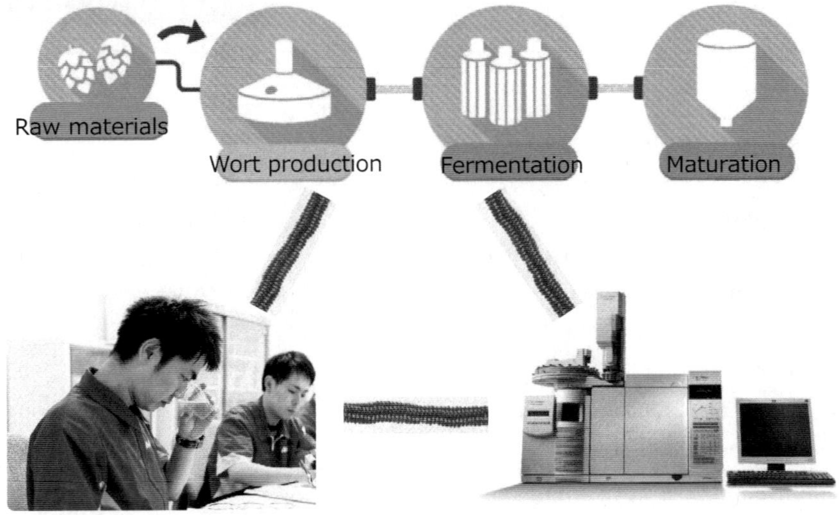

Research scheme

Exploration (sensory evaluation, GC-O)

↓

Identification (GC/MS, GC/QTOF)

↓

Quantification (GC/MS, LC/MS/MS)

↓

Factor analysis

↓

Control

Compounds that are difficult to analyze due to low threshold and low concentration in beer

RI	Compounds	Odor description	Odor threshold
1088	3-methyl-2-butene-1-thiol (MBT)	Burnt	0.002
1293	2-methyl-3-furanthiol	Yeast extract	(0.1)
1409	2-furfurylthiol	Burnt, Roasted	2.8
1377	4-mercapto-4-methylpentan-2-one	Black currant	0.0015
1414	mercaptoethyl acetate	Rubber	6.3
1593	Benzenemethanethiol (BMT)	Burnt	0.002
1640	2-mercapto-3-methylbutanol (2M3MB)	Onion	0.280
1719	3-mercaptohexyl acetate	Fruity	0.005
1840	3-mecaptohexanol	Fruity	0.055

Kobayashi, ASBC meeting (2007), Kishimoto, JAFC (2006)

Identification of burnt flavor by heart-cut GC/MS

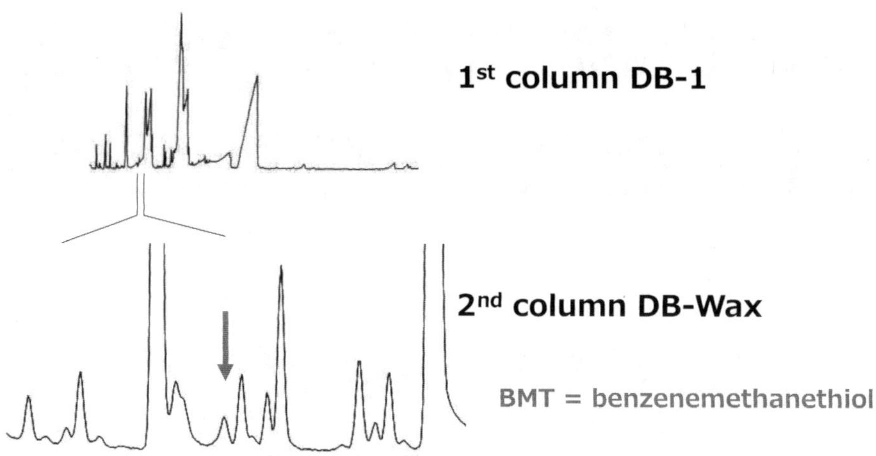

Kobayashi, ASBC meeting (2007)

Identification of burnt flavor by heart-cut GC/MS

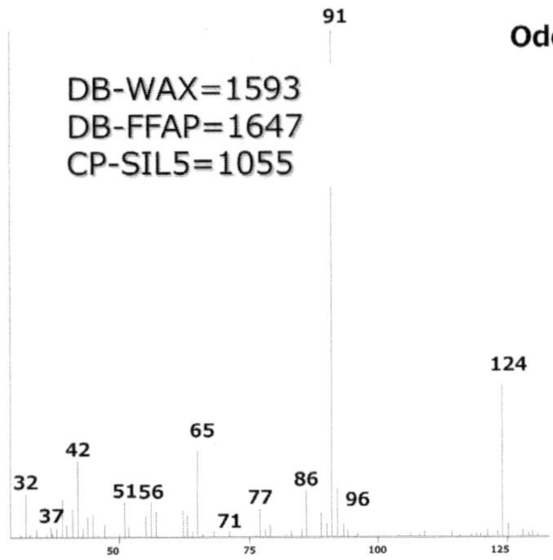

Odor: gun flint, smoky

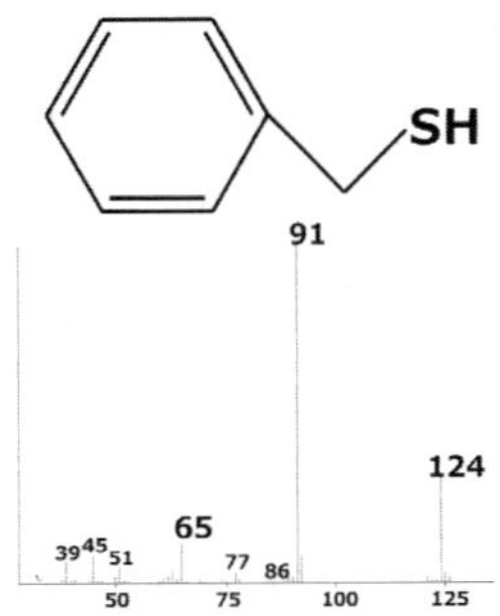

BMT = benzenemethanethiol

Kobayashi, ASBC meeting (2007)

Identification of a precursor of onion off-flavor in beer

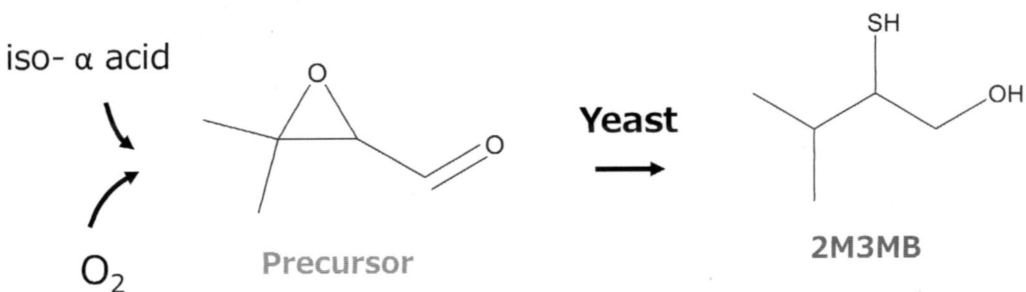

- ✓ As a result of our ability to analyze the precursors in the brewing process, it became possible to identify control points and to hypothesize a way to reduce the onion-like off-flavor.

Noba, WBC (2016)

To achieve consistency

- **Technological-innovation capability**
  - Identification and quantification of trace compounds
  - Pursuit of fundamental principles/understandings
- **Cross-divisional coordination**
  - Transnational Management
  - Quality Conference
- **Improvement capability on the manufacturing site**
  - "Thinking" operators: self-modification of the working process
  - Process-capability-index (Cp)-based approach

Transnational Organization

~~Global,~~ ~~International,~~ and ~~multi-domestic~~

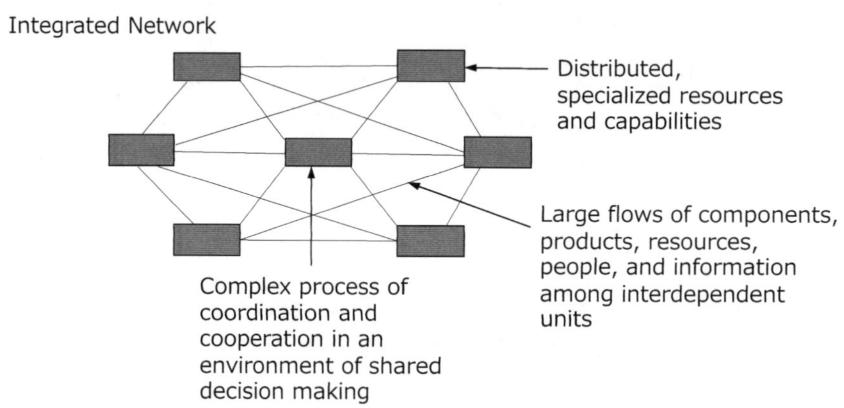

Managing Across Borders: The Transnational Solution
Christopher A. Bartlett, Sumantra Ghoshal (1989)

Quality Conference

1) Vision of the brand
2) Concrete goal in achievement of brand vision
3) Mass-production technology for the goal
4) Specific issues in each brewery

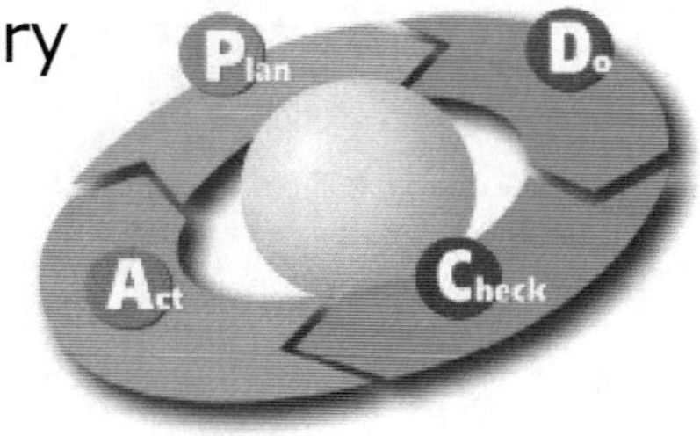

"Concept-based" approach to beer tasting

- Brand concept

↓

- Concrete sensory characteristics

Brew masters meeting
→share information
and discussions

"Concept-based" approach to beer tasting

Brand concept

Target character of the brand

Sensory characteristics

Necessary condition

a	Overall characteristic, first impression
b	Character particularly important in (a)
c	Aromatic impression and its strength
d	Overall taste, after taste impression
e	Sufficient condition for full mark

Sufficient condition

Analysis of sensory-evaluation data

Sensory-profile-analysis method

+

Concept-based sensory evaluation

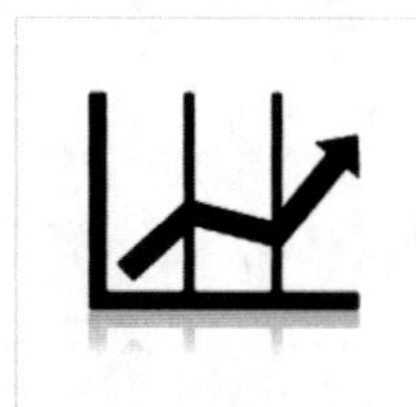

Trend management

Principal-component analysis (PCA)

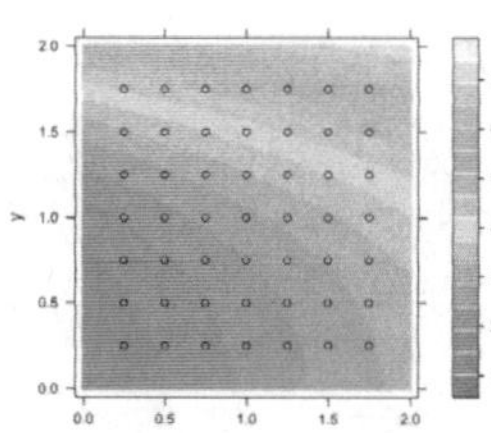

Preference mapping

Iguchi, ASBC meeting (2013)

Positioning breweries

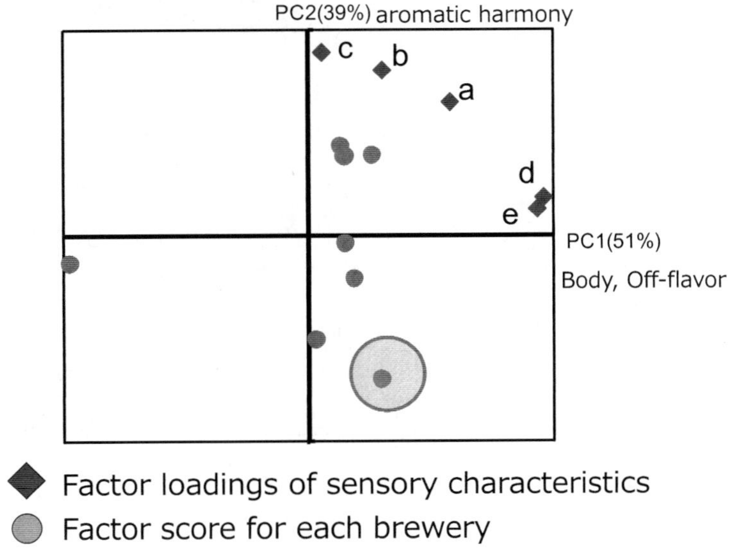

Sensory characteristics and overall score

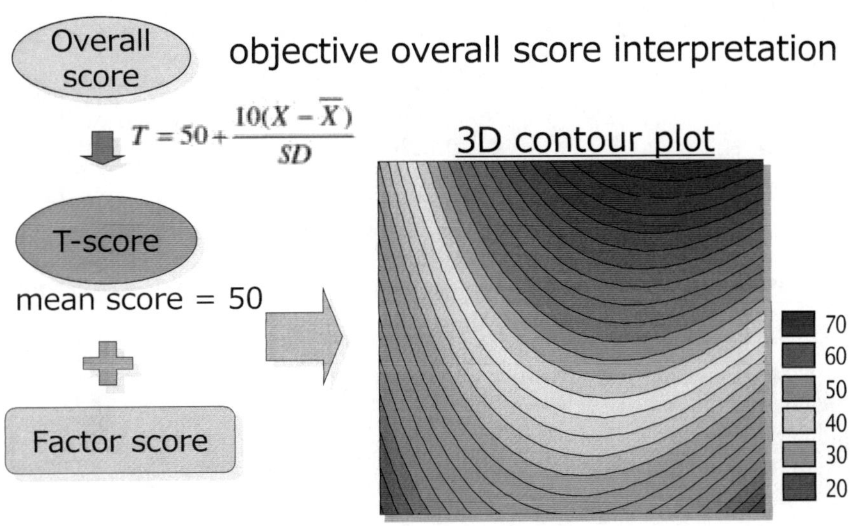

Sensory characteristics and overall score

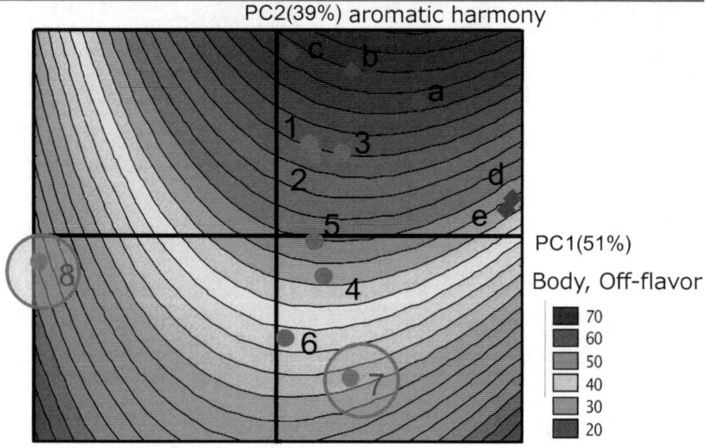

◆ Factor loadings of sensory characteristics

● Factor score for each brewery (number indicating the rank of overall score)

Off-flavor reduction process through quality conference

R & D center
- Development of analytical methods
- Lab-scale test
- Building up a hypothesis

Breweries
- Monitoring in practical brewing
- In-process control
- Narrowing down possible causes
- Verifying a hypothesis

Lab-scale test : precursor of onion off-flavor

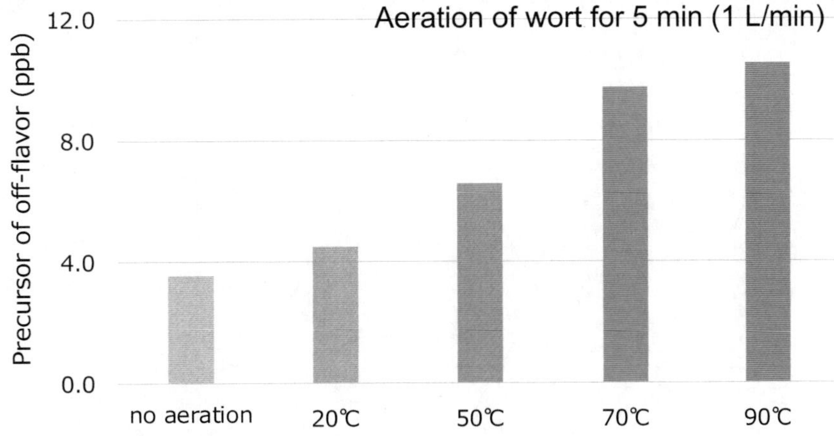

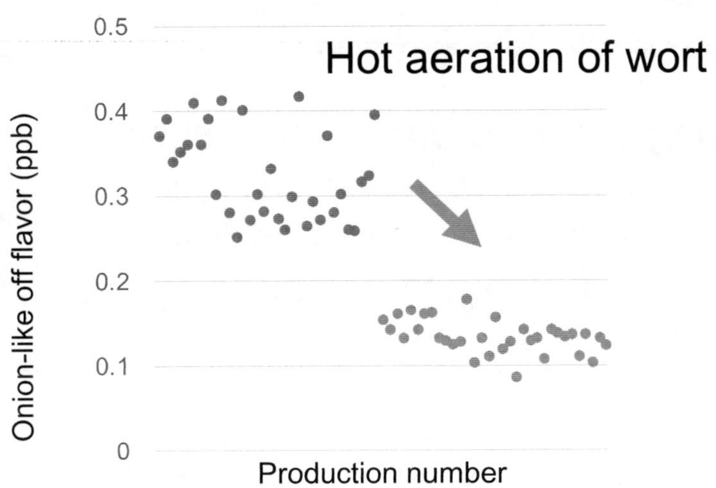

Kobayashi, ASBC Meeting (2006)

Trend of onion off-flavor in each brewery

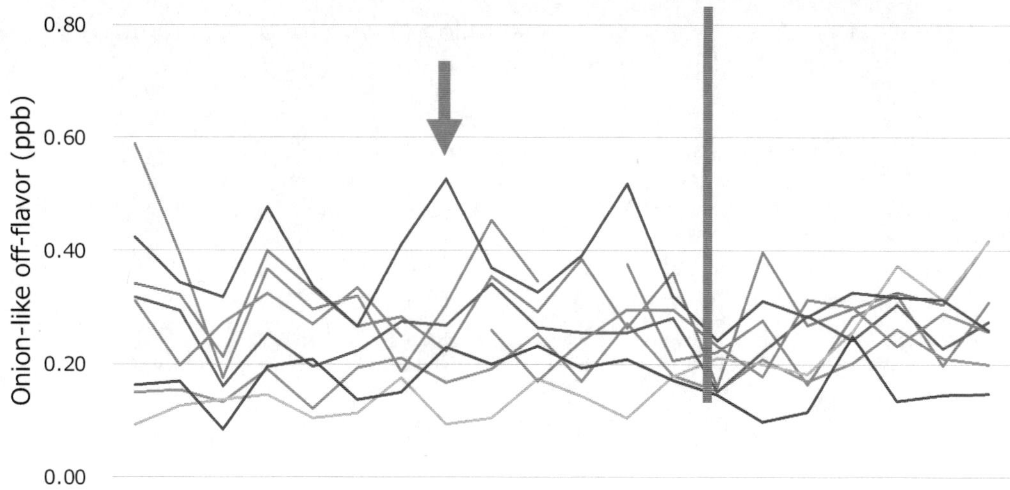

To achieve consistency

- **Technological-innovation capability**
  - Identification and quantification of trace compounds
  - Pursuit of fundamental principles/understandings
- **Cross-divisional coordination**
  - Transnational Management
  - Quality Conference
- **Improvement capability on the manufacturing site**
  - "Thinking" operators: self-modification of the working process
  - Process-capability-index (Cp)-based approach

"Thinking" operators : self-modification of working process

In order to minimize the variation of raw materials,

- feedback-control based on analysis values even at midnight on holidays
  (brewing section: 3 teams in 8-hour shifts, QC section: day shift)
- reduce waste time to transport samples from brewhouse to QC section

→ Create a system for on-site analysis

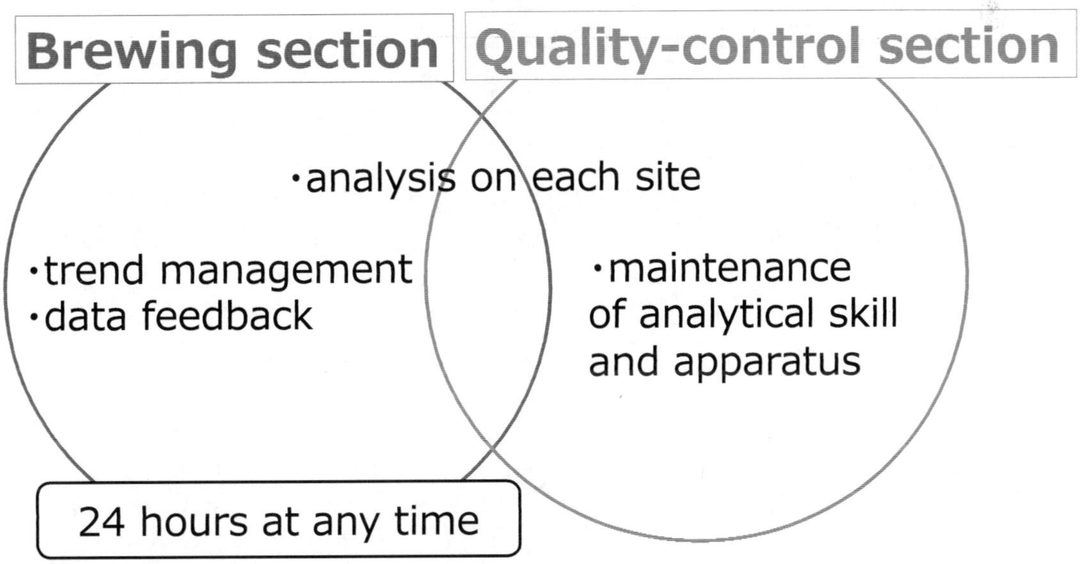

Sample transportation

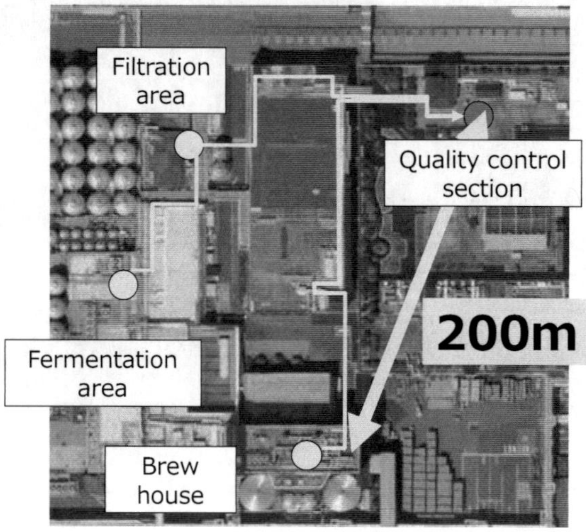

Color degree (EBC°)

Before
Cp : 0.84

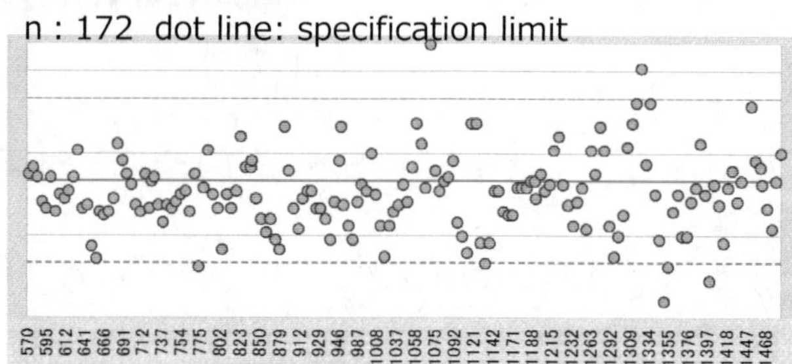

After
Cp : 1.53

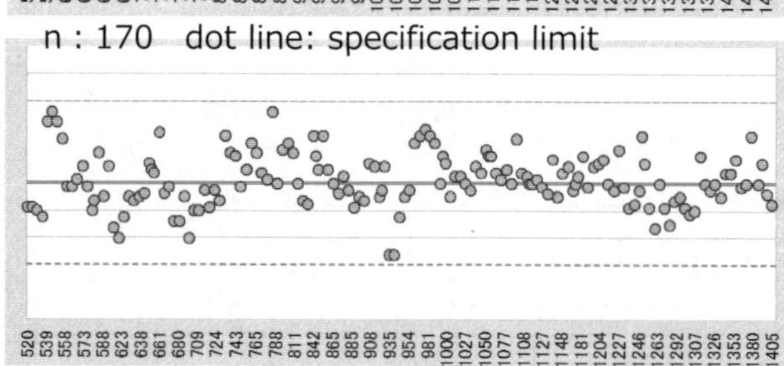

Attenuation limit (%)

Before
Cp : 0.94

n : 252 dot line: specification limit

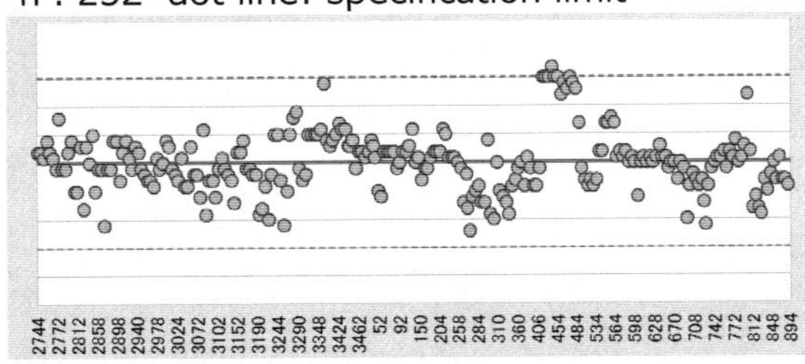

After
Cp : 1.57

n : 282 dot line: specification limit

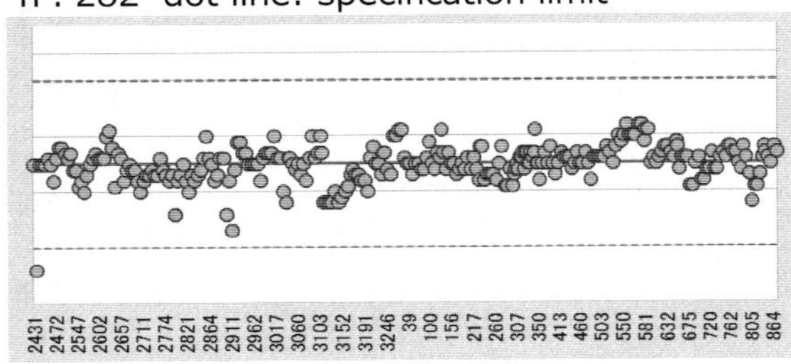

Trend of sensory evaluation

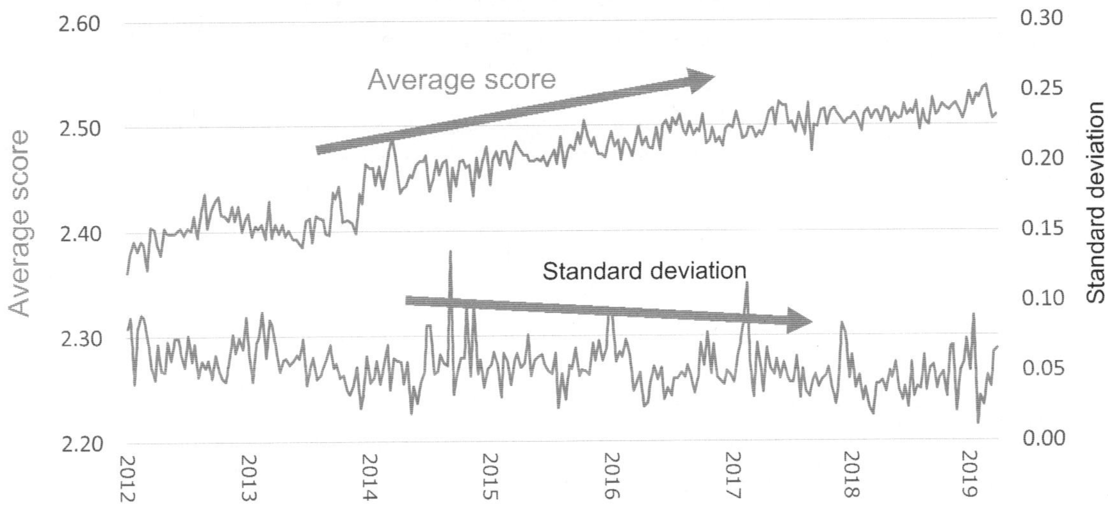

Flavor stability : Trans-2-nonenal range trend

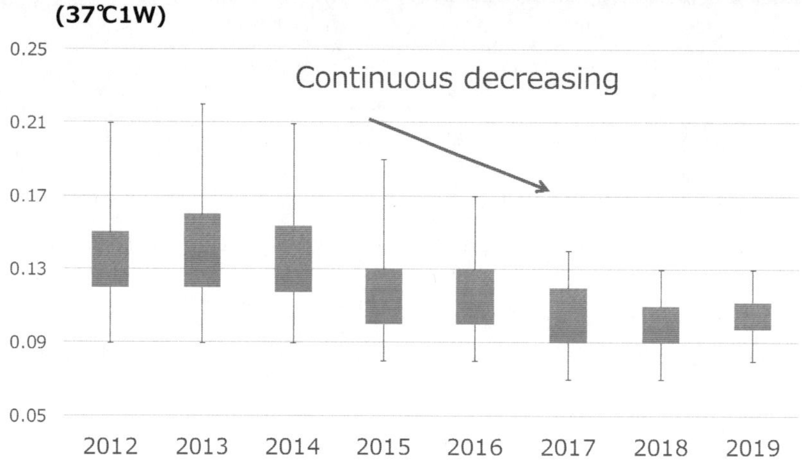

Asahi Group Philosophy

Our Mission

Deliver on our great-taste promise and bring more fun to life

Our Vision

Be a value creator globally and locally, growing with high-value-added brands

Our Values

Challenge and innovation
Excellence in quality
Shared inspiration

Our Principles

Building value together with all our stakeholders

Customers: Win customer satisfaction with products and services that exceed expectations
Employees: Foster a corporate culture that promotes individual and company growth
Society: Contribute to a sustainable society through our business
Partners: Build relationships that promote mutual growth
Shareholders: Increase our share value through sustainable profit growth and shareholder returns

Summary

- No effort is spared to serve our customers **the best SUPER DRY every time, everywhere**
- Technological-innovation capability
  - Identification and quantification of trace compounds
  - Pursuit of fundamental principles/understandings
- Cross-divisional coordination
  - Transnational Management
  - Quality Conference
- Improvement capability on the manufacturing site
  - "Thinking" operators: self-modification of the working process
  - Process-capability-index (Cp)-based approach

Future plan

Pilsner Urquell

Transnational

PERONI NASTRO AZZURRO

SUPER "DRY" Asahi

Consistent taste of SUPER DRY in all breweries
different climates/facilities/materials

L11 - IDENTIFICATION OF FACTORS AT THE ORIGIN OF THE PHENOMENON OF BEER GUSHING IN ORDER TO DEVELOP A PREMATURE DETECTION TECHNIQUE OF THE GUSHING RISK

Julien Billard[1]; Sophie Schwebel[1]; Romain Kapel[2]; Marc Schmitt[1]

1 - IFBM; 2 - University of Lorraine

Abstract

Beer gushing is a sometimes explosive over-generation of foam following the opening of a bottle. It is highly dependent on barley growing and harvesting climatic conditions and has been observed more frequently over the past years.

Current techniques, such as the modified Carlsberg test, only provide information on predictive gushing potential of malt with a big uncertainty. A rapid screening test on raw material leading to the identification of actual factor(s) responsible for the gushing is consequently highly demanded by the brewing industry.

The initial goal of this study was to determine the compounds involved in beer gushing. This approach has been applied to both naturally gushing positive malts and artificially contaminated barleys. Barley artificially contaminated was used as model to develop the extraction of gushing compounds. Then this method was used to screen and compare gushing positive and negative industrial malts in order to try to identify compounds with possible gushing potential.

The methodology and the results will be presented and discussed. The final objective of the work is the development of a simple test to quantify the gushing potential of raw material.

Short biography

PhD in food process and biotechnology, he works in the R&D department at IFBM on barley-to-beer themes. He also has worked on the identification of malt spoiling moulds by MALDI-TOF analyses during his MSC specialized in food process and biotechnology. He have worked 2 years with AEB group France on the screening and characterization of yeast strain for the fermentation of beer wort and the impact of the glutathione on the growth of brewing yeast.

VTT

Sourdough cultures as a source of speciality yeast for the brewing industry

Brian Gibson

29/05/2019 VTT – beyond the obvious

VTT

Bioprospecting for brewers

29/05/2019 VTT – beyond the obvious

Bioprospecting for brewers

- The attempt to discover living organisms, or biological processes, that have medical, agricultural, or industrial value
- Allows creation of novel, speciality beers
  - Wild lager (*S. eubayanus*)
  - Sour beer (*Lachancea thermotolerans*)
  - Low-alcohol beer (*Pichia*, *Torulaspora*, etc.)
- Yeast can be sourced from:
  - Nature
  - Natural fermentations

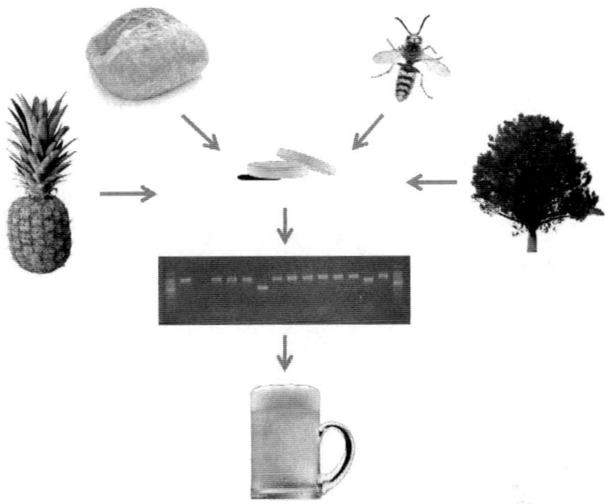

Cubillos et al. (2019)

Sourdough cultures

- A fermented flour culture used to leaven dough
- Complex fermentation involving mix of yeasts and bacteria
- Propagated using a back-slopping technique (repitching)
- Often maintained for extended periods (generations)

Sourdough as a reservoir of novel yeast strains

VTT

- Advantages:
  - Accessibility of material
  - Abundance and diversity of yeasts
  - Yeasts with fermentative potential (Ascomycota)
  - Qualified Presumption of Safety (QPS/GRAS)
  - Ease of isolation
    - Absence of mycelial fungi
    - Simple steps for removal of bacteria

29/05/2019 VTT – beyond the obvious

VTT

Sourdough yeasts for low-alcohol brewing

VTT

Sourdough preparation

- 10 different sourdough culture prepared in duplicate
  - 4 x wheat
  - 4 x rye
  - 1 x barley
  - 1 x oat

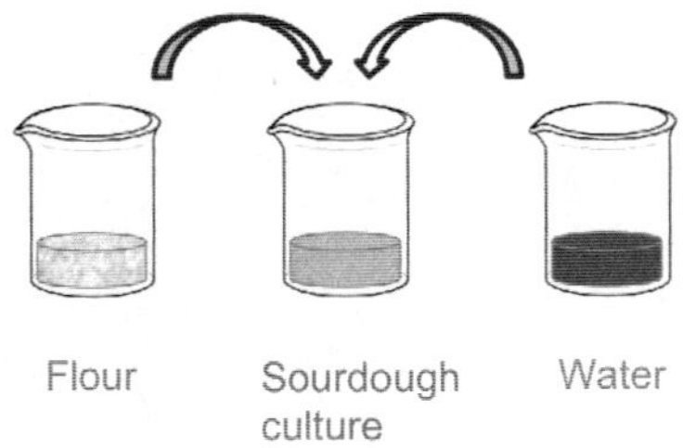

- Flour:water paste prepared aseptically
- Incubate at 25°C
- Fed daily for 1 week
- Weekly thereafter

VTT

Sourdough preparation

- Yeast abundant after only 1 week
- Sequence analysis shows dominance of Ascomycota

Culture	Flour	Yeast count Cells/ml
1.	Wheat	3.0×10^5
2.	Wheat	c
3.	Wheat	3.1×10^4
4.	Wheat	6.5×10^3
5.	Rye	1.0×10^6
6.	Rye	2.5×10^5
7.	Rye	1.0×10^7
8.	Rye	3.1×10^6
9.	Barley	2.0×10^4
10.	Oat	3.0×10^2

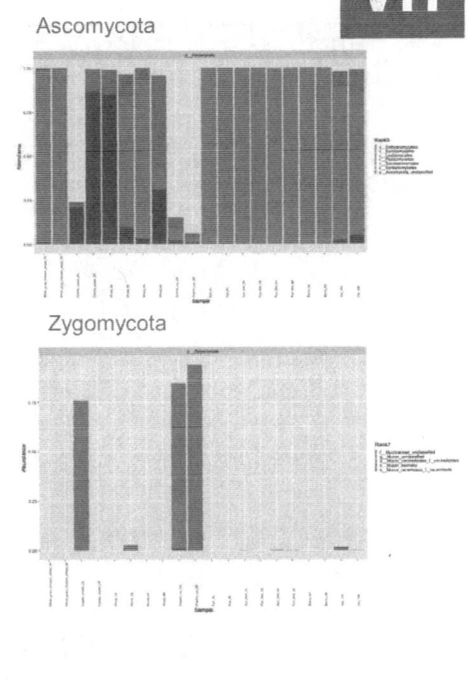

Yeast isolation

- Cultures plates on YPD + chloramphenicol
- 10 yeast species isolated
- 1-4 species per culture

	Flour	*Pichia fermentans*	*P. kudriavzevii*	*Hanseniaspora uvarum*	*Kluyveromyces marxianus*	*Hyphopichia burtonii*	*Wickerhamomyces anomalus*	*W. ciferii*	*Cyberlindnera fabianii*	*Torulaspora delbrueckii*	*Kazachstania servazzi*
1.	Wheat	✓	✓	✓	✓						
2.	Wheat	✓									
4.	Wheat					✓	✓				
5.	Rye									✓	
6.	Rye	✓	✓					✓	✓		
7.	Rye		✓						✓		✓
8.	Rye		✓							✓	
9.	Barley		✓		✓		✓				

Screening for fermentation performance (alcohol yield)

- 10 isolates screened for fermentation performance
- 12°Plato, all-malt wort, 20°C, 6d
- Reference strains
  - *Saccharomycodes ludwigii*
  - *S. cerevisiae (ale)*
- All strains MAL-negative
- Similar alcohol production (<1.0%)
- Which are most suitable?

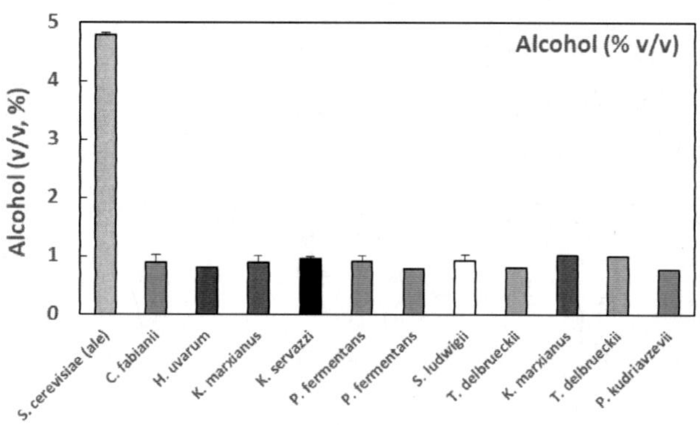

Screening for fermentation performance (aldehyde reduction at 24h)

- Aldehydes contribute 'worty' character
- Normally reduced by yeast after pitching

- Simple aldehydes fully reduced
  - Furfural, methional, hexanal, benzaldehyde

- Reduction variable with complex aldehydes
  - Decrease (*S. cerevisiae, H. uvarum, K. servazzi, T. delbrueckii*)
  - No change *(Pichia spp.)*
  - Increase *(K. marxianus, S. ludwigii)*

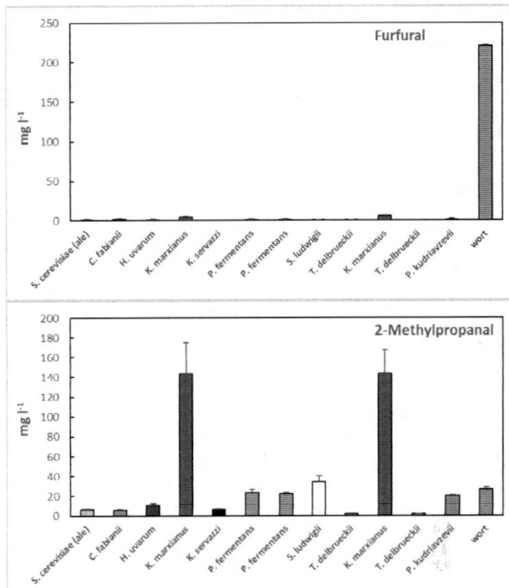

Screening for fermentation performance

- Aroma analysis facilitates selection
- Excessive volatile production:
  - *C. fabianii, K. marxianus*

- *Kazachastania servazzi*
  - 3-methylbutanol (banana)
  - Ethyl acetate (fruit)
  - Phenylethyl acetate (rose)

  - Aldehyde reduction

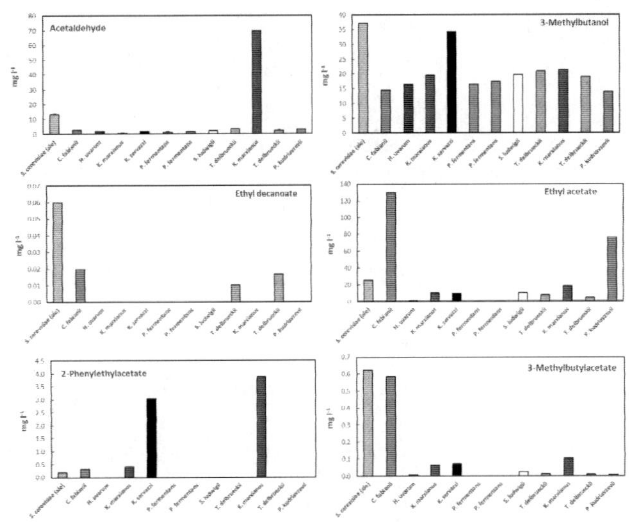

Isolate selection

- Heat tolerance
  - *C. fabianii*
  - *K. marxianus*
  - *P. kudriavzevii*
- Cold tolerance
  - *H. uvarum*
  - *K. servazzi*
  - *P. fermentans*
- Cold tolerance an advantage:
  - Safety
  - Compatibility with lager brewing

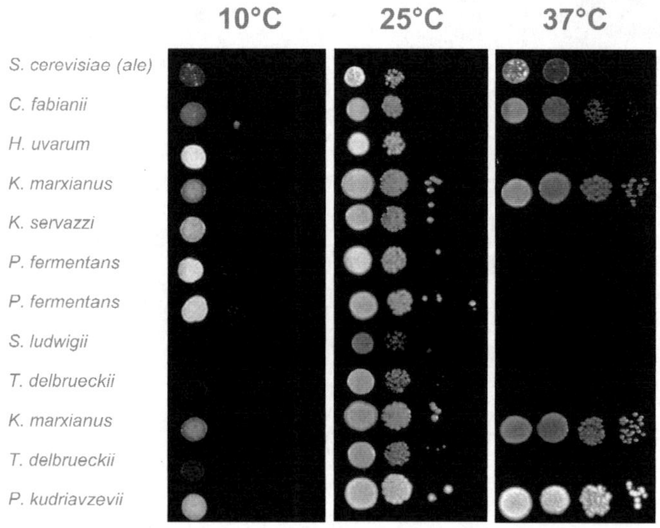

29/05/2019 VTT – beyond the obvious

Isolate selection

- Variable ethanol tolerance
- *Sensitive strains*
  - *Kazachstania servazii*
  - *Torulaspora delbrueckii*

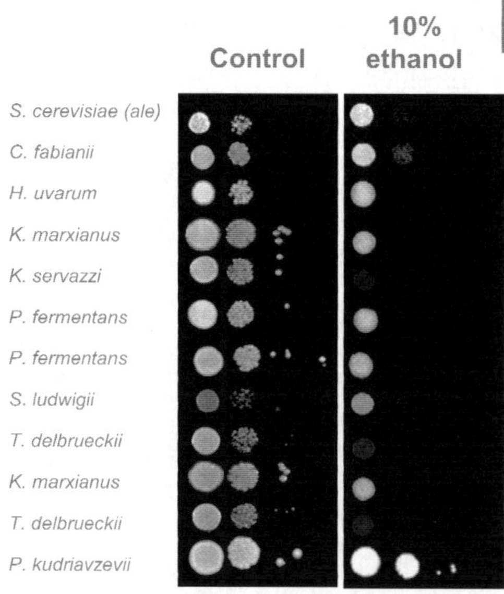

29/05/2019 VTT – beyond the obvious

Fermentation with *Kazachstania servazzi*

- *K. servazzi* features
  - MAL-negative
  - POF-negative
  - Good aldehyde reduction
  - Interesting flavour attributes (2-PEA)
- 10L fermentation: *K. servazzi* performing well relative to *S. ludwigii*
- 2-PEA = 7.4mg l^{-1} (2x flavour threshold)

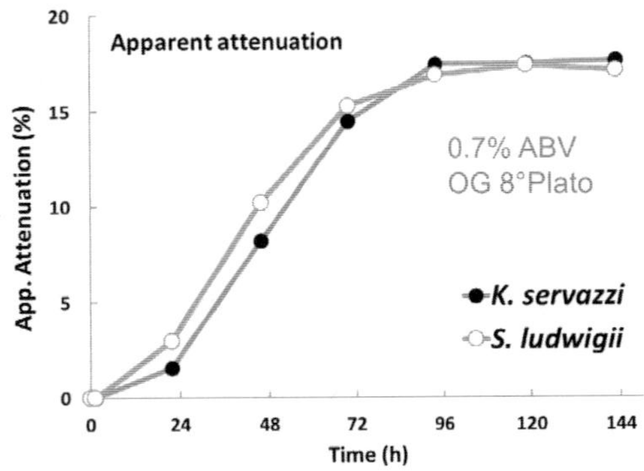

29/05/2019 VTT – beyond the obvious

Sourdough *S. cerevisiae* strains for brewing

Selected strains

	Strain	Species	Notes
1	Suomen Hiiva	*S. cerevisiae*	Commercial Finnish baker's yeast and standard sahti brewing strain
2	VTT C-81117	*S. cerevisiae*	Finnish sourdough isolate from 1981 (VTT Culture Collection)
3	VTT C-81118	*S. cerevisiae*	Finnish sourdough isolate from 1981 (VTT Culture Collection)
4	VTT C-94207	*S. cerevisiae*	Finnish sourdough isolate from 1994 (NCYC 2937 and VTT Culture Collection)
5	M	*S. cerevisiae*	Finnish sourdough isolate 2019 of M. Laitinen
6	Y15	*S. cerevisiae*	Traditional Sicilian sourdough isolate (University of Catania)
7	Y17	*S. cerevisiae*	Traditional Sicilian sourdough isolate (University of Catania)
8	Y23	*S. cerevisiae*	Traditional Sicilian sourdough isolate (University of Catania)

29/05/2019 VTT – beyond the obvious

Temperature tolerance

- Heat tolerant
- Cold sensitive

25°C 37°C*

Suomen Hiiva
VTT C-81117
VTT C-81118
VTT C-94207
M
Y15
Y17
Y23
S. pastorianus
S. cerevisiae (ale)

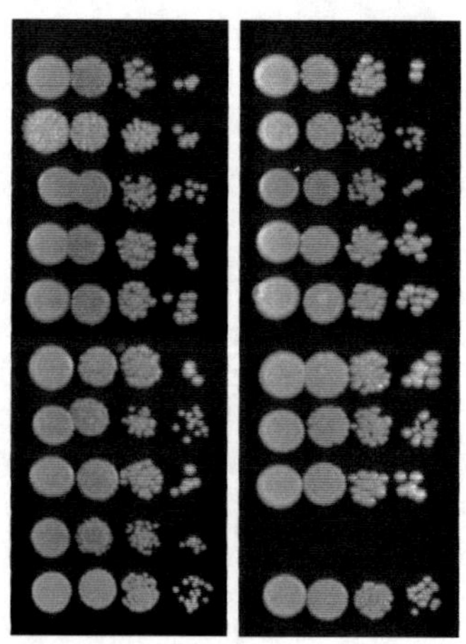

29/05/2019 VTT – beyond the obvious

*3 days' growth at 25°C and 37°C, 6 days' growth at 12°C

Phenolic flavour notes

- All strains POF+
- POF status and temperature tolerance limit application to certain styles
  - Wheat beer
  - Saison, other Belgian ale styles
  - Farmhouse ales…

Strain	Species	POF status*
Suomen Hiiva	*S. cerevisiae*	Positive
VTT C-81117	*S. cerevisiae*	Positive
VTT C-81118	*S. cerevisiae*	Positive
VTT C-94207	*S. cerevisiae*	Positive
M	*S. cerevisiae*	Positive
Y15	*S. cerevisiae*	Positive
Y17	*S. cerevisiae*	Positive
Y23	*S. cerevisiae*	Positive
Negative reference	*S. pastorianus*	**Negative**
Positive reference	*S. cerevisiae*	Positive
Control	None	**Negative**

29/05/2019 VTT – beyond the obvious

Sahti

- An 'ancient' (pre-hop) beer style from Finland
- Raw farmhouse ale
- Juniper-flavoured
- Strong (>20°P OG)
- Brewed with **baking yeast**
- Farmhouse yeast maintenance no longer practised
  - **Strain diversity limited**
- Restoration of diversity via bioprospecting strategy?

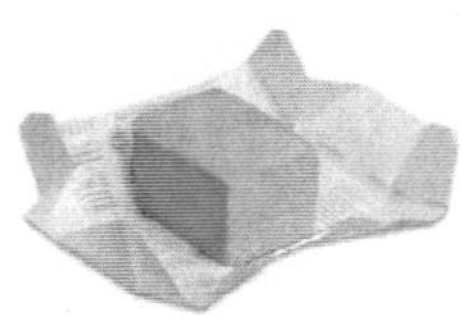

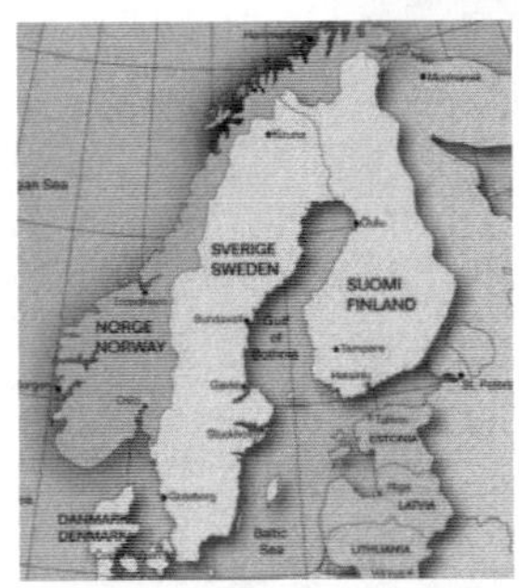

29/05/2019 VTT – beyond the obvious

Fermentation performance

VTT

- Duplicate 2L, 20°P, 20°C sahti wort fermentations monitored over 8 days
- Differences in fermentation rate and yield (6.6 - 7.9% ABV)
- Fastest initial fermentation rate with Suomen Hiiva (SH) reference strain
- Highest ethanol values with Italian isolates

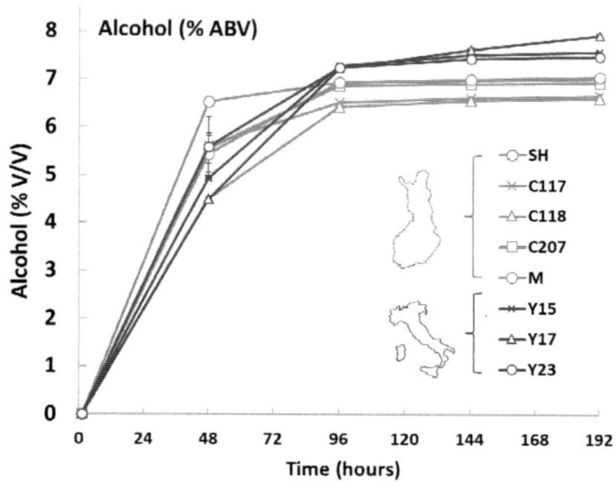

Fermentation performance

VTT

- Yield related to maltotriose use

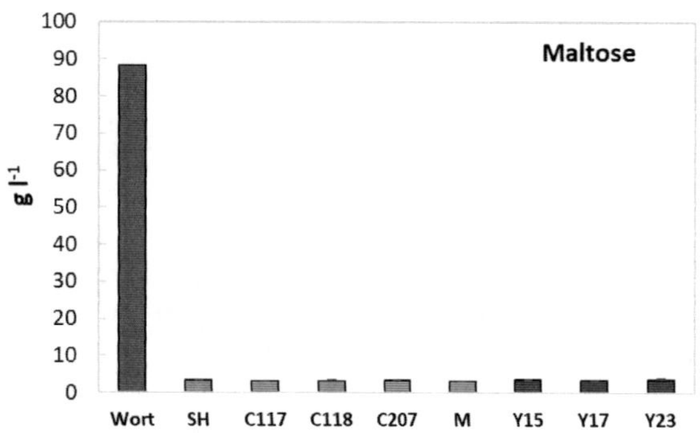

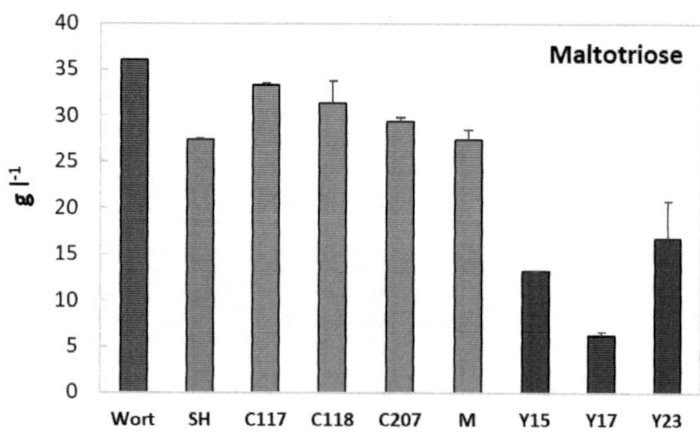

Viability

- Low viability in all yeast cultures after 1 week secondary fermentation at 5°C
- Most tolerant strains are:
  - Finnish C118 strain
  - Italian Y17 strain
- Repitching not typical in sahti brewing

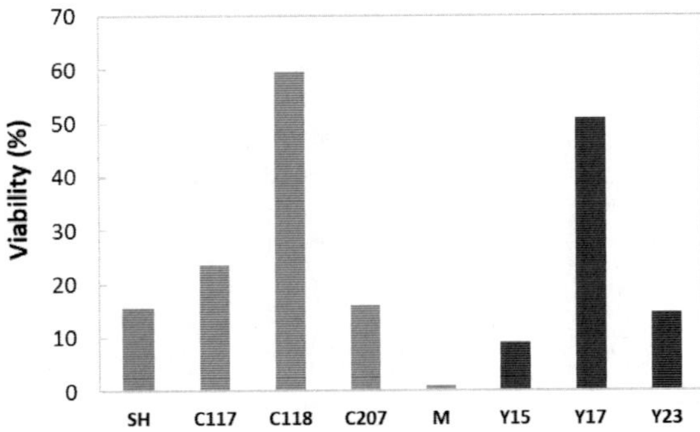

29/05/2019 VTT – beyond the obvious

Aroma volatiles

- Very high levels of ethyl acetate and 3-methylbutyl acetate in certain strains (typical for sahti)
- Low-ester strains for less 'yeast-forward' beers:
  - C117, C118

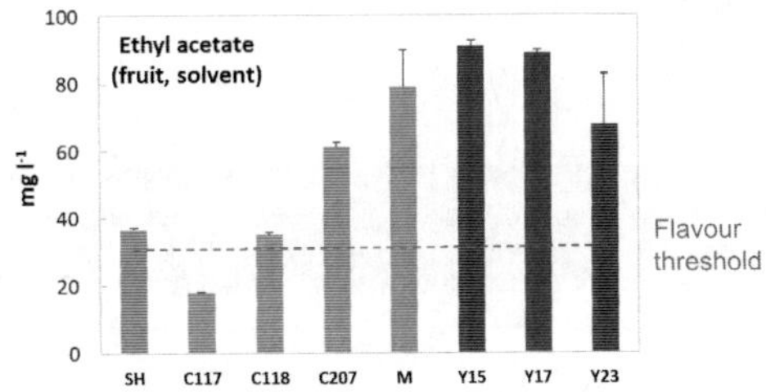

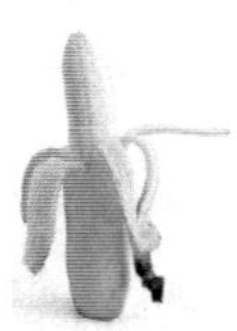

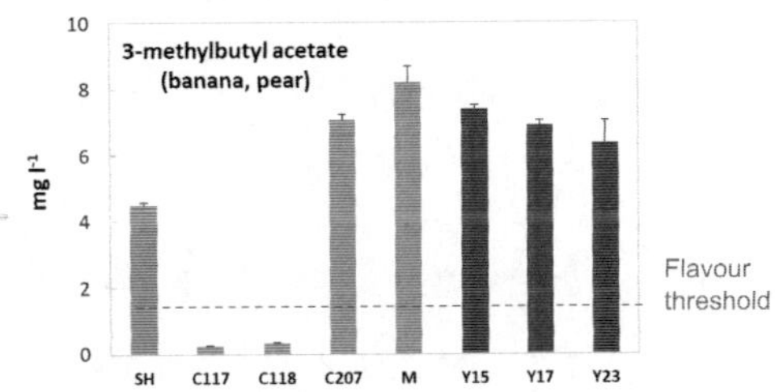

29/05/2019 VTT – beyond the obvious

Aroma volatiles

- Ethyl esters high in certain strains
  - Ethyl hexanoate in Y15 and Y17
  - Ethyl octanoate in 'M' strain

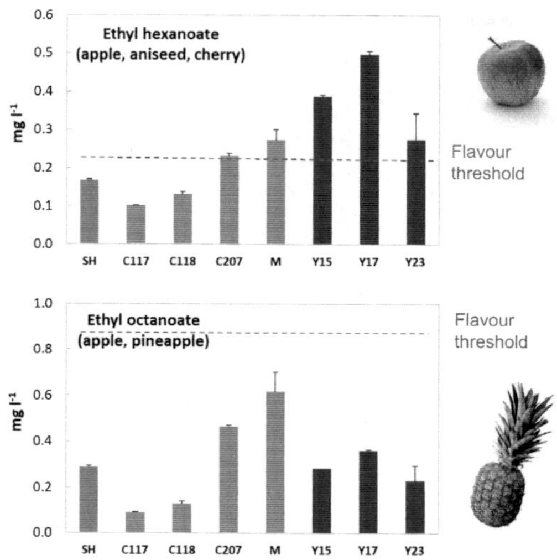

29/05/2019 VTT – beyond the obvious

Features of selected strains

	Low Yield	POF-	High yield	Tolerance	Low ester	3-methyl butyl acetate	Ethyl hexanoate	Ethyl octanoate
C118	✓	✓		✓	✓			
Maisa						✓		✓
Y15			✓			✓	✓	
Y17			✓	✓		✓	✓	

- Distinct differences in fermentation performance and beer quality

29/05/2019 VTT – beyond the obvious

Conclusions

VTT

- Sourdough is a potentially valuable reservoir of new brewing yeast strains
- Non-*Saccharomyces* species show good potential for low alcohol brewing
  - Absence of off-flavours, aldehyde reduction
  - Cold tolerance
  - Stress tolerance
  - Bioflavouring potential (esp. *K. servazzi*)
- *S. cerevisiae* strains suitable for alcoholic beer production
- May be suitable only for particular styles (cold sensitive, POF+)
- Proof-of-Concept sahti study shows potential for introducing diversity
  - Yield
  - Flavour profile
  - Tolerance

29/05/2019 VTT – beyond the obvious

*3 days' growth at 25°C and 37°C, 6 days' growth at 12°C

Acknowledgements

VTT

- Martina Catallo, University of Modena and Reggio Emilia
- Lisa Solieri, University of Modena and Reggio Emilia
- Cinzia Randazzo, University of Catania
- Linnea Johansson, Metropolia University of Applied Sciences
- Marjokaisa Piironen, Sahtipäät ry
- Jarkko Nikulin, VTT
- Kristoffer Krogerus, VTT
- Frederico Magalhães, VTT
- Atte Mikkelson, VTT
- Paula Jouhten, VTT
- Maija Nuppunen-Puputti, VTT
- Elina Sohlberg, VTT

UNIMORE
UNIVERSITÀ DEGLI STUDI DI MODENA E REGGIO EMILIA

Fund Fonds
Fonds Baillet Latour

Metropolia

29/05/2019 VTT – beyond the obvious

L13 - EXPLOITING YEAST DIVERSITY FOR PRODUCT AND PROCESS OPTIMISATION

Katherine Smart[1]

1 - University of Cambridge

Abstract

Yeast species and strain diversity has been poorly exploited by the brewing industry. This is potentially due to the lack of effective assays to enable the selection of candidate strains for expensive larger scale trial work to be conducted. An assay specifically designed to assess the phenotypic diversity of yeast strains has been developed. This assay permits the examination of differential responses of yeast strains to new wort recipes for product development or enhanced stresses that might be encountered during brewing fermentation optimisation. The assay comprising a bespoke 96-well plate, with wort and stress effector challenge media, is assessed using a Phenotypic Microarray Omnilog reader. A redox reporter within the assay permits analysis of metabolic activity in aerobic and/or anaerobic conditions. Results from the assay compare well with traditional growth on spot plates and tolerance assays in which maintenance of viability was assessed. Perhaps more critically the data obtained appears to provide a predictive assessment when compared to larger scale fermentations. For the first time a high throughput assessment of yeast strain fitness for fermentation has been developed.

Short biography

Katherine Smart holds the role of University Lecturer in Brewing and Distilling at the University of Cambridge. She is a Co-Founder and the Distiller of The Surrey Copper Distillery Limited. Until October 2017, Katherine was the Global Chief Brewer of ABInBev and SABMiller PLC. For SABMiller PLC she was responsible for research, innovation, brewing strategy, raw material strategy and technical stewardship of global brands. Katherine has 21 years' academic experience at the Universities of Cambridge, Oxford Brookes and Nottingham. At Nottingham she was the SABMiller Professor of Brewing Science from 2005 to 2012 and she is currently an honorary Professor. She founded the National Brewing Library at Oxford Brookes University and the Bioenergy and Brewing Science Centre at the University of Nottingham. Katherine has holds three doctorates and successfully completed her DSc by publication in December 2016. She is an author of scientific papers and technical books.

EBC, 2019

IMPACT OF FERMENTATION PARAMETERS AND HOPS ON THE FLAVOUR EXPRESSION OF BEER YEASTS

THE OBVIOUS CHOICE FOR BEVERAGE FERMENTATION

GABRIELA MONTANDON, PHD

Fermentis
LESAFFRE FOR BEVERAGES

THERE ARE NO BREWERS

BUT 'YEAST SHEPHERDS'

(MAKING THE BEST 'CULTURE MEDIA AND CONDITIONS' FOR THE YEAST TO PRODUCE BEERS)

Fermentis
LESAFFRE FOR BEVERAGES

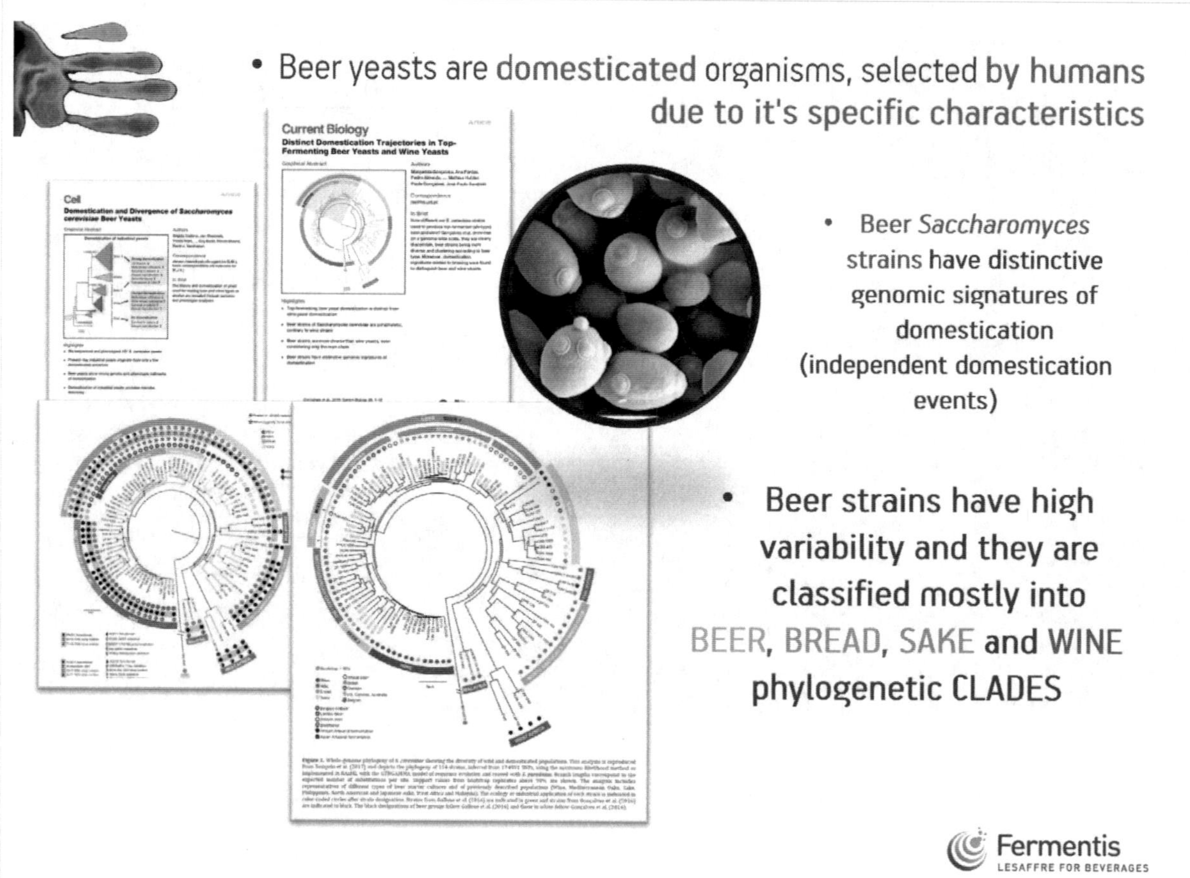
Beer yeasts are domesticated organisms, selected by humans due to it's specific characteristics
Beer *Saccharomyces* strains have distinctive genomic signatures of domestication (independent domestication events)
Beer strains have high variability and they are classified mostly into BEER, BREAD, SAKE and WINE phylogenetic CLADES
Current Biology
Distinct Domestication Trajectories in Top-Fermenting Beer Yeasts and Wine Yeasts
Cell
Domestication and Divergence of Saccharomyces cerevisiae Beer Yeasts
Fermentis
LESAFFRE FOR BEVERAGES

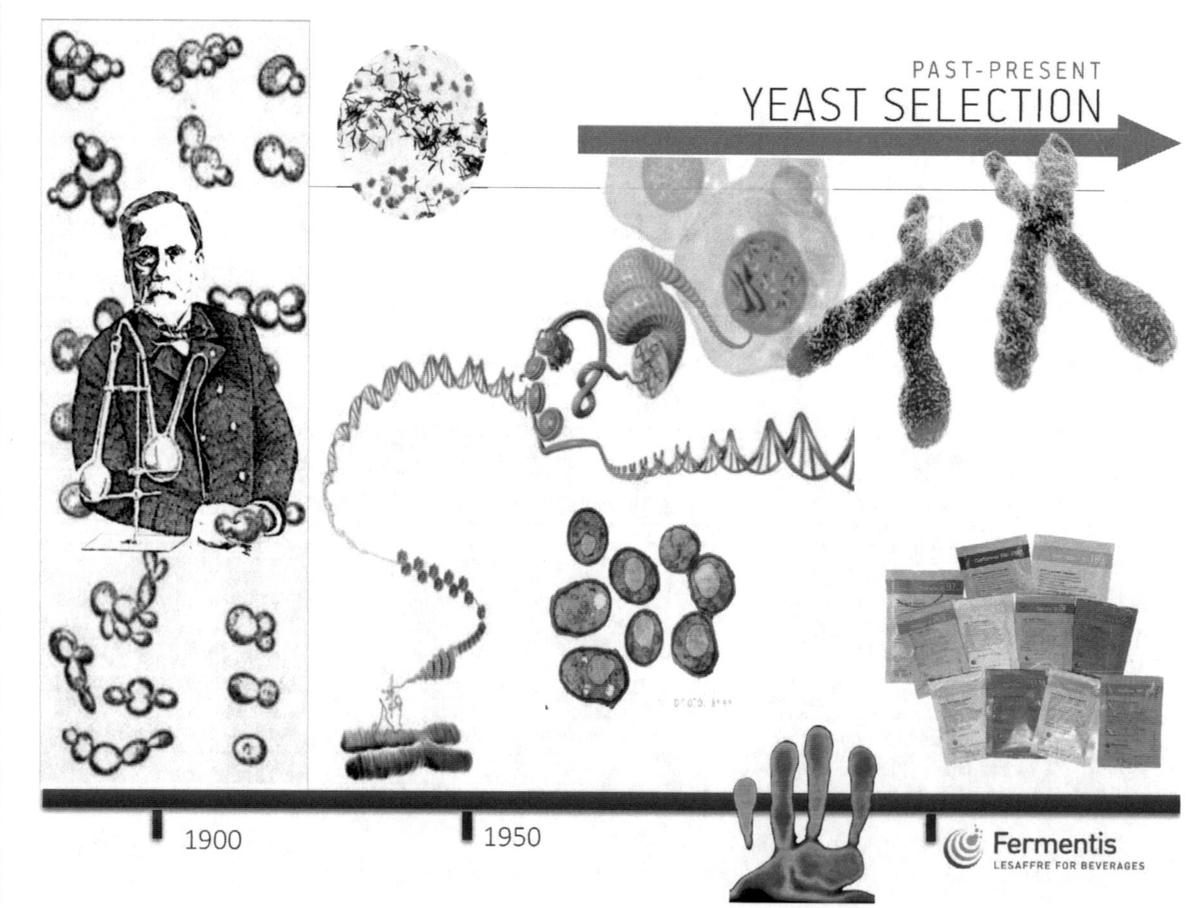
PAST-PRESENT
YEAST SELECTION
1900
1950
Fermentis
LESAFFRE FOR BEVERAGES

PAST-PRESENT-FUTURE
MARKET SELECTION
Its all about (consumer)
HUMAN PLEASURE!
GOT A THIRST FOR
MAN SIZE PLEASURE?
Pleasures
Are followed by consequences
I HAVE A PLEASURE ROOM
DO YOU WANT TO SEE IT?
1900
2000
2100
Fermentis
LESAFFRE FOR BEVERAGES

= Pleasure
Flavors
Fermentis
LESAFFRE FOR BEVERAGES

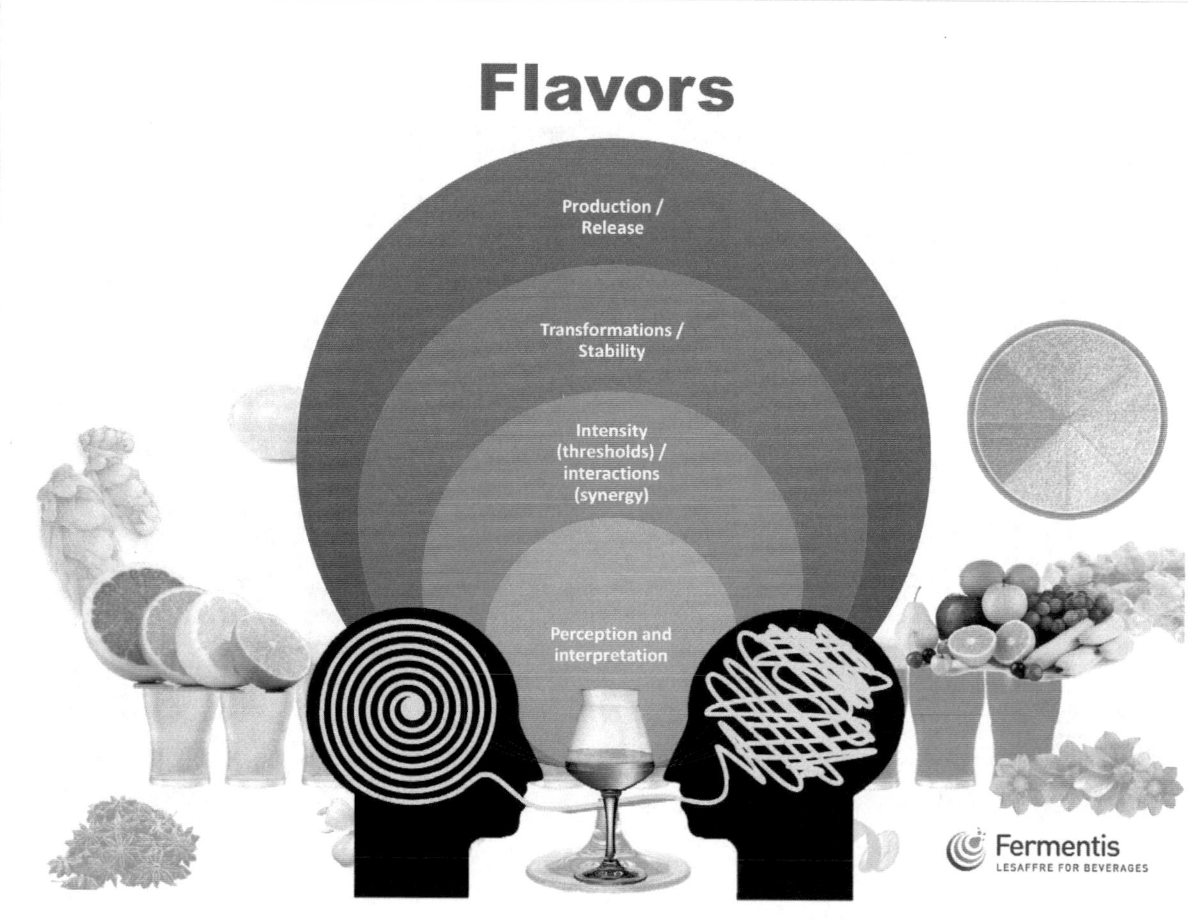
Flavors
Production / Release
Transformations / Stability
Intensity (thresholds) / interactions (synergy)
Perception and interpretation
Fermentis
LESAFFRE FOR BEVERAGES

= Pleasure Flavors
YES
MAYBE
NO
How important is the yeast to produce/ contribute to 'pleasure notes' in beer?
Fermentis
LESAFFRE FOR BEVERAGES

YEAST & PLEASURE

OBJECTIVES OF OUR MULTIPLE STUDIES

- ✓ **Designing of experiment for phenotype expression**
- ✓ Focus on <u>application</u>
- ✓ Sensory oriented

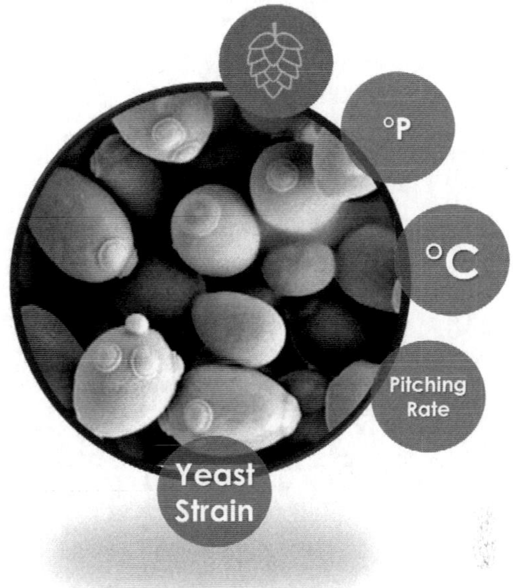

Fermentis
LESAFFRE FOR BEVERAGES

Fermentis Expert Beer Panel (2y)

Weekly Sessions	40 panelists	<u>Ramdomly, blindly, repetitions, statistics tests</u>

Fermentis
LESAFFRE FOR BEVERAGES

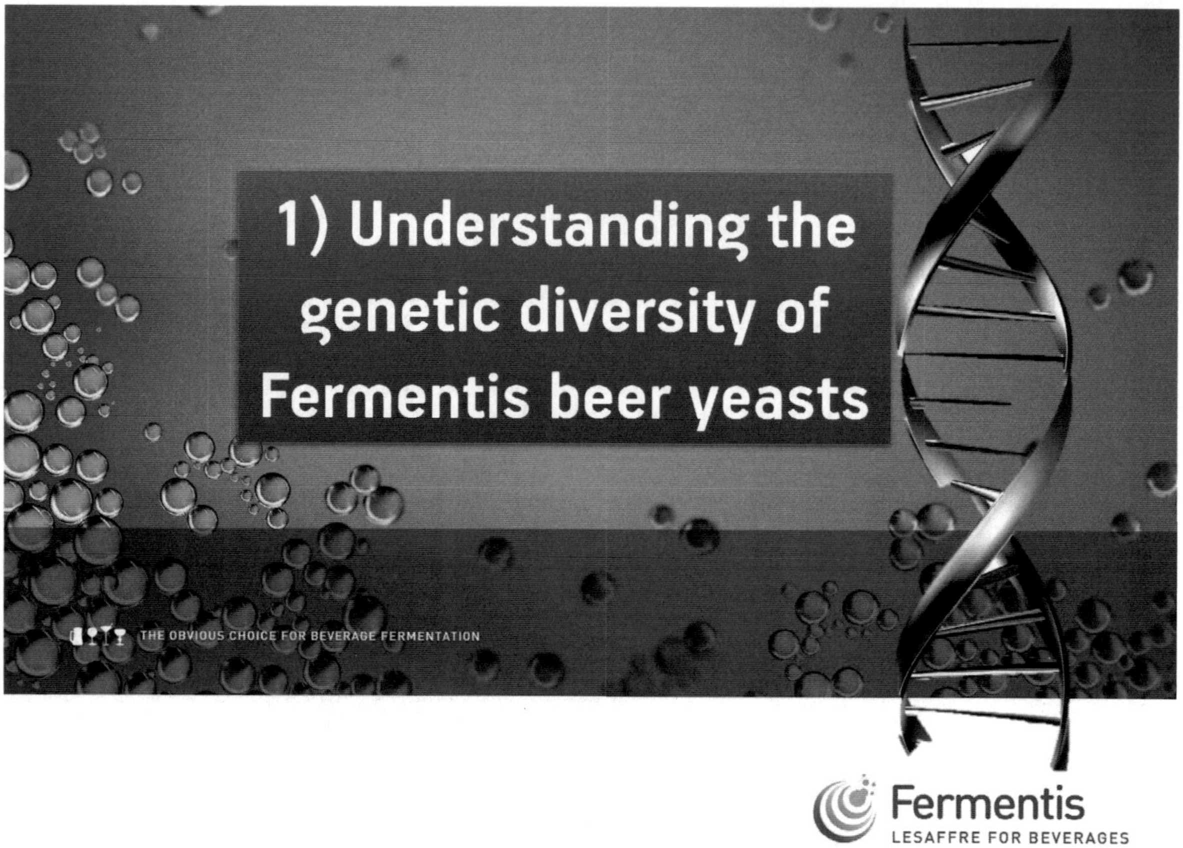

PHYLOGENETIC PLACEMENT OF FERMENTIS BEER STRAINS

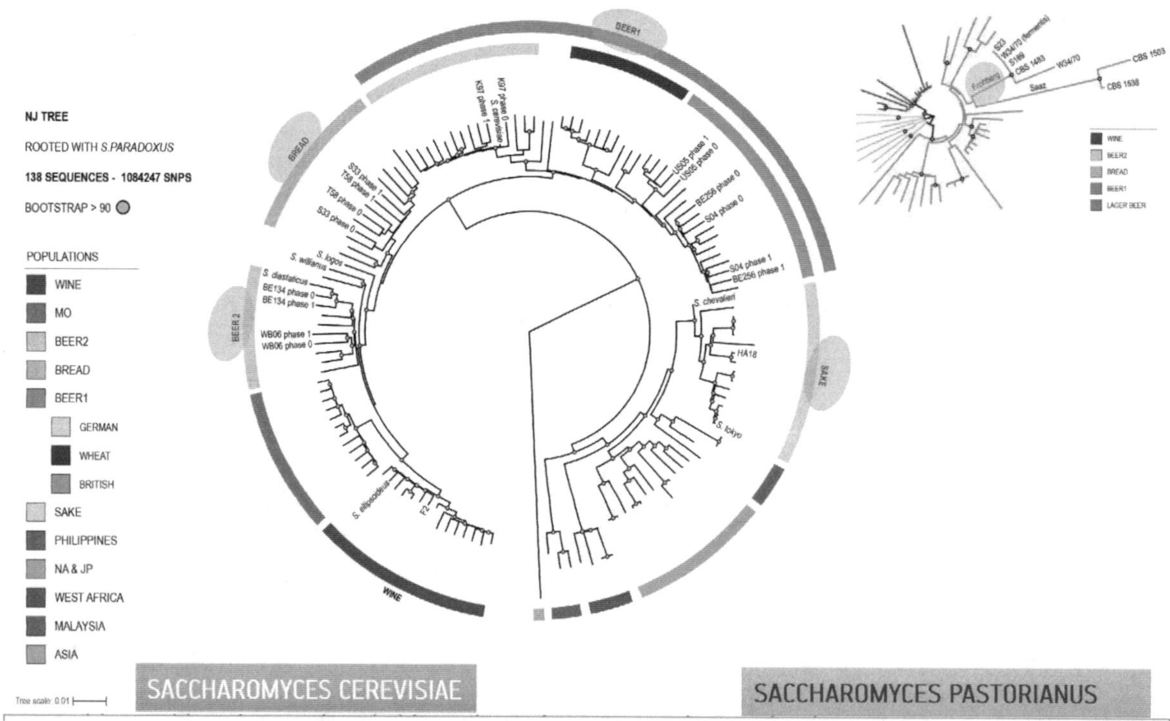

Remarkable genetic diversity of the portfolio of *Fermentis* strains **that covers basically the entire spectrum of industrial relevant strains.**

GENETIC DIVERSITY = FLAVOUR DIVERSITY?
Fermentis
LESAFFRE FOR BEVERAGES

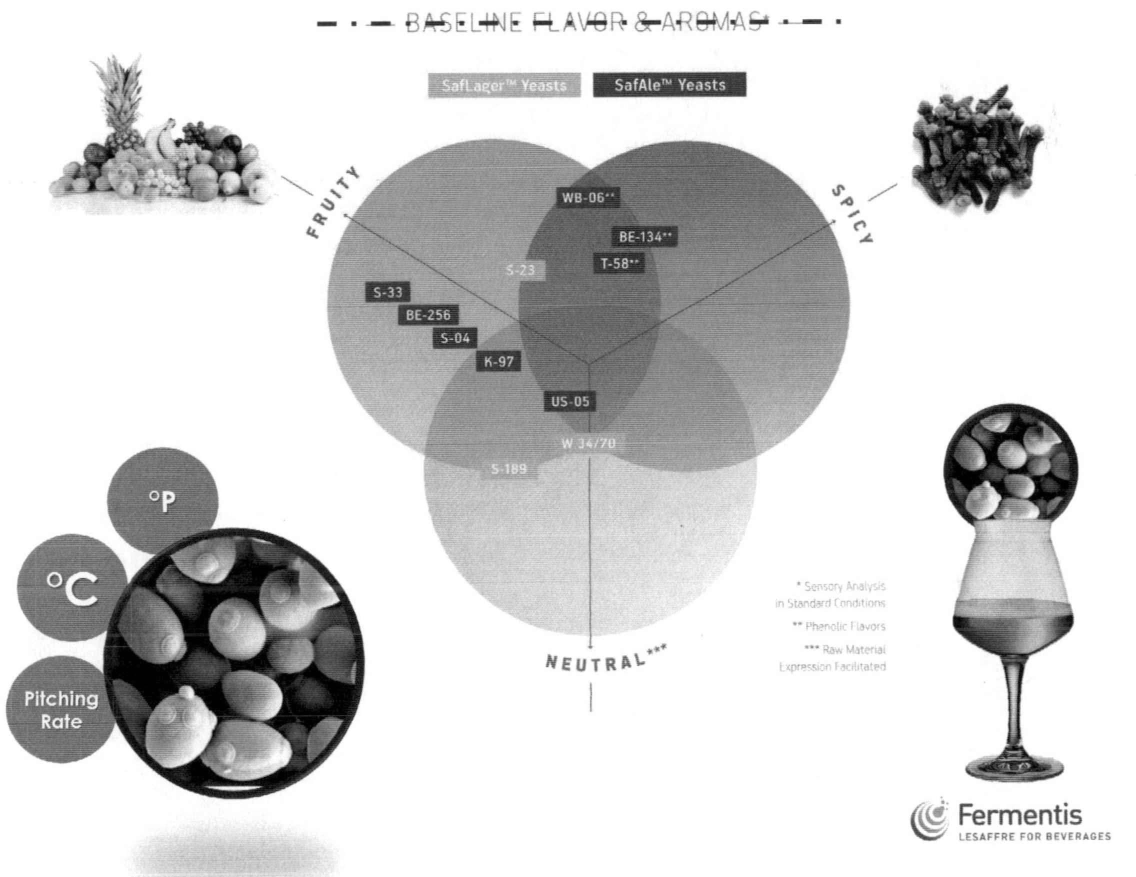
BASELINE FLAVOR & AROMAS*
SafLager™ Yeasts
SafAle™ Yeasts
FRUITY
SPICY
NEUTRAL***
WB-06**
BE-134**
T-58**
S-23
S-33
BE-256
S-04
K-97
US-05
W 34/70
S-189
°P
°C
Pitching Rate
* Sensory Analysis in Standard Conditions
** Phenolic Flavors
*** Raw Material Expression Facilitated
Fermentis
LESAFFRE FOR BEVERAGES

PROTOCOLS

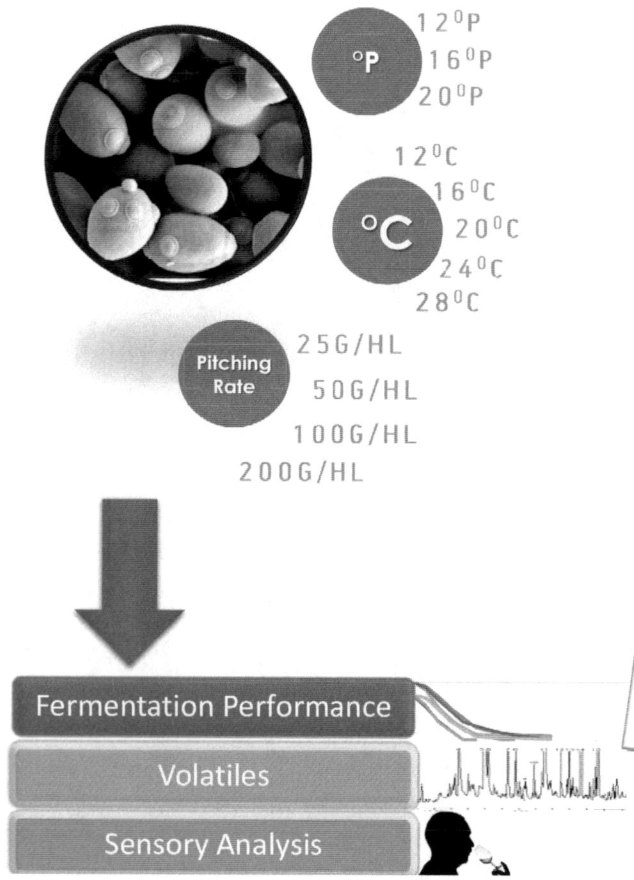

✓ **LAB TRIALS** 0,2L

- Beer wort
- Multiple conditions for phenotype expression

Fermentis
LESAFFRE FOR BEVERAGES

Phenotype / Flavor Expression
Screening of important fermentation flavor components in different conditions

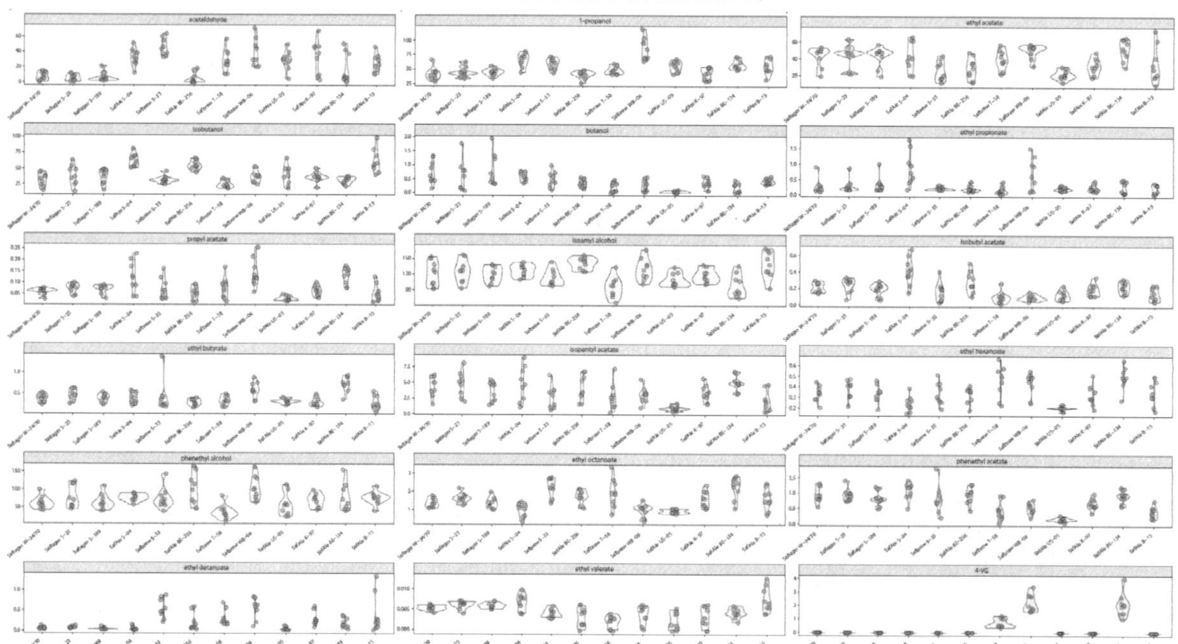

Fermentis
LESAFFRE FOR BEVERAGES

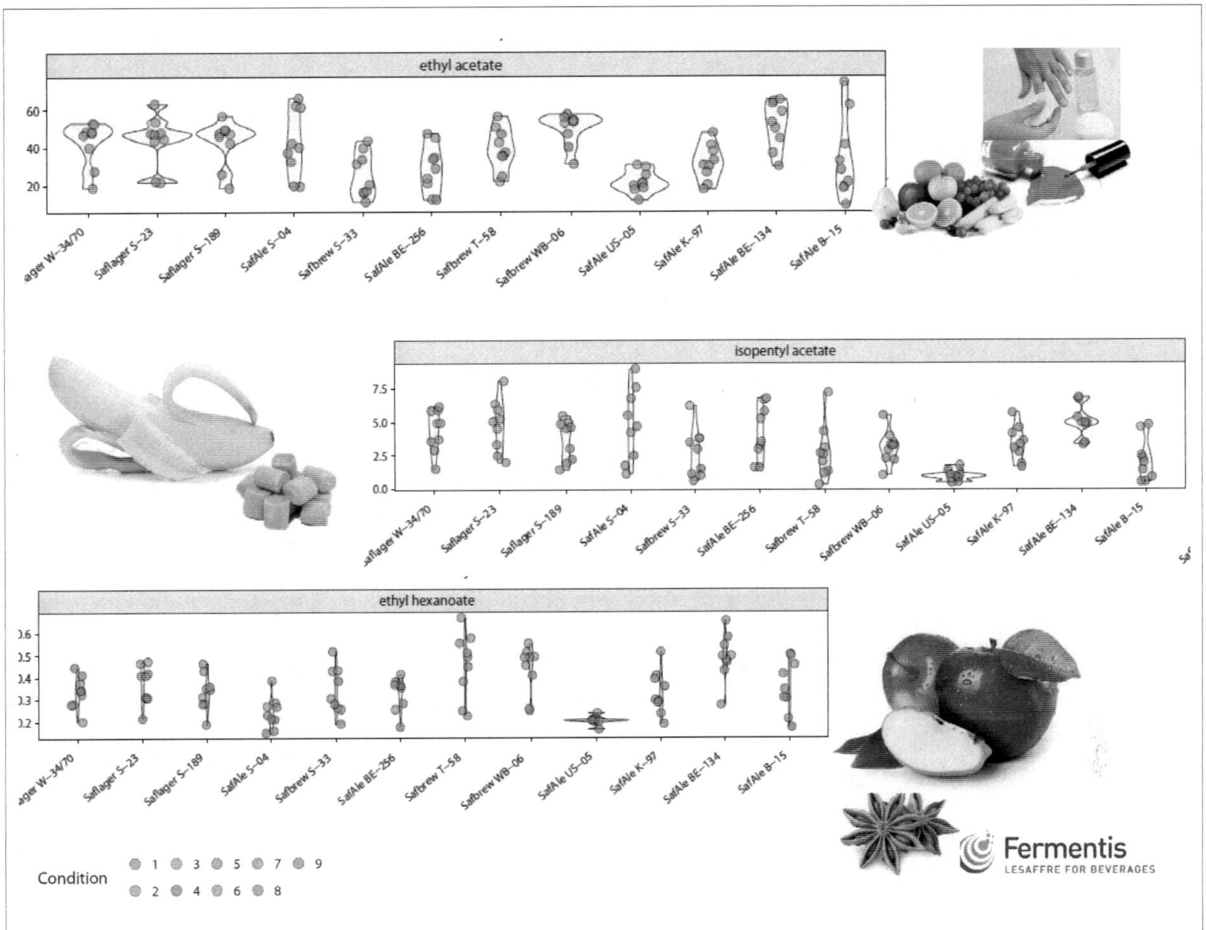
ethyl acetate
60
40
20
Saflager S-23
Saflager S-189
SafAle S-04
Safbrew S-33
SafAle BE-256
Safbrew T-58
Safbrew WB-06
SafAle US-05
SafAle K-97
SafAle BE-134
SafAle B-15
isopentyl acetate
7.5
5.0
2.5
0.0
ethyl hexanoate
Condition
1 2 3 4 5 6 7 8 9
Fermentis
LESAFFRE FOR BEVERAGES

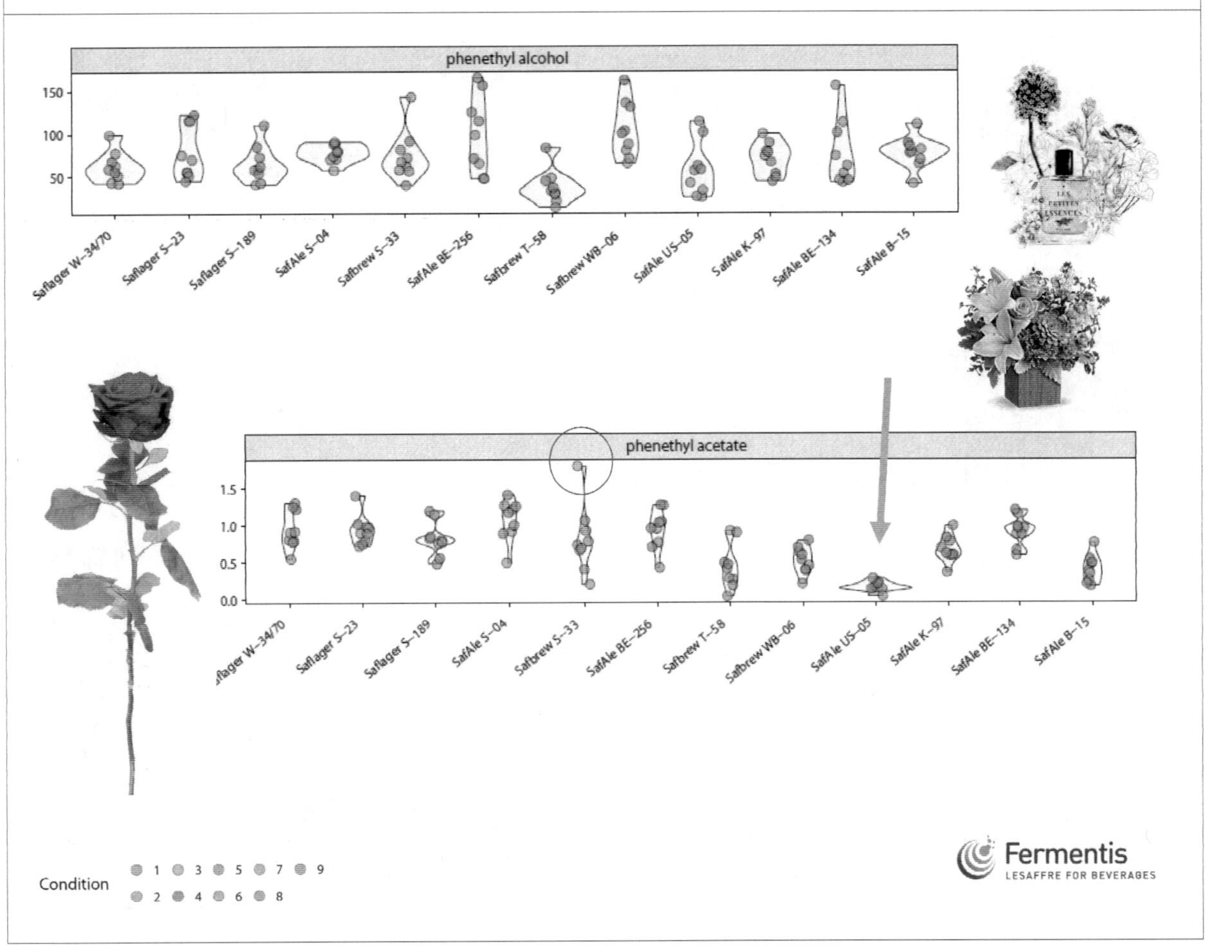
phenethyl alcohol
150
100
50
Saflager W-34/70
Saflager S-23
Saflager S-189
SafAle S-04
Safbrew S-33
SafAle BE-256
Safbrew T-58
Safbrew WB-06
SafAle US-05
SafAle K-97
SafAle BE-134
SafAle B-15
phenethyl acetate
1.5
1.0
0.5
0.0
Condition
1 2 3 4 5 6 7 8 9
Fermentis
LESAFFRE FOR BEVERAGES

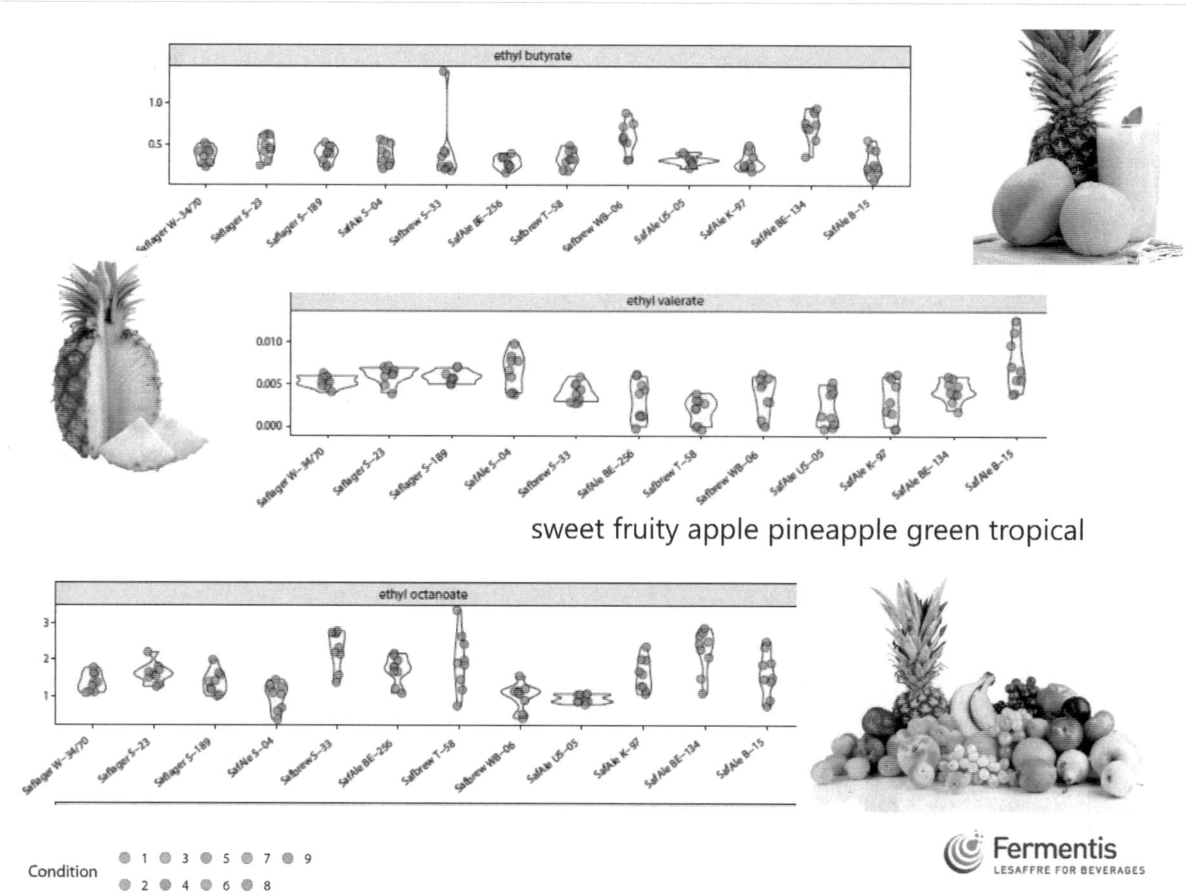
ethyl butyrate
ethyl valerate
sweet fruity apple pineapple green tropical
ethyl octanoate
Saflager W-34/70
Saflager S-23
Saflager S-189
SafAle S-04
Safbrew S-33
SafAle BE-256
Safbrew T-58
Safbrew WB-06
SafAle US-05
SafAle K-97
SafAle BE-134
SafAle B-15
Condition
1 3 5 7 9
2 4 6 8
Fermentis
LESAFFRE FOR BEVERAGES

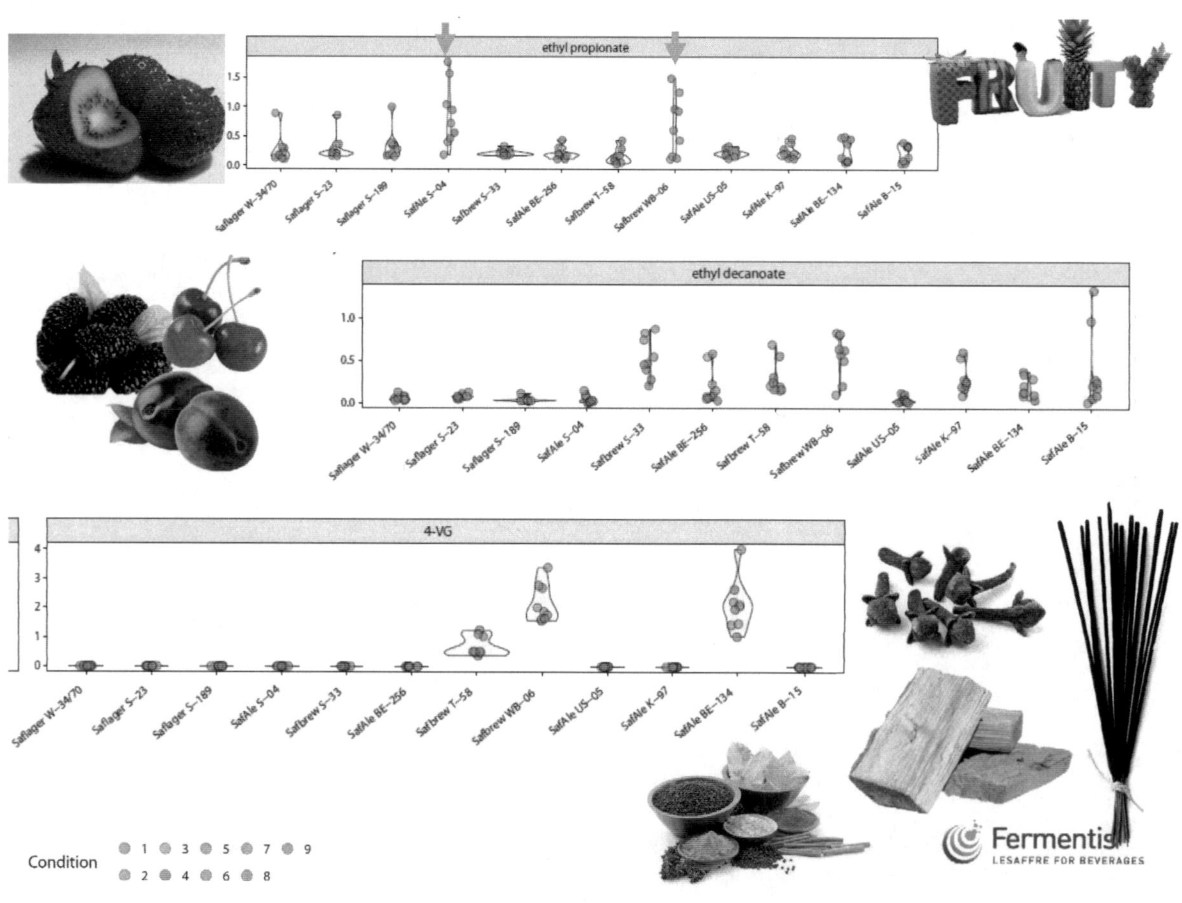
ethyl propionate
FRUITY
ethyl decanoate
4-VG
Saflager W-34/70
Saflager S-23
Saflager S-189
SafAle S-04
Safbrew S-33
SafAle BE-256
Safbrew T-58
Safbrew WB-06
SafAle US-05
SafAle K-97
SafAle BE-134
SafAle B-15
Condition
1 3 5 7 9
2 4 6 8
Fermentis
LESAFFRE FOR BEVERAGES

Lab x Pilot trials can possibly affect:

The high **FRUITY** flavor profile produced of the ale yeast **BE256**?

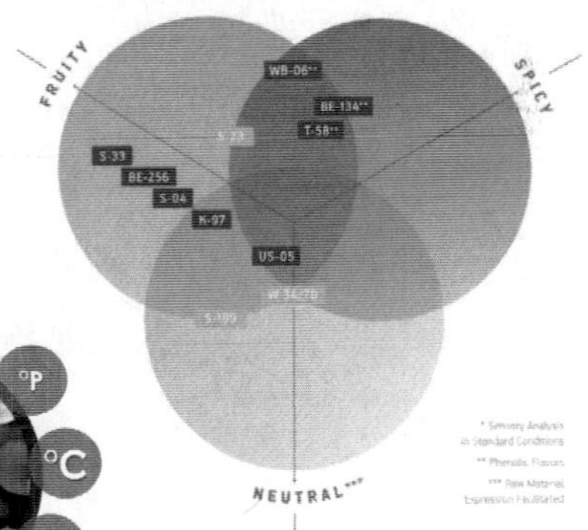

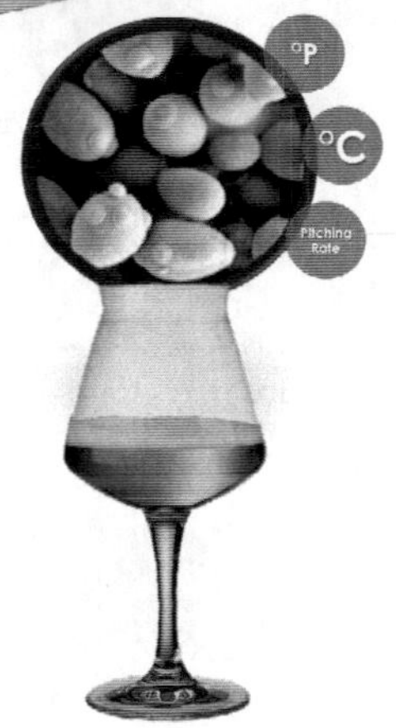

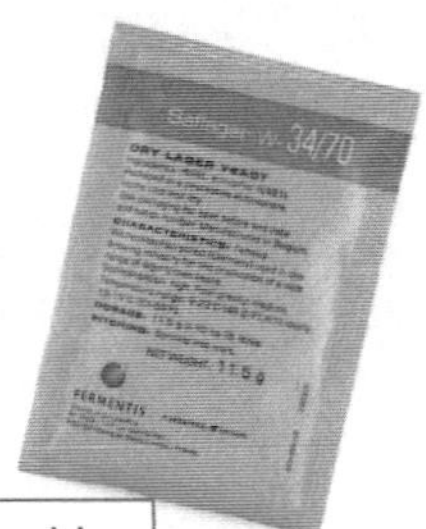

Or the **NEUTRAL** flavor profile produced by the lager yeast **W-3470**?

Fermentis
LESAFFRE FOR BEVERAGES

CLUSTER LAB X PILOT TRIALS

BE256

W34/70

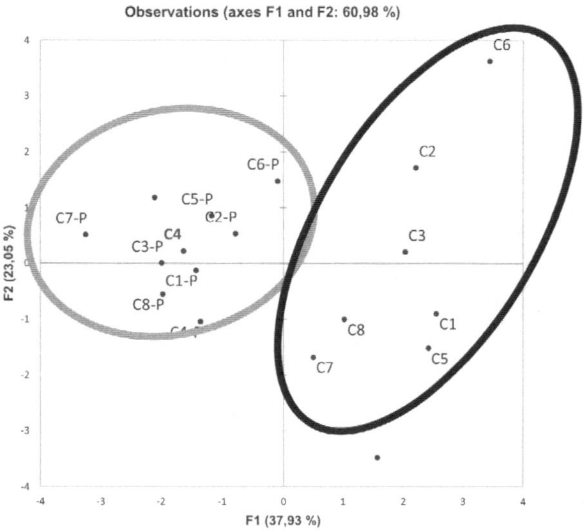

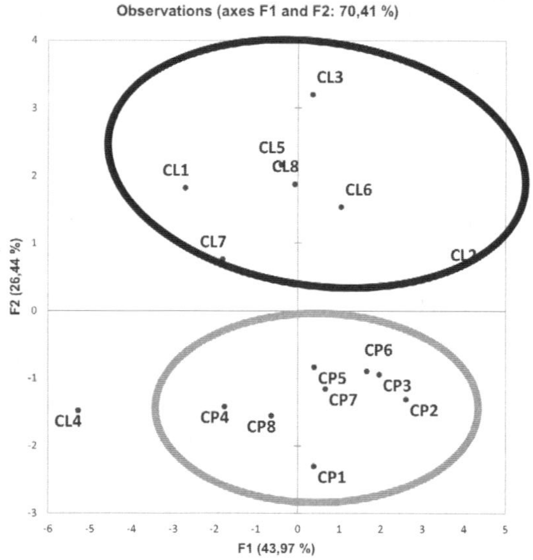

- Lab trials and pilot trials clusters separately
- REAL LIMITATION IN APPLICATION X PREDICTION OF THE FLAVOR EXPRESSION OF LAGER AND ALE YEASTS UNDER DIFFERENT CONDITIONS

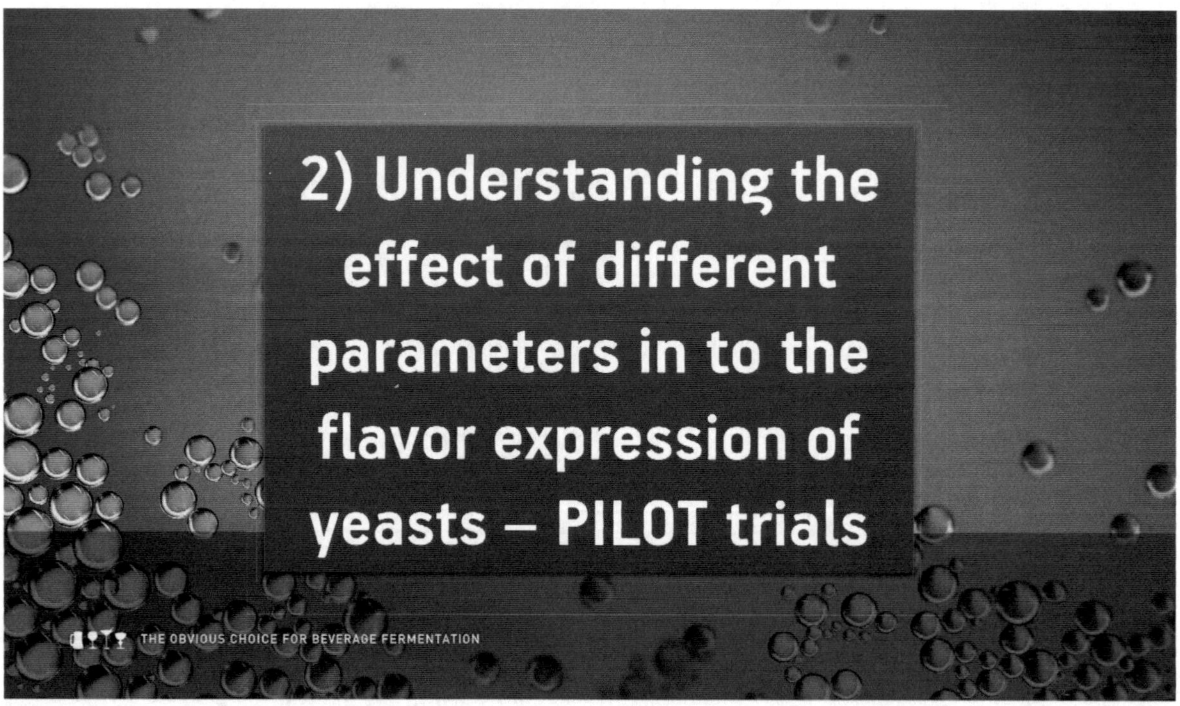

Fermentis
LESAFFRE FOR BEVERAGES

STUDIED CONDITIONS (14)

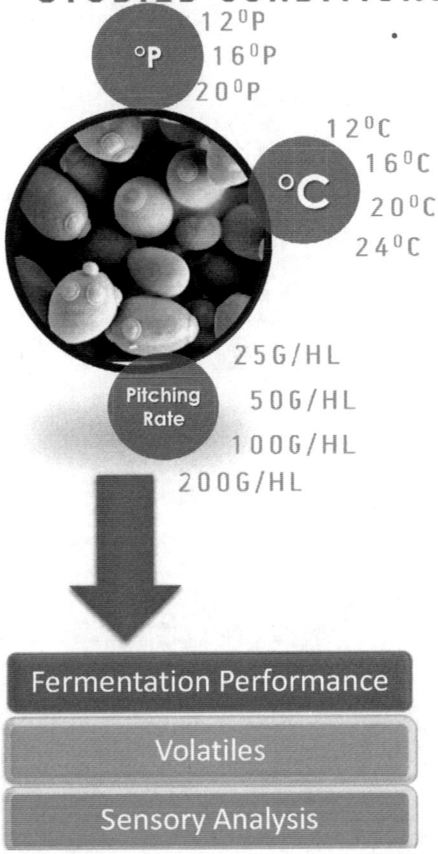

✓PILOT TRIALS

- 14 conditions (MATRIX) for phenotype expression

W34/70

Condition	Density (°P)	Temperature (°C)	Pitching Rate (g/hL)
C1	16	12	50
C2	20	16	100
C3	20	16	200
C4	12	20	50
C5	16	20	50
C6	20	20	50
C7	16	20	25
C8	16	20	100
C9	16	16	100
C10	12	20	100
C11	12	12	25
C12	12	16	50
C13	20	16	25
C14	20	12	100

BE256

Condition	Density (°P)	Temperature (°C)	Pitching rate (g/hL)
C1	16	12	50
C2	20	24	50
C3	20	24	100
C4	12	20	50
C5	16	20	50
C6	20	20	50
C7	16	20	25
C8	16	20	100
C9	12	20	100
C10	16	16	100
C11	12	12	25
C12	12	16	50
C13	20	16	25
C14	20	12	100

Analysed Fermentation volatiles / flavors

Component	Diacetyl	4-VG	n-propanol	Ethyl-acetate	Iso-butanol	2-methyl-butanol	3-methyl-butanol	Isobutyl-acetate	Ethyl-butanoate	Isoamyl-acetate	Ethyl-hexanoate	Phenylethyl-alcohol	Ethyl-octanoate	Phenylethyl-acetate	Ethyl-decanoate
Synonym	2,3 butanodion e					Amyl Alcohol	Isoamyl Alcohol		Ethyl butyrate	Isopentyl Acetate	Ethyl Caproate		Ethyl Caprylate		Ethyl Caprate
Threshold in ppm	0,15	0,3	800	33	1500	65	70	1,6	0,4	1,6	0,23	125	0,9	3,8	1,5
Flavour Perception		Phenolic, bitter, cloves	alcohol	fruity, solvent, sweetish	alcohol	Alcohol, banana, medicinal, solvent	Aromatic, Alcohol, banana, medicinal, solvent	banana, sweet, fruity	tropical, papaya, butter, cheesy, sweetish, perfumed	Banana, candies, pear	Aniseed, Apple, Fruiy, Sweetish	parfum, roses, sweetish	Apple, sweetish, fruity	sweetish, roses, honey	apple, fatty acids, fruity, solvent

ppm, ppb

Concentration / Threshold = Flavour units

'Sensory relevancy'

Different for every component and matrix

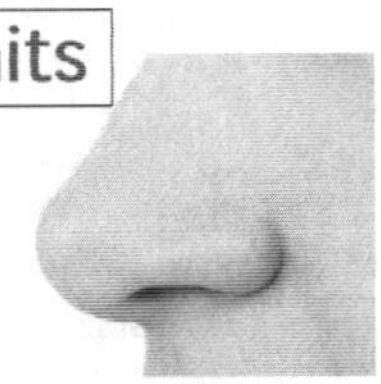

Fermentis
LESAFFRE FOR BEVERAGES

W34/70 STUDIED CONDITIONS (14)

Condition	Density (°P)	Temperature (°C)	Pitching Rate (g/hL)
C1	16	12	50
C2	20	16	100
C3	20	16	200
C4	12	20	50
C5	16	20	50
C6	20	20	50
C7	16	20	25
C8	16	20	100
C9	16	16	100
C10	12	20	100
C11	12	12	25
C12	12	16	50
C13	20	16	25
C14	20	12	100

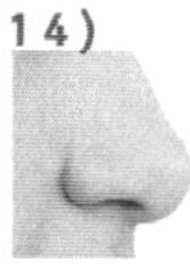

SENSORY PROFILE

Statistically relevancy: density related
(sweetness, alcohol + warmth)

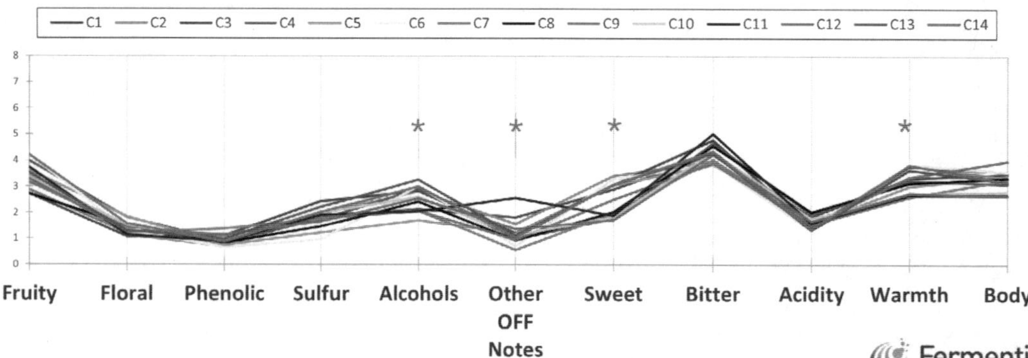

Fermentis
LESAFFRE FOR BEVERAGES

W3470 - PILOT TRIALS

VOLATILES ALCOHOLS in W3470

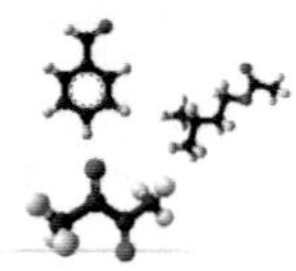

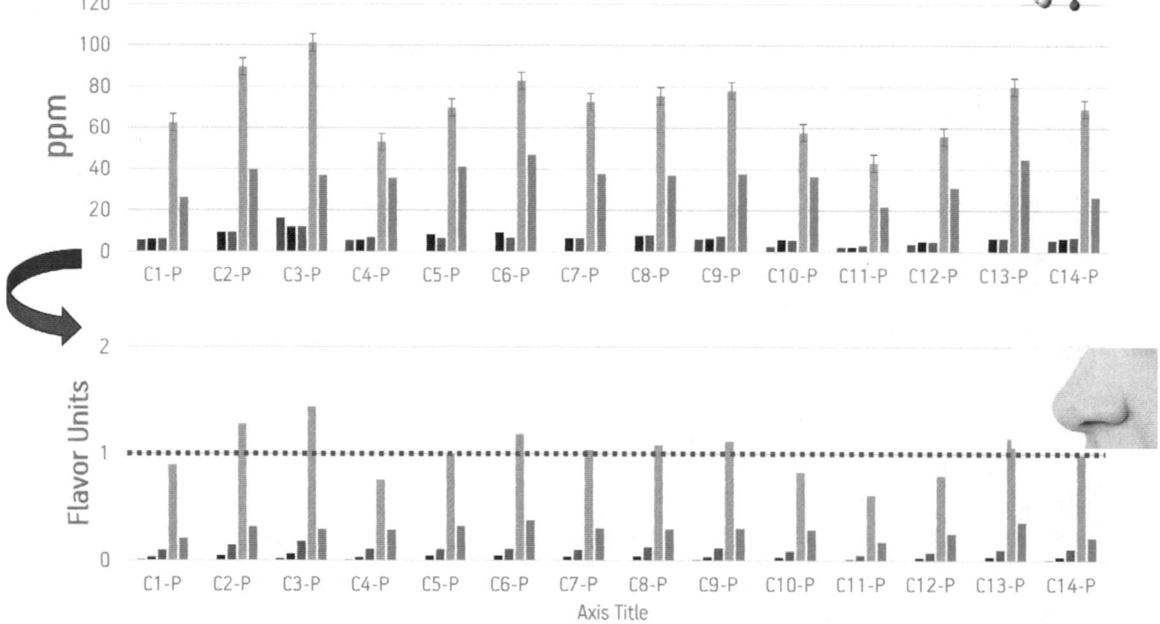

VOLATILES ESTERS (+4VG) in W3470

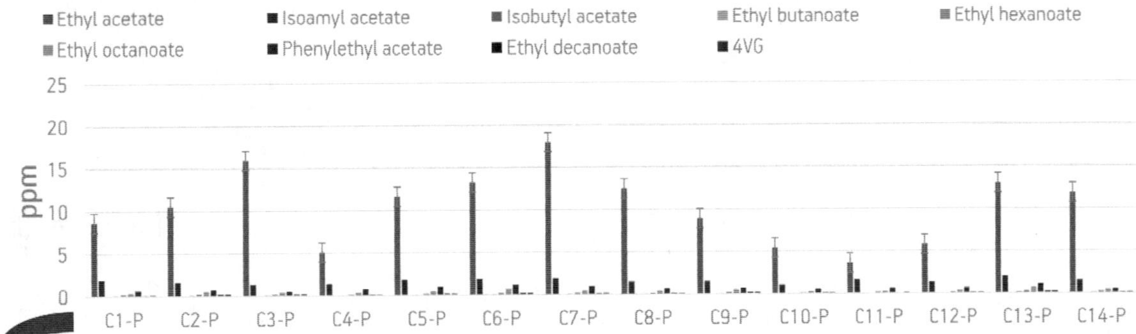

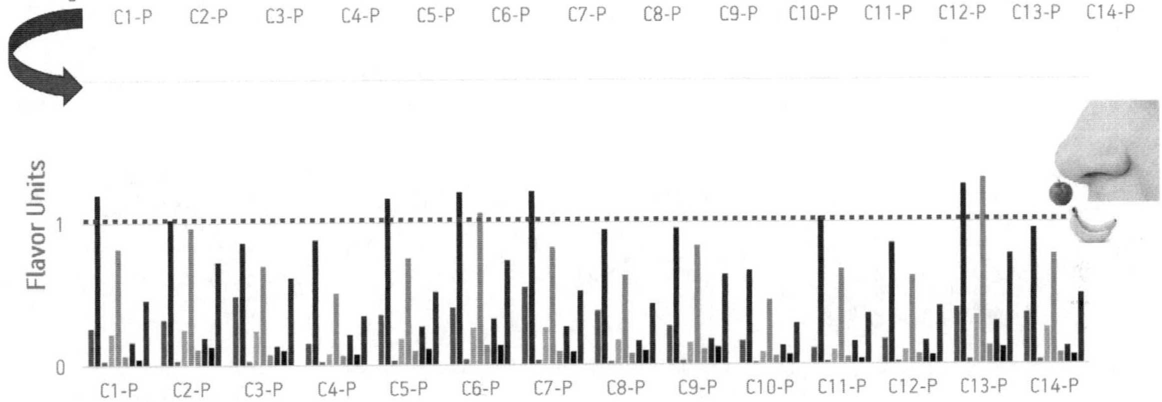

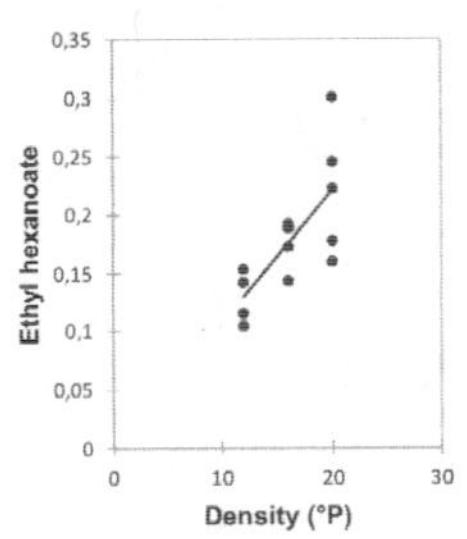

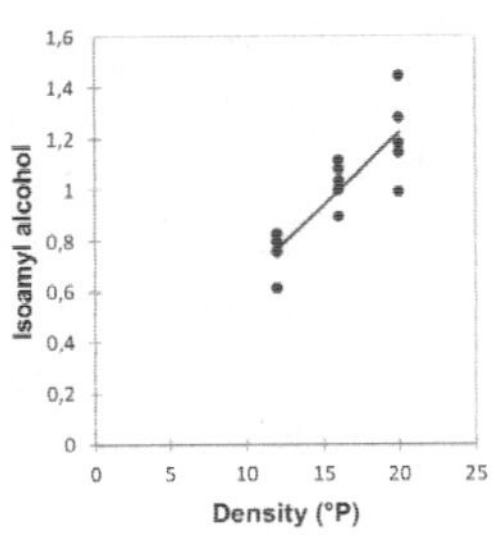

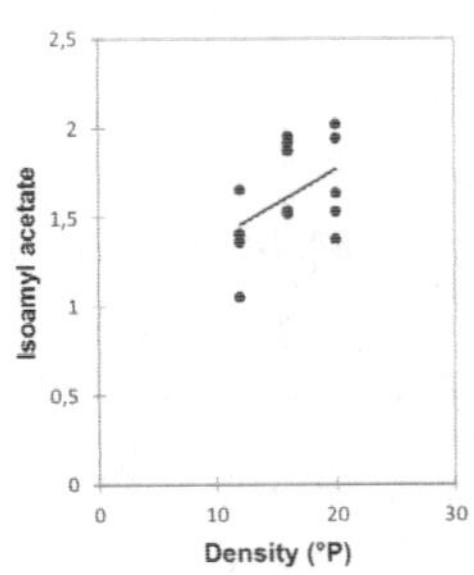

correlation coefficients **r**

W3470 - PILOT TRIALS

Variables	Density (°P)	Temperature (°C)	Pitching rate (g/hL)
Density (°P)	**1**	-0,141	0,345
Temperature (°C)	-0,141	**1**	-0,040
Pitching rate (g/hL)	0,345	-0,040	**1**
Duration (h)	**0,575**	**-0,799**	0,026
Ethanol (ABV)	**0,986**	-0,117	0,280
n-propanol	0,098	-0,334	**0,726**
Isobutanol	**0,726**	0,321	**0,648**
Amyl alcohol	**0,641**	0,152	**0,808**
Isoamyl alcohol	**0,853**	0,148	**0,605**
Phenyl ethyl alcohol	0,448	**0,718**	0,046
Ethyl acetate	**0,747**	0,246	0,243
Isoamyl acetate	0,460	-0,081	**-0,566**
Isobutyl acetate	**0,766**	0,027	-0,142
Ethyl butanoate	**0,910**	-0,132	0,042
Ethyl hexanoate	**0,739**	-0,160	-0,259
Ethyl octanoate	**0,706**	0,274	-0,131
Phenyl ethyl acetate	0,294	0,488	**-0,569**
Ethyl decanoate	**0,635**	0,496	0,210
4VG	**0,855**	-0,024	0,104
Fruity	0,050	-0,171	-0,395
Floral	-0,312	0,316	-0,203
Phenolic	-0,193	0,056	0,055
Sulfur	0,121	-0,492	0,434
Alcohols	0,463	-0,015	0,215
Other OFF Notes	-0,283	**-0,696**	-0,151
Sweet	**0,891**	-0,231	0,259
Bitter	-0,323	-0,326	0,199
Acidity	**-0,707**	0,245	-0,091
Warmth	**0,848**	-0,224	0,446
Body	**0,597**	0,104	0,356

*Values in **bold** are different from 0 with a significance level alpha=0,05* (p-values that correspond to each correlation coefficient).

BE256 STUDIED CONDITIONS (14)

SENSORY PROFILE

Condition	Density (°P)	Temperature (°C)	Pitching rate (g/hL)
C1	16	12	50
C2	20	24	50
C3	20	24	100
C4	12	20	50
C5	16	20	50
C6	20	20	50
C7	16	20	25
C8	16	20	100
C9	12	20	100
C10	16	16	100
C11	12	12	25
C12	12	16	50
C13	20	16	25
C14	20	12	100

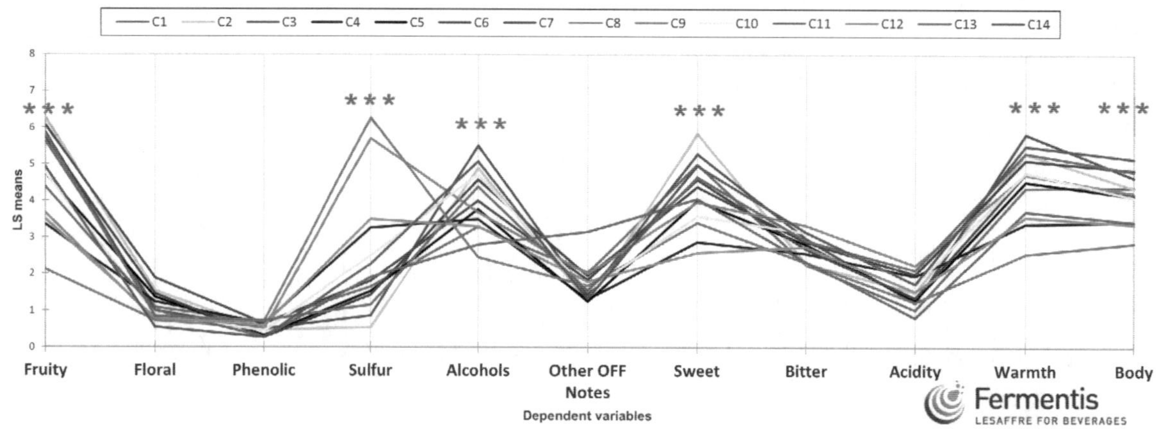

Fermentis
LESAFFRE FOR BEVERAGES

BE256 - PILOT TRIALS

VOLATILES in BE256

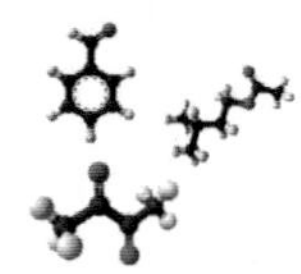

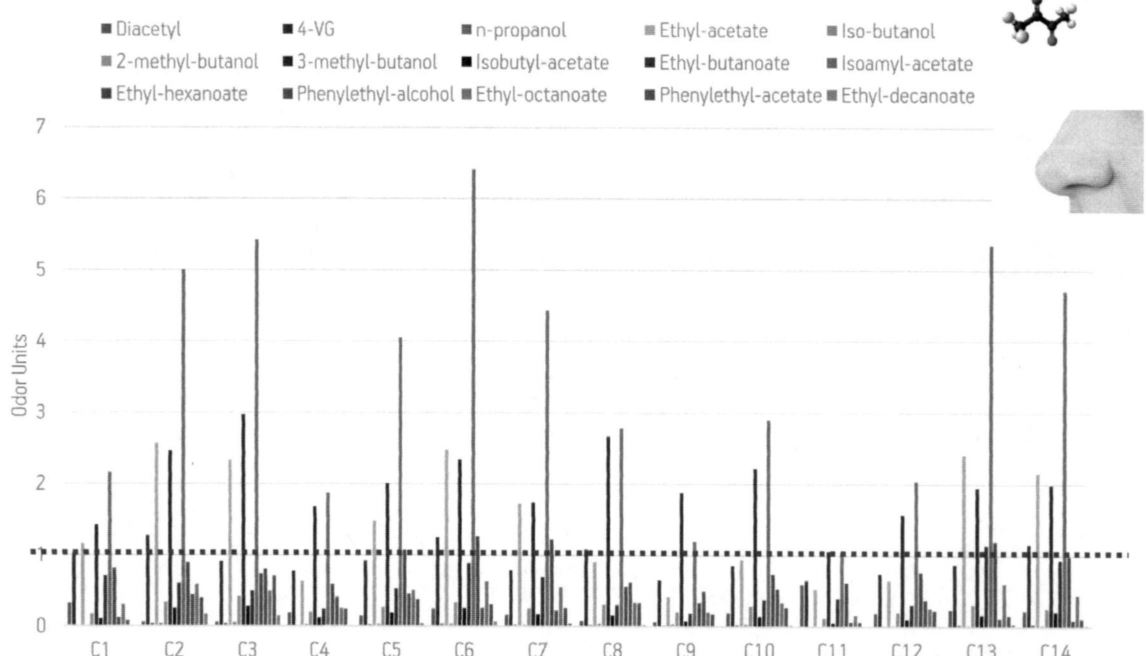

Fermentis
LESAFFRE FOR BEVERAGES

VOLATILES - ODOUR UNITS

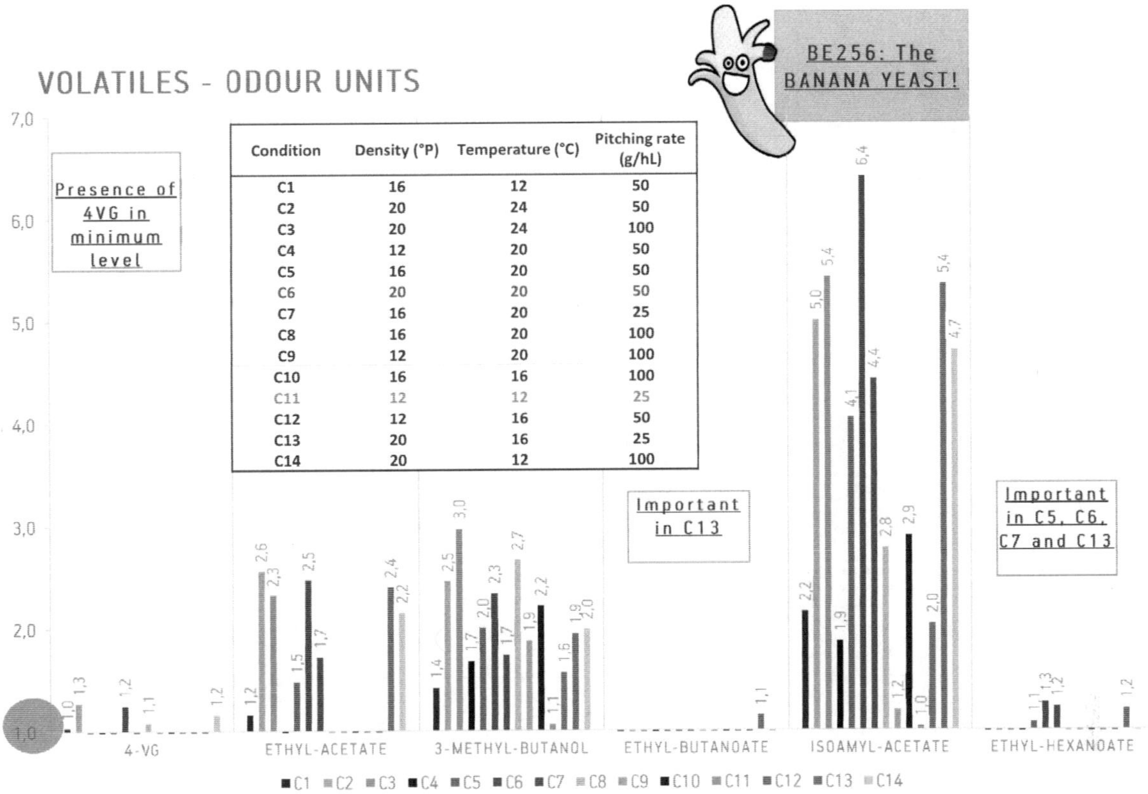

Condition	Density (°P)	Temperature (°C)	Pitching rate (g/hL)
C1	16	12	50
C2	20	24	50
C3	20	24	100
C4	12	20	50
C5	16	20	50
C6	20	20	50
C7	16	20	25
C8	16	20	100
C9	12	20	100
C10	16	16	100
C11	12	12	25
C12	12	16	50
C13	20	16	25
C14	20	12	100

Fermentis
LESAFFRE FOR BEVERAGES

BE256 - PILOT TRIALS

'NOISE of ESTERS' in BE256
Below threshold – importance?

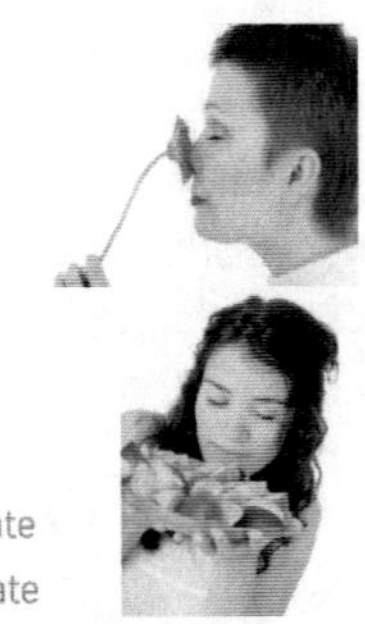

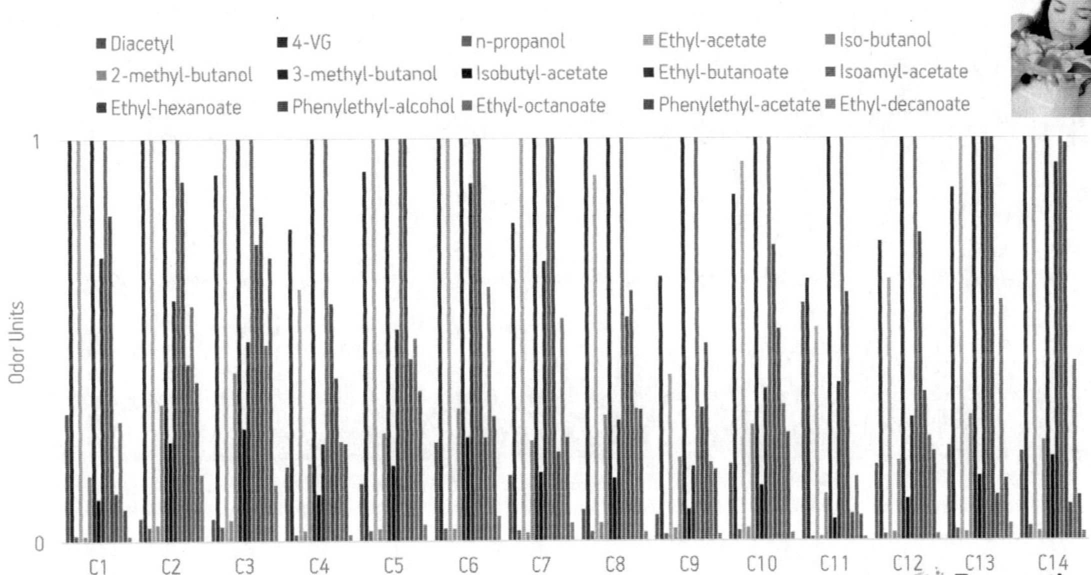

Fermentis
LESAFFRE FOR BEVERAGES

correlation coefficients **r**

Variables	Density (°P)	Temperature (°C)	Pitching rate (g/hL)
Temperature	0,228	**1**	0,154
Plato	**1**	0,228	0,114
Pitching rate (g / hL)	0,114	0,154	**1**
Fruity	**0,833**	0,132	-0,371
Floral	0,092	0,384	**-0,672**
Phenolic	-0,193	-0,192	0,116
Sulfur	**-0,594**	-0,087	0,491
Alcohols	**0,899**	0,239	0,186
Other OFF Notes	-0,358	-0,502	-0,220
Sweet	**0,832**	0,164	-0,078
Bitter	0,230	0,047	0,459
Acidity	0,386	0,094	0,063
Warmth	**0,945**	0,183	0,092
Body	**0,914**	-0,025	0,023
Diacetyl	-0,275	**-0,759**	-0,517
4-VG	**0,780**	0,191	0,164
n-propanol	**0,897**	0,452	0,308
Ethyl-acetate	**0,952**	0,318	-0,105
Iso-butanol	0,413	**0,808**	**0,622**
2-methyl-butanol	**0,777**	**0,683**	0,351
3-methyl-butanol	**0,645**	**0,701**	**0,585**
Isobutyl-acetate	**0,867**	**0,571**	0,185
Ethyl-butanoate	**0,770**	-0,238	-0,336
Isoamyl-acetate	**0,920**	0,390	-0,053
Ethyl-hexanoate	**0,645**	-0,001	-0,503
Phenylethyl-alcohol	0,008	**0,742**	**0,618**
Ethyl-octanoate	**0,861**	0,419	-0,209
Phenylethyl-acetate	0,401	**0,825**	0,328
Ethyl-decanoate	**0,626**	**0,712**	0,036

Values in bold are different from 0 with a significance level alpha=0,05

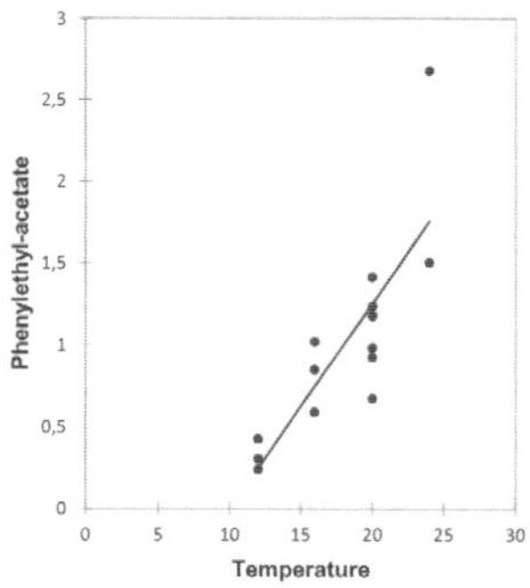

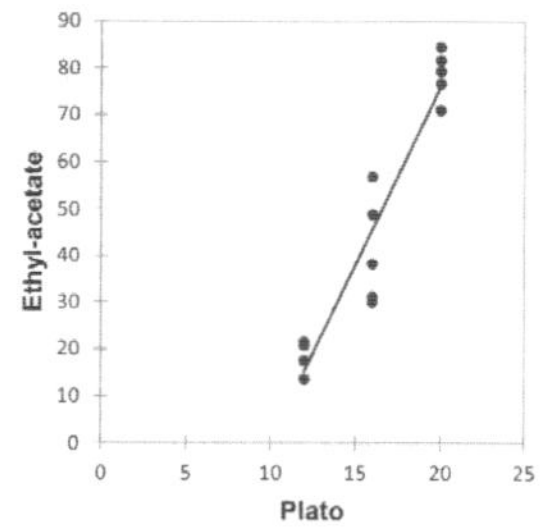

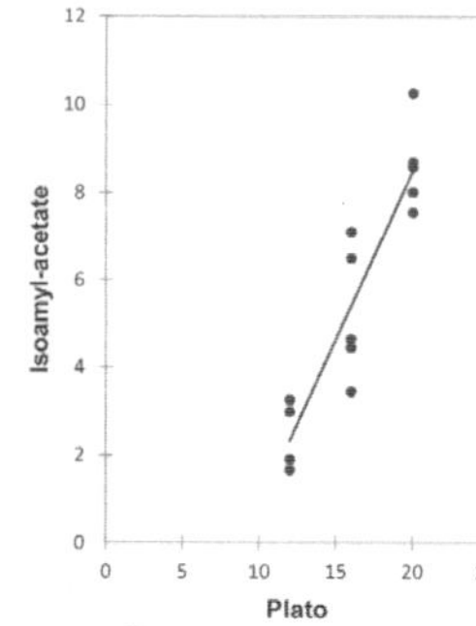

Fermentis
LESAFFRE FOR BEVERAGES

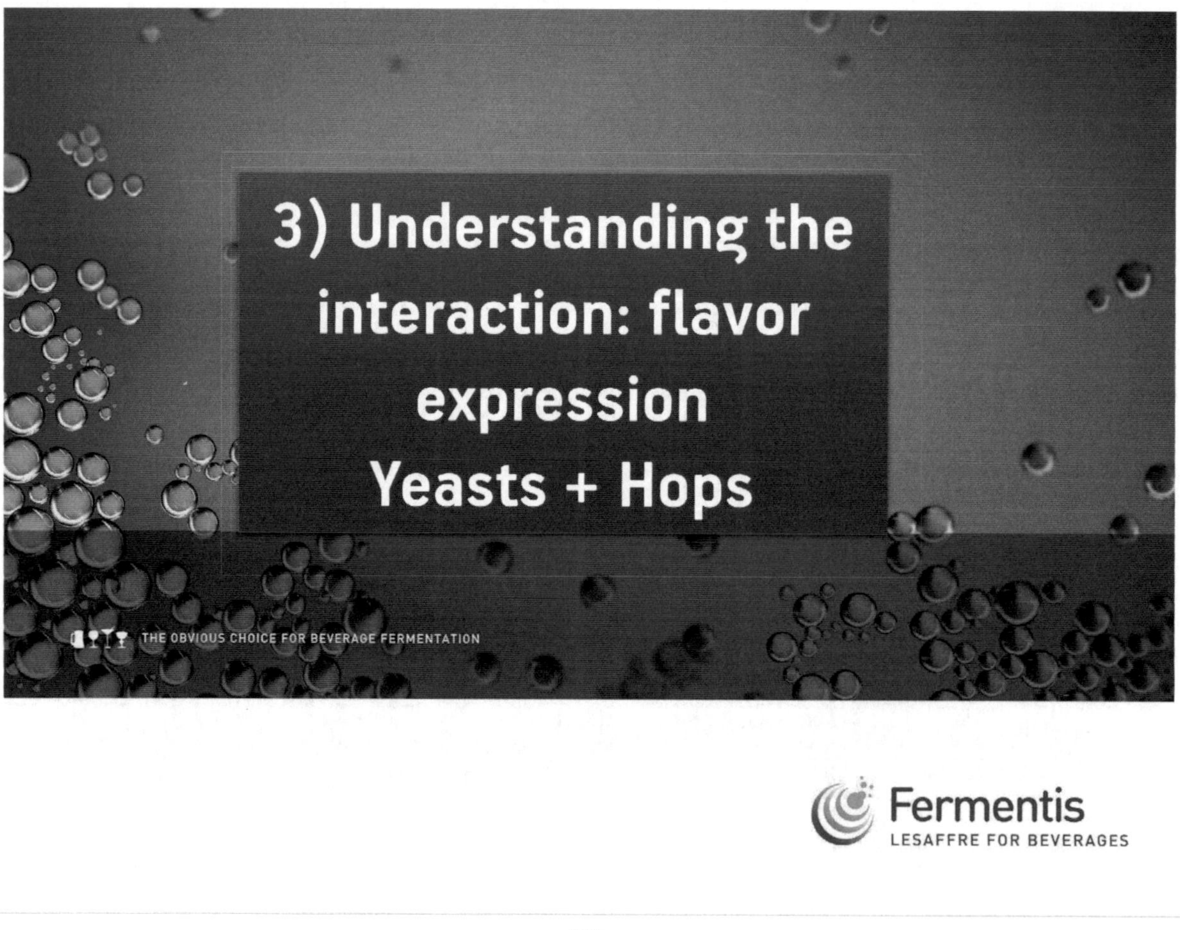

Fermentis
LESAFFRE FOR BEVERAGES

New England IPA
Champagne IPA
BREWERS ASSOCIATION RELEASES 2018 BEER STYLE GUIDELINES
New Categories Include Three Styles of Juicy or Hazy Ales
Beer Judge
21B. Specialty IPA: New England IPA
Overall Impression
Aroma
Appearance
The Birth of the Brut IPA
An enzyme long used to help make big imperial stouts a little easier on the palate has found a new purpose in an emerging style of IPA. The Brut IPA is a dry–0° Plato–version of the style that was created just months ago and is now spreading like wildfire.
Is Brut IPA the Newest Hop Trend?
Beer News
Juicy Beer?
NEIPA Yeast??
"Think of a sparkling glass of dry Champagne but with fruit-forward hop aromas"
Fermentis
LESAFFRE FOR BEVERAGES

NEW BEER STYLE = NEW YEAST?
Fermentis
LESAFFRE FOR BEVERAGES

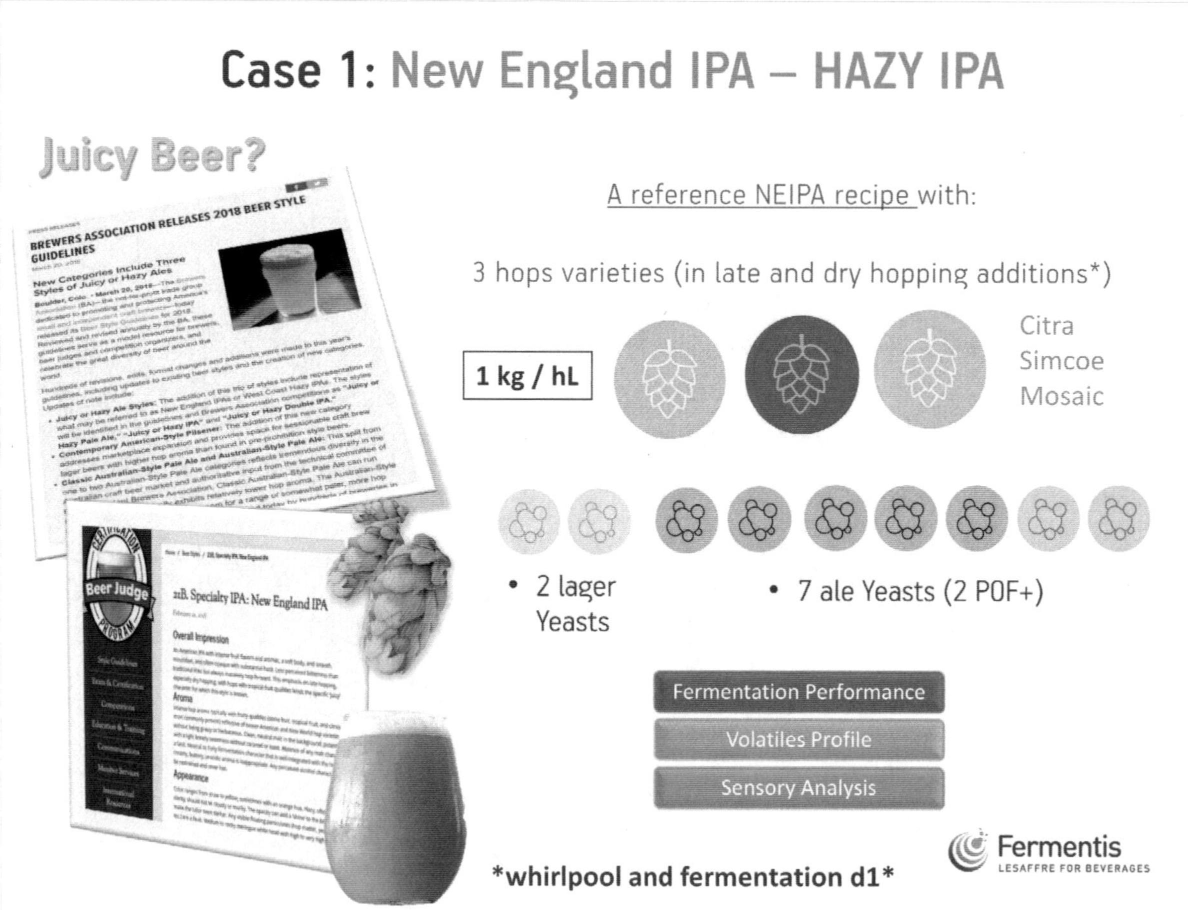
Case 1: New England IPA – HAZY IPA
Juicy Beer?
BREWERS ASSOCIATION RELEASES 2018 BEER STYLE GUIDELINES
New Categories Include Three Styles of Juicy or Hazy Ales
Beer Judge
21B. Specialty IPA: New England IPA
Overall Impression
Aroma
Appearance
A reference NEIPA recipe with:
3 hops varieties (in late and dry hopping additions*)
1 kg / hL
Citra
Simcoe
Mosaic
• 2 lager Yeasts
• 7 ale Yeasts (2 POF+)
Fermentation Performance
Volatiles Profile
Sensory Analysis
whirlpool and fermentation d1
Fermentis
LESAFFRE FOR BEVERAGES

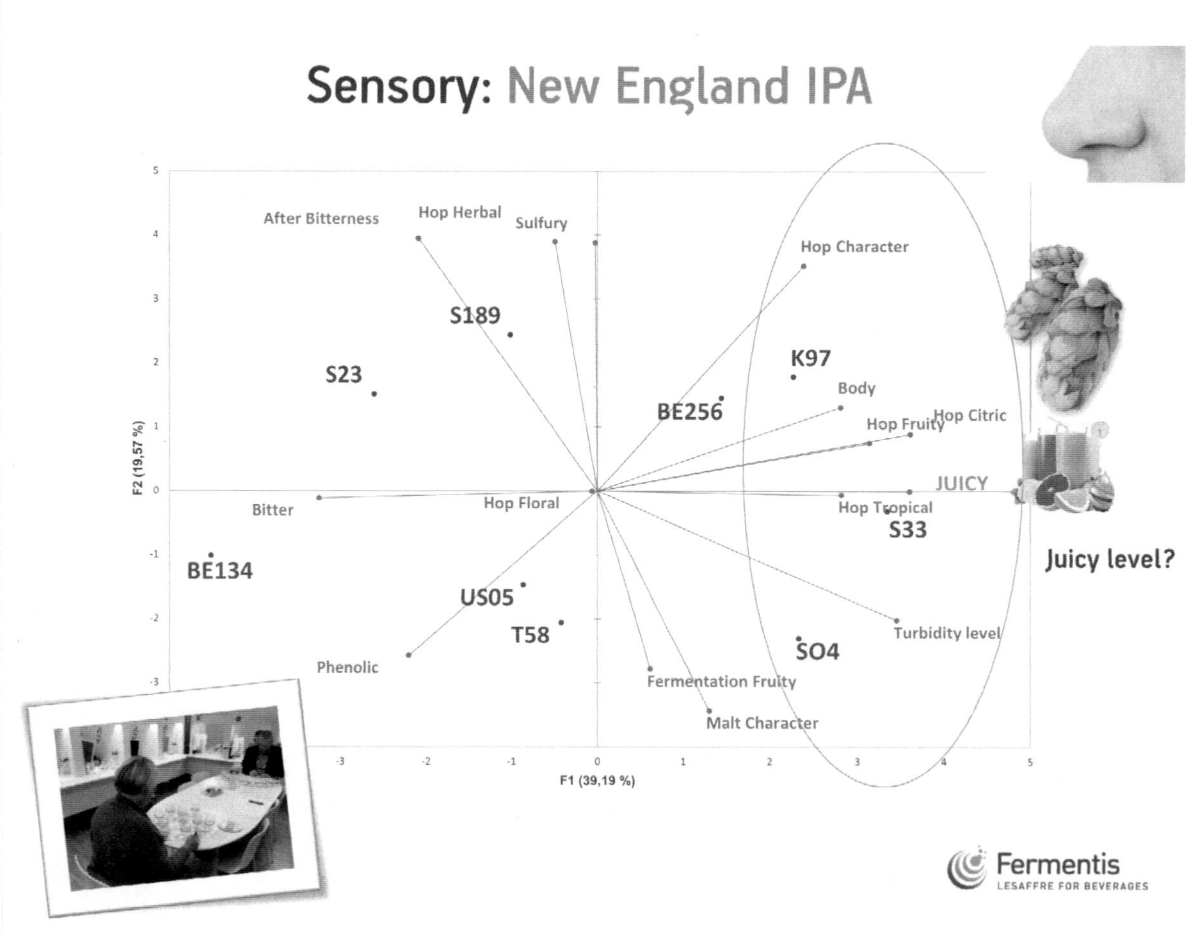
Sensory: New England IPA
After Bitterness
Hop Herbal
Sulfury
Hop Character
S189
S23
K97
Body
BE256
Hop Fruity
Hop Citric
JUICY
Bitter
Hop Floral
Hop Tropical
S33
BE134
Juicy level?
US05
T58
Turbidity level
Phenolic
SO4
Fermentation Fruity
Malt Character
F2 (19,57 %)
F1 (39,19 %)
Fermentis
LESAFFRE FOR BEVERAGES

Hop Oils: New England IPA

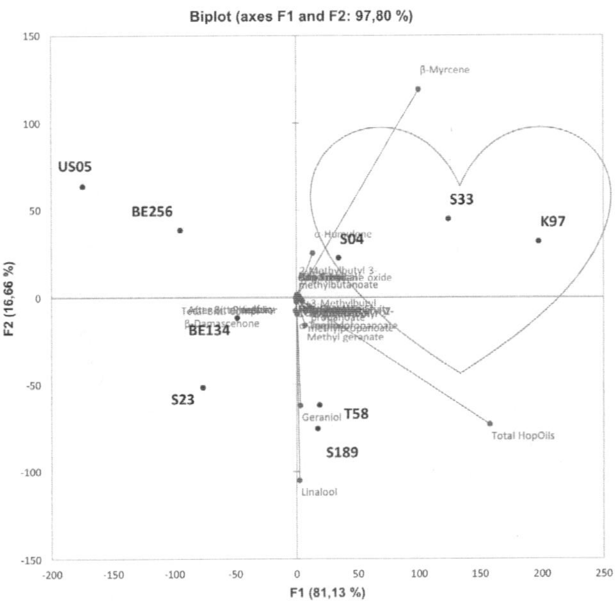

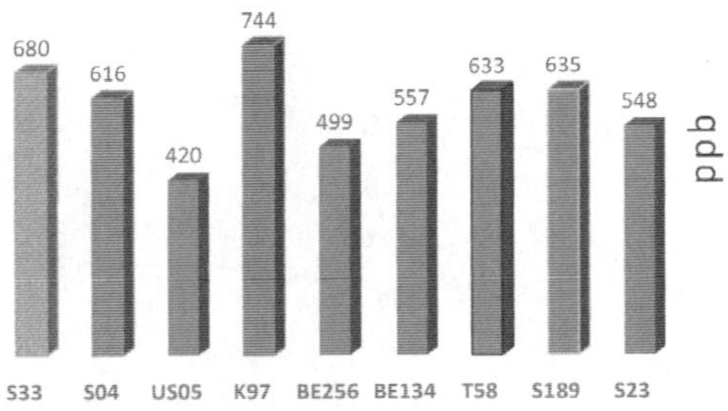

Fermentis
LESAFFRE FOR BEVERAGES

Case 2: Brut IPA – Champagne IPA

"Think of a sparkling glass of **dry Champagne** but with fruit-forward hop aromas"

Fermentis
LESAFFRE FOR BEVERAGES

Case 2: Brut IPA – Champagne IPA

1 type of BRUT IPA reference recipe
+ 6 Fermentis Yeast (1 POF+)

700 g / hL
whirlpool

Cascade
Mosaic
Citra

+ AMG 50 g/hl

Fermentation Performance

Volatiles Profile

Sensory Analysis

BRUT IPA

HOP VOLATILES

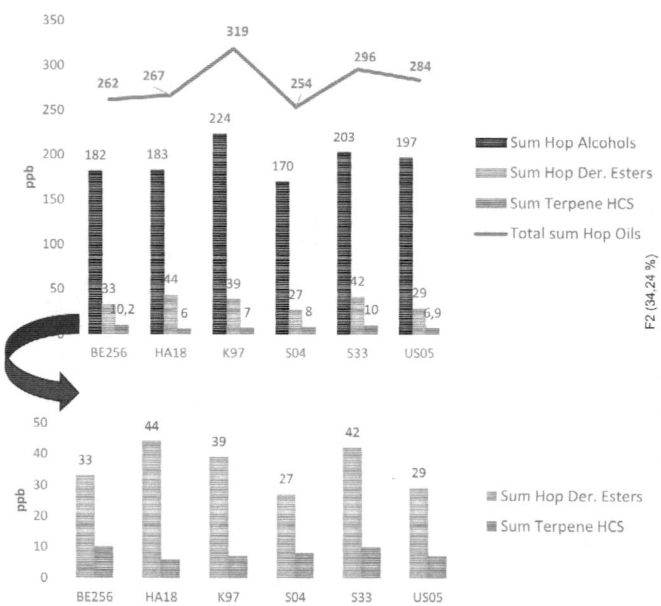

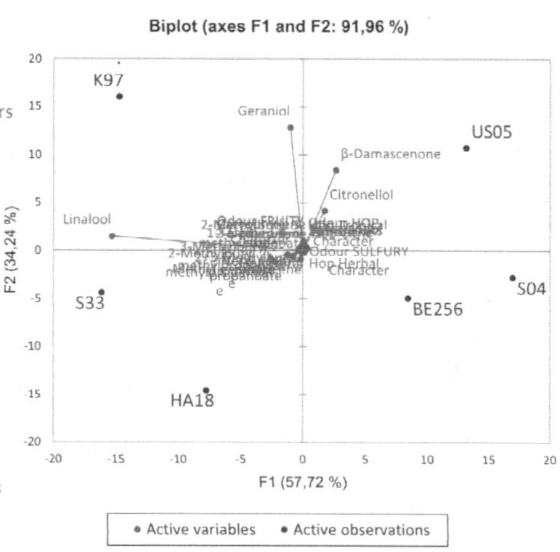

Fermentis
LESAFFRE FOR BEVERAGES

Sensory: BRUT IPA

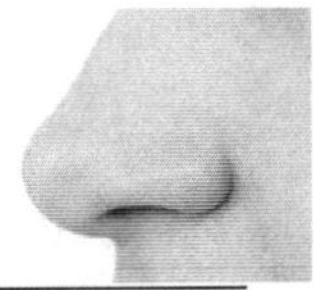

	Odour MALT Character	Odour HOP Character	Odour FRUITY Character	Odour SULFURY Character	Hop Citric	Hop Tropical	Stone Fruits	Hop Herbal	Hop Floral	Fermentation Fruity	Phenolic	Sulfury	Bitterness	After Bitterness	Body
			*			***				^	***				
Pr > F(Yeast)	0,706	0,602	0,046	0,665	0,344	< 0,0001	0,755	0,498	0,704	0,131	< 0,0001	0,525	0,422	0,154	0,721
Significant	No	No	Yes	No	No	Yes	No	No	No	No	Yes	No	No	No	No

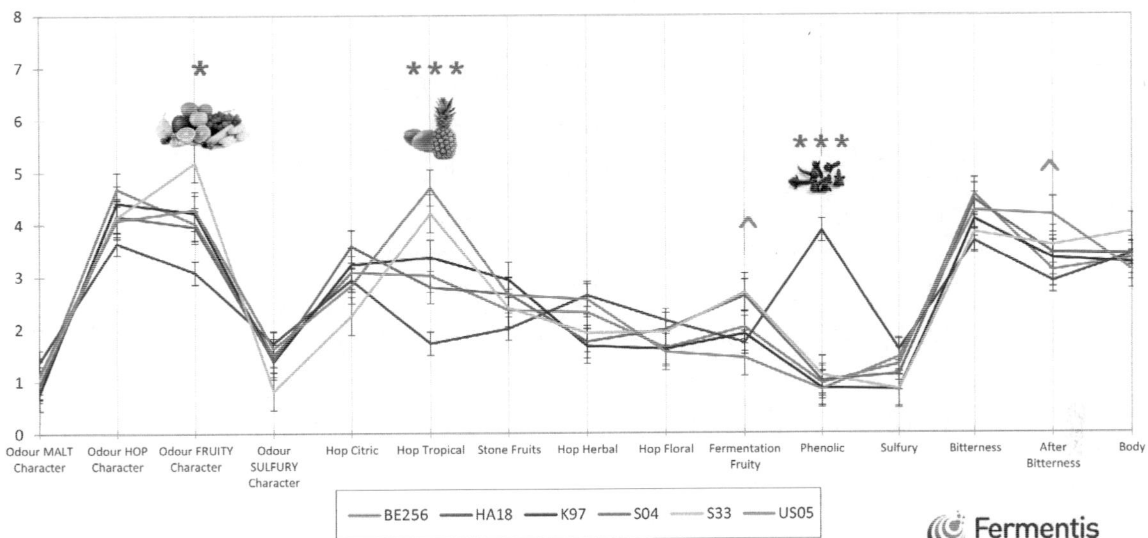

Fermentis
LESAFFRE FOR BEVERAGES

*(statistically relevant) ^(sensory tendency)

BRUT IPA

FERMENTATION VOLATILES

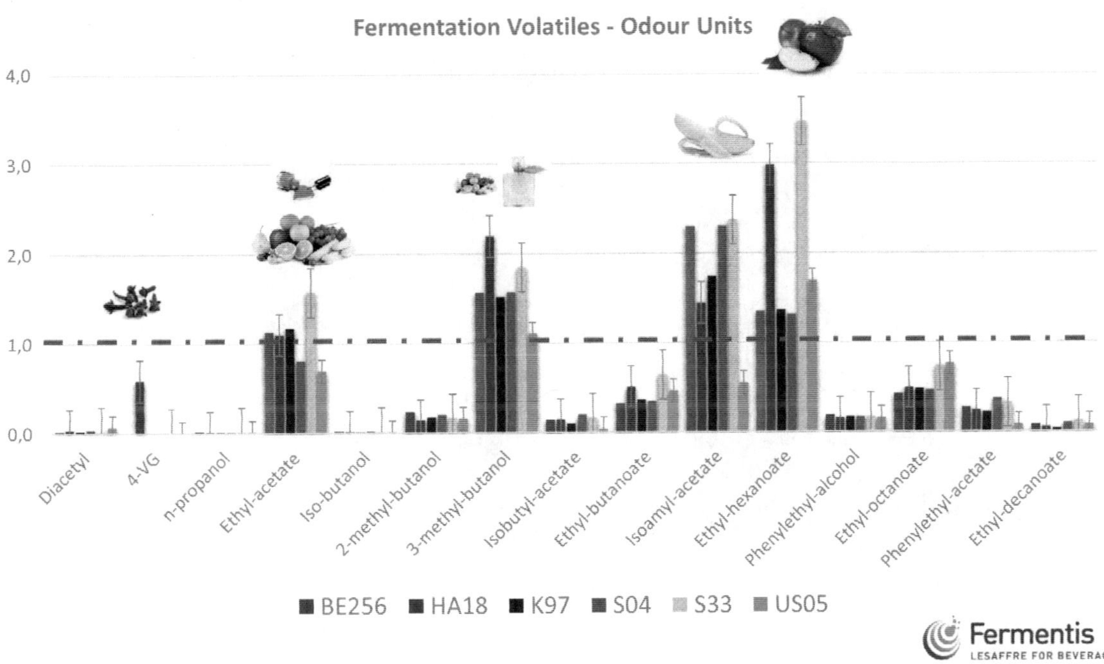

Fermentis
LESAFFRE FOR BEVERAGES

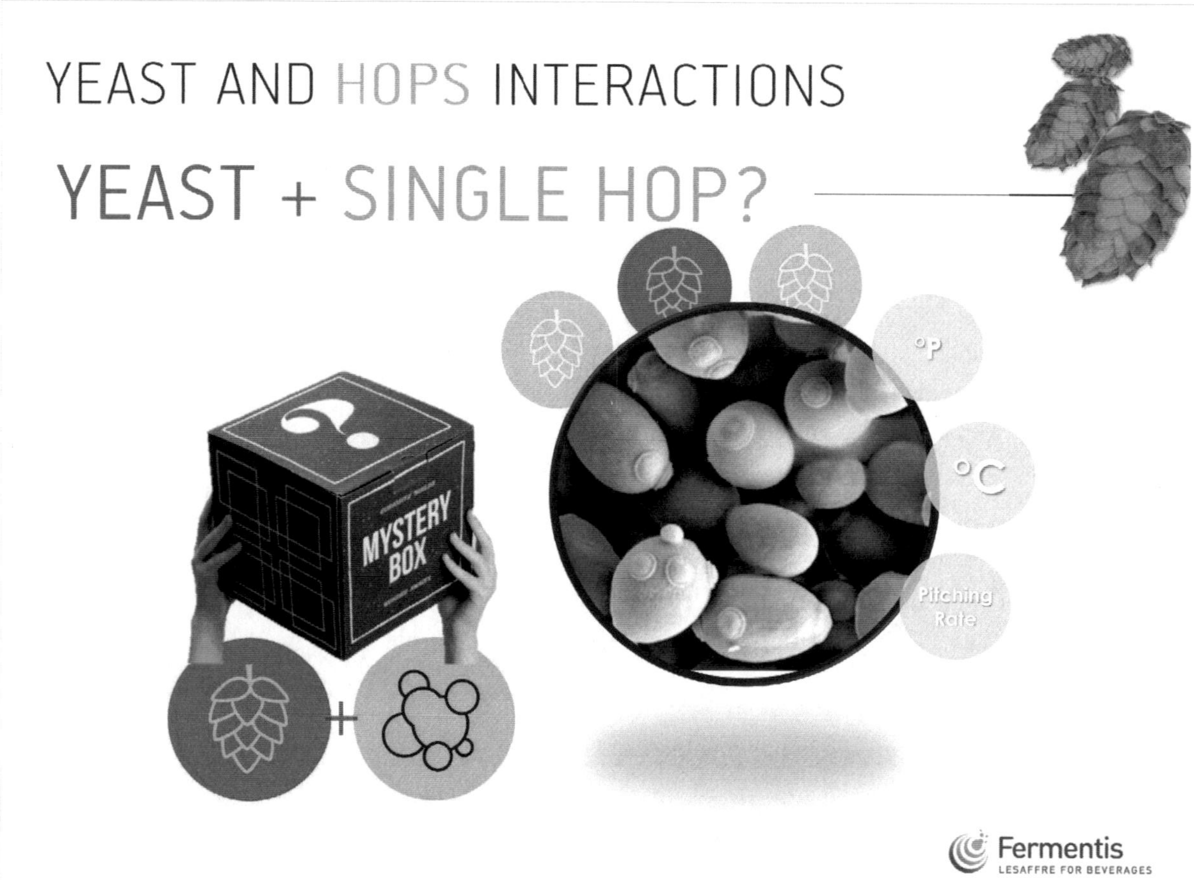
YEAST AND HOPS INTERACTIONS
YEAST + SINGLE HOP?
°P
°C
Pitching Rate
MYSTERY BOX
Fermentis
LESAFFRE FOR BEVERAGES

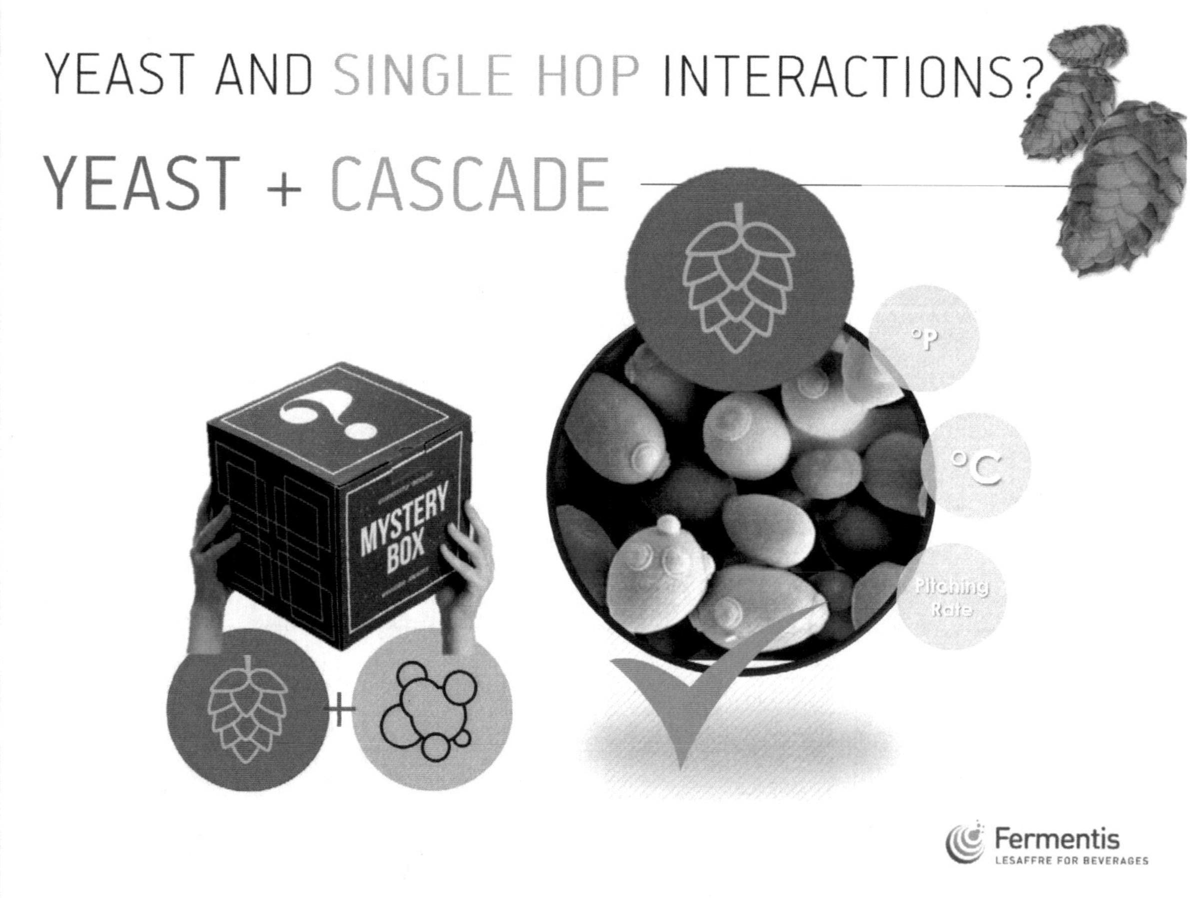
YEAST AND SINGLE HOP INTERACTIONS?
YEAST + CASCADE
°P
°C
Pitching Rate
MYSTERY BOX
Fermentis
LESAFFRE FOR BEVERAGES

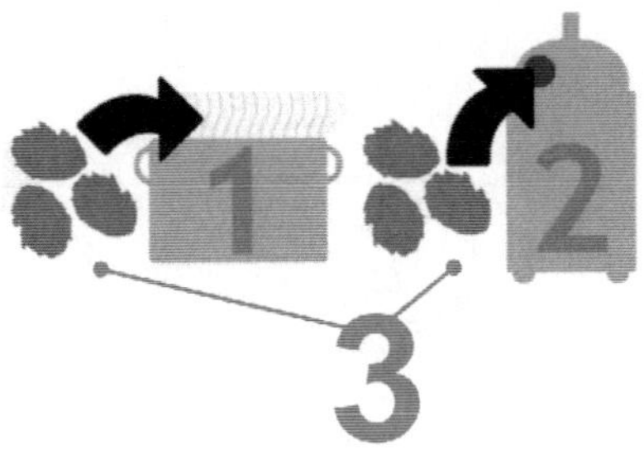

CASCADE HOP

PROTOCOLS

(3 late hopping regimes)

	Code	Hop Addition	Amount (g/hL)
1	**Lh**	Late Hopping	300
2	**Dh**	Dry Hopping**	300
3	**LDh**	Late Hopping + Dry Hopping	150+150

+ 11 Fermentis yeasts
(3 lagers, 7 ales (3 POF+)

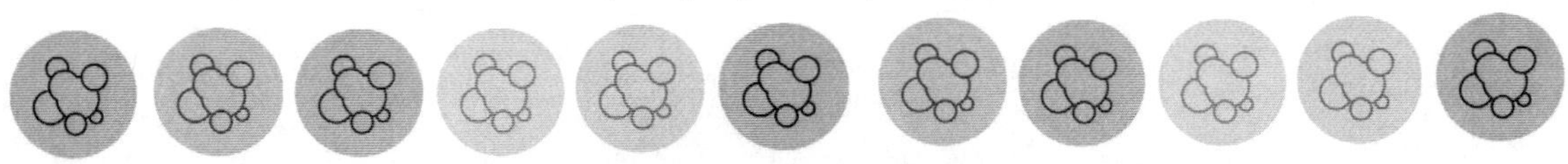

Late Hopping - Added at beginning of pumping the wort in the whirlpool (pumping takes 10 minutes, then 15 minutes of settling), around 90oC
**Dry Hopping - added 24h after filling of vessel, Fermentation day 1 (24 °C).

Fermentation Performance

Volatiles Profile

Sensory Analysis

General flavor families

Fermentis
LESAFFRE FOR BEVERAGES

CLUSTERING

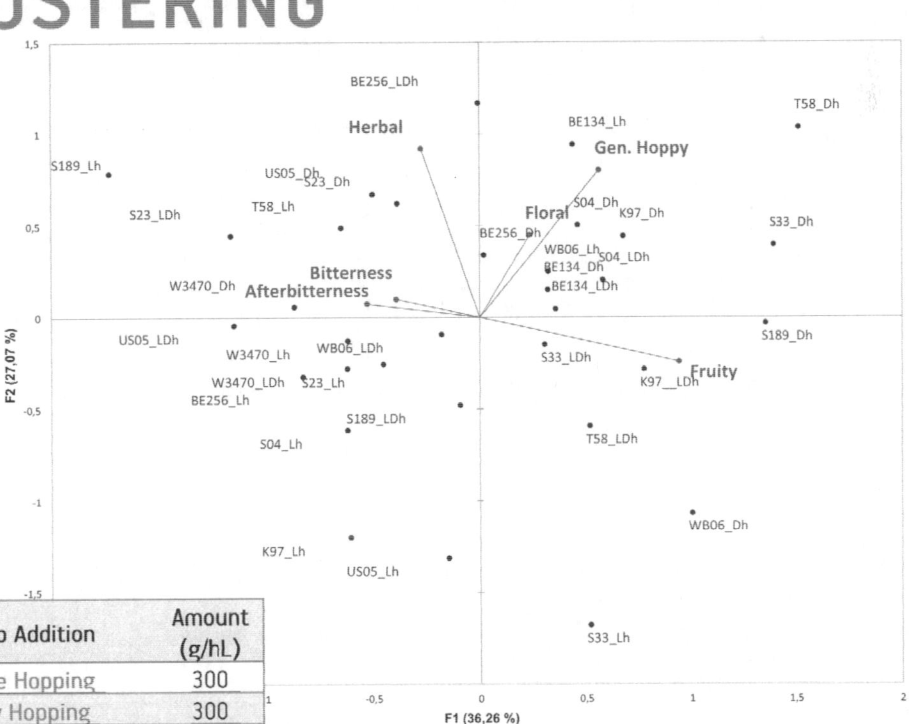

Code	Hop Addition	Amount (g/hL)
Lh	Late Hopping	300
Dh	Dry Hopping	300
LDh	Late Hopping + Dry Hopping	150+150

Fermentis
LESAFFRE FOR BEVERAGES

The cases of W3470 and BE256

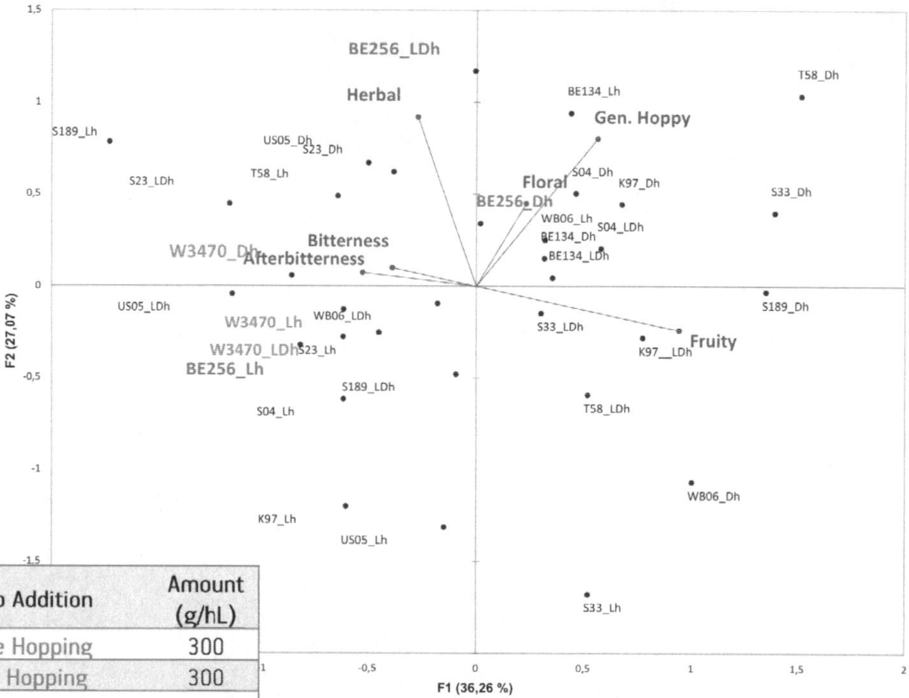

Code	Hop Addition	Amount (g/hL)
Lh	Late Hopping	300
Dh	Dry Hopping	300
LDh	Late Hopping + Dry Hopping	150+150

Fermentis
LESAFFRE FOR BEVERAGES

The gypsies S189 (lager), K97 and S33 (ales)

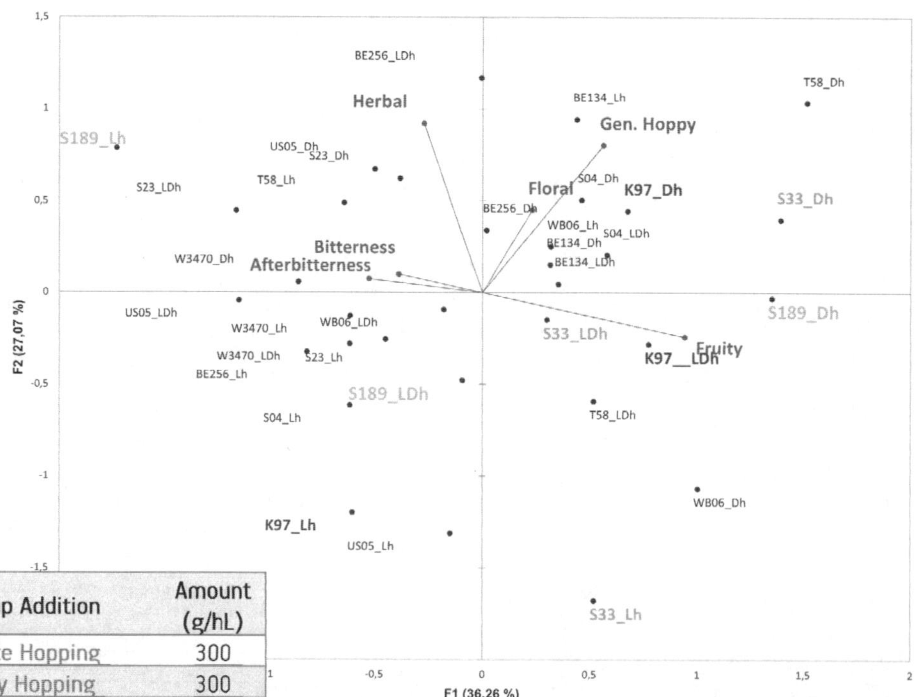

Code	Hop Addition	Amount (g/hL)
Lh	Late Hopping	300
Dh	Dry Hopping	300
LDh	Late Hopping + Dry Hopping	150+150

Fermentis
LESAFFRE FOR BEVERAGES

YEASTS + CASCADE

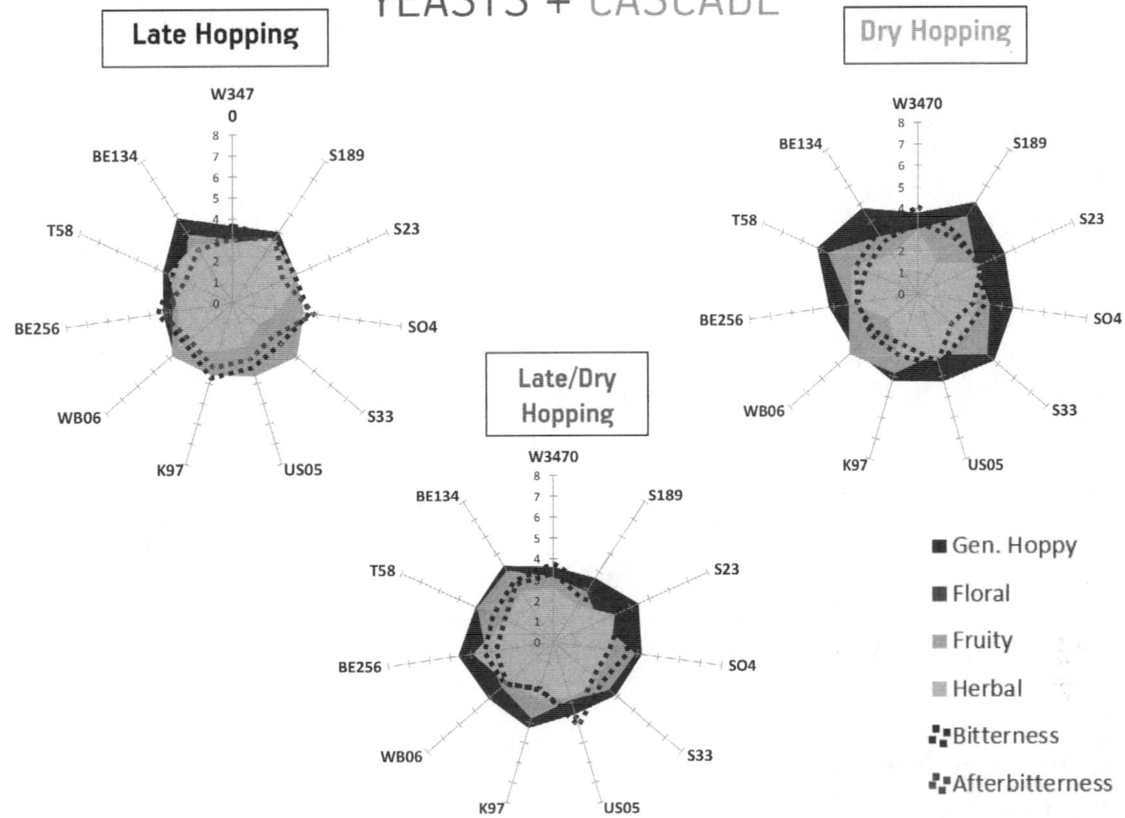

The yeasts "MOVE" into the hop flavors perception possibilities.
But what would be the impact of primary **fermentation flavors** in the overall hop-fruity perceptions?

Fermentis
LESAFFRE FOR BEVERAGES

❑ Integrating a '*Blanco*' (non hopped)

BLANCO + 1 LATE HOPPING REGIME

Hop Addition	Amount (g/hL)
NO LATE Hopping	0
Late Hopping + Dry Hopping*	200+200

Late Hopping - Added at beginning of pumping the wort in the whirlpool (pumping takes 10 minutes, then 15 minutes of settling), around 90oC
Dry Hopping* - added 24h after filling of vessel, Fermentation day 1 (24 °C).

- + Detailed Sensory (discrimination fermentation X hop fruity flavors)
- + Analytics

Fermentis
LESAFFRE FOR BEVERAGES

YEAST & CASCADE

LAGERS

SafLager W3470, S23 and S189

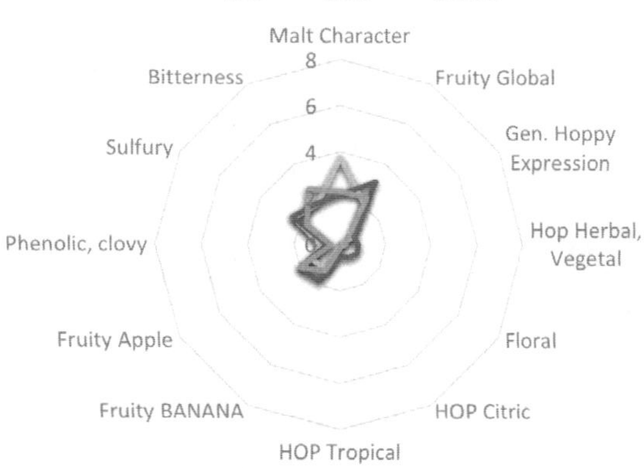

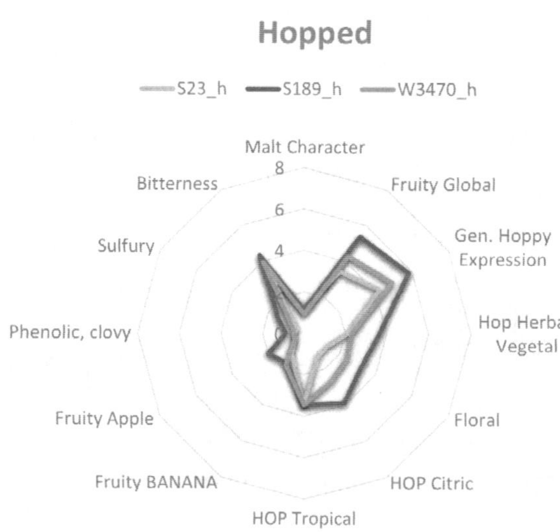

Fermentis
LESAFFRE FOR BEVERAGES

YEAST & CASCADE

ALES (POF-)

SafAles BE256, K97, S04, S33, US05

Non Hopped

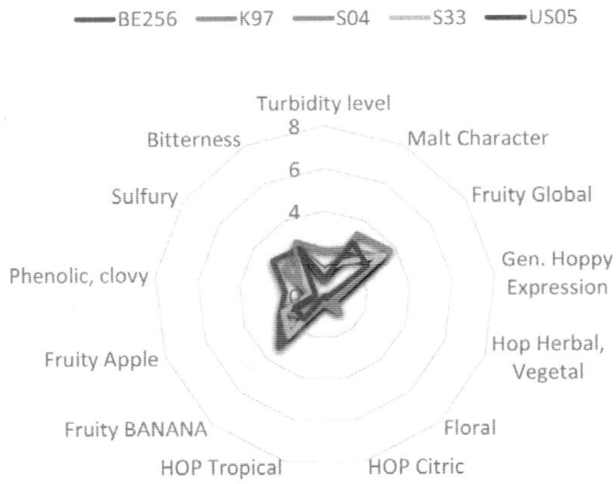

Hopped

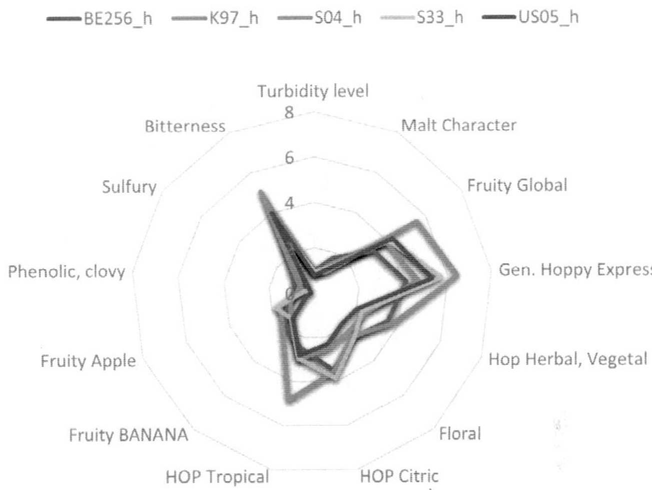

Fermentis
LESAFFRE FOR BEVERAGES

YEAST & CASCADE

ALES (POF+)

SafAles BE134, T58 and WB06

Non hopped POF+

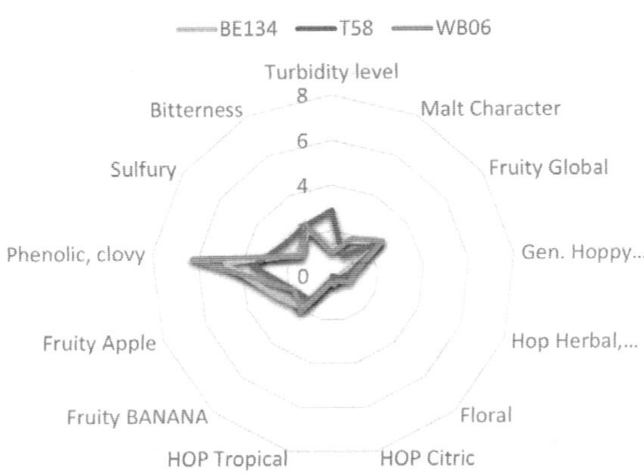

Hopped POF+

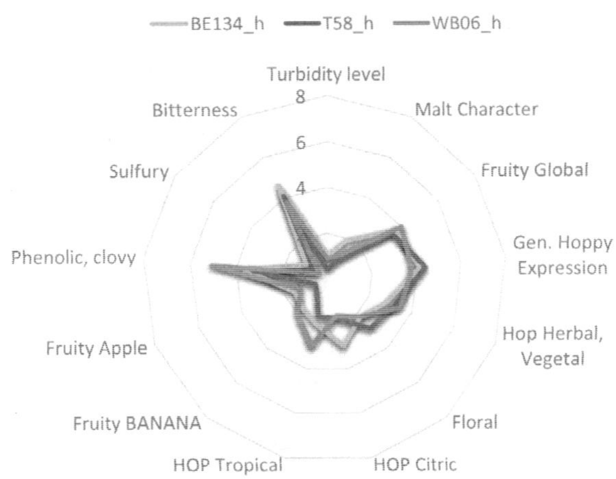

Fermentis
LESAFFRE FOR BEVERAGES

Clustering

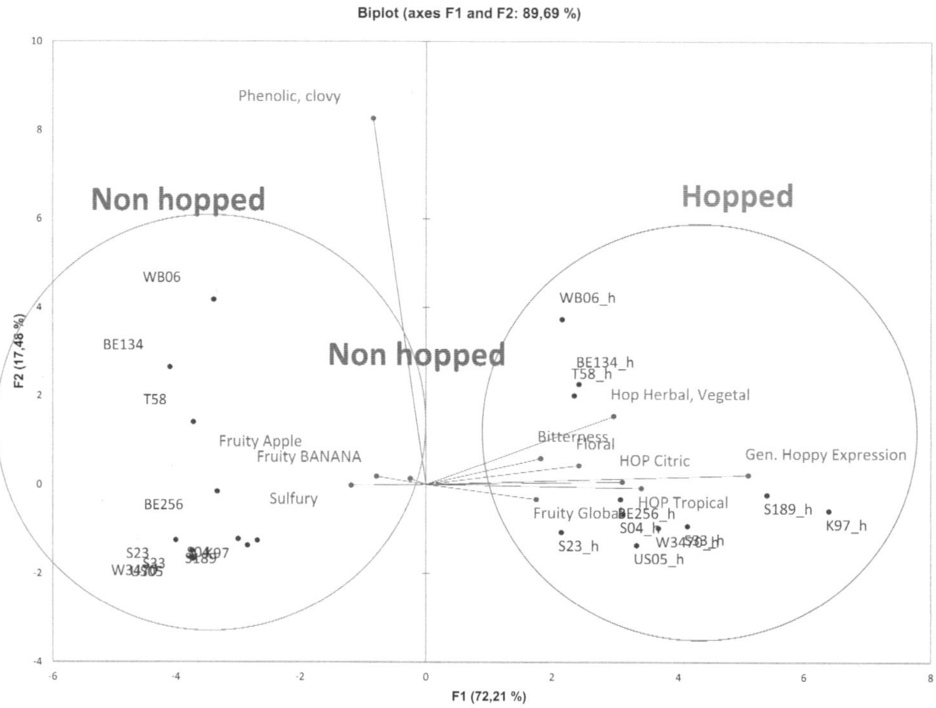

Fermentis
LESAFFRE FOR BEVERAGES

Clustering

Phenolic zone
Masking compound

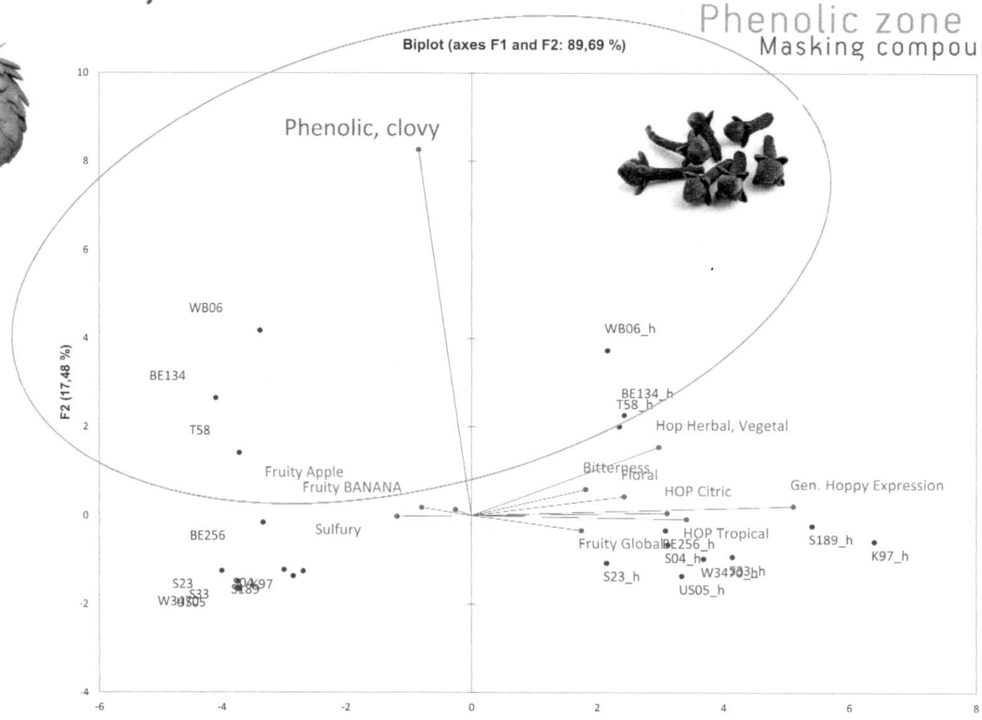

Fermentis
LESAFFRE FOR BEVERAGES

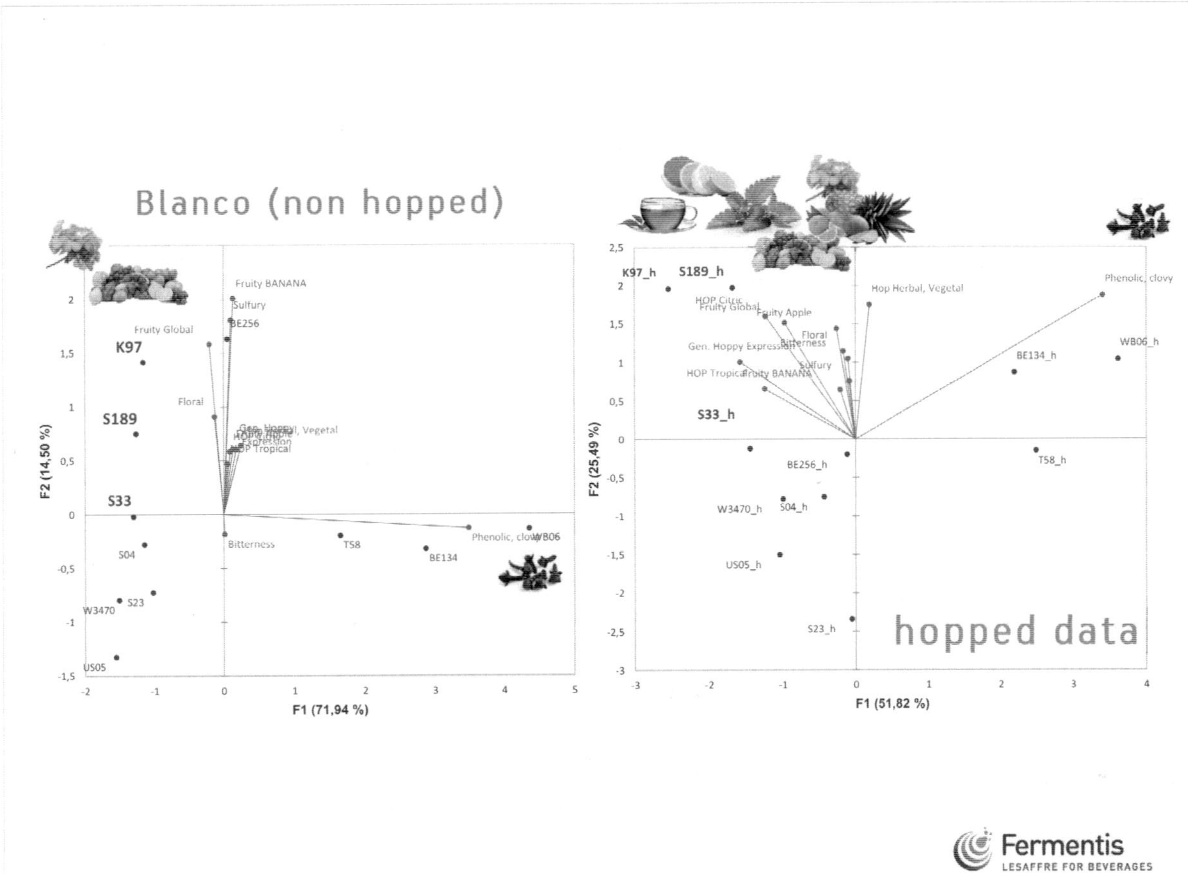
Blanco (non hopped)
K97
S189
S33
S04
S23
W3470
US05
Fruity BANANA
Sulfury
BE256
Fruity Global
Floral
Bitterness
T58
BE134
Phenolic, clovy
WB06
F2 (14,50 %)
F1 (71,94 %)
K97_h
S189_h
S33_h
HOP Citric
Fruity Global
Fruity Apple
Floral
Bitterness
Gen. Hoppy Expression
Sulfury
HOP Tropical
Fruity BANANA
Hop Herbal, Vegetal
Phenolic, clovy
WB06_h
BE134_h
T58_h
BE256_h
W3470_h
S04_h
US05_h
S23_h
hopped data
F2 (25,49 %)
F1 (51,82 %)
Fermentis
LESAFFRE FOR BEVERAGES

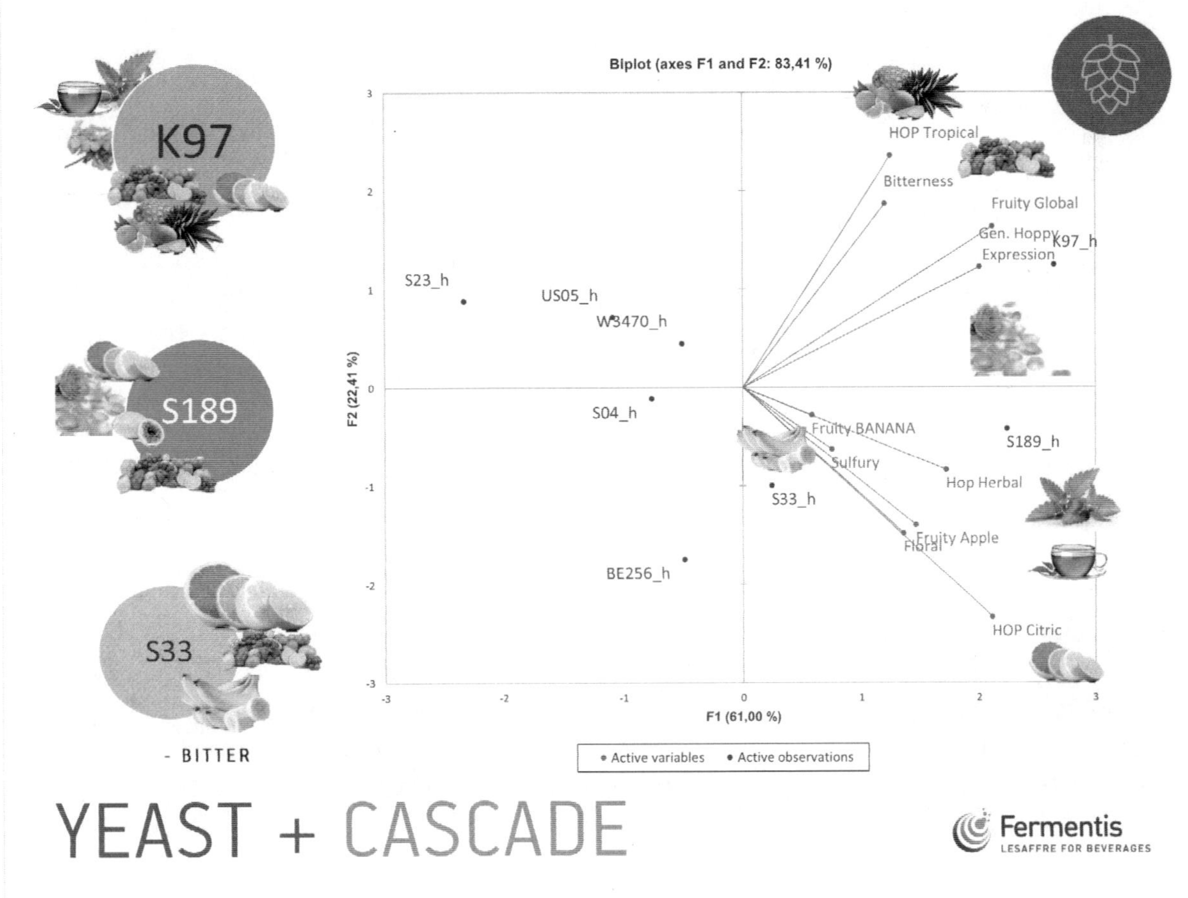
Biplot (axes F1 and F2: 83,41 %)
K97
S189
S33
- BITTER
HOP Tropical
Bitterness
Fruity Global
Gen. Hoppy Expression
K97_h
S23_h
US05_h
W3470_h
S04_h
Fruity BANANA
Sulfury
S189_h
Hop Herbal
S33_h
Fruity Apple
Floral
BE256_h
HOP Citric
F2 (22,41 %)
F1 (61,00 %)
Active variables
Active observations
YEAST + CASCADE
Fermentis
LESAFFRE FOR BEVERAGES

KEY CONCLUSION POINTS

- Scale up X yeast fermentation flavor prediction. Limitation over application of lab size trials
- Density has a higher impact (compared with fermentation temperature and pitching rate) in the formation of fermentation flavor components
- Fermentation flavors may interact and positively contribute to hop fruity / floral perceptions
- Flavor expression and yeasts can be more or less impacted by different hop regimes
- Fruity / floral as well phenolics fermentation flavors must not be underestimated when choosing a specific yeast strain – they can rather facilitates or masks important sensory perceptions

Fermentis
LESAFFRE FOR BEVERAGES

PAST-PRESENT-FUTURE

HUMAN FACTOR

Its all about feelings... and

INTERPRETATION!

Above all, the human factor must not be underestimated in flavor formation studies.

A flavor-rich or new (hopped?) beer style may not necessarily need to be along with new yeasts nor raw materials: but from the better understanding about raw materials and process / sensory interactions

Fermentis
LESAFFRE FOR BEVERAGES

ACKNOWLEDGMENTS
KU LEUVEN
GHENT TECHNOLOGY CAMPUS
LABORATORY OF ENZYME, FERMENTATION AND BREWING TECHNOLOGY (EFBT)
KU LEUVEN
LAB FOR SYSTEMS BIOLOGY, CENTER FOR MICROBIOLOGY
FACULDADE DE CIÊNCIAS E TECNOLOGIA
UNIVERSIDADE NOVA DE LISBOA
Departamento de Ciências da Vida
And
Technical Team of
Fermentis
LESAFFRE FOR BEVERAGES
Fermentis
LESAFFRE FOR BEVERAGES

Thank you!
Fermentis.com
THE OBVIOUS CHOICE FOR BEVERAGE FERMENTATION
Fermentis
LESAFFRE FOR BEVERAGES

Rapid Assessment of Yeast Phenotypic Diversity via the Application of Raman Spectroscopy

Yang He
State Key Laboratory of Biological Fermentation Engineering of Beer, Tsingtao Brewery Co. Ltd

Presentation Overview

Ramanome

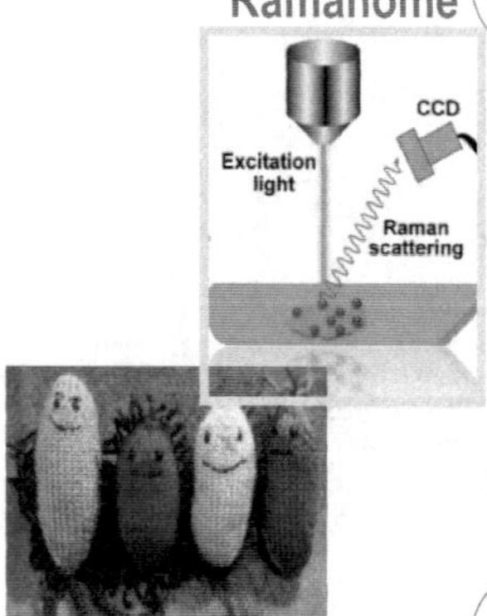

Phenotype

I Introduction

II What's the relationship between Ramanome & Phenome?

III How to select strains via Ramanome?

IV Application: Screen candidates with low acetaldehyde

V Summary

I II III IV V

Yeast heterogeneity

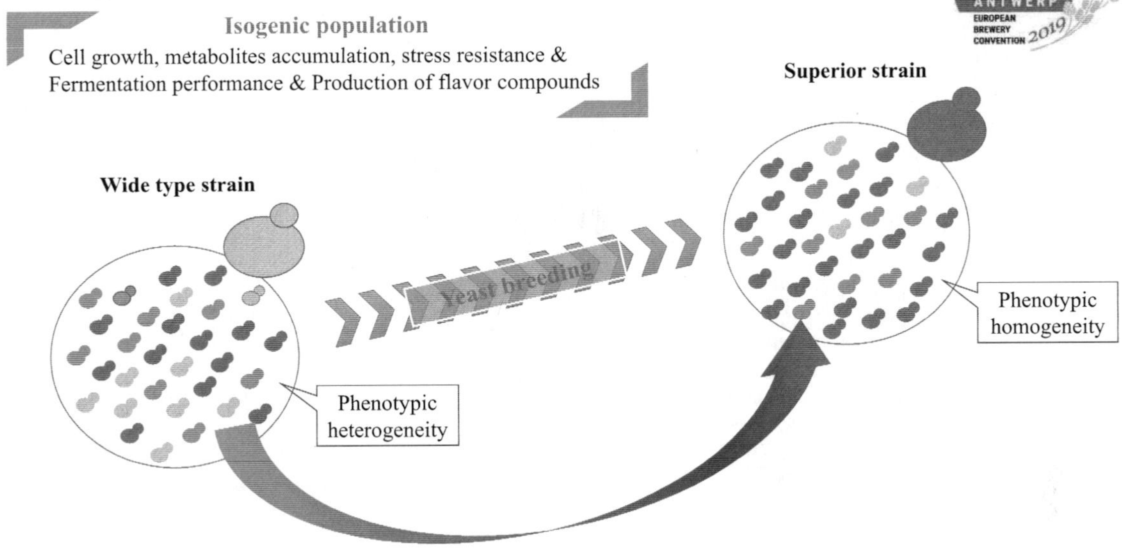

I II III IV V

Strategies for yeast phenotypic measurement

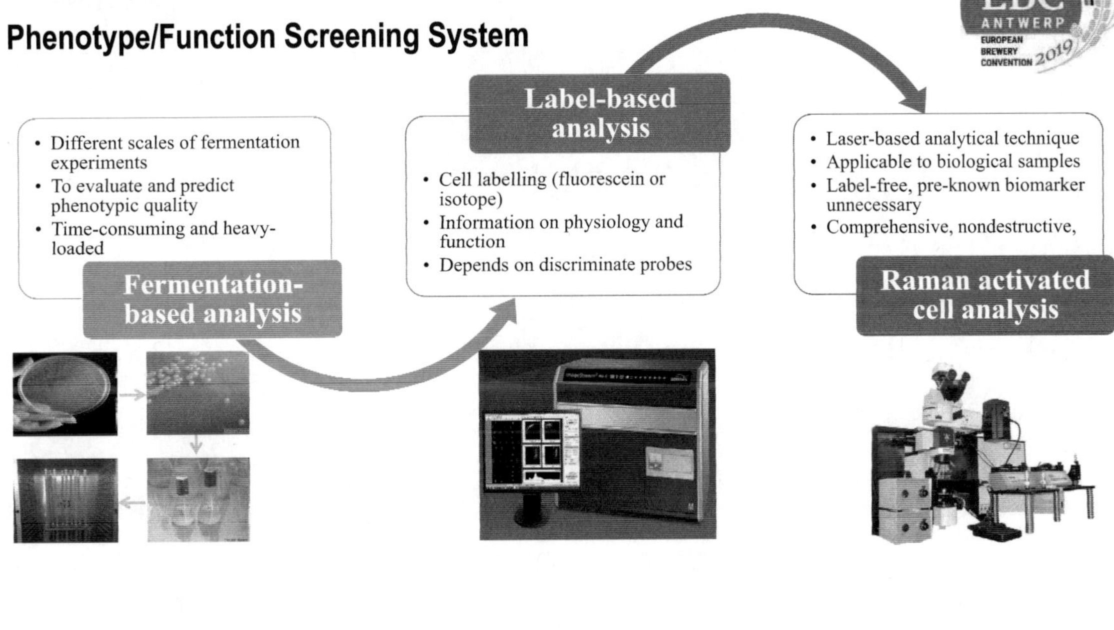

I II III IV V

Single cell Raman spectrum (SCRS)

- Specific molecules have unique Raman spectral signatures
- Provide rich and intrinsic information of the cell
- Ramanome: the collection of SCRS from an isogenic population
- Biochemical ID card

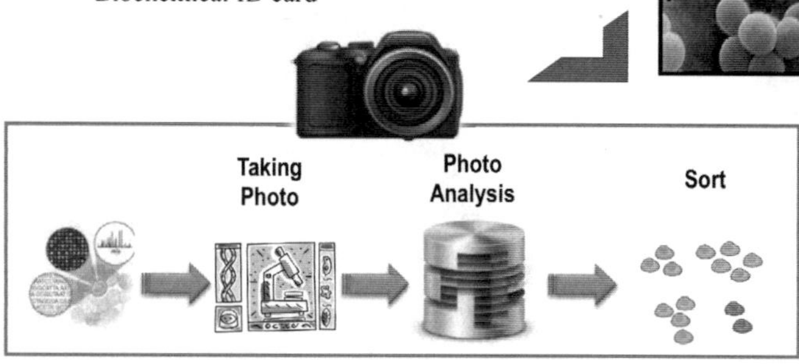

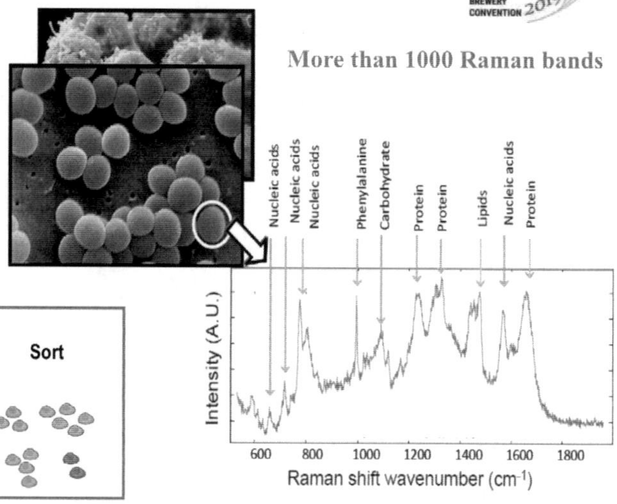

I II III IV V

Research outline

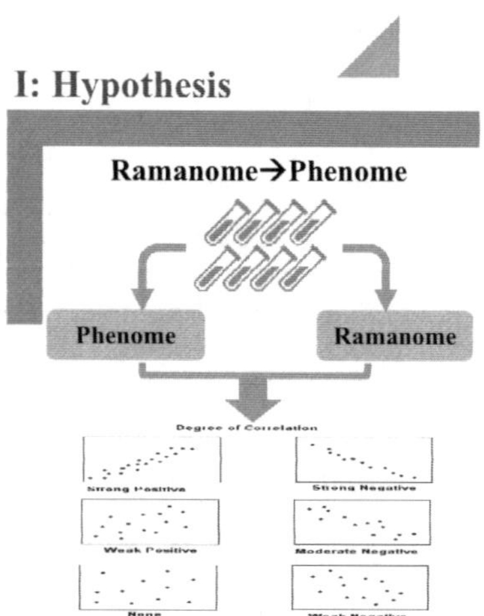

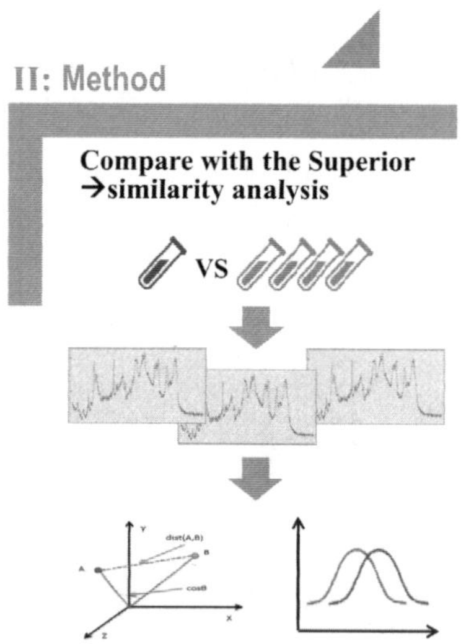

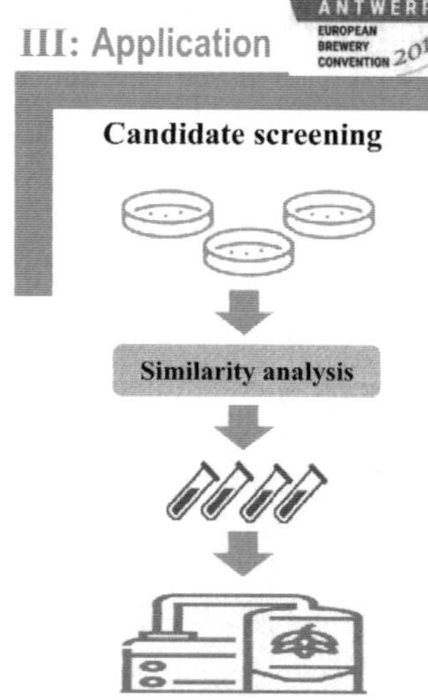

I II III IV V

What's the relationship between Ramanome & Phenome?

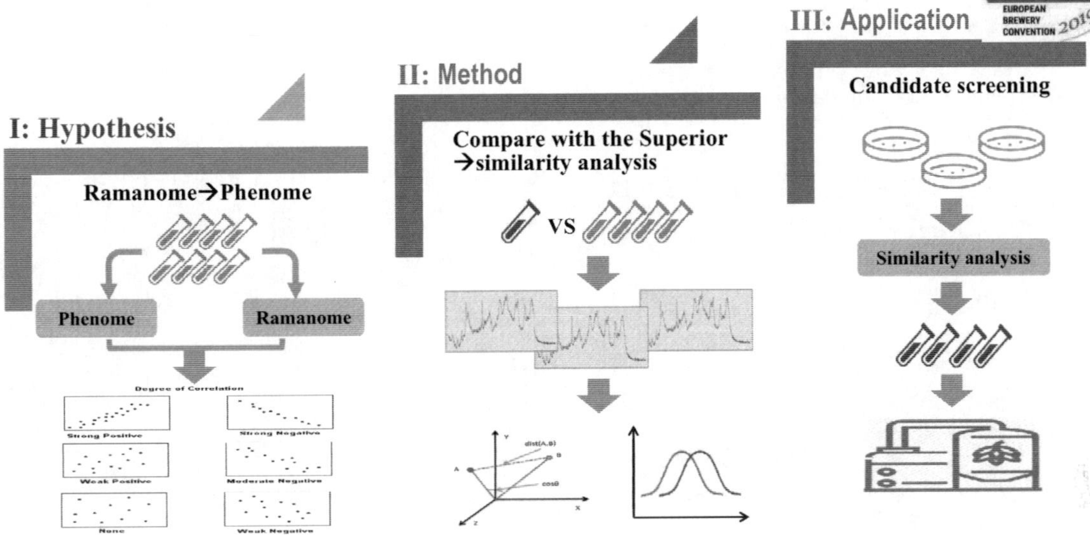

I II III IV V

Data acquisition

Different yeast strains
(1 control strain + 8 test strains)

Yeast characteristcs

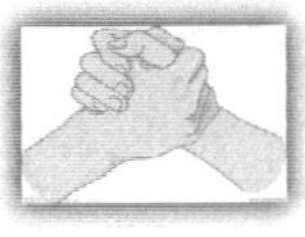

Phenome

- Brewing properties
  (cell growth, flocculation, sugar utilization, RDF, alcohol production)
- Flavour compounds formation
  (acetaldehyde, higher alcohols, esters)

Ramanome

- Signals of biochemical compounds
  (nucleic acid, protein, polysaccharide, carbohydrate, lipid)
- Relative abundances

I II III IV V

Phenotypic diversity of yeast cells

- Fractional phenome data : 15 representative phenotypes
- Strains exhibited **phenotypic diversity**
- PCA discrimination: distinct difference between the control strain and S7, followed by S5 and S8

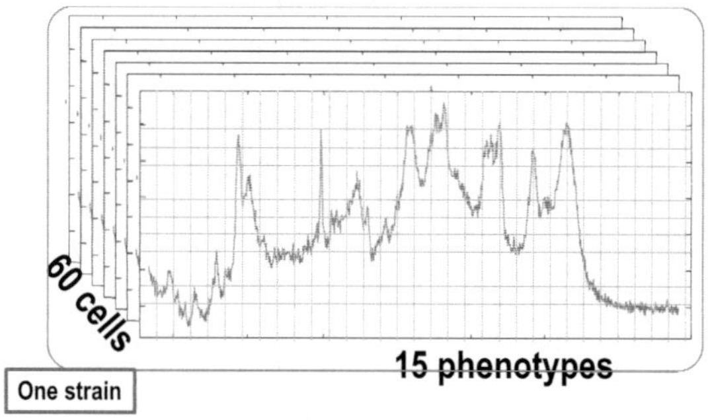

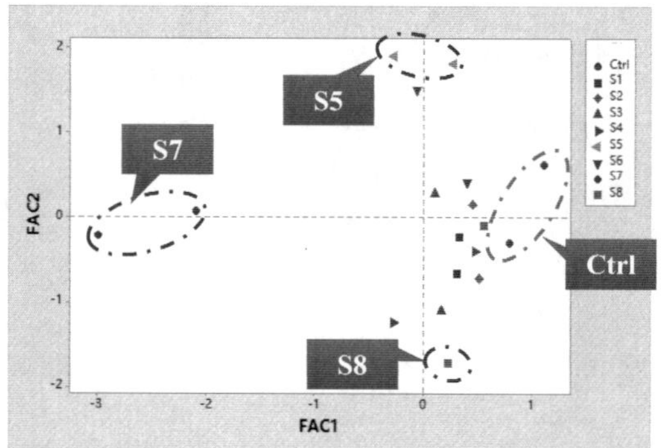

I II III IV V

Ramanome of yeast cells

- Raman profiling of different strains shared similar peaks
- Panels of test strains exist differences in intensity with certain bands
- Group I: S1-S4 & **Group II: S5, S7 and S8**
- Group II : great differences with the control

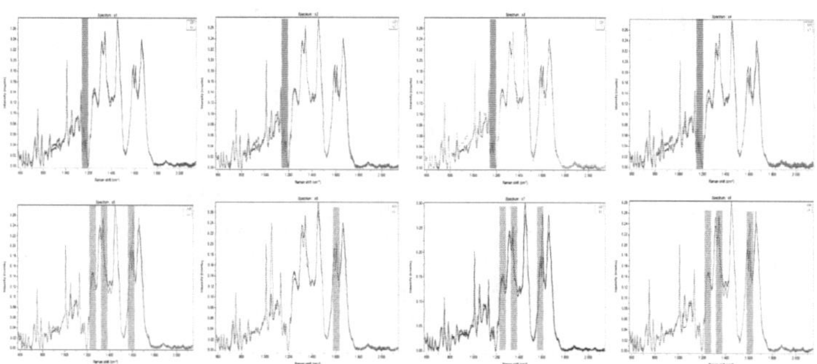

Hypothesis: The Ramanome of yeast correlates with their Phenome

Raman band (cm^{-1})	Assignment
749	Symmetric breathing of Try
1006	Symmetric ring breathing mode of Phe
1130	n(C-C) skeletal of acyl backbone in lipid
1155	Proteins (C–C/C–N str)
1258	Adenine, amide III
1312	Amide III, C–H def
1341	CH stretching of adenine
1450	Lipids and proteins (CH def)
1587	C=C olefinic stretch
1605	Phenylalanine
1650	Unsaturated lipids

I II III IV V

Correlation between Ramanome and Phenome

37th CONGRESS EBC ANTWERP EUROPEAN BREWERY CONVENTION 2019

- Ramanome & phenome data : multivariate, high-dimensional
- Mantel test : comparing the Ramanome and phenome matrices of data
- **A linear regression relationship** was found with Pearson correlation coefficient was 0.797 ($p < 0.01$)

Ramanome was strongly positive correlated with these representative phenotypes

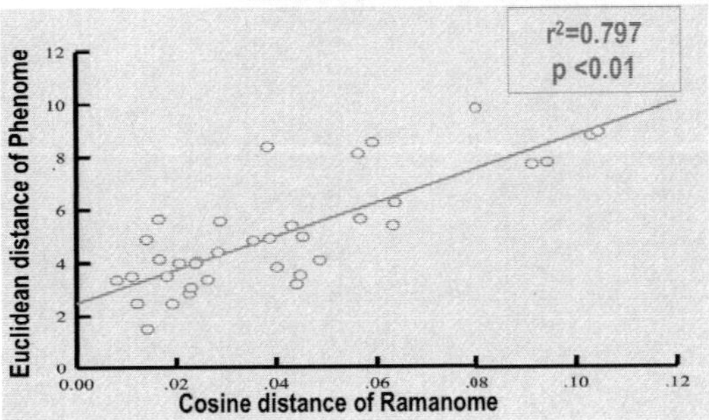

Ramanome could be used as a global evaluation for phenome and having the potential to resolve the unmeasurable or unknown phenotypes

I II III IV V

How to select strains via Ramanome?

37th CONGRESS EBC ANTWERP EUROPEAN BREWERY CONVENTION 2019

I: Hypothesis

Ramanome→Phenome

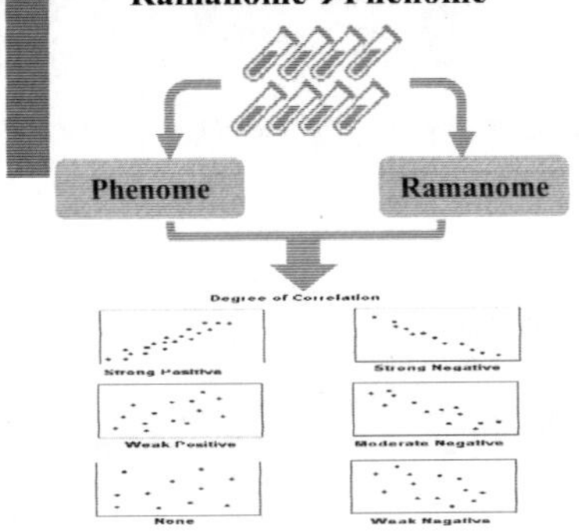

II: Method

Compare with the Superior →similarity analysis

III: Application

Candidate screening

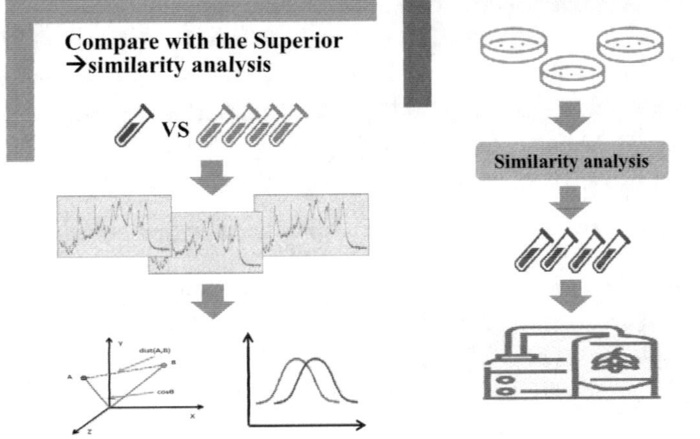

I II III IV V

Similarity analysis of Ramanome

➢ PLS-DA & PDF

- To highlight the cluster pattern between groups
- PLS-DA: the control and sample groups were visibly separated
- PDF: significant dissimilarities were detected between strain S5, S7, S8 and the control
- **Difficult to quantify the difference**

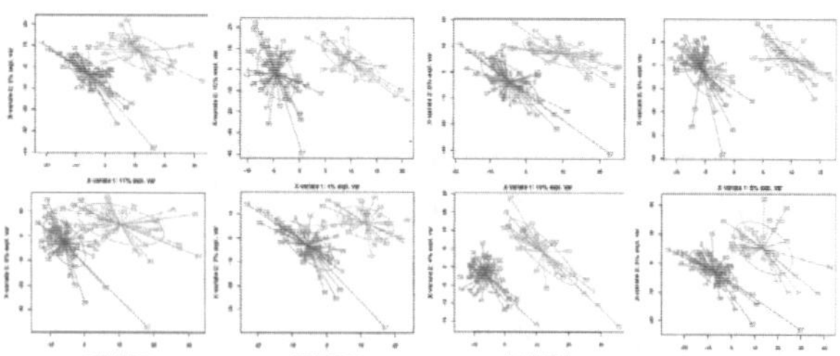
PLS-DA

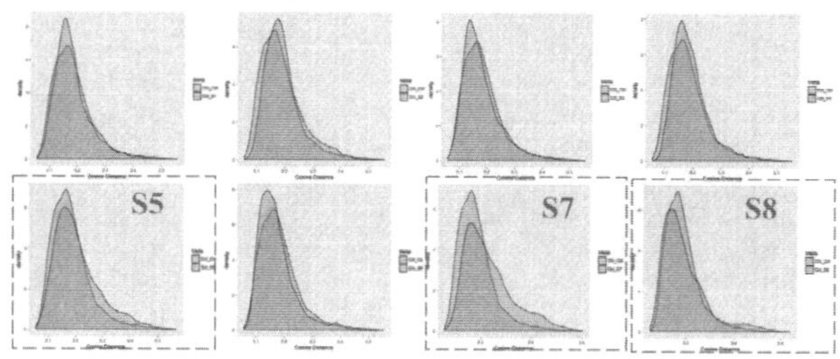

PDF

I II III IV V

Similarity analysis of Ramanome

➢ ANOSIM

- To quantify the degree of similarity between groups
- R values gives a picture of distances between strains
  near "0" similar; close to "1" strong differences
- Significant dissimilarities were detected between S5, S7, S8 and the control strain than others

Compare with Ctrl	S1	S2	S3	S4	S5	S6	S7	S8
R	0.023	0.06	0.022	0.006	0.265	0.031	0.398	0.158
P	0.247	0.083	0.237	0.368	0.001	0.171	0.001	0.002

ANOSIM

A quantifiable indicator for Ramanome comparison

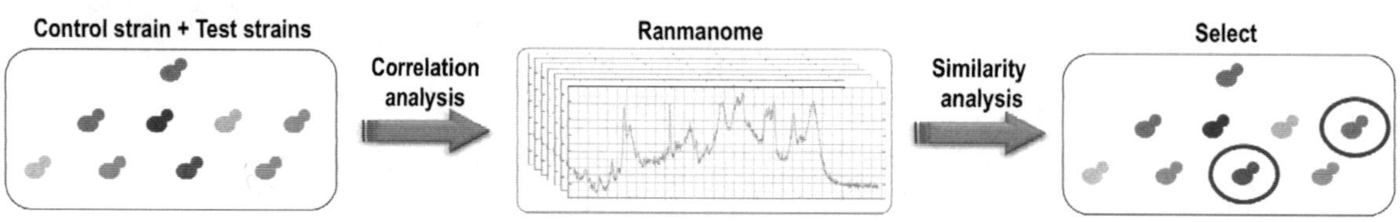

I II III IV V

Application: Screen candidates with low acetaldehyde

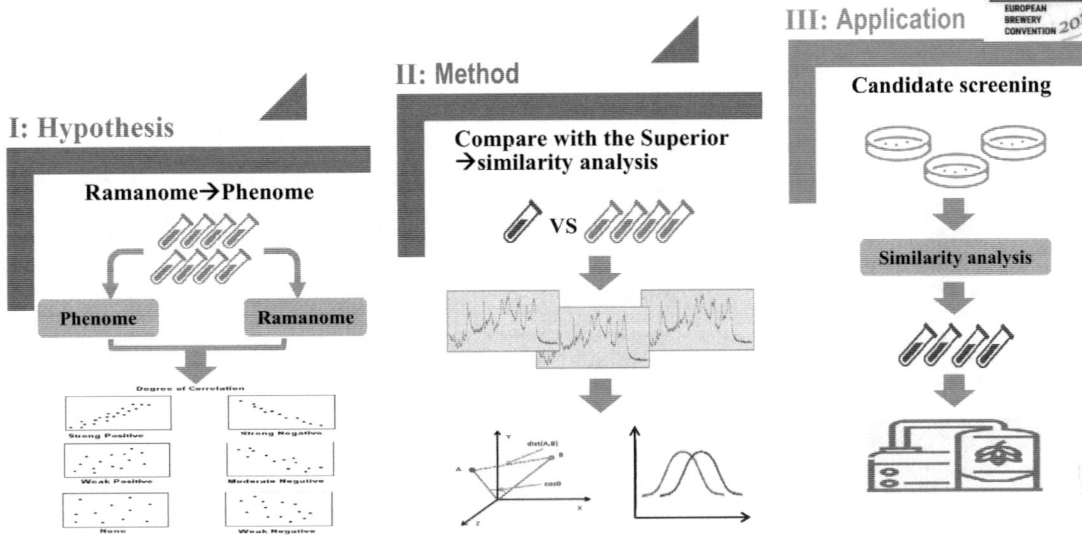

I II III IV V

Yeast acetaldehyde metabolism

Acetaldehyde: Green Beer Flavor

- Natural by-product of fermentation
- Exceeded amount: pungent aroma
  Appropriate amount: green apple aroma
- Formation occurs during the first three days of fermentation
- Peaks during primary fermentation, declines during maturation
- Depends also on yeast strain

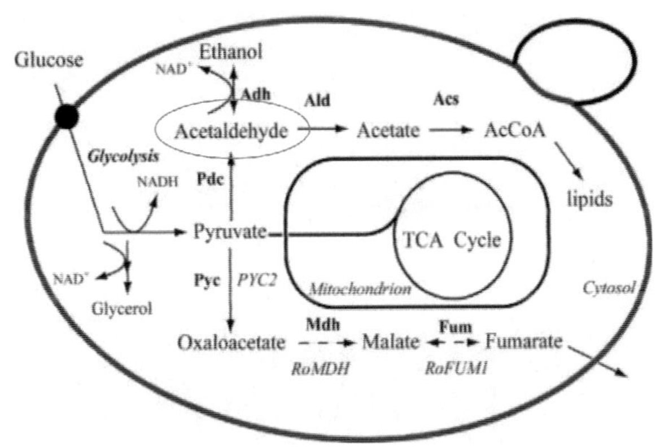

Pathways for acetaldehyde metabolism in *S. cerevisiae*

I II III IV V

Acetaldehyde metabolism in brewing yeast

Acetaldehyde production

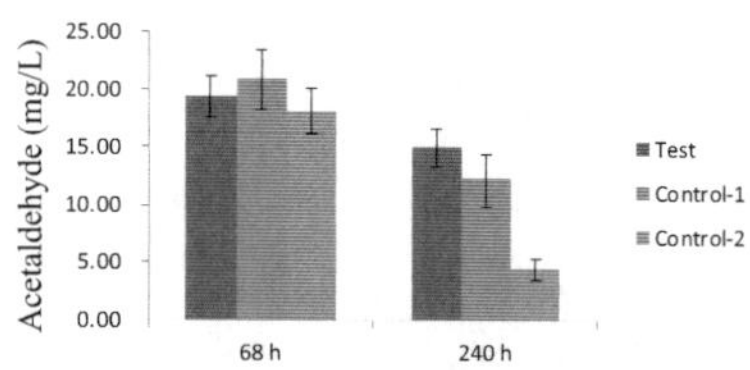

Concentration of acetaldehyde during fermentation with different strains

Final : Test > Control-1 > Control-2
Peak: Control-1 >Test > Control-2

Yeast growth

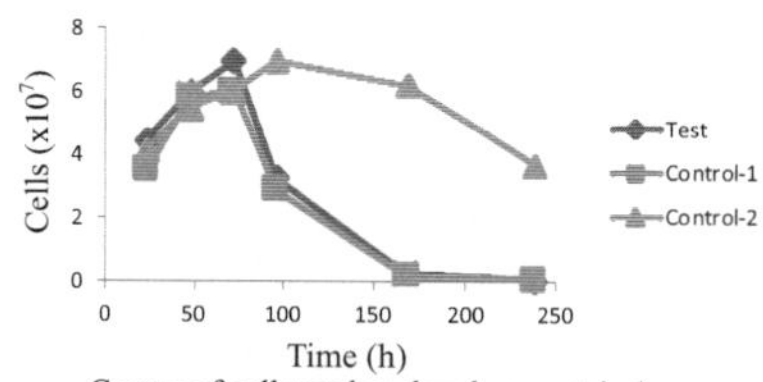

Course of cell number development during fermentation with different strains

Control-2 > Test & Control -1

Acetaldehyde reducing ability

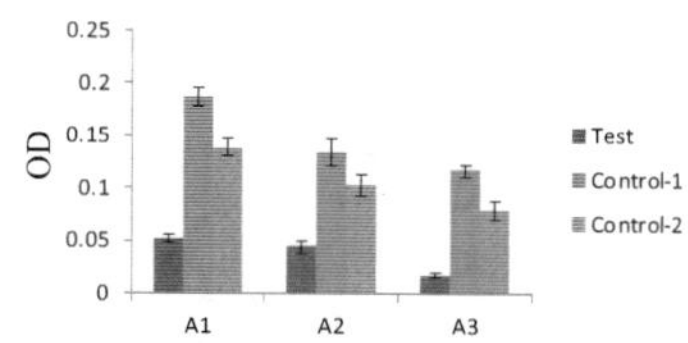

Yeast growth with acetaldehyde as sole carbon source in different levels of acetaldehyde addition (A1-10mg/L; A2-20mg/L; A3:30mg/L)

Control-1> Control-2>Test

- ✓ Test: High concentration of final acetaldehyde production
- ✓ Control-1: Low acetaldehyde production with high acetaldehyde reducing ability
- ✓ Control-2: Low acetaldehyde production with high suspending yeast during maturation

I II III IV V

Strategies

Screening of yeast cells with low acetaldehyde production

- Yeast with enhanced suspension potential (Control-2)
  - Prediction the density of yeast flocculin based on SCRS
- Yeast with strong acetaldehyde reducing ability (Control-1)
  - Prediction of yeast similarity based on SCRS

Methods

ANOSIM

I II III **IV** V

Prediction the density of yeast flocculin

Flocculins on the yeast cell surface
- Fluorescent probe (FITC-Avidin) measured by flow cytometer
- 20,000 cells were analyzed/sample

Different strains

Single cell Raman spectroscopy
- Frequency-doubled 532 nm Nd:YAG laser
- Exposure time 10 s
- 60 cells were analyzed/sample

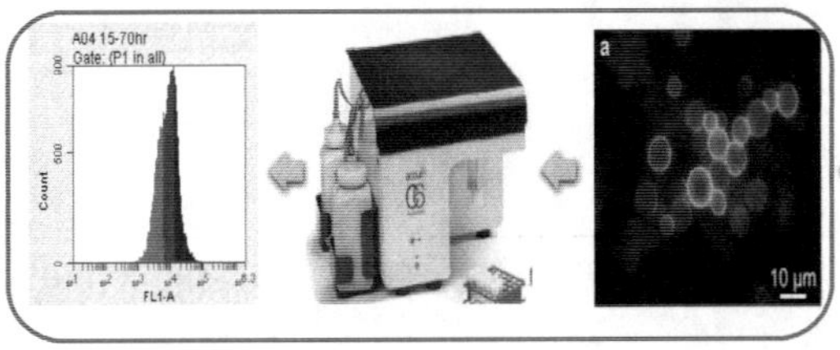

ABCD

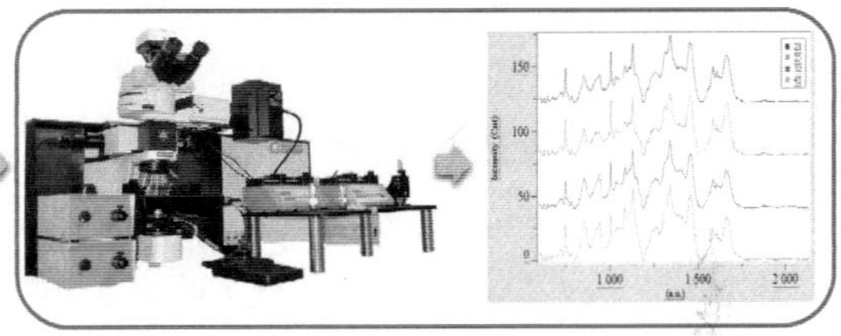

I II III **IV** V

Prediction model of yeast flocculin densities

- Flocculins on the yeast cell surface were different with strains
- Random forest regression: prediction model of the yeast flocculin densities with R-square value at 0.87

Strains	Flocculin			
A	60102	60737	59292	55660
B	49582	44988	49319	46841
C	55962	58034	52919	54328
D	40580	41125	41163	41349

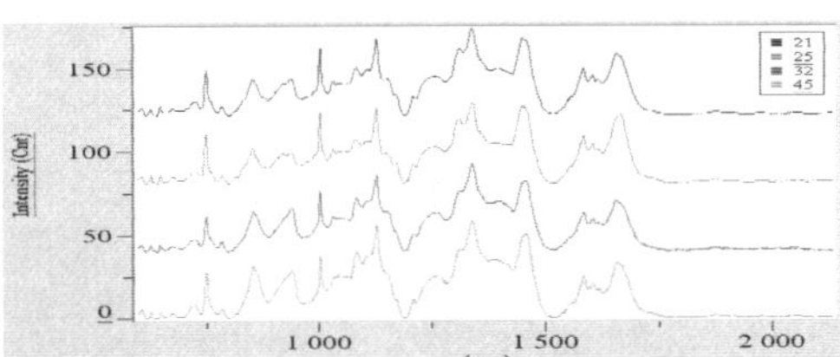

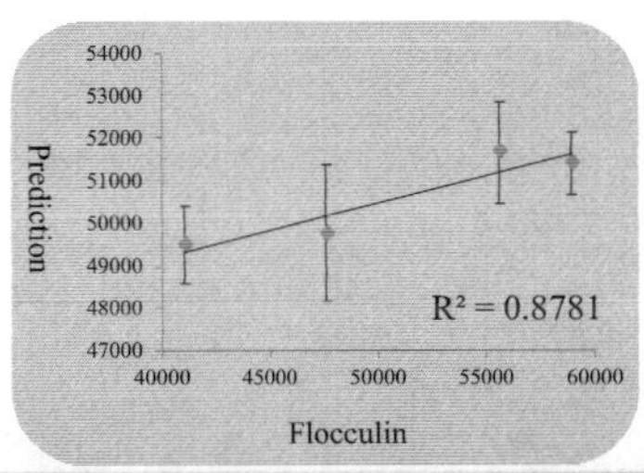

Ramanome could be used to predict yeast surface flocculins

I II III **IV** V

Screening of cells based on flocculin and similarity prediction

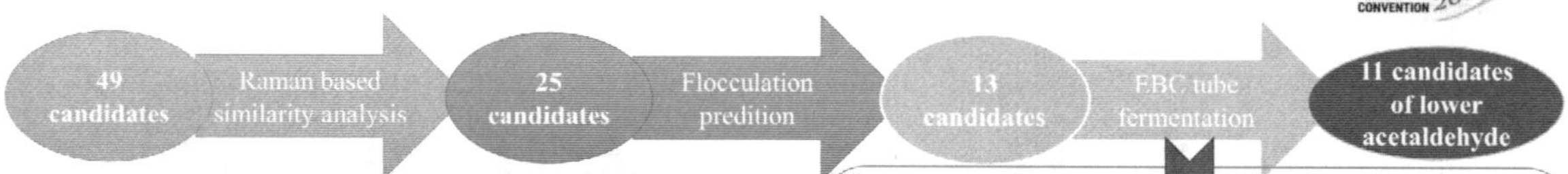

vs Control-1	1	2	3	4	5	6	7
R	0.13	0.125	0.006	0.09	0.116	0.102	0.072
F	0.99	0.96	0.99	0.98	0.98	0.98	0.99
vs Control-1	8	9	10	11	12	13	14
R	0.05	0.087	0.07	0.068	0.108	0.068	0.154
F	0.97	0.96	0.98	0.98	0.98	0.98	0.98
vs Control-1	15			18	19		21
R	0.054	0.033	0.011	0.051	0.094	0.047	0.115
F	1.00	1.01	1.00	1.00	1.01	1.00	1.00
vs Control-1	22	23	24	25	26	27	28
R	0.084	0.044	0.04	0.021	0.002	0.019	0.059
F	0.98	0.97	0.98	0.98	0.98	0.97	1.00
vs Control-1	29	30		32		34	35
R	0.091	0.169	0.041	0.076	0.039	0.152	0.16
F	1.06	1.06	1.03	1.00	1.05	1.01	1.03
vs Control-1		37	38		40	41	
R	0.064	0.168	0.126	0.09	0.08	0.125	0.083
F	1.00	1.00	1.01	1.00	0.98	[illegible]	[illegible]
vs Control-1				46	47	[illegible]	[illegible]
R	0.02	0.06	0.02	0.01	0.27	[illegible]	[illegible]
F	1.00	1.00	1.01	0.97	0.98	[illegible]	[illegible]

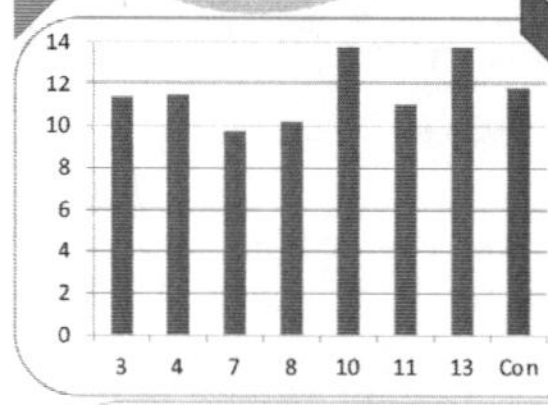

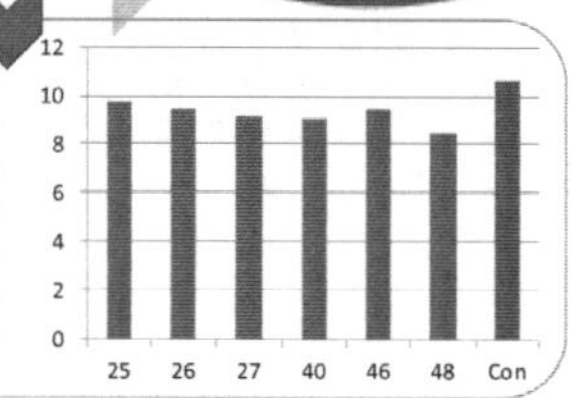

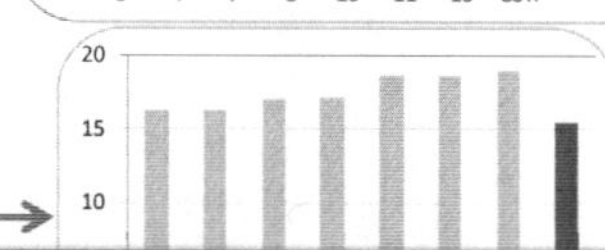

Raman spectroscopy could be used as a rapid assessment for screening of yeast cells with low acetaldehyde production

I II III IV **V**

Summary

➢ **Omics**

Genomics

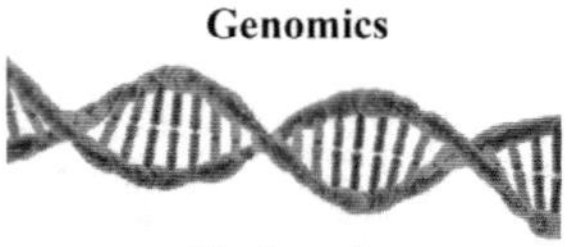

Proteomics

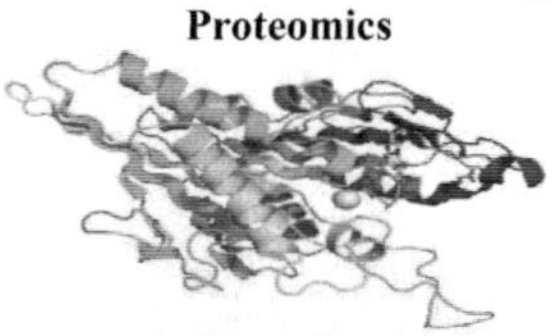

Metabolomics

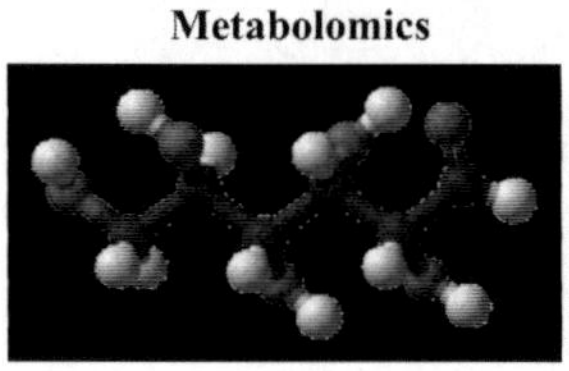

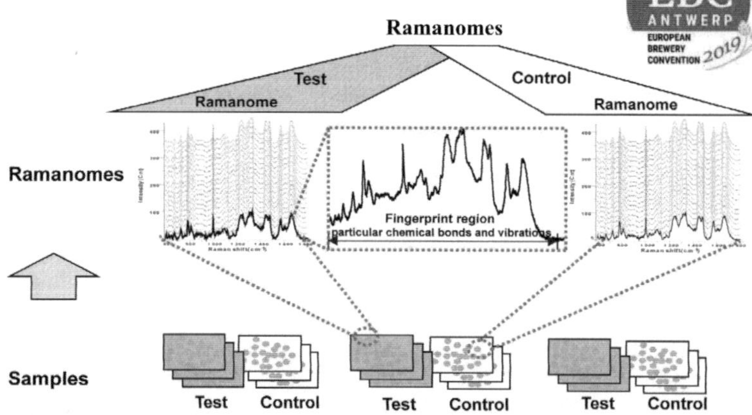

- Genomics, Proteomics & Metabolomics: disruptive nature precludes rapid detection
  Limited by the inability to amplify single-cell proteome and metabolome
- Ramanome: provides a multiplex and comprehensive metabolic profile at the single-cell level
  Label-free, non-disruptive, simple yet sensitive

I II III IV V

Summary

- There exist a linear correspondence between Ramanome and Phenome
- Ramanome could be used to as a global evaluation for predicting yeast Phenome & predict yeast similarity & yeast surface flocculins
- A rapid assessment for screening of yeast cells with low acetaldehyde production has been established through the application of Raman spectroscopy

Raman spectra can be directly linked to cellular omics information

Huge thanks

QIBEBT Single Cell Center One cell, one world

- Methodology and analytical software of Raman spectroscopy

TSINGTAO 青岛啤酒

- Support

THANK YOU FOR YOUR KIND ATTENTION!

Ongoing

➢ **Acetaldyhete utilization rate**

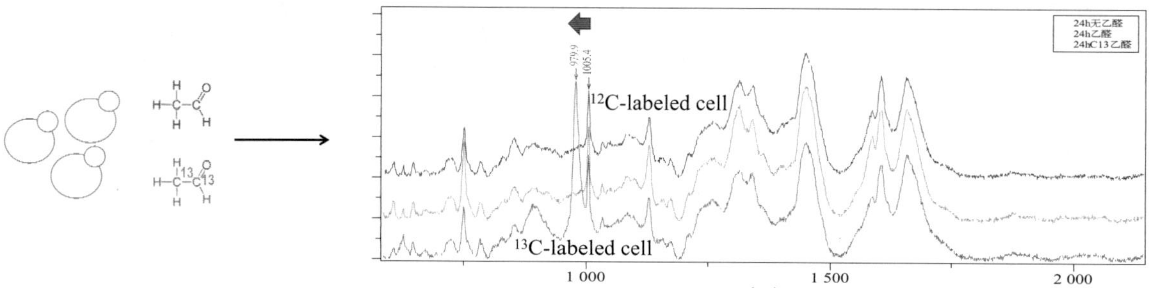

Ongoing

➢ **Deuterated water utilization rate**

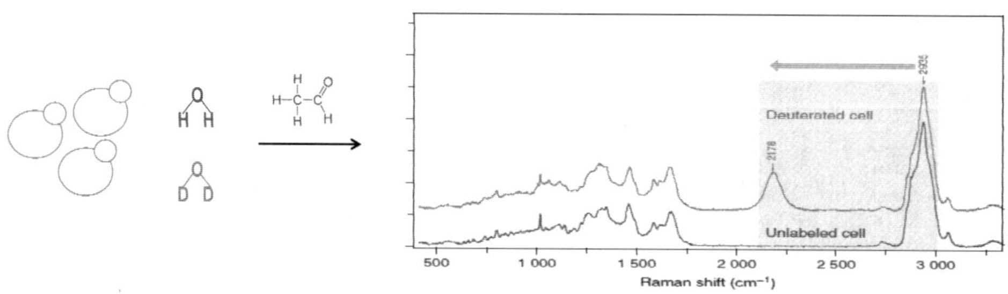

Ongoing

➤ **Deuterated water utilization rate**

Dynamic change of C-D peak

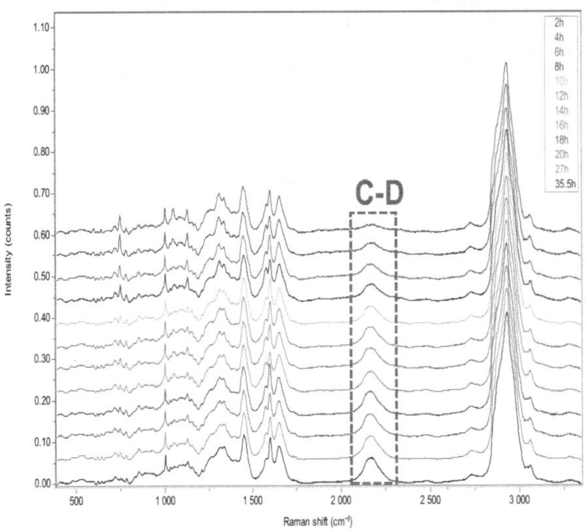

Changes in yeast metabolic activity

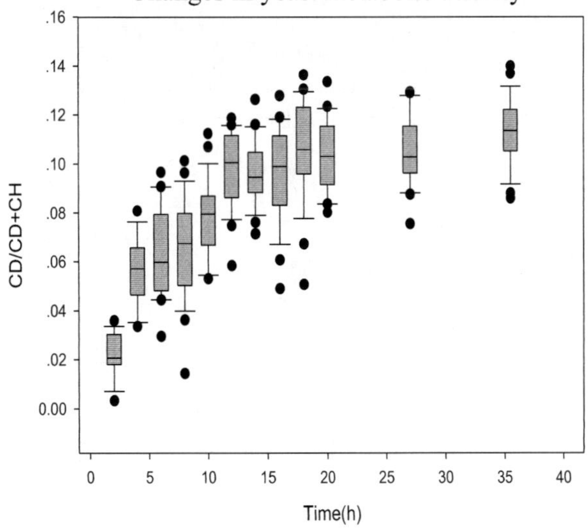

European Joint Doctorate Food Science

European Union

HORIZON 2020

The copper binding affinity of wort during the brewing process

Marcus Pagenstecher

University of Copenhagen
University of Agriculture in Krakow

1

EJD Food Sci

Introduction – Flavor Stability

- Beer flavor stability is a result of the whole brewing process

2

EJD Food Sci

Spontaneous chemical changes of beer quality

Oxidative reactions

- The presence of oxygen
- Radical reactions
- Formation of off-flavors
- Loss of bitterness
- Haze formation

Thermal reactions

- Maillard chemistry
- Formation of flavor active aldehydes

3

EJD Food Sci

Introduction – Beer oxidation

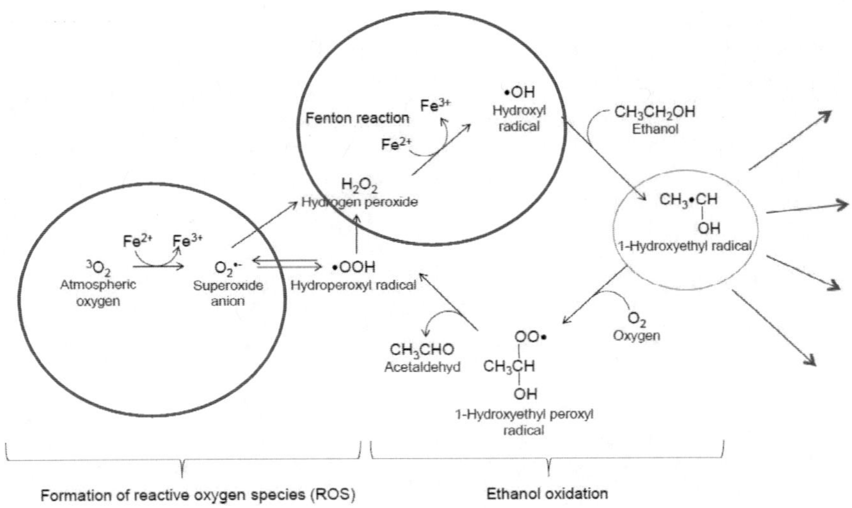

Huvaere & Andersen **2009** In: Beer in Health and Disease Prevention, Preedy, V.R. (Ed), Burlington, Academic Press.

4

EJD Food Sci

Background – Metal sources

Fe & Cu uptake and removal during the early brewing stages

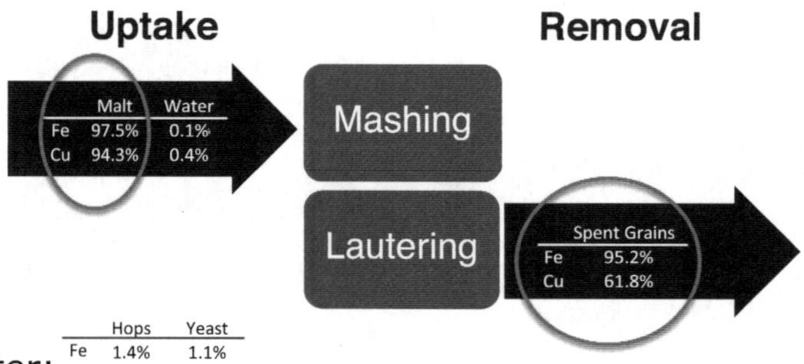

Later:

	Hops	Yeast
Fe	1.4%	1.1%
Cu	5.0%	0.4%

Wietstock, Kunz, Waterkamp & Methner **2015**: *J. Am. Soc. Brew. Chem.* 73(2): 179–184.

5

EJD Food Sci

Background - Roasting

Effect of malt roasting on Fe & Cu levels in sweet wort

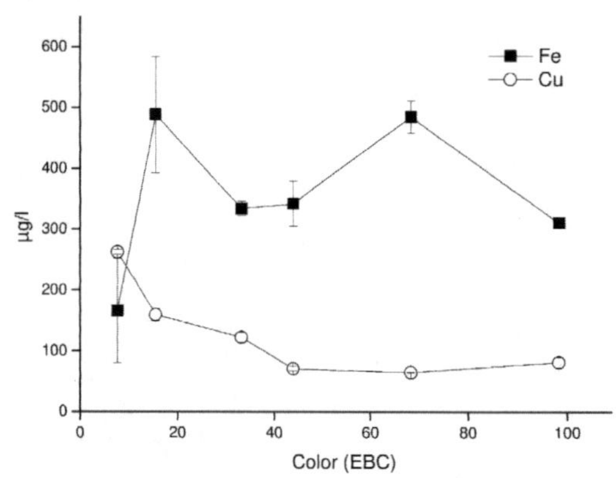

Figure 2. Fe and Cu content in sweet worts 1 to 6 plotted against color (EBC). Results are shown as mean values, and standard deviations are given as error bars ($n = 2$).

Hoff, Lund, Petersen, Jespersen & Andersen **2012**: *J. Agric. Food Chem.* 60: 5652–5659

6

EJD Food Sci

Background - Copper levelling during mashing

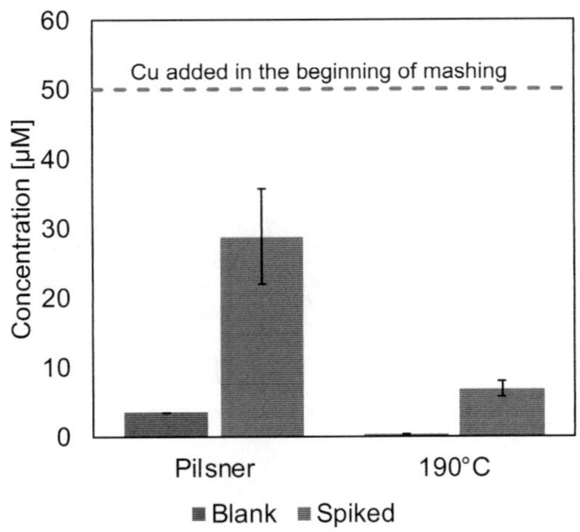

Cu (wort) ⇌ Cu (spent grains)

Malt	Spiked with 50 µM of	Wort pH	Wort Extract [°Plato]	Color [EBC]
Pilsner	-	5.8 ± 0.3	8.2 ± 0.1	6.9 ± 2.6
Pilsner	Cu	5.8 ± 0.1	8.3 ± 0.4	7.4 ± 1.3
Roasted at 190°C	-	4.8 ± 0.3	2.5 ± 0.2	96.6 ± 23.4
Roasted at 190°C	Cu	4.5 ± 0.0	3.1 ± 0.7	78.0 ± 1.9

Manuscript in progress 7

EJD Food Sci

Project Aim:

- What determines how much copper ends up in beer?
  - Roasting dependent copper affinity
  - pH effect
  - Copper affinity during the brewing process
  - Oxidative stability

KØBENHAVNS UNIVERSITET

UNIVERSITY OF AGRICULTURE IN KRAKOW

8

EJD Food Sci

Roasting dependent copper affinity

Congress Mashing of malts made from the same base Pilsner malt

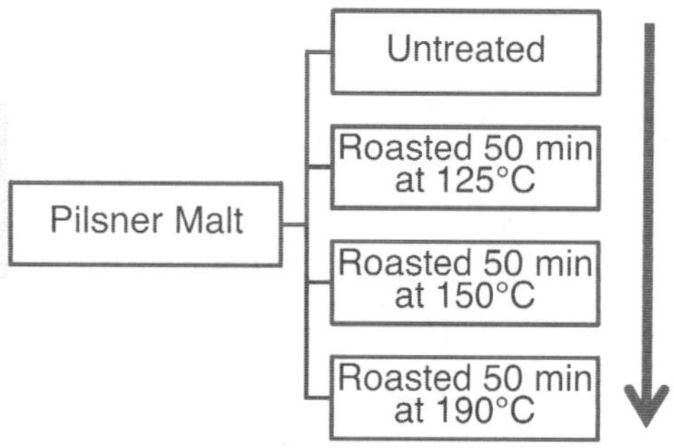

- Decreasing pH
- Increasing Color
- Decreasing total copper

Sample	Wort pH	Color [EBC]	Total Cu [μM]
Pilsner	5.84 ± 0.01	13.2 ± 0.2	2.83 ± 0.05
Roasted at 125°C	5.62 ± 0.01	19.4 ± 0.8	1.78 ± 0.03
Roasted at 150°C	5.39 ± 0.01	53.8 ± 0.5	0.89 ± 0.04
Roasted at 190°C	4.83 ± 0.01	89.2 ± 1.6	0.91 ± 0.01

9

EJD Food Sci

Copper(II) binding affinity

Measured by Cu(II) ion selective electrode in connection with a reference electrode.

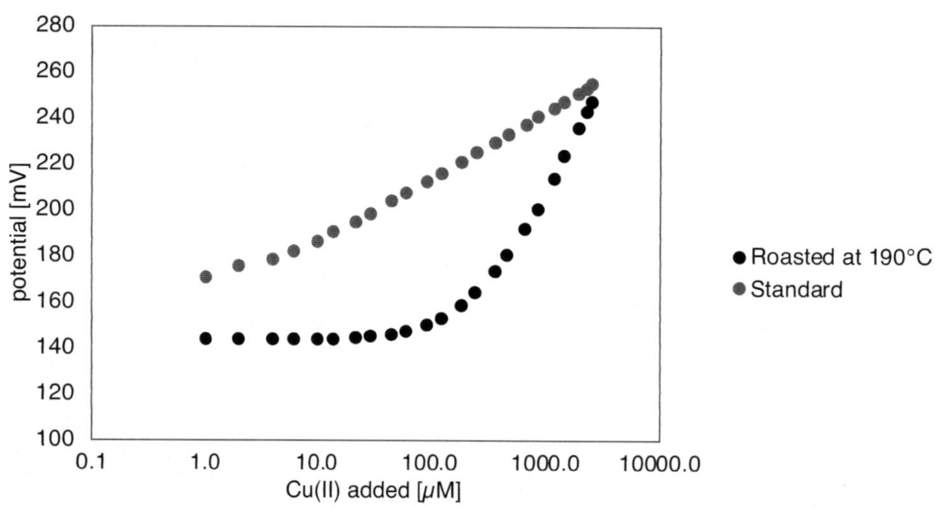

10

EJD Food Sci

Cu (II) binding affinity

Roasting dependent copper(II) affinity

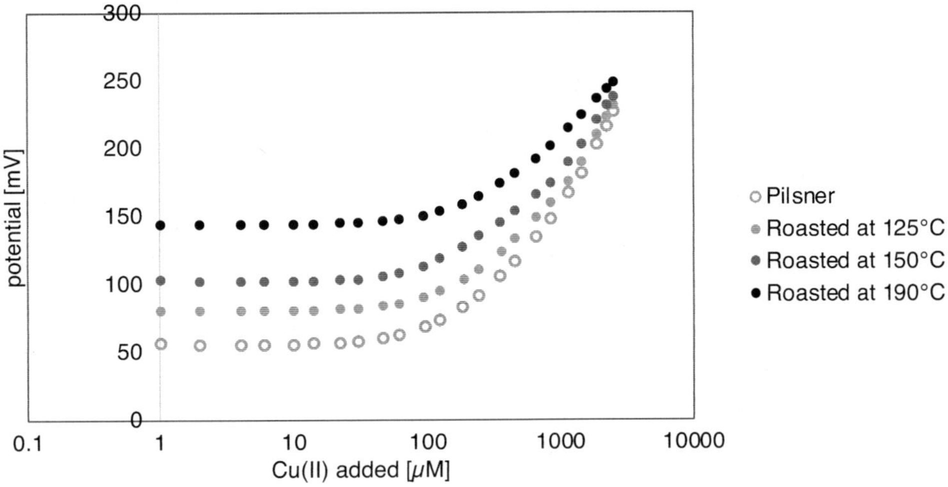

11

EJD Food Sci

Copper(II) binding affinity

Relative amount of added copper bound by different sweet worts.

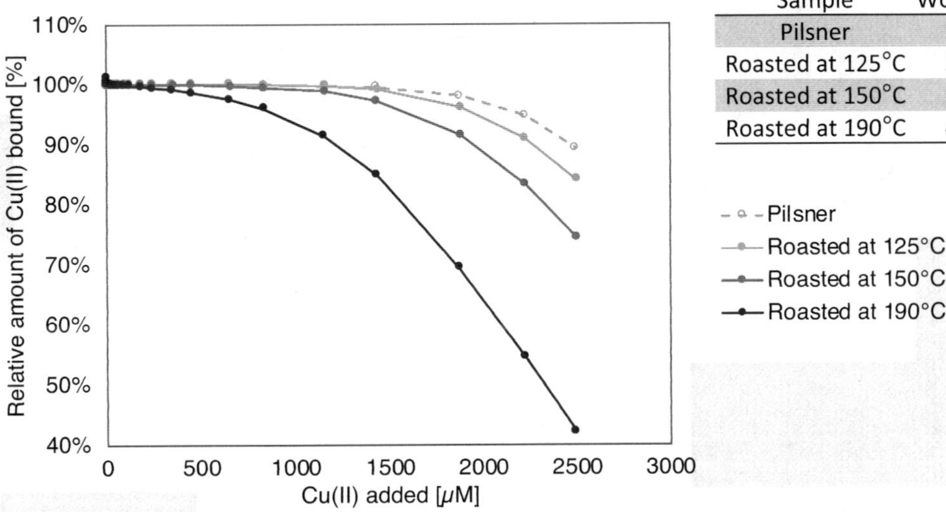

Sample	Wort pH
Pilsner	5.8
Roasted at 125°C	5.6
Roasted at 150°C	5.4
Roasted at 190°C	4.8

12

EJD Food Sci

Copper(II) binding affinity

Relative amount of added copper bound by different sweet worts – pH dependence.

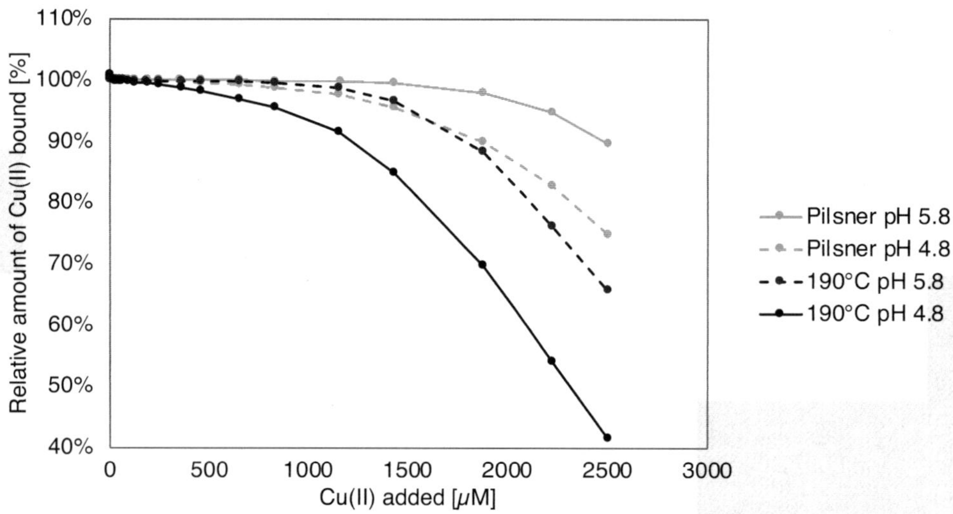

➢ pH has a strong influence, but wort made from roasted mal still has a lower copper(II) binding affinity

13

EJD Food Sci

Copper(II) binding affinity

Relative amount of added copper bound by different sweet worts.

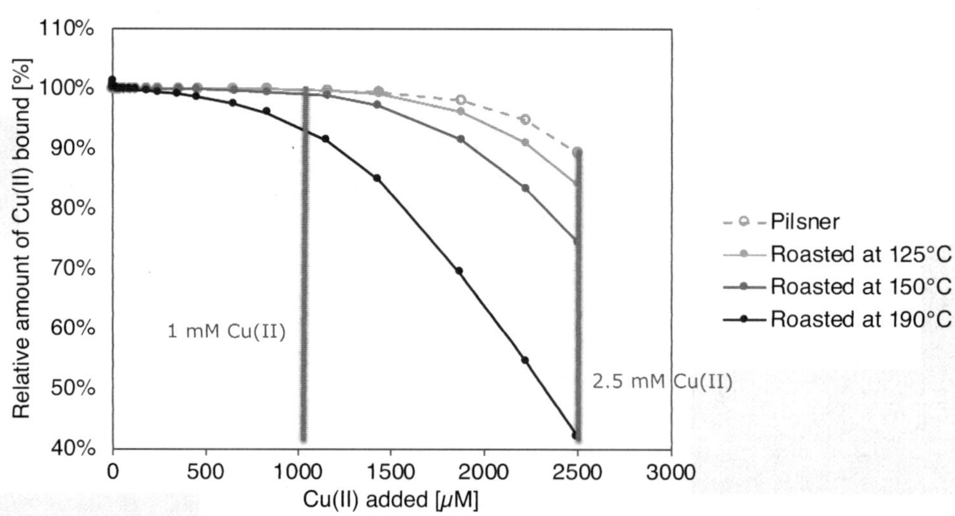

14

EJD Food Sci

Structual Analysis by ESR Spectroscopy

ESR spectra of 1 mM Copper(II)(aq.) and Pilsner wort with addition of 1 mM Copper(II).

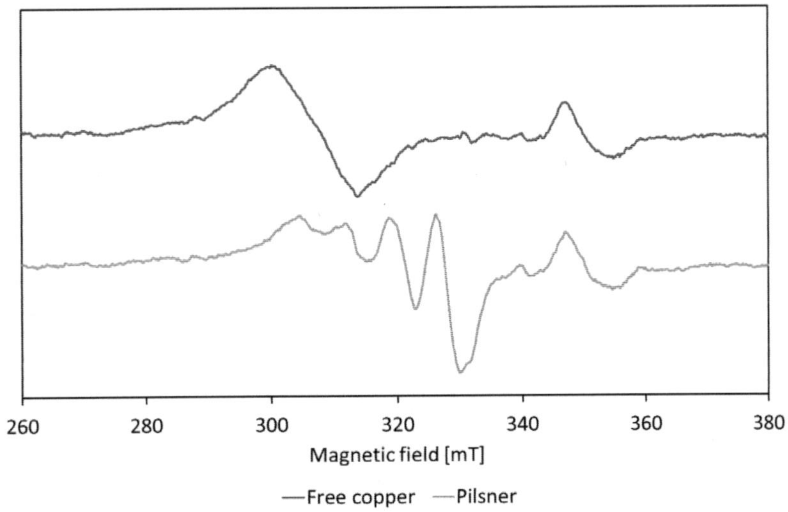

15

EJD Food Sci

Structual Analysis by ESR Spectroscopy

ESR spectra of 1 mM Cu(II) in Pilsner wort, Glycine and a simple dipeptide.

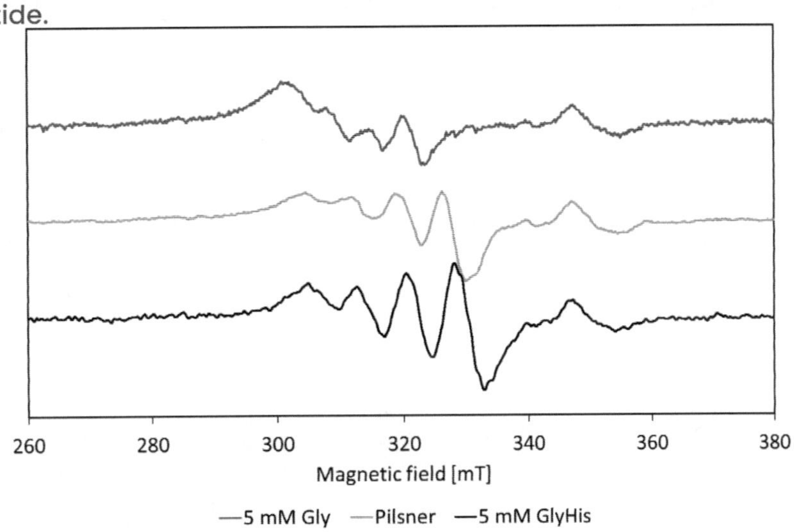

- Similar structure, but different binding constants
- Tested for different amino acids/dipeptides

16

EJD Food Sci

Structual Analysis by ESR Spectroscopy

ESR spectra of roasted malt sweet wort with addition of 1 mM Cu(II) and 1 mM GlyHis.

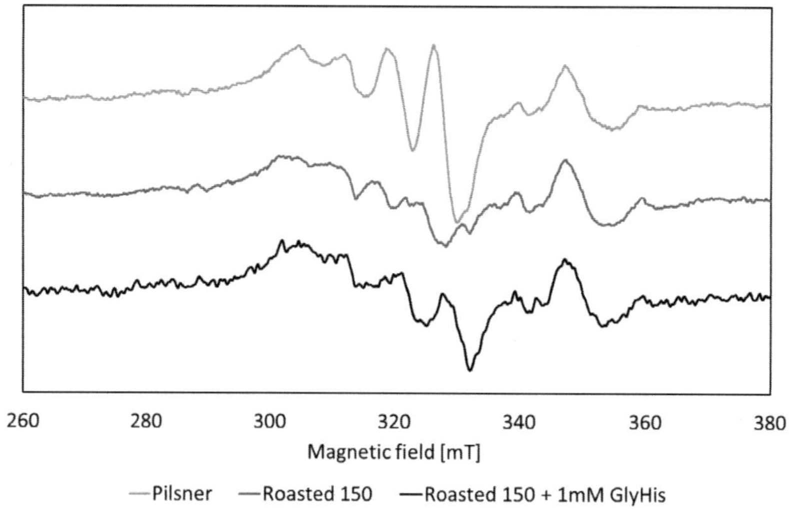

- Weaker affinity in sweet wort made from roasted malt
- Addition of dipeptide males it more similar to Pilsner

17

EJD Food Sci

Free amino nitrogen (FAN)

FAN levels in sweet worts vs. total copper

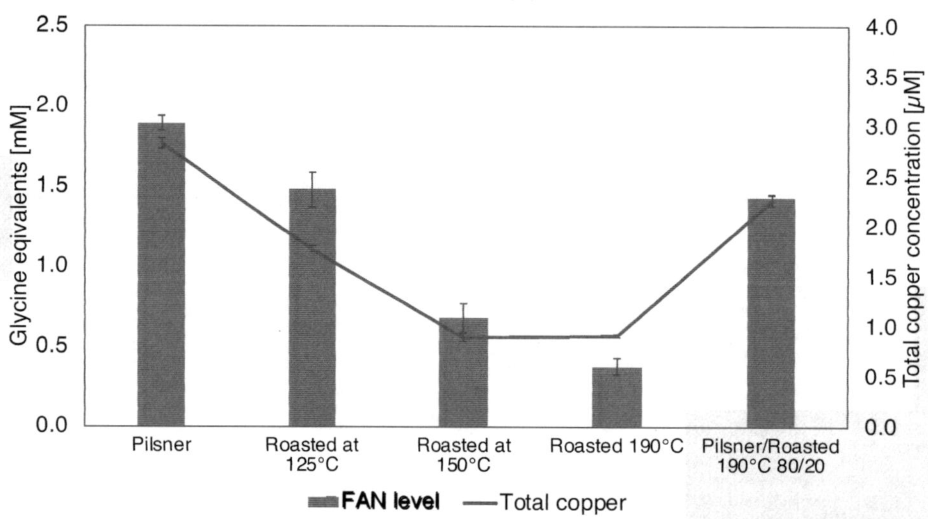

18

EJD Food Sci

Copper(II) binding affinity during the brewing process

Spring bock, with a around 40 IBU bitterness.

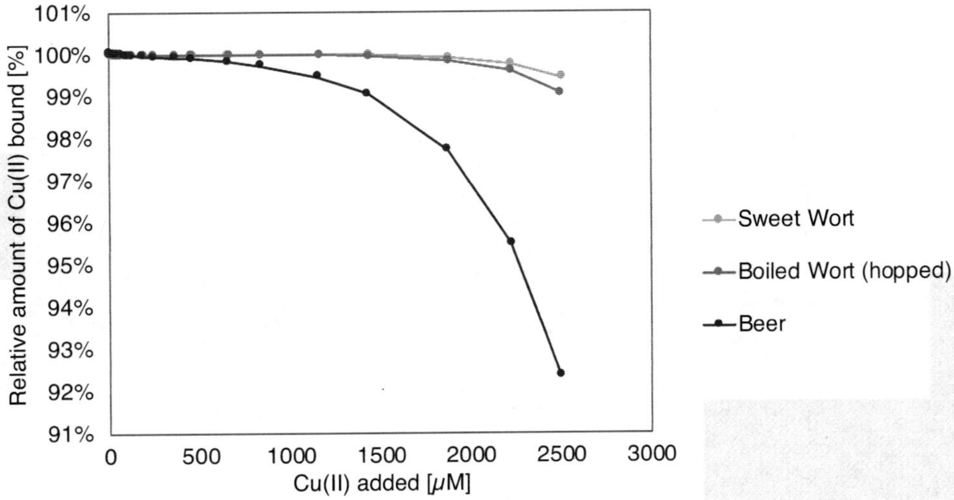

19

EJD Food Sci

Structual Analysis by ESR Spectroscopy

ESR spectra during the brewing process with 1 mM Cu(II) added.

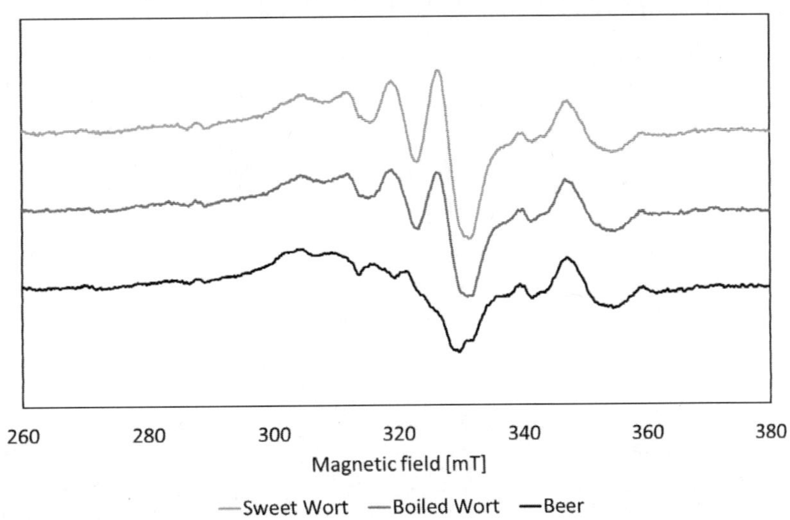

- No difference between sweet wort and boiled wort (hopped)
- Lower binding affinity in beer

20

EJD Food Sci

Free amino nitrogen (FAN)

FAN levels during the brewing process vs total copper

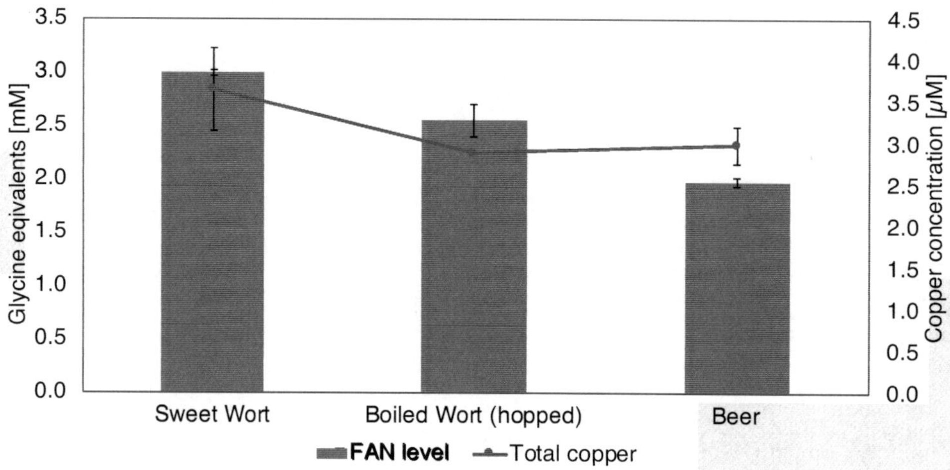

21

EJD Food Sci

Oxidative Stability

Forced ageing of sweet words with the addition of copper(II).

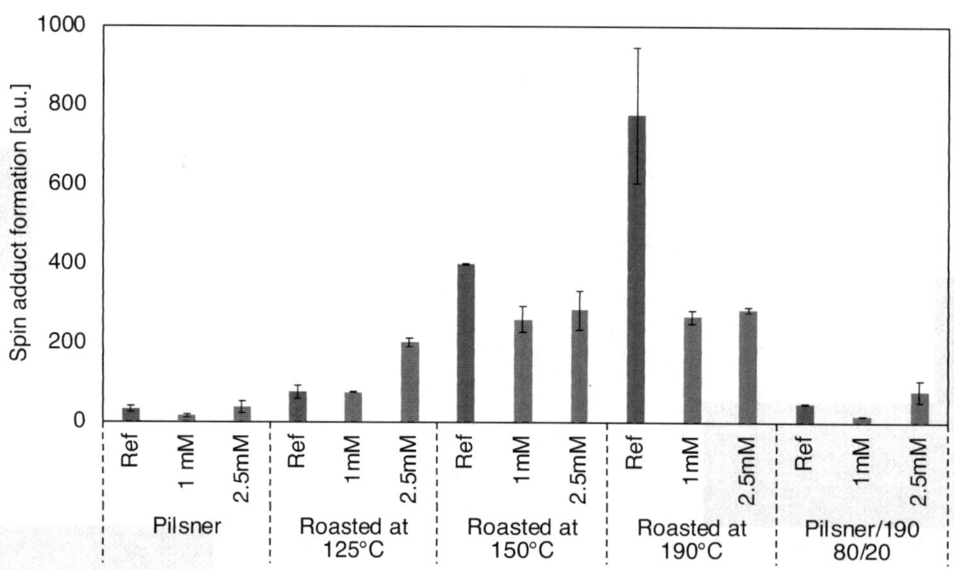

22

EJD Food Sci

Conclusions

- Roasting malt decreases the copper(II) affinity of sweet wort, resulting in lower total copper levels
- Copper(II) binding affinity…
  - decreases with decreasing pH
  - decreases with decreasing FAN
- Amino acids/dipeptides are efficient copper(II) chelators during the brewing process
- Increased copper(II) levels up to 1 mM have a stabilizing effect on oxidation rates in sweet wort

23

EJD Food Sci

Acknowledgements

- **Mogens L. Andersen**, University of Copenhagen
- **Morten J. Bjerrum,** University of Copenhagen
- EU H2020 Integrated Training Network **EJDFoodScience**

24

European
Joint Doctorate
Food Science
Thank you for your attention!
Any questions?
Marcus Pagenstecher
University of Copenhagen
University of Agriculture in Krakow
This project has received funding from the European Union's Horizon 2020 research and innovation programme under the Marie Skłodowska-Curie grant agreement No 722166

School of Food and Nutritional Sciences
University College Cork
Ireland

University College Cork, Ireland
Coláiste na hOllscoile Corcaigh

Unlocking the potential of lactic acid bacteria and their application in malting and brewing

Professor Elke Arendt, *MSc, PhD, DSc,*
School of Food and Nutritional Sciences,
University College, Cork.

Lactic acid bacteria (LAB) - Introduction

University College Cork, Ireland
Coláiste na hOllscoile Corcaigh

Gram +
Organic Acids (e.g. Lactic Acid)
Acid Tolerant (< pH 5)

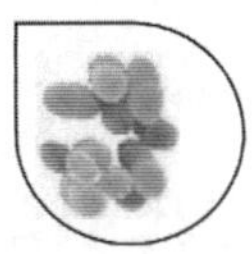

LAB & Food

Traditional
Sourdough
Cheese
Wine
Beer
Etc.

Food Grade
GRAS Status

Indigenous Microbiota
Ubiquitous

Traditional use in brewing:

- Acid malt
- Biological acidification
- Sour beer

28/06/2019 2

Unlocking the potential of Lactic acid bacteria

UCC
University College Cork, Ireland
Coláiste na hOllscoile Corcaigh

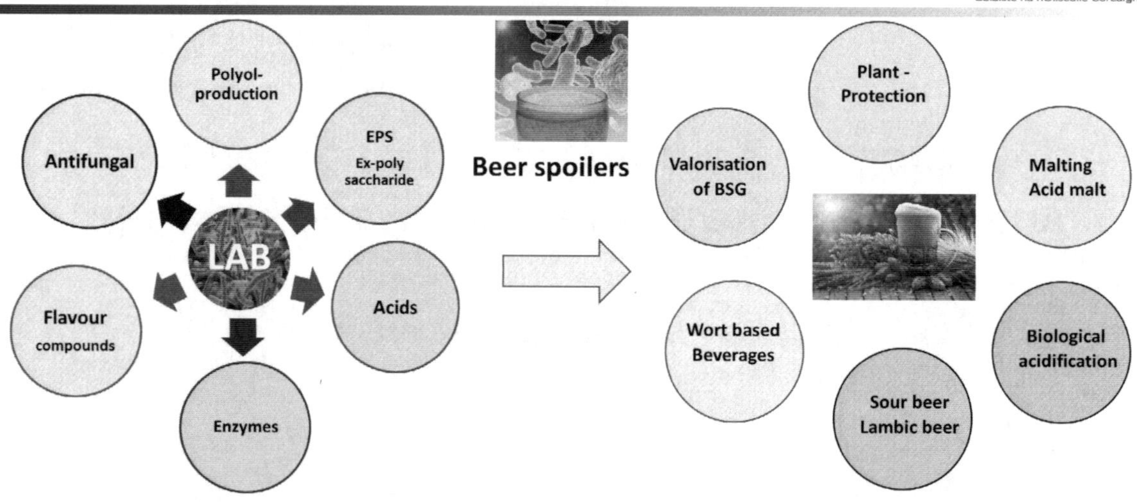

28/06/2019 3

School of Food and Nutritional Sciences
University College Cork
Ireland

UCC
University College Cork, Ireland
Coláiste na hOllscoile Corcaigh

Lactic acid bacteria with antifungal activity

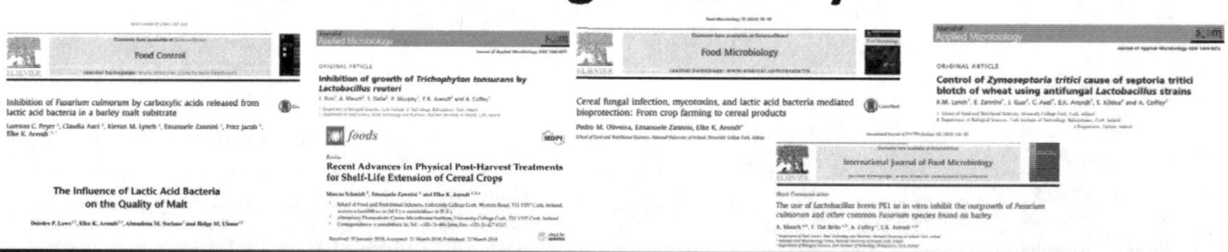

Why are fungi a problem ?

University College Cork, Ireland
Coláiste na hOllscoile Corcaigh

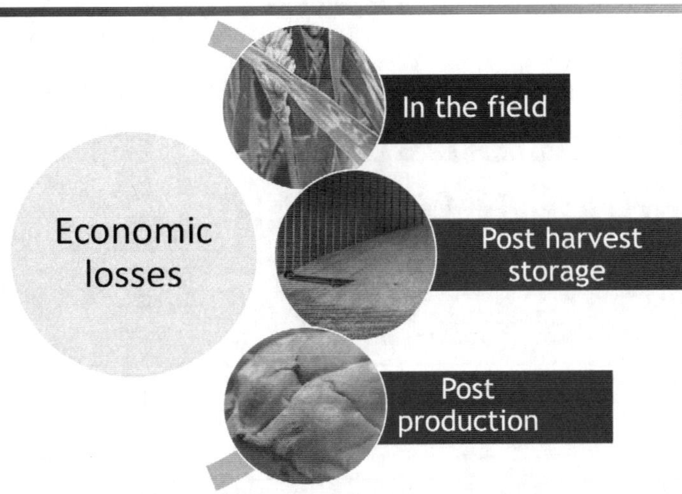

Approximately **15% of all cereals** worldwide are **lost** due to **microbial spoilage**

Approximately **25% of the global cereal crops**, equivalent to over 500 million tons per annum, are **contaminated with mycotoxins** and thus present a potential consumer health hazard (Cheli et al., 2013)

5

Research approach – LAB with antifungal activity

University College Cork, Ireland
Coláiste na hOllscoile Corcaigh

Isolation of lactic acid bacteria
Isolation of LAB from cereal source
Microbial ecology
16S rRNA identification

Screening und characterisation of the isolates
Screening and Characterization,
Antifungal and anti-microbial properties

Isolation and characterisation of antifungal commpounds
Bio-assays, GC / MS, HPLC etc.

Processing
Pilot-scale Processing (laboratory scale and pilot scale)

Bio-protection | **Malting / Beverages** | **Cereal processing**

28/06/2019

6

School of Food and Nutritional Sciences
University College Cork
Ireland

University College Cork, Ireland
Coláiste na hOllscoile Corcaigh

Isolation and characterization of antifungal compounds from LAB

ORIGINAL PAPER

Rapid identification, by use of the LTQ Orbitrap hybrid FT mass spectrometer, of antifungal compounds produced by lactic acid bacteria

Anal Bioanal Chem (2014) 406:2435–2444
DOI 10.1007/s00216-014-7626-3

RESEARCH PAPER

Quantification of cyclic dipeptides from cultures of *Lactobacillus brevis* R2Δ by HRGC/MS using stable isotope dilution assay

Claudia Axel · Emanuele Zannini · Elke K. Arendt · Deborah M. Waters · Michael Czerny

Analytical Methods

PAPER

A comprehensive investigation into sample extraction and method validation for the identification of antifungal compounds produced by lactic acid bacteria using HPLC-UV/DAD

The Use and Effects of Lactic Acid Bacteria in Malting and Brewing with Their Relationships to Antifungal Activity, Mycotoxins and Gushing: A Review

Deirdre P. Lowe and Elke K. Arendt

Food Control

Inhibition of *Fusarium culmorum* by carboxylic acids released from lactic acid bacteria in a barley malt substrate

Talanta

The QuEChERS approach in a novel application for the identification of antifungal compounds produced by lactic acid bacteria cultures

Brid Brosnan, Aidan Coffey, Elke K. Arendt, Ambrose Furey

Antifungal screening

University College Cork, Ireland
Coláiste na hOllscoile Corcaigh

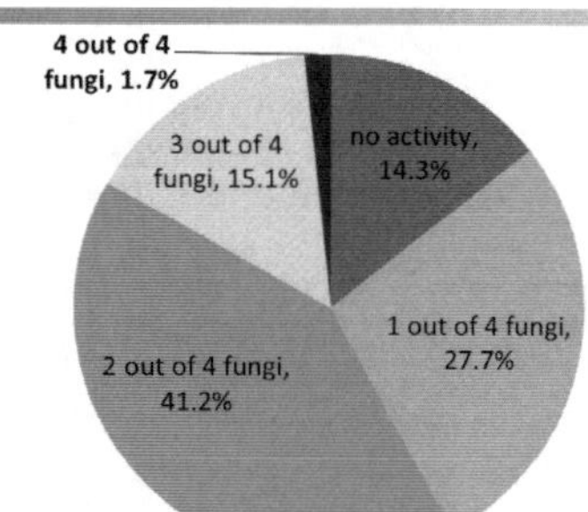

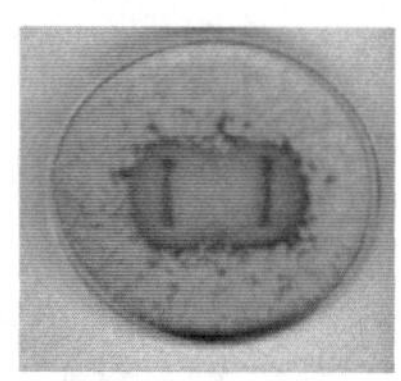

LAB inhibiting *Fusarium* species

- LAB were screened against four species of *Fusarium*, *Apergillus niger*, *Penecillium expansum* and *roforteii* ;
- *Phytophthora infestans* (Potato blight),
- *Zymoseptoria tritici* (Septoria tritici blotch)
- 1 % of the strains showed anti-fungal activity against all 12 fungal strains

28/06/2019 8

Antifungal compounds from LAB

University College Cork, Ireland
Coláiste na hOllscoile Corcaigh

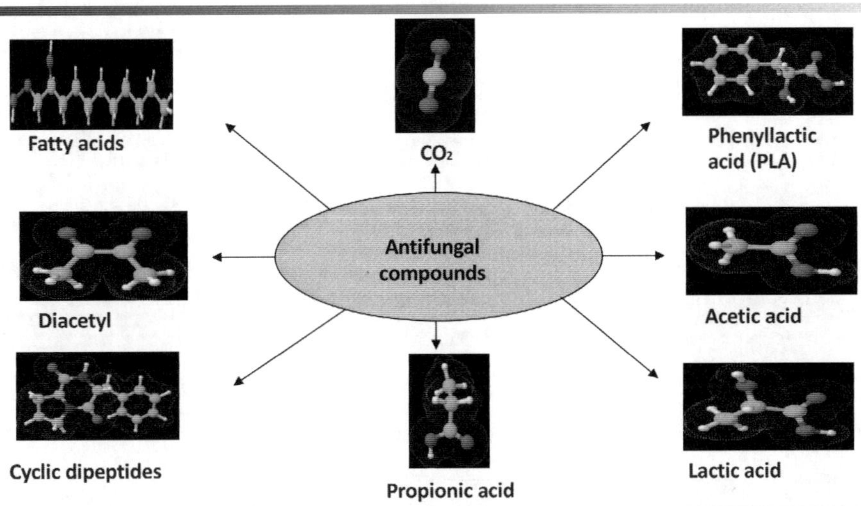

28/06/2019 9

LAB with antifungal activity

University College Cork, Ireland
Coláiste na hOllscoile Corcaigh

***L. amylovorus* FST 2.1**

15 different compounds,
9 new for LAB

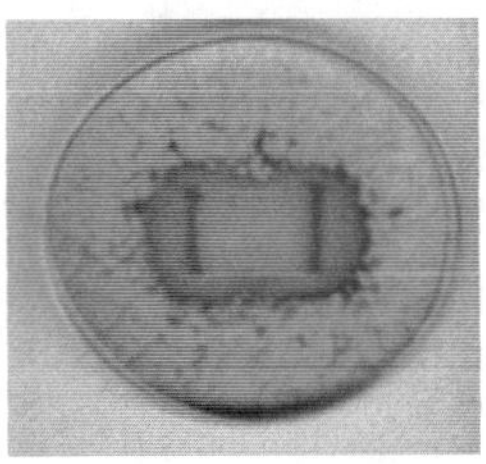

Code	Compound	Chemical formula	MW	Fragments (MW)	Inhibition (day)
1	Glucuronic acid	$C_6H_{10}O_7$	194.14	105.1, 91.1	>10
2	Cytidine	$C_9H_{13}N_3O_5$	243.22	109.1, 91.1	1
3	Deoxycytidine	$C_9H_{13}N_3O_4$	227.22	183.0, 93.1	1
4	Sodium decanoate	$C_{10}H_{19}O_2Na$	194.25	153.3, 103.0	>10
5	Hydrocinnamic acid	$C_9H_{10}O_2$	150.17	105.1	>10
6	Phenyllactic acid	$C_9H_{10}O_3$	166.17	149.0, 119.8	>10
7	OH- Phenyllactic acid	$C_9H_{10}O_4$	182.17	163.1, 135.2	>10
8	Coumaric acid	$C_9H_8O_3$	164.16	119.2	>10
9	Methylcinnamic acid	$C_{10}H_{10}O_2$	162.19	147.1, 117.1	>10
10	Salicylic acid	$C_7H_6O_3$	136.12	93.2	>10
11	Cyclo His – Pro	$C_{11}H_{14}N_4O_2$	234.26	207.1,166.0	2
12	Cyclo Pro – Pro	$C_{10}H_{14}N_2O_2$	194.23	151.2, 70	4
13	Cyclo Met –Pro	$C_{10}H_{16}N_2O_2S$	228.3	181.2,	3
14	Cyclo Tyr – Pro	$C_{14}H_{16}N_2O_3$	260.3	233.0, 136.1	3
15	Cyclo Leu – Pro	$C_{11}H_{18}N_2O_2$	210.13	183.2, 155.0	3

28/06/2019 10

School of Food and Nutritional Sciences
University College Cork
Ireland

University College Cork, Ireland
Coláiste na hOllscoile Corcaigh

Lactic acid bacteria and their role in plant protection / barley protection - Application

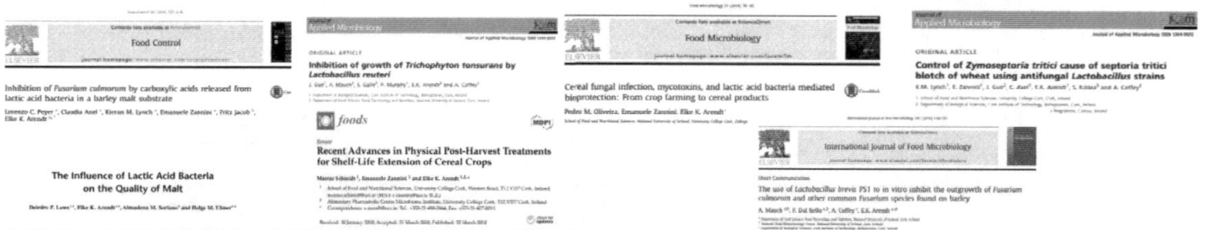

LAB against *Zymoseptoria tritici* – *Treatment of Septoria tritci boltch*

University College Cork, Ireland
Coláiste na hOllscoile Corcaigh

Anti- *Zymoseptoria tritici* plate assay

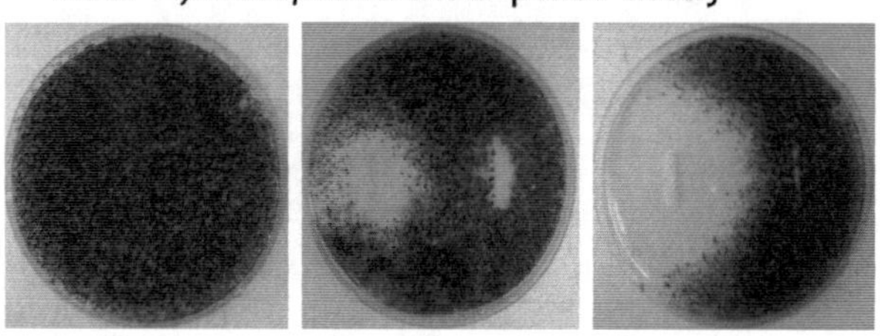

Fungal plate with no LAB.

Fungal plate with *Lb. brevis* (left) *Lb. brevis* L1105 (right)

Fungal plate with *Lb. reuteri* (left) *Lb. reuteri* M13 (right)

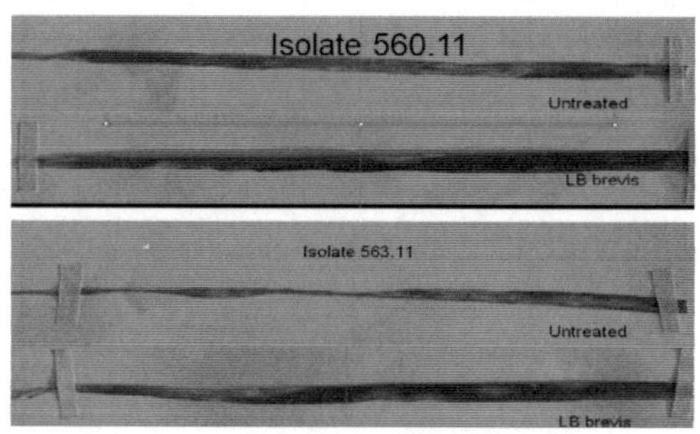

The CFS of *Lactobacillus brevis* JJ2P significantly reduced the diseased leaf area caused by STB strain (a) *Zymoseptoria tritici* 56011 and (b) *Z. tritici* 56311 compared to the control plants (diseased wheat leaf without application of CFS)

LAB with antifungal activity can be as effective as fungicides and can be applied for plant protection as alternative treatments to fungicides – Example *Zymoseptoria tritici*

28/06/2019 *In collaboration with Prof Aidan Coffey, CIT* 12

LAB treatment of Fusarium head blight

UCC
University College Cork, Ireland
Coláiste na hOllscoile Corcaigh

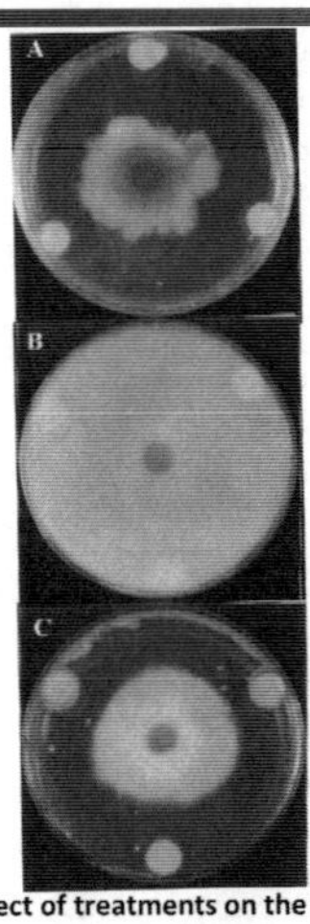

Lactobacillus amylovorous **antifungal**

Negative Control strain

Lactobacillus brevis **antifungal**

Effect of treatments on the radial growth of *F. graminearum* on PDA plates

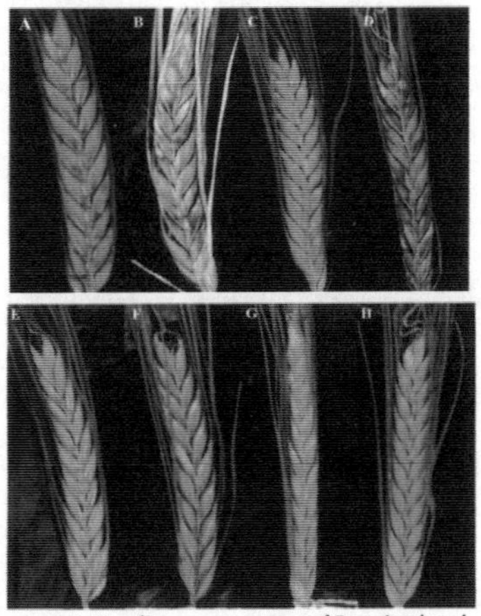

Representative disease symptoms of Fusariun head blight disease on spikelet of two row barley cultivar

A: Positive Control 5 days
B: Positive Control 10 days
C: *L. amylovorous* (neg. control) 5 days
D: *L. amylovorous* (neg. control) 10 days
E: *L. brevis* (antifungal) 5 days
F: *L. brevis* (antifungal) 10 days
G: *L. brevis R 2* (antifungal) 5 days
H: *L. brevis R 2* (antifungal) 10 days

- Treatment with antifungal LAB was as effective as treatment with fungicides
- Barley also expressed genes which are related to increased fungal resistance

Work was performed in collaboration with Prof Jimmy Burke, UCD 16

School of Food and Nutritional Sciences
University College Cork
Ireland

UCC
University College Cork, Ireland
Coláiste na hOllscoile Corcaigh

Lactic acid bacteria and their role in malting – antifungal activity

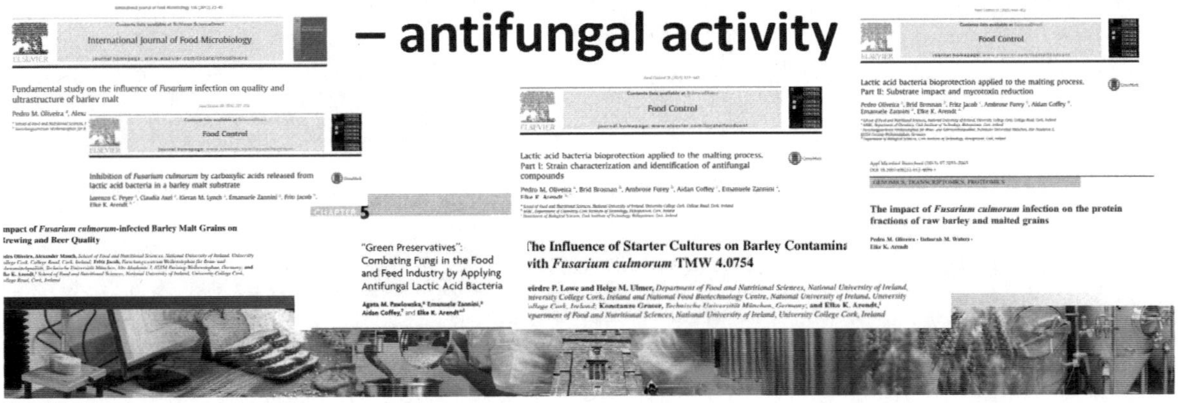

Mycotoxins in brewing industry

UCC
University College Cork, Ireland
Coláiste na hOllscoile Corcaigh

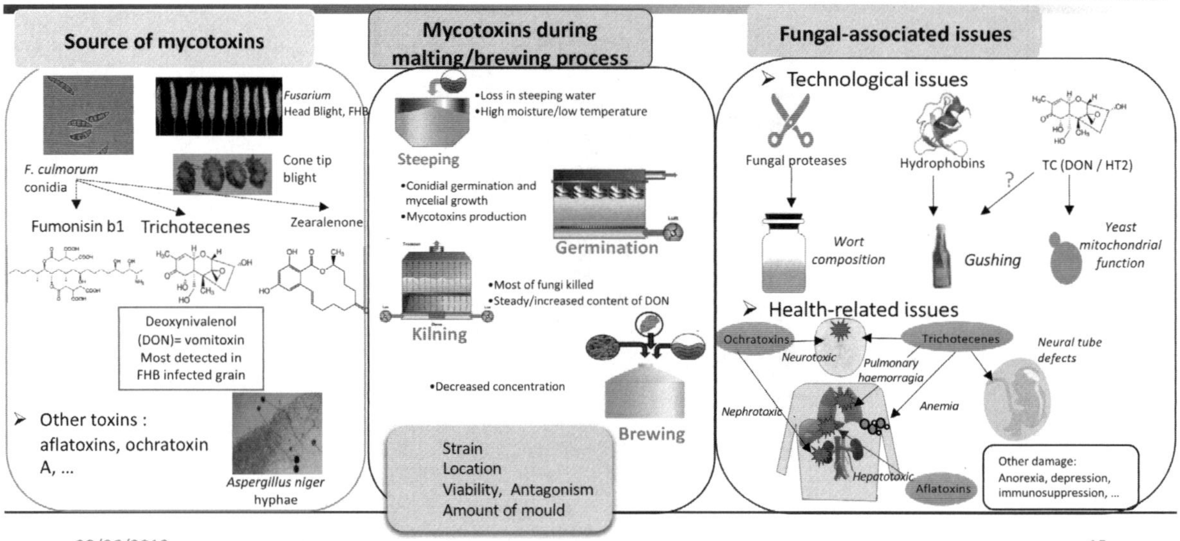

28/06/2019 15

Fungal infection: technological complications

UCC
University College Cork, Ireland
Coláiste na hOllscoile Corcaigh

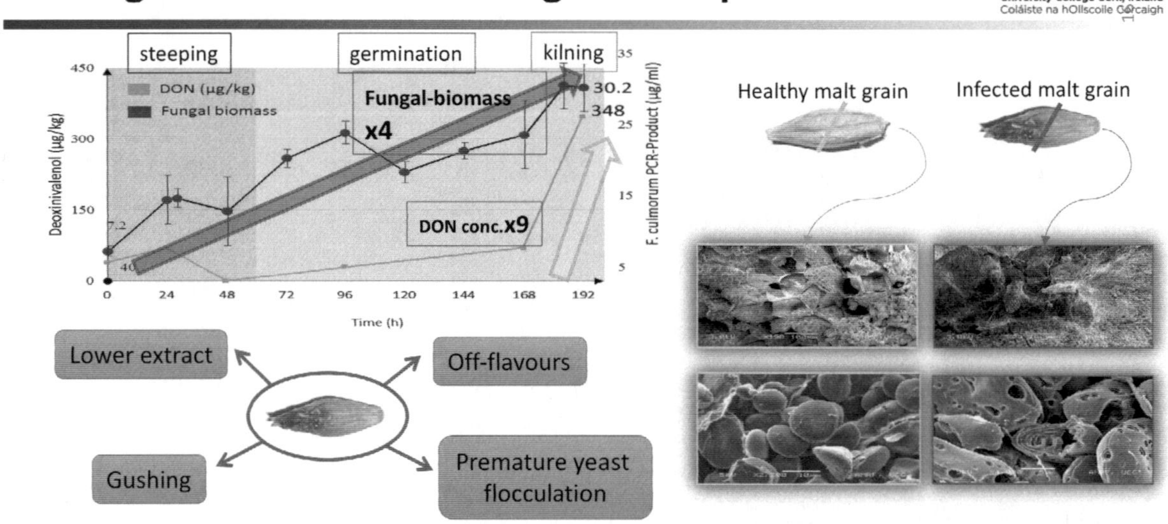

In collaboration with Prof Jacob and Dr Zarnkow 19

Fungal infection: technological complications

University College Cork, Ireland
Coláiste na hOllscoile Corcaigh

Beer – Color / Foam

Standard EBC/MEBAK methods

- Higher coloration (67%)
- Tendency for better foam stability (8%)

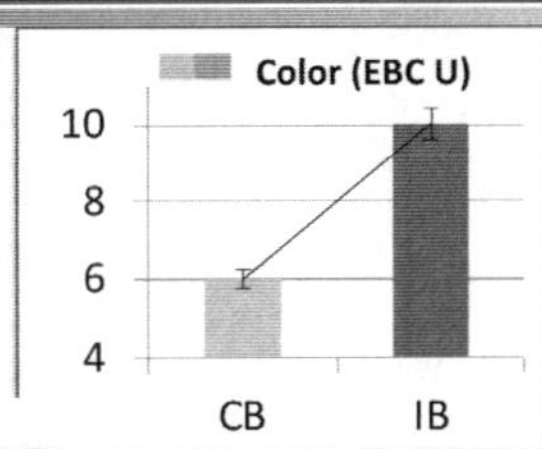

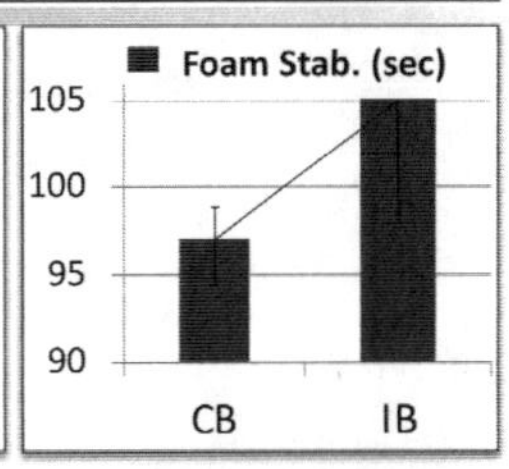

Beer – DON mycotoxin

HPLC-UV

78% of initial mycotoxin from infected grains accumulated in final beer

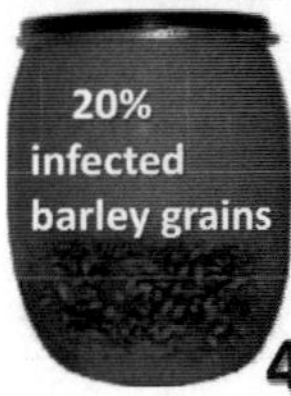

40 ppb

Malting

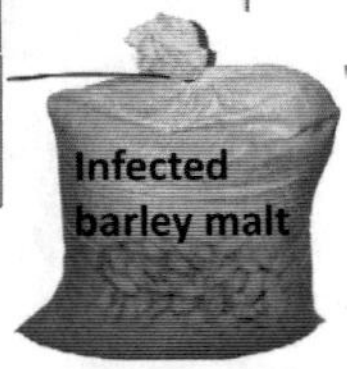

348 ppb

Brewing

271 ppb

DON - Concentration

In collaboration with Prof Jacob and Dr Zarnkow

20

Anti-fungal LAB in malting –Experimental approach

University College Cork, Ireland
Coláiste na hOllscoile Corcaigh

Fonds InBev-Baillet Latour Fund

→ Release of carboxylic acids (antifungal compounds) in wort from antifungal LAB and efficiency against *F. culmorum*

Overlay assay method
(10^4 *Fusarium culmorum* spores/ml)

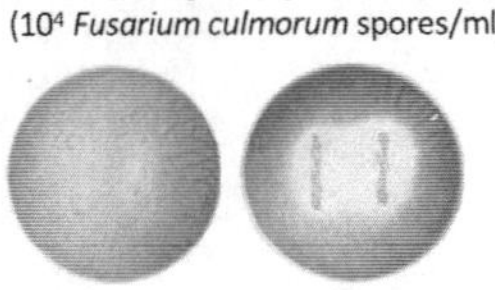

Lb. brevis L1105 *Lb. brevis* R2Δ

Screening for AF +
Lb. plantarum FST1.7
Lb. brevis R2Δ
(*Lb. brevis* L1105 as AF -)

Microtiter plate assay
(10^4 *Fusarium culmorum* spores/ml)

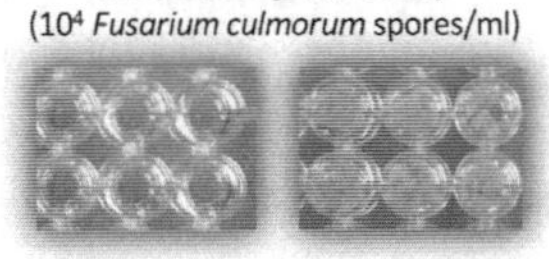

Lb. plantarum FST1.7 Unfermented control

Production of wort

Inoculation of wort

Investigation of anti-fungal compounds
QuEChERS cleanup for phenolic acids
HPLC-UV/PDA

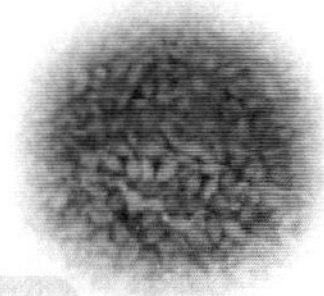

Different fermentation conditions

- Extract content: 3, 6, 9, 12 % w/w
- Temperature: 30, 37°C
- Time: 24, 48, 72, 120 h

Spiking
Mimic concentrations

Minimal inhibitory concentration
pH corrected and uncorrected

28/06/2019

18

Experimental design

University College Cork, Ireland
Coláiste na hOllscoile Corcaigh

→ **Bioprotection using fermented wort as antifungal ingredient and influence on malt quality**

Strain: ***Lb. brevis* R2Δ**

Substrate: **Wort 3% w/w**

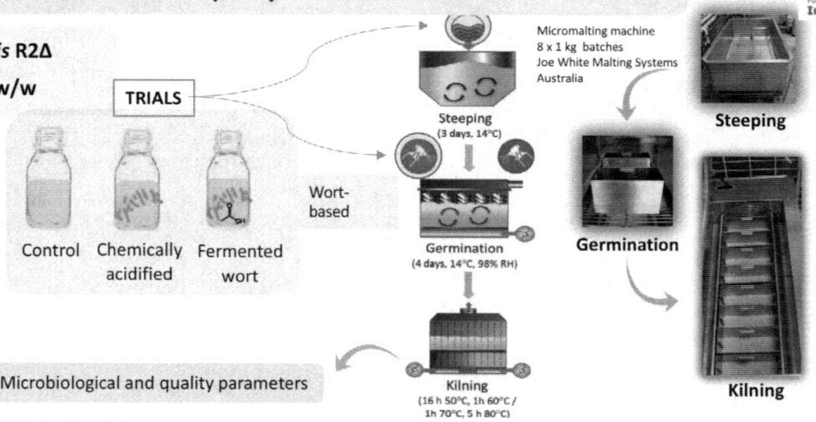

Microbiological and quality parameters

Fonds InBev-Baillet Latour Fund

In collaboration with Prof Jacob and Dr Zarnkow 22

Impact of antifungal LAB on the Barley microflora

University College Cork, Ireland
Coláiste na hOllscoile Corcaigh

Fonds InBev-Baillet Latour Fund

- LAB grew well on malt
- Aerobic heterotrophic bacteria when acidified wort or fermented wort was added
- Addition of fermented wort increased yeast growth

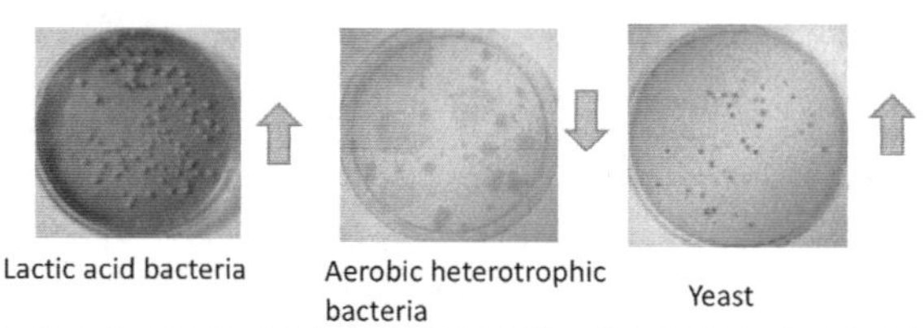

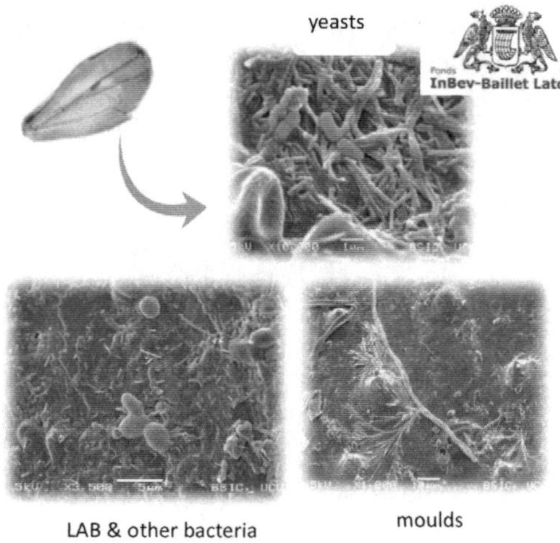

In collaboration with Prof Jacob and Dr Zarnkow 23

Fusarium spp. infection and mycotoxins

	Control	Chemical acidified	LAB fermented wort
Red kernels %	31 ± 3	10 ± 1	3 ± 1
Red halos	1 ± 1	1 ± 1	1 ± 1
***Fusarium* DNA** (ng/g)	78 ± 13	<LOD^{a}	<LOD
Deoxynivalenol-3-Glucoside *(μg/kg)*	19 ± 3	19 ± 2	<LOD
Enniatins A+A1 *(μg/kg)*	52 ± 2	<LOD	<LOD
Enniatins B+B1 *(μg/kg)*	120 ± 9	<LOD	24 ± 3

*Targeted gene for qPCR was **ZEA** (synthesis of Zearalenone toxin in *F. culmorum* and *F. graminearum*) according to [8]^{a} LOD < 20 ng/g

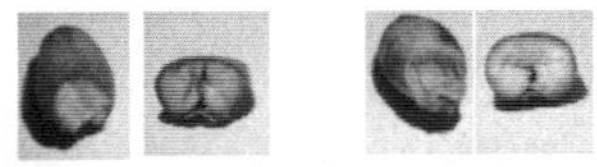

- **Every treatment performed better than control**
- **LAB cells can help adsorption of mycotoxins**
- **Wort fermented with anti-fungal LAB performed best**

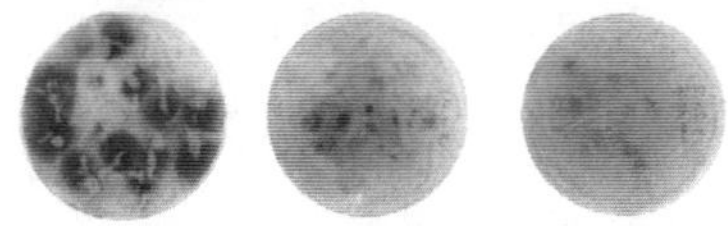

Modified Czapek-Dox agar selective for *Fusarium* spp.

In collaboration with Prof Jacob and Dr Zarnkow

24

Malt quality of treated samples

	Contorl	Chemical acidification	LAB fermented wort
Malting loss (% w/w, d.m.)	**10.81 ± 0.57**^{a}	**9.65 ± 0.92**^{a}	**7.37 ± 0.23**^{b}
α-amylase (units/g)	290.0 ± 10.1^{a}	287.4 ± 8.1^{a}	227.0 ± 18.2^{b}
Friability (%)	90.8 ± 0.3^{a}	86.0 ± 0.2^{b}	88.8 ± 1.4^{ab}
Total protein (% w/w, d.m.)	9.46 ± 0.16^{a}	9.60 ± 0.11^{a}	9.84 ± 0.14^{a}
Sol. nitrogen(g/100 g malt, d.m.)	0.78 ± 0.01^{a}	0.82 ± 0.02^{a}	0.79 ± 0.02^{a}
Extract (% d.m.)	**85.0 ± 0.3**^{a}	**84.9 ± 0.1**^{a}	**84.4 ± 0.1**^{bc}

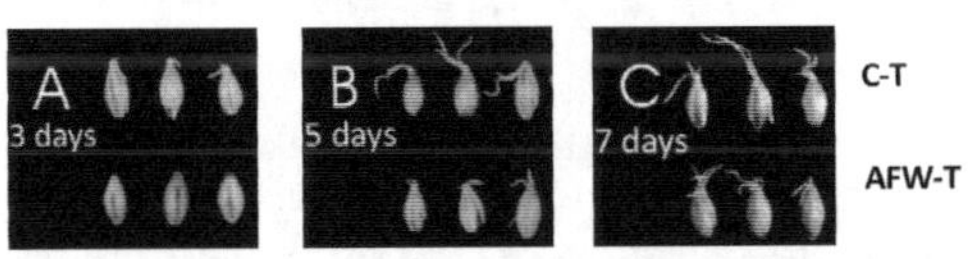

- **Decreased rootlet growth**
  → lower malting losses (- 32%)
- **Reduction in α-amylase activity which did not effect the extract content of the wort**

$^{a-c}$ Grouping according to Tukey's post hoc test after one-way ANOVA (significance at 5%)

In collaboration with Prof Jacob and Dr Zarnkow

25

Conclusion – LAB and Malting

Fonds InBev-Baillet Latour Fund

University College Cork, Ireland
Coláiste na hOllscoile Corcaigh

- **Promotion of yeast** and **decrease of aerobe bacteria and *Fusarium*** species
- **Bacterial metabolites** are primarily **responsible** for inhibition
- **LAB treatment of malt had a minimal affect on malt, wort and beer quality**
- **No** evidence on **negative effect** on yeast fermentation (e.g. stuck fermentation)

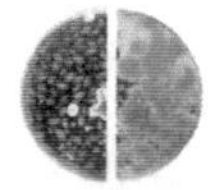

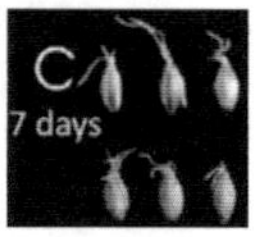

Most effective treatment of malt was the inoculation with LAB (antifungal activity) fermented wort – reduced malting loss and prevented formation of mycotoxins and reduced fungal growth

28/06/2019 — *In collaboration with Prof Jacob and Dr Zarnkow* — 23

School of Food and Nutritional Sciences
University College Cork
Ireland

University College Cork, Ireland
Coláiste na hOllscoile Corcaigh

Lactic acid bacteria and their role in malting – anti-rootlet activity

Part I. The Use of *Lactobacillus plantarum* Starter Cultures to Inhibit Rootlet Growth During Germination of Barley, Reducing Malting Loss, and Its Influence on Malt Quality

Alexander Mauch, *Department of Food Science, Food Technology and Nutrition and National Food Biotechnology Centre, National University of Ireland, Cork, Ireland;* **Fritz Jacob,** *Forschungszentrum Weihenstephan für Brau- und Lebensmittelqualität, Technische Universität München, Freising-Weihenstephan, Germany;* **Aidan Coffey,** *Department of Biological Sciences, Cork Institute of Technology, Bishopstown, Cork, Ireland;* **and Elke K. Arendt,**[1] *Department of Food Science, Food Technology and Nutrition, National University of Ireland, Cork, Ireland*

Part II. The Use of Malt Produced with 70% Less Malting Loss for Beer Production: Impact on Processability and Final Quality

Alexander Mauch and Sascha Wunderlich, *School of Food and Nutritional Sciences, University College Cork, and National Food Biotechnology Centre, National University of Ireland, Cork, Ireland;* **Martin Zarnkow and Thomas Becker,** *Lehrstuhl für Brau- und Getränketechnologie, Technische Universität München, Freising-Weihenstephan, Germany;* **Fritz Jacob,** *Forschungszentrum Weihenstephan für Brau- und Lebensmittelqualität, Technische Universität München, Freising-Weihenstephan, Germany;* **and Elke K. Arendt,**[1] *School of Food and Nutritional Sciences, University College Cork, National University of Ireland, College Road, Cork, Ireland*

Experimental design – anti-rootlet LAB

Fonds InBev-Baillet Latour Fund

UCC University College Cork, Ireland
Coláiste na hOllscoile Corcaigh

- Barley
- Steeping ← Addition of fermentation broth or chemically acidified solution
- Germination
- Kilning → Rootlet removal
- Malt → Determination of malting loss and malt quality including enzyme activities
- Beer → Design of an adequate mashing regime, determination of process- and product quality

Pre screening of LAB for reduction of malting loss ← **How ?**

Reduction of malting loss using LAB

UCC University College Cork, Ireland
Coláiste na hOllscoile Corcaigh

Malting loss due to rootlet removal 4 to 10%

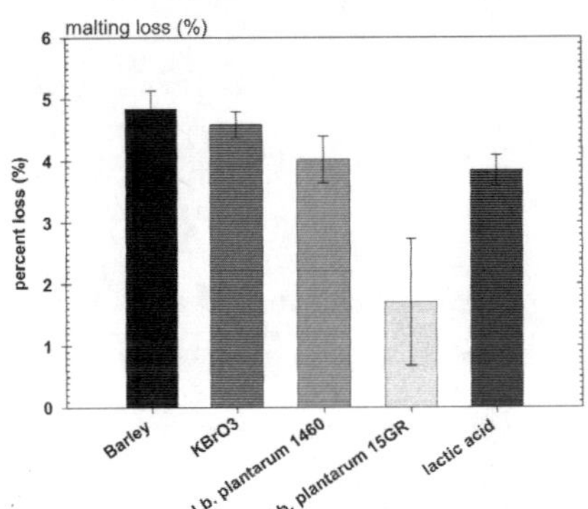

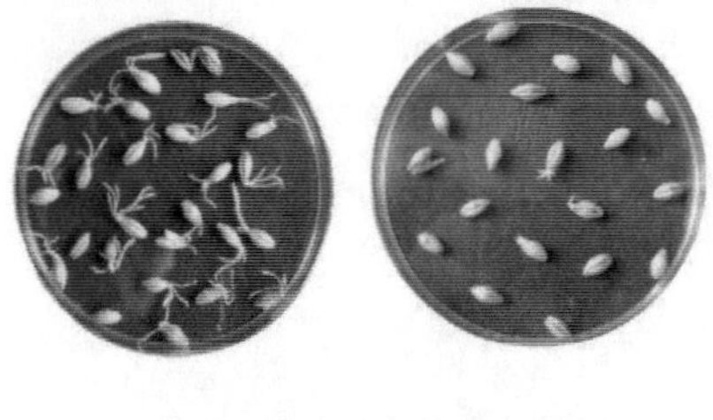

Malt treated with LAB (right), no treatment (left)

➡ **Application of LAB producing anti-rootlet compounds leads to 50% - 70 % reduction of malting loss**

PhD Alex Mauch *In collaboration with Prof Jacob and Dr Zarnkow* 29

School of Food and Nutritional Sciences
University College Cork
Ireland

University College Cork, Ireland
Coláiste na hOllscoile Corcaigh

Lactic acid bacteria and their role in beer – Sour beers

Sour brewing: impact of *Lactobacillus amylovorus* FST2.11 on technological and quality attributes of acid beers

Lorenzo C. Peyer, Martin Zarnkow, Fritz Jacob, David P. De Schutter and Elke K. Arendt

***Accepted** in the Journal of American Society of Brewing Chemists (March 2017)*

The Science of Beer

Sour beer interest

University College Cork, Ireland
Coláiste na hOllscoile Corcaigh

- Interest in sour beer at an all-time high
- Increasing number of new sour beer brands
- Annual Growth Rate of 62% (as of 05/2017)

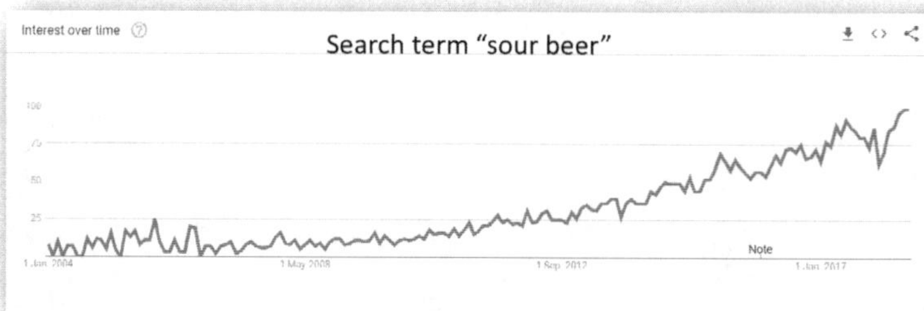

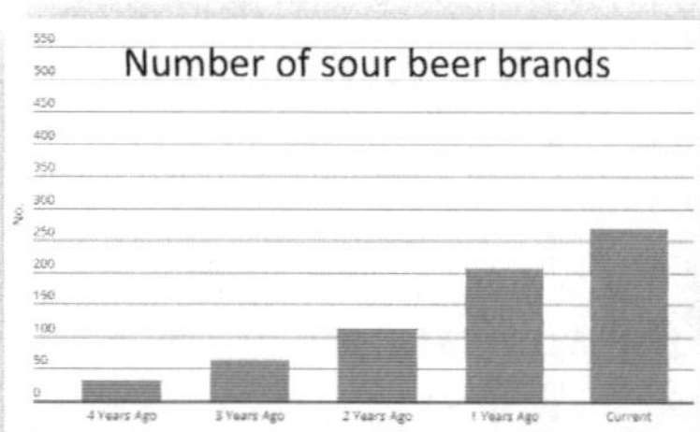

Nielsen Retail Measurement Services, 52 weeks ended May 20, 2017
Google Trends, search term "sour beer", worldwide interest, accessed: 06/07/2018

Sour beer: traditional styles

University College Cork, Ireland
Coláiste na hOllscoile Corcaigh

- Spontaneous fermentation
- Yeasts and bacteria
- Barley malt and unmalted wheat

Lambic

- Inoculated, mixed fermentation
- *Saccharomyces* and bacteria
- All barley malt and red malt

Flanders Red Ale

- Inoculated, mixed fermentation
- *Saccharomyces* and *Lactobacillus*
- Wheat malt and barley malt

Berliner Weisse

- Inoculated, mixed fermentation
- *Saccharomyces* and *Lactobacillus*
- Wheat malt, barley malt and salt

Gose

© 2016 Analytical Flavor Systems

Experimental design –sour beer production

Fonds InBev-Baillet Latour Fund

University College Cork, Ireland
Coláiste na hOllscoile Corcaigh

→ Investigate differences between souring done at different time points along the brewing process

Brewing recipe

- 100% Pilsner malt (grist:liquor 1:5)
- 20' @ 50°C; 40' @ 62°C; 20' @ 72°C; 5' @ 78°C
- Lautering 2 x 5 L / Boiling 15 min
- No hopping

Mash souring (MS)

Preboil wort souring (preBWS)

Postboil wort souring (postBWS)

Yeast control (YC)

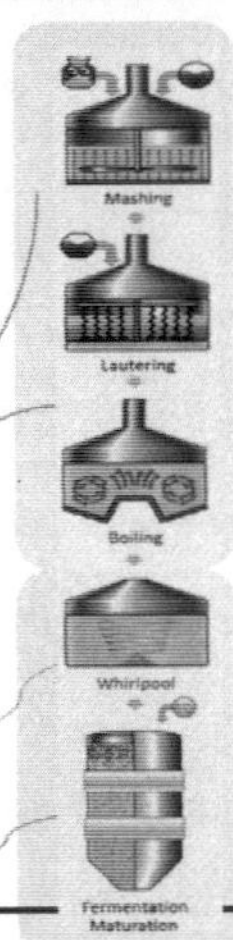

Lactic acid fermentation (LF)

- *Lb. amylovorus* FST2.11
- 7 log CFU/ml
- 40°C for 18 h → 5-6 g/L lactic acid

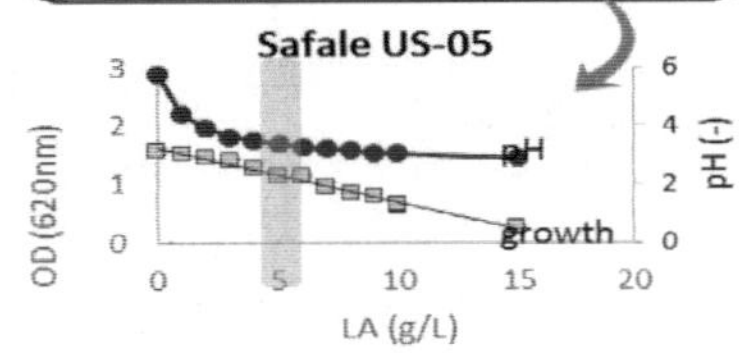

Alcoholic fermentation (AF)

- Safale US-05 (*S. cerevisiae*)
- 7 log CFU/ml
- 20°C for 2 weeks (+ 2 maturation)

Experimental design and analysis

UCC
University College Cork, Ireland
Coláiste na hOllscoile Corcaigh

Measurements:
- pH, TTA
- CFU
- Acids
- Sugars

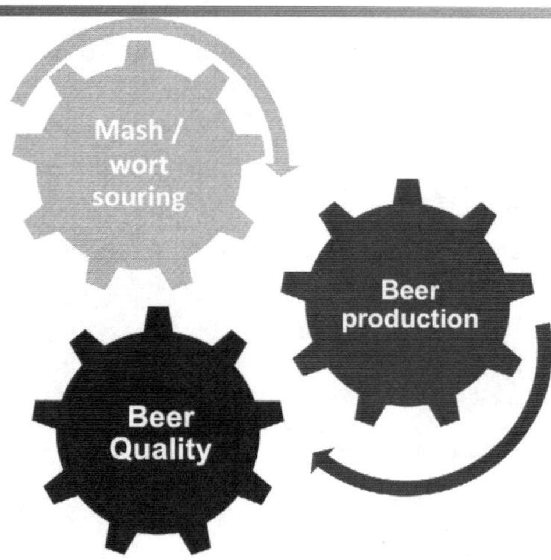

Measurements on beer:
- Soluble Amino Nitrogen
- Sugars and Acids
- Polyphenols
- Color
- Foam stability
- Flavour analysis (Esters, Fusel-alcohols, Fatty acids, Ketons)
- Sensory: trained panel

Measurements on mash and wort:
- Wort viscosity
- Beta glucan
- buffering capacity
- pH, TTA
- FAN
- Sugars
- Acids
- CFU – Yeast and LAB
- Color

Fermentation profile – sour beer

Fonds InBev-Baillet Latour Fund

UCC
University College Cork, Ireland
Coláiste na hOllscoile Corcaigh

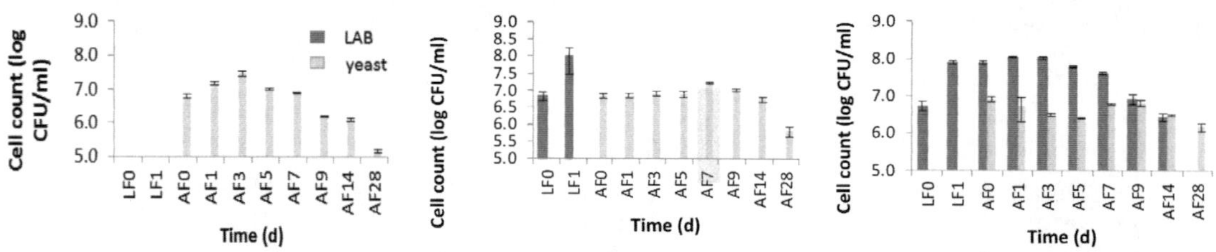

Control **Pre-boil wort souring** **Post-boil wort souring**

LF: lactic fermentation AF: alcoholic fermentation

- Yeast growth was delayed by 2 to 4 days if souring was applied
- Yeast counts slightly lower in soured samples (MS and Pre-boil souring); most obvious when yeast and LAB are part of the same fermentation
- Alcohol and extract content at the end of fermentation in depend of the treatments

pH of Beer at end of fermentation

Treatment	pH
Control	4.13
Mash acidification	3.75
Pre-boil acidification	3.44
Post-boil acidification	3.44

Quality attributes of wort and beers

Fonds InBev-Baillet Latour Fund

UCC University College Cork, Ireland
Coláiste na hOllscoile Corcaigh

Comparison of metabolites found in the final beer

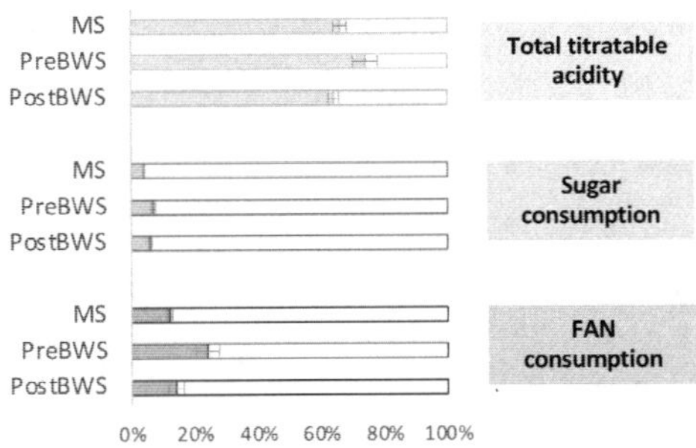

*glucose, fructose, maltose and maltotriose

During lactic acid fermentation
- Majority of acids formed
- Little consumption of sugars and FAN

Treatment	Haze EBC formazin units	Colour EBC units	Foam stability $t_{1/2}$sec
Control	1.6 ± 0.2^{b}	6.4 ± 0.1^{a}	189 ± 10^{ab}
Mash souring	0.4 ± 0.2^{c}	5.2 ± 0.0^{c}	224 ± 44^{a}
Pre-boiled wort souring	0.4 ± 0.0^{c}	4.9 ± 0.1^{b}	161 ± 14^{b}
Post- boil wort souring	3.4 ± 0.6^{a}	5.8 ± 0.0^{ab}	206 ± 33^{ab}

- Mash souring and Preboil wort souring → Light and clear beers
- Different foam stability
  → proteolytic damage not evident

a-c Grouping according to Tukey's post hoc test after one-way ANOVA (significance at 5%)

Flavour profile – analytical results

Fonds InBev-Baillet Latour Fund

UCC University College Cork, Ireland
Coláiste na hOllscoile Corcaigh

	Isoamyl alcohol	Isovaleric acid	Diacetyl (total)	Acetoin	Acetaldehyde
Control	56.5 ± 0.2^{a}	0.83 ± 0.06^{b}	0.10 ± 0.01^{b}	2.2 ± 0.2^{b}	5.4 ± 0.5^{c}
Mash souring	76.7 ± 15.8^{a}	0.72 ± 0.02^{b}	0.07 ± 0.01^{bc}	2.5 ± 0.1^{b}	40.3 ± 10.0^{a}
Pre-boiled wort souring	59.7 ± 3.8^{a}	0.50 ± 0.03^{c}	0.05 ± 0.00^{c}	1.4 ± 0.3^{c}	14.3 ± 0.5^{b}
Post-boiled wort souring	60.2 ± 0.7^{a}	1.15 ± 0.07^{a}	0.28 ± 0.00^{a}	22.6 ± 0.2^{a}	14.1 ± 0.4^{b}

- Diacetyl level highest using post-boiled wort souring - due to impact on yeast growth
- Acetaldehyde level highest using Mash souring
- **Best overall result: Pre-boiled wort souring**

A-c Grouping according to Tukey's post hoc test after one-way ANOVA (significance at 5%)

Sensory analysis of sour beers

University College Cork, Ireland
Coláiste na hOllscoile Corcaigh

	Sensory Trained panel of 10
Control	Typical (unhopped lager)
Mash souring	Untypical / not clean
Pre boiled wort souring	Slight plum aroma
Post boiled wort souring	Untypical / astringent

Spider chart in logarithmic scale for better representation

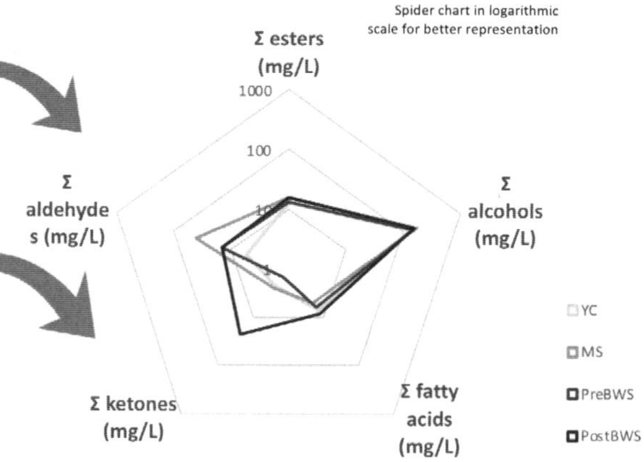

Best overall flavour
Pre-boiled wort souring

Conclusion – sour-beer

Fonds InBev-Baillet Latour Fund

University College Cork, Ireland
Coláiste na hOllscoile Corcaigh

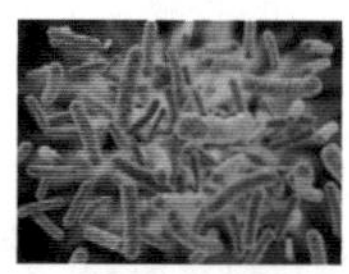

- Lactic acid bacteria for sour beer production need to be **specially selected**
- Souring done at different points gives rise to **different final products**
- Timing of acidification influences both **microbiological quality and process ability of beer**
- Exposure of yeast to acid condition **delayed fermentation** by ca. 2-4 days
  - Final attenuation reached by all beer samples
- Untypical smells of soured beers related to **off-flavours - need to redefine flavour evaluation of sour beer**
- LAB-own flavour compounds are present at **higher levels**
  - Acetaldehyde, diacetyl, and acetoin
- Souring of wort before boiling (→ **preBWS**) as best souring method, as:
  - Avoids LAB in the cold side of brewing process
  - Leads to a bright and light product
  - Limited off-flavours

ASBC *The Science of Beer*

Sour brewing: impact of *Lactobacillus amylovor* on technological and quality attributes of acid b

Lorenzo C. Peyer, Martin Zarnkow, Fritz Jacob, David P. De Schutter a

***Accepted** in the Journal of American Society of Brewing Chemi*

School of Food and Nutritional Sciences
University College Cork
Ireland

UCC
University College Cork, Ireland
Coláiste na hOllscoile Corcaigh

Lactic acid bacteria: the creation of novel wort based beverages

Journal of Applied Microbiology

ORIGINAL ARTICLE

Barley malt wort fermentation by exopolysaccharide-forming *Weissella cibaria* MG1 for the production of a novel beverage

E. Zannini[1,2], A. Mauch[1,2], S. Galle[1,3], M. Gänzle[3], A. Coffey[4], E.K. Arendt[1], J.P. Taylor[1] and D.M. Waters[1]

Growth Study, Metabolite Development, and Organoleptic Profile of a Malt-Based Substrate Fermented by Lactic Acid Bacteria

Lorenzo C. Peyer and Emanuele Zannini, *School of Food and Nutritional Sciences, National University of Ireland, University College Cork, College Road, Cork, Ireland;* **Fritz Jacob,** *Research Center Weihenstephan for Brewing and Food Quality, Technische Universität München, Alte Akademie 3, 85354 Freising-Weihenstephan, Germany;* **and Elke K. Arendt,**[1] *School of Food and Nutritional Sciences, National University of Ireland, University College Cork*

J. Am. Soc. Brew. Chem. 73(4):303-313, 2015

Trends in Food Science & Technology

Lactic acid bacteria as sensory biomodulators for fermented cereal-based beverages

Lorenzo C. Peyer, Emanuele Zannini, Elke K. Arendt*

Growth Study, Metabolite Development, and Organoleptic Profile of a Malt-Based Substrate Fermented by Lactic Acid Bacteria

ANNUAL REVIEWS

Annual Review of Food Science and Technology

Lactic Acid Bacteria Exopolysaccharides in Foods and Beverages: Isolation, Properties, Characterization, and Health Benefits

Kieran M. Lynch,[1] Emanuele Zannini,[1] Aidan Coffey,[2] and Elke K. Arendt[1]

Mauch, A., Galle, S., Schwab, C., Gänzle, M., Jacob, F., Coffey, A., Zannini, E., Arendt, E. K.

The use of exopolysaccharides (EPS) forming *Weissella* strains for fermentation of barley malt derived wort: Potential basis for a functional beverage

International Journal of Food Microbiology (2011)

ASBC
The Science of Beer

Lactic acid bacteria and novel beverages

UCC
University College Cork, Ireland
Coláiste na hOllscoile Corcaigh

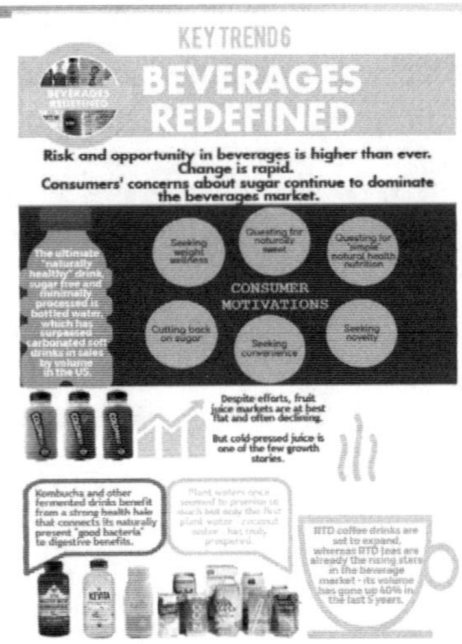

Wort based beverages – unlocking the potential of LAB

UCC
University College Cork, Ireland
Coláiste na hOllscoile Corcaigh

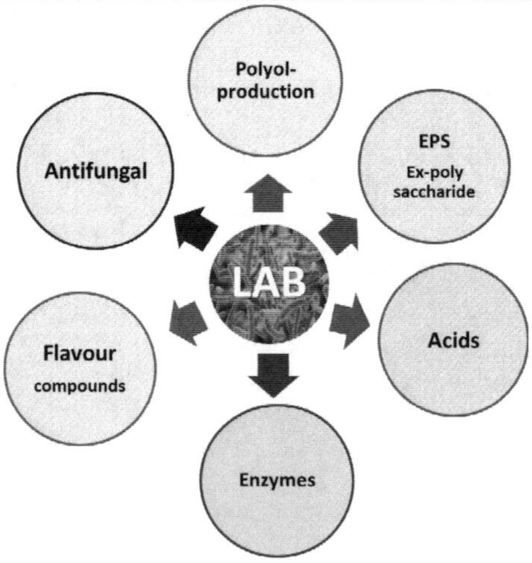

Modulating flavour through LAB fermentation - wort based beverages and sour-beer production

Increasing shelf-life through anti-fungal LAB

Reducing sugar through polyol production

Influencing the mouth feel through EPS production by Lactic acid bacteria

Wort base - Flavour modulation

InBev-Baillet Latour Fund
UCC
University College Cork, Ireland
Coláiste na hOllscoile Corcaigh

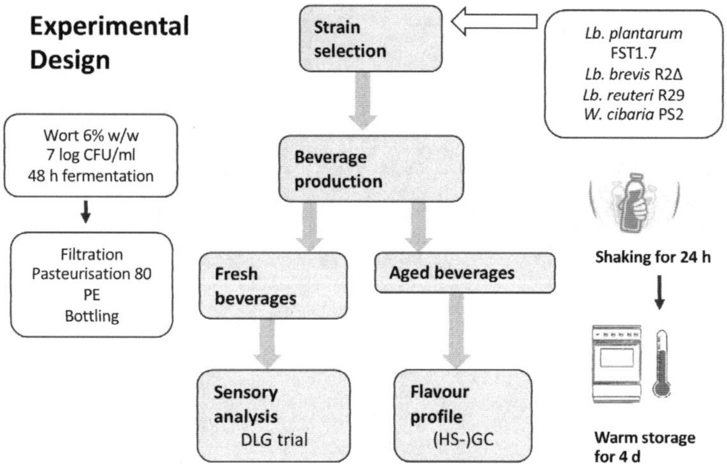

Objective: Identify the LAB species-dependent flavour formation in wort

Application:

- Wort based beverages
- Base for sour-beer

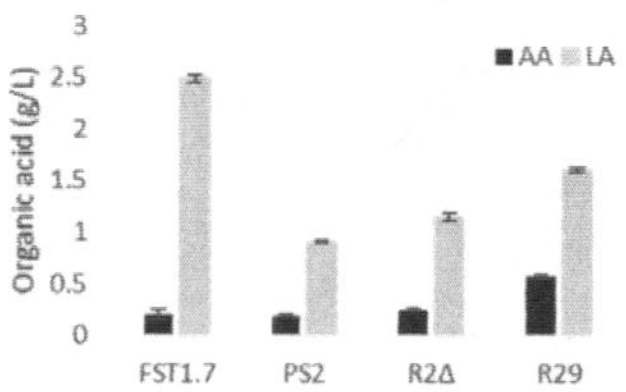

Wort based beverage sensory and stability

	Strecker aldehydes	γ-nonalactone	2-furfural	Total staling indicators	Total oxygen indicators	Total thermal indicators
Unfermented	77.0 (+44%)	10.5 (-32%)*	25.5 (+6%)	212 -> 255	167 -> 214	40 -> 36
Lb. plantarum FST1.7	30.0 (+50%)	38.0 (+33%)	118.0 (+219%)*	13 -> 266	55 -> 105	66 -> 156
W. cibaria PS2	59.5 (+260%)*	13.0 (-37 %)	65.0 (+1200%)*	68 -> 287	42 -> 193	21 -> 78
Lb. brevis R2Δ	31.0 (+94%)*	68.0 (+100%)*	46.0 (+820%)*	76 -> 225	36 -> 96	34 -> 114
Lb. reuteri R29	45.5 (+29%)	19.0 (+6%)	47.5 (+850%)*	108 -> 201	84 -> 130	18 -> 67

* Concentration increment (in %) significant after student t-test (significance at 5%)

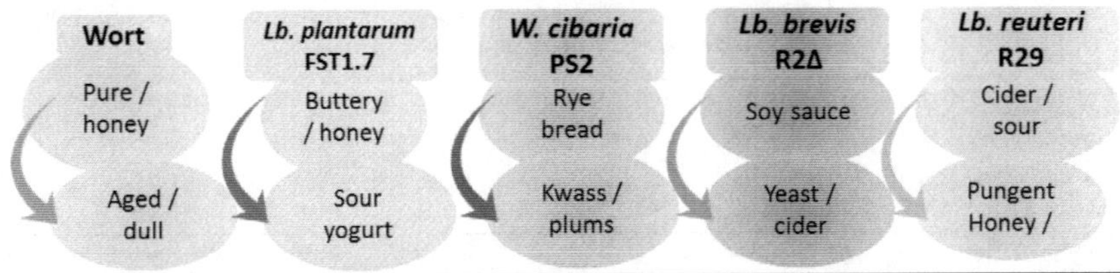

Sensory and flavor profile

Sensory analysis
DLG – German Agricultural Society

"honey" "malty" — Unfermented control

"bready" "yogurty" "appely" "buttery" — Fermented samples

Flavor and aroma profiles
GC-FID (flame ionisation detector)

Strecker aldehydes ↓
(2-, 3-methylbutanal, 2-phenylacetaldehyde)
→ **"malty"** ↓

Diacetyl / acetoin / acetaldehyde ↑
→ "dairy" ↑

Fermentation of wort by LAB cultures resulted in a beverage with a **mild sourness, reduced wort and increased dairy flavour**

Flavour is very LAB strain depended

Reconsideration of 'OFF-Flavours' for novel beverages

Both **sensorial and analytical analysis** needs to be approached simultaneously

School of Food and Nutritional Sciences
University College Cork
Ireland

University College Cork, Ireland
Coláiste na hOllscoile Corcaigh

Lactic acid bacteria producing polyols – reducing sugar in wort based beverages

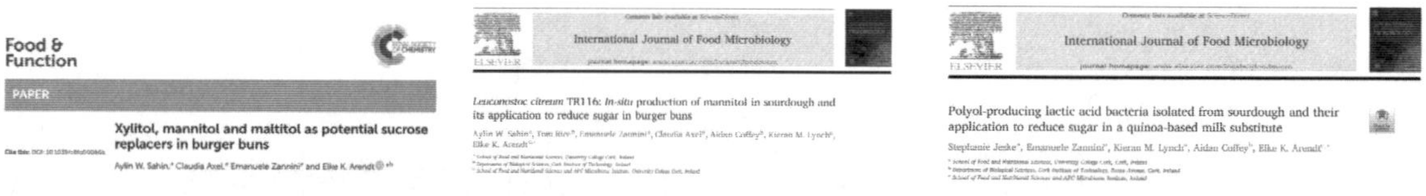
Food & Function
PAPER
Xylitol, mannitol and maltitol as potential sucrose replacers in burger buns

International Journal of Food Microbiology
Leuconostoc citreum TR116: In-situ production of mannitol in sourdough and its application to reduce sugar in burger buns

International Journal of Food Microbiology
Polyol-producing lactic acid bacteria isolated from sourdough and their application to reduce sugar in a quinoa-based milk substitute

Sources of added sugar

University College Cork, Ireland
Coláiste na hOllscoile Corcaigh

38.6 pounds of added sugar from sugar drinks are consumed by the average American each year
(Center for science in the public interest, USA, 2013)

"What is the Soft Drinks Industry Levy (sugar tax)?" by Jamie Oliver, April 2018

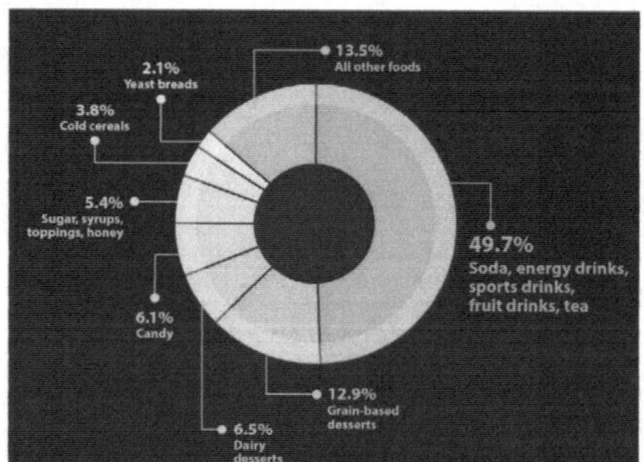

(Center for science in the public interest, USA, 2013)

What is sugar ?

UCC
University College Cork, Ireland
Coláiste na hOllscoile Corcaigh

- **Carbohydrates (mono- and disaccharides [EU Regulation 2011/1169])**
- **Convertible saccharides ↔ Unchangeable saccharides**

Sweeteners

Carbohydrates				Sweetening agents	
Monosaccharides	Disaccharides	Blends of mono- and disaccharides	Polyols	Artificial	Natural
Glucose Fructose Galactose	Sucrose Maltose Lactose	Glucose syrup Isoglucose Starch sugar Invert sugar syrup	Sorbitol Mannitol Xylitol Maltitol Lactitol Erythritol	Saccharin Cyclamate Aspartame Acesulfame-k Sucralose	Stevia Thaumatin Inulin Tagatose Glycyrrhizin

Nabors (2002), Gomez (2008)

28/06/2019 45

Sugar in Beverages

UCC
University College Cork, Ireland
Coláiste na hOllscoile Corcaigh

- **Effects taste, texture and colour**
- **Sucrose or HFCS in most beverages**
- **Carbonated beverages contain about 10% sugar**
- **Corn syrup or invert syrup were often used as a sugar replacer, but limited sweetening power**
- **Polyols and artificial sweetners**
- **Inversion of sucrose**

Sucrose → Invertase → **Glucose** + **Fructose**

– The lower the pH and the higher the processing or storage temperature, the faster the inversion

R.J. Alexander (1998), Nordic Sugar

28/06/2019 46

Polyols – Example Mannitol

UCC
University College Cork, Ireland
Coláiste na hOllscoile Corcaigh

- D-Mannitol is a naturally occurring six-carbon sugar alcohol
  - Used in confectionery (42%), food (13.3%)
  - Mannitol most abundant polyol in nature
    - produced by bacteria, yeasts, fungi, algae, lichens, and many plants
  - 1.6 calories per gram (sucrose has 4)
  - 50 – 70 % of sweetness of sugar
  - Cooling effect on the tongue
  - Low hygroscopicy
  - In oral doses larger than 20 g, mannitol acts as an osmotic laxative

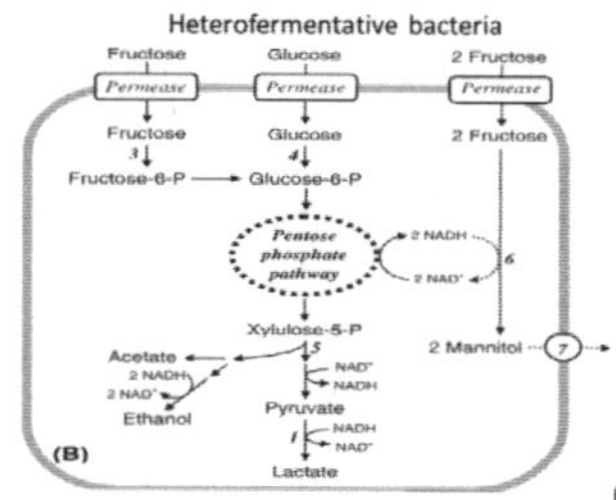

Manitol

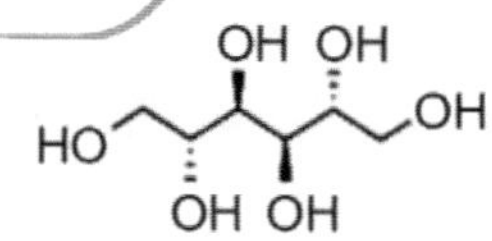

Mannitol pathway of LAB

UCC
University College Cork, Ireland
Coláiste na hOllscoile Corcaigh

Heterofermentative LAB

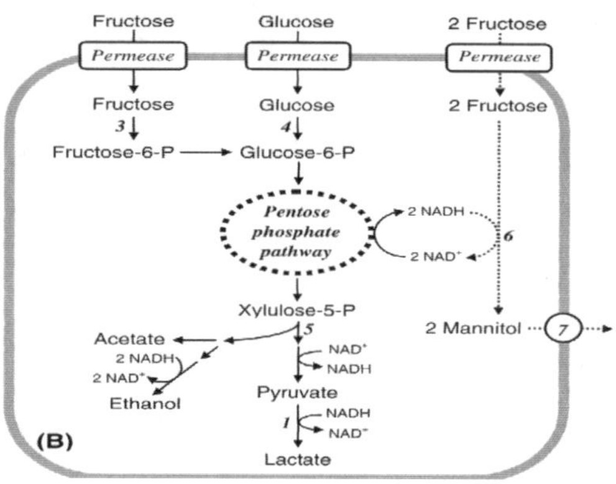

Homofermentative LAB

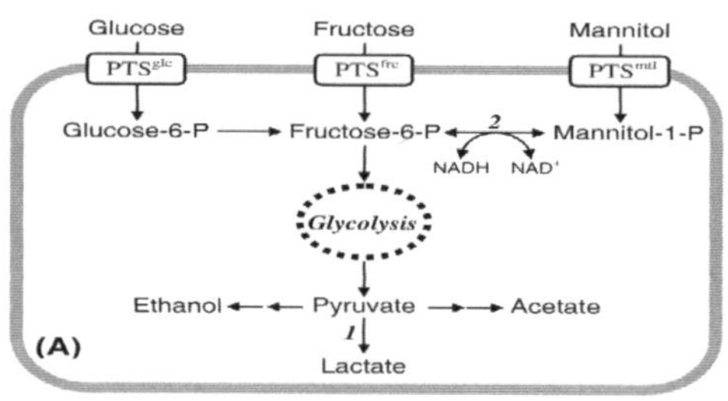

Wisselink et al., (2002); Mannitol production by lactic acid bacteria: a review; Int Dairy J, 12 (2002), pp. 151–161

Research approach & results - polyol producers

University College Cork, Ireland
Coláiste na hOllscoile Corcaigh

Screening for LAB	Isolation of polyol producers from cereal and pulse based sourdough
Screening results	196 strains were isolated 25 % produced polyols
Strain selection	3 LAB strains were selected, sequenced, pathway of production identified
Fermentation	Wort production, Fructose additions 48 h fermentation, pH 4, 10^9CFU/ml
Quality evaluation	7.5 g/l to 25g/l of Mannitol production (depending on fructose level) → low sugar beverage Sensory: pleasant sourness, vinegar flavour

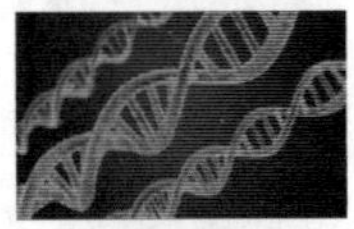

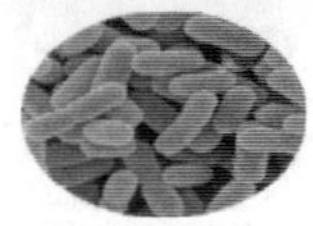

28/06/2019 49

School of Food and Nutritional Sciences
University College Cork
Ireland

University College Cork, Ireland
Coláiste na hOllscoile Corcaigh

Lactic acid bacteria the creation of novel beverages – Canvas

ORIGINAL PAPER

Fibre, protein and mineral fortification of wheat bread through milled and fermented brewer's spent grain enrichment

Deborah M. Waters · Fritz Jacob · Jean Titze · Elke K. Arendt · Emanuele Zannini

WO 2018/033521 A1

A PROCESS FOR PREPARING A BEVERAGE OR BEVERAGE COMPONENT, BEVERAGE OR BEVERAGE COMPONENT PREPARED BY SUCH PROCESS, AND USE OF BREWER'S SPENT GRAINS FOR PREPARING SUCH BEVERAGE OR BEVERAGE COMPONENT

ABInBev

Research Article

Wheat bread biofortification with rootlets, a malting by-product

Deborah M Waters, Wilma Kingston, Fritz Jacob, Jean Titze, Elke K Arendt and Emanuele Zannini

Review article

Brewers' spent grain: a review with an emphasis on food and health

Kieran M. Lynch, Eric J. Steffen and Elke K. Arendt

Brewers spent grain

UCC
University College Cork, Ireland
Coláiste na hOllscoile Corcaigh

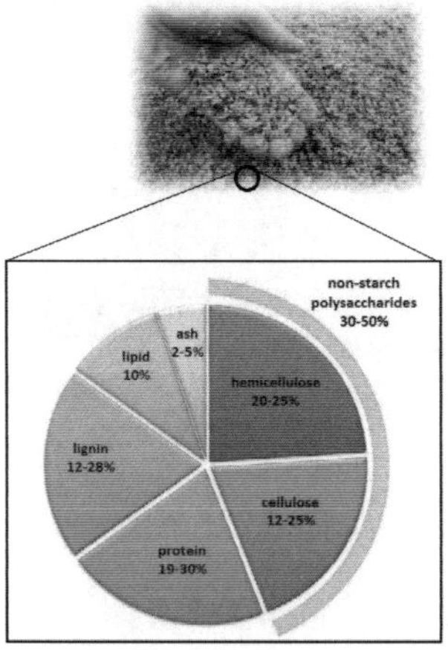

Most abundant **by-product** generated from the brewing process, ~85% of the total by-products

Average annual **global production** is estimated to be ~39 million tonnes

Majority goes for **animal feed** or **to landfill**

BSG is rich in *fibre* and *protein*...

...locked up in a complex lignocellulosic plant structure

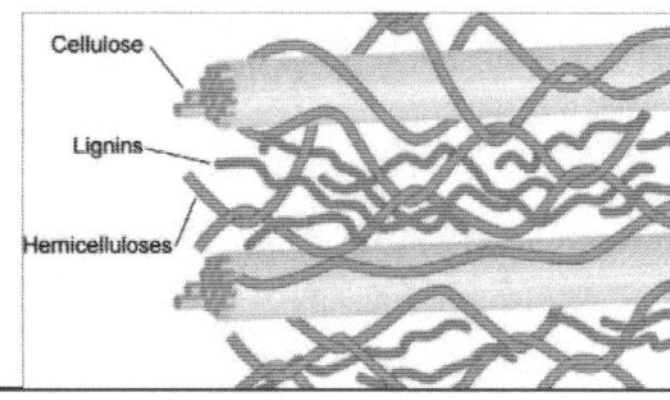

Complex, structure: cellulose forms a skeleton which is surrounded by hemicellulose and lignin

Is it possible to *unlock* the nutritional value in BSG for use in human nutrition?

Lynch, K. M., Steffen, E. J., & Arendt, E. K. (2016). Brewers' spent grain: a review with an emphasis on food and health. *Journal of the Institute of Brewing*, *122*(4), 553-568.

Lactic acid bacteria fermentation of BSG

ABInBev

UCC
University College Cork, Ireland
Coláiste na hOllscoile Corcaigh

Patented fermentation process developed between UCC and AB-InBev

WO2018033522A1: A process for preparing a beverage or beverage component from brewer's spent grains

Enzymatic hydrolysis
- ✓ Nutrient release for LAB fermentation
- ✓ Unlocking of functional fibres and protein

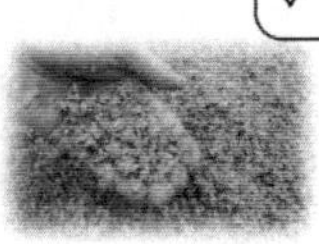

Fresh BSG

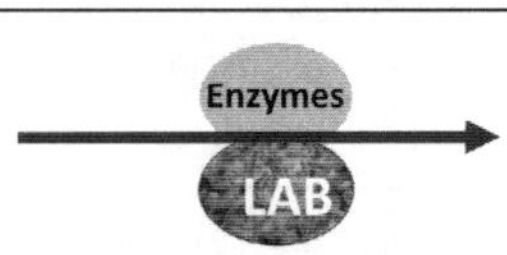

Simultaneous saccharification and fermentation (SSF) process

Fermented BSG beverage base

APPLICATIONS

canvas

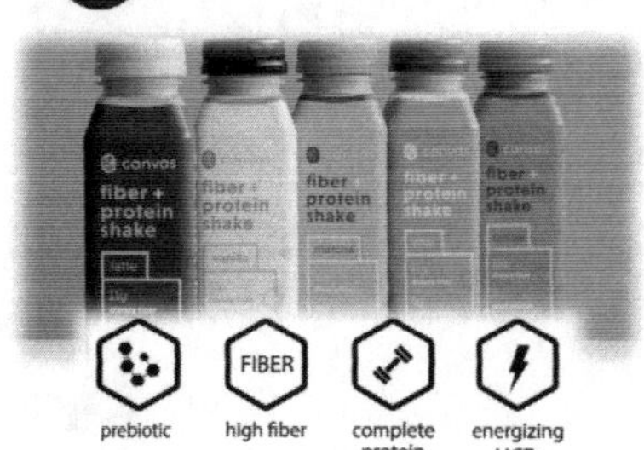

LAB fermentation
- ✓ Stabilisation of product
- ✓ Aroma and flavour formation

"a prebiotic fiber and plant protein shc

https://drinkcanvas.com/

https://patentscope.wipo.int/search/docservicepdf_pct/id00000041321219/PAMPH/WO2018033522.pdf?psAuth=6IAYf3F34lg56zmTRTXjLTbqjisdNJ_ngUZPruzx74I

Canvas First nutritional beverage from Brewer's ~~Spent~~ Saved Grains

UCC
University College Cork, Ireland
Coláiste na hOllscoile Corcaigh

ABInBev

Saved Grains are combined with other remarkable ingredients like **coconut**, **cashews**, **herbs** and **spices**. Coming in 5 different versions: Original (vanilla), Cocoa, Cold Brew Latte, Matcha, and Turmeric Chai. We crafted and perfected each and every detail of our beverage, from flavor, viscosity, efficaciousness, serving size, nutrition, and packaging. As a result, plant-based delicious nourishment for everyone, with **11 g of dietary fiber and 8 g of protein per serving**

28/06/2019 79

Unlocking the potential of Lactic acid bacteria

UCC
University College Cork, Ireland
Coláiste na hOllscoile Corcaigh

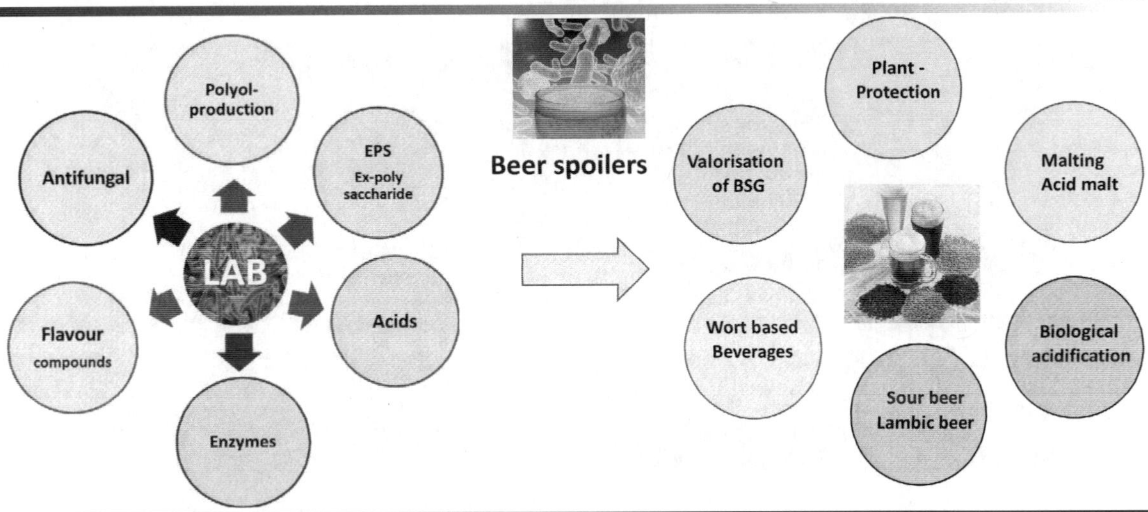

28/06/2019 54

Acknowledgements

UCC
University College Cork, Ireland
Coláiste na hOllscoile Corcaigh

Cereal and Beverage research group at UCC

University College Cork
Dr Alex Mauch
Dr Pedro Olivera
Dr Lorenzo Peryer
Konstatin Bellut
Aylin Sahin

TUM, Germany
Prof Fritz Jacob
Dr Martin Zarnkow

CIT, Ireland
Dr Aidan Coffey
Dr Ambrose Fury
Tom Rice

ABInBev

Fonds InBev-Baillet Latour Fund

Dr Patrick O'Riordan
Dr Jorge Gil-Martinez
Dr Gregory Belt
Dr David DeSchutter
Dr Jerome Pellaut
Dr Luk Daeneen
Dr Philippe Malcorps

Department of Agriculture, Food and the Marine
An Roinn Talmhaíochta, Bia agus Mara
Seventh Framework Programme
Horizon 2020
sfi
JPI a healthy diet for a healthy life
PROTEIN 2FOOD

28/06/2019 90

Thank you for your attention

UCC
University College Cork, Ireland
Coláiste na hOllscoile Corcaigh

28/06/2019 56

Challenging the assumptions around the pasteurisation requirements of beer - spoilage bacteria

Dr Grzegorz Rachon
Campden BRI
Brewing Services (Nutfield)
T: +44(0)1737824298
E: Grzegorz.Rachon@campdenbri.co.uk

37th EBC Congress - 2019

Pasteurisation in brewing

Product	Typical Minimum PU	Typical Maximum PU
Pilsner and Lager Beer	15	25
Ales and Stout	20	35
Low Alcohol Beer	40	60
Non Alcoholic Beer	80	120
Lemonade	300	500
Fruit Juices	3,000	5,000

* EBC Technology and Engineering Forum (1995) *Beer Pasteurisation: Manual of Good Practice*,

Pasteurisation/sterilisation in food industry

- Target microorganism is known
  - Sterilisation - *Bacillus stearothermophilus*
  - Pasteurisation - *L. monocytogenes*, *S.* Senftenberg or *S.* Enteritidis, *E. coli* O157
- Use of surrogates
  - *Salmonella / Listeria / E.coli - E. faecium*
  - *Cl. botulinum - Cl. sporogenes*
  - *L. monocytogenes - L. innocua*
  - *E. coli* O157 VTEC+ve - *E. coli* O157 VTEC-ve

Campden BRI
food and drink innovation

Pasteurisation in brewing

- Tunnel/Flash pasteurisation
- Pasteurisation validation
  - Target microorganism is unknown (NPD, low alcohol beverages)
  - Current approach - trial and error
  - Lethality of the process can't be correctly calculated

Campden BRI
food and drink innovation

Pasteurisation units (PU)

- 60°C for 1 min = 1 PU
- 60°C for 5 min = 5 PU
- 70°C for 1 min = 27.6 PU
- 85°C for 15 sec = 1000 PU

$$PU_{tot} = \sum_{0}^{t} L_T \times \Delta t_T$$

$$L_T = 10^{(T - T_{Ref})/z}$$

Campden BRI
food and drink innovation

Pasteurisation units (PU) – whole process

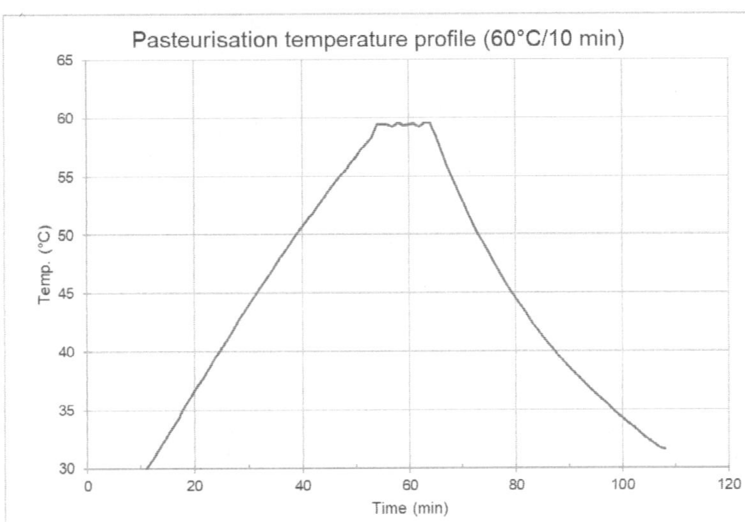

Campden BRI
food and drink innovation

Pasteurisation units (PU) – whole process

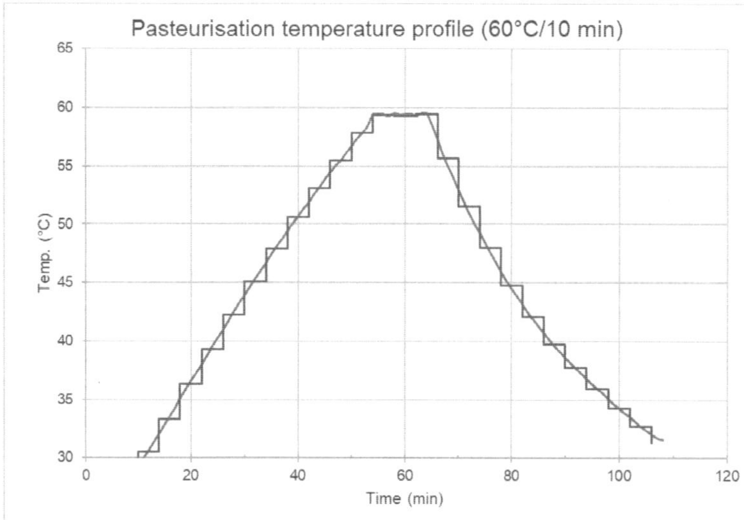

Section	PU
Ramp time	3.3
Hold time	9.2
Cooling	2.0
Total	14.5

Campden BRI
food and drink innovation

PU of the process and Process Lethality

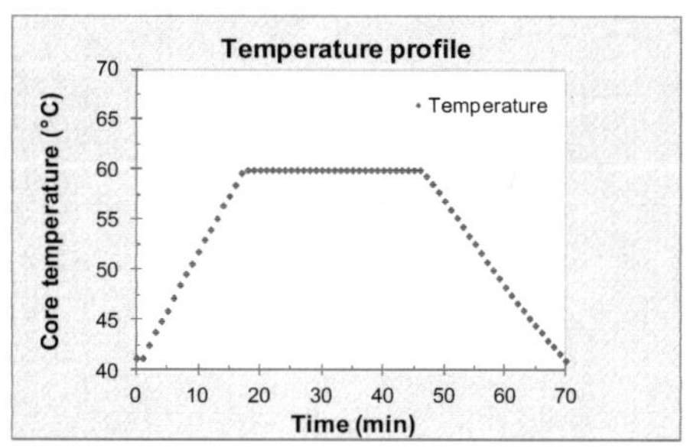

$$PU_{tot} = \sum_{0}^{t} L_T \times \Delta t_T$$

$$L_T = 10^{(T - T_{Ref})/z}$$

z-value = 6.94°C for Del Vecchio (1951)

$$L_{Tproc} = \frac{PU_{tot}}{D_{REF}}$$

D_T-value – decimal reduction time
z-value - measure of the change of the D-value with varying temperature

* Del Vecchio, H. W., Dayharsh, C. A., and Baselt, F. C. (1951) Thermal death time studies on beer spoilage organisms,

Campden BRI
food and drink innovation

z-value = 6.94°C for Del Vecchio (1951)*

- Used beer, supplemented with five percent boiled wort (**higher sugar content**)
- Inactivation kinetics were based on an unidentified (and **uncharacterised**) so-called "**abnormal yeast**".
- This work was never intended to provide a catch-all model for pasteurisation. Indeed, the authors themselves reported in the study that:

  "*...It should not be concluded that the resistance found for the organisms in these tests are necessarily the maximum existing in the beer industry. Only numerous tests with different types and strains and on different beers and ales can determine this point*"

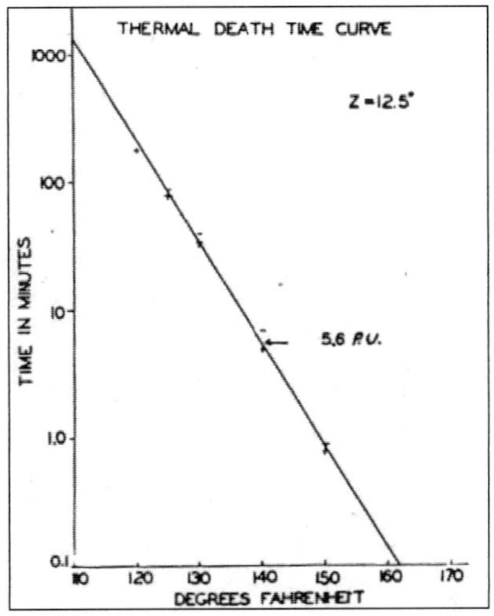

- z = 6.94°C (12.5°F)
- D_{60} = 5.6 min (140°F)

Campden BRI
food and drink innovation

* Del Vecchio, H. W., Dayharsh, C. A., and Baselt, F. C. (1951) Thermal death time studies on beer spoilage organisms,

Campden BRI study - method

- Preliminary screen
  - 14 microorganisms
  - growth check
  - heat resistance

Gluconobacter oxydans
Gluconacetobacter saccharivorans
Acetobacter pasteurianus
Kocuria kristinae
Obesumbacterium proteus
Enterobacter kobei
Bacillus megaterium
Lactobacillus brevis
Lactobacillus paracasei
Lactobacillus brevis
Lactobacillus hilgardii
Pediococcus cerevisiae
Pediococcus pentosaceus
Pediococcus damnosus

- Determination of D- and z-values
  - three most heat resistant strains
  - capillary tube method

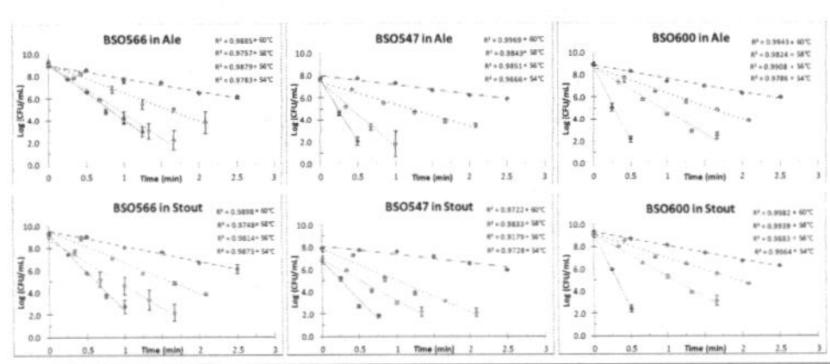

- Pasteurisation validation
  - tunnel pasteuriser
  - bottles
  - range of PU settings

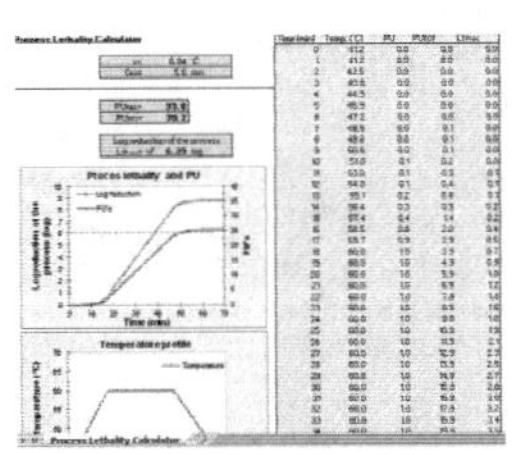

Campden BRI
food and drink innovation

Campden BRI study - results

Growth (ale and stout) and heat resistance (0.5 PU)

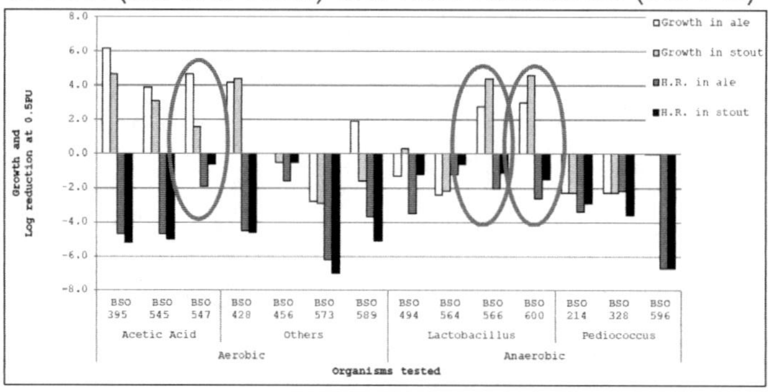

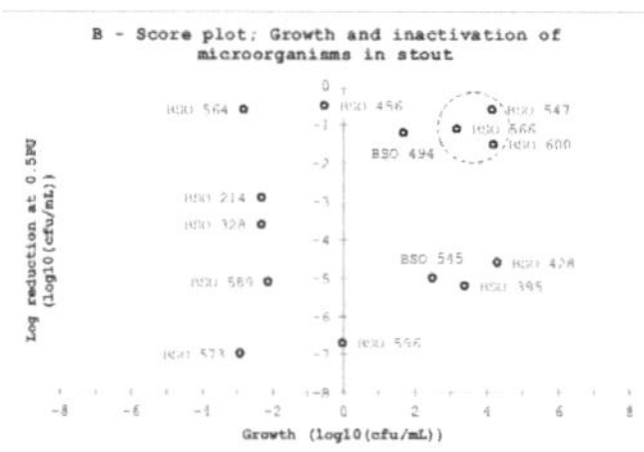

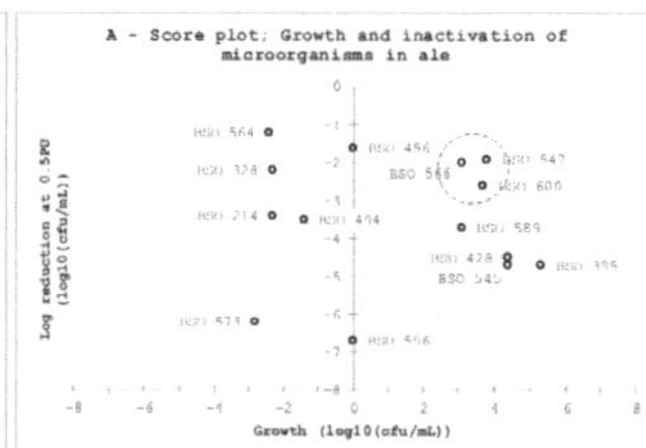

3 the most heat resistant microorganisms were:

- ***Lactobacillus brevis***
- ***Lactobacillus hilgardii***
- ***Acetobacter pasteurianus***

Campden BRI
food and drink innovation

Campden BRI study - results

D- and z-values

Strain	Beer	D-values (min) 54°C	56°C	58°C	60°C	z-value (°C)
Acetobacter pasteurianus (BSO547)	Ale	1.20 ± 0.08	0.49 ± 0.01	0.17 ± 0.02	0.09 ± 0.02	5.17 ± 0.32
	Stout	1.30 ± 0.07	0.35 ± 0.03	0.27 ± 0.03	0.14 ± 0.01	6.71 ± 0.14
Lactobacillus brevis (BSO566)	Ale	0.85 ± 0.03	0.39 ± 0.07	0.22 ± 0.02	0.20 ± 0.02	9.48 ± 0.37
	Stout	0.74 ± 0.07	0.36 ± 0.01	0.24 ± 0.01	0.15 ± 0.01	8.68 ± 0.39
Lactobacillus hilgardii (BSO600)	Ale	0.81 ± 0.03	0.42 ± 0.01	0.25 ± 0.01	0.07 ± 0.00	5.97 ± 0.19
	Stout	0.80 ± 0.03	0.46 ± 0.01	0.27 ± 0.01	0.08 ± 0.00	6.08 ± 0.19

- *Acetobacter pasteurianus;* **D_{60}= 0.09 – 0.14 min** and z-values (**z = 5.17 - 6.71°C**)
- *Lactobacillus brevis;* **D_{60}= 0.15 – 0.20 min** and z-values (**z = 8.68 – 9.48°C**)
- *Lactobacillus hilgardii;* **D_{60}= 0.07 – 0.08 min** and z-values (**z = 5.97 – 6.08°C**)
  - z-value for Del Vecchio abnormal yeast; **z = 6.94°C**

Spoilage bacteria	Medium	D_{60}	z - value	Reference
Lactobacillus paracasei	Pilsner	0.02	6.5	Reveron et al. (2005)
Lactobacillus sp.	Lager (4.4%v/v)	0.31	10.7	Adams et al. (1989)
Lactobacillus delbrueckii	Lager	0.64*	7.2	L'Anthoen et al. (1996)
Lactobacillus sp.	Lager (low alcohol)	1.57	9.2	Adams et al. (1989)
Lactobacillus delbrueckii	alcohol free beer	4.01	15.8	L'Anthoen et al. (1996)

*extrapolation

Campden BRI
food and drink innovation

z - values

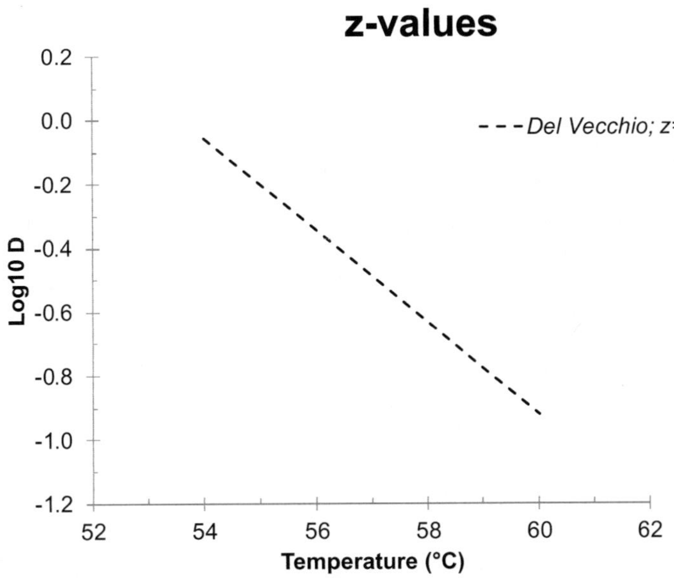

Campden BRI
food and drink innovation

z - values

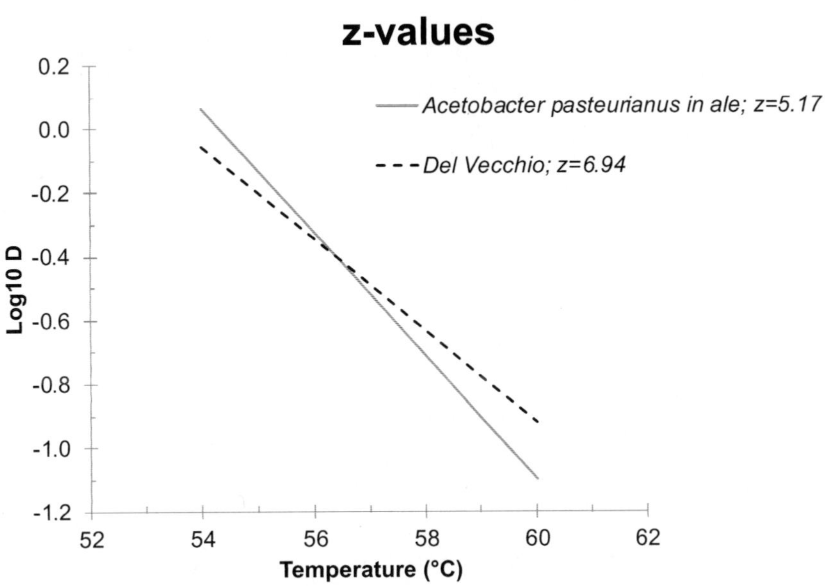

Campden BRI
food and drink innovation

z - values

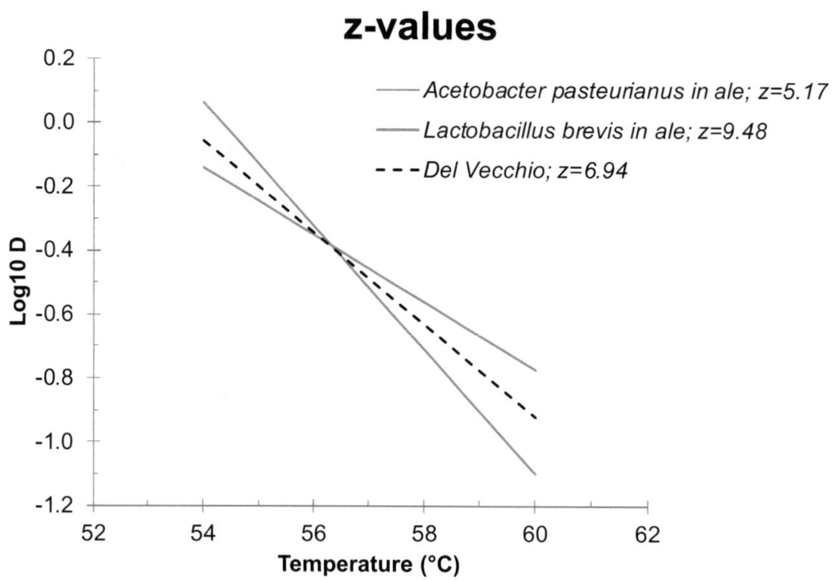

Campden BRI
food and drink innovation

Pasteurisation validation

- Beer inoculated at high level (10^7 CFU/mL) with cocktail of *Lactobacillus* spp *and Acetobacter* sp

- *Lactobacillus brevis* sp
  - D_{60} = 0.2 min; z = 9.48 °C
  - 0.2 min X 6 = 1.2 min → 1.2 PU

- Three trials (one fail and two successful)
  - 1st at 0.75 PU - achieved **0.6 PU**;
    45 min ramp time
    1 min holding time at 52°C
    cooling to 35°C for 25 min
  - 2nd at 1.5 PU - achieved **1.6 PU**;
    45 min ramp time
    1 min holding time at 54°C
    cooling to 35°C for 25 min
  - 3rd at 3 PU - achieved **3 PU**;
    45 min ramp time
    1 min holding time at 56°C
    cooling to 35°C for 25 min

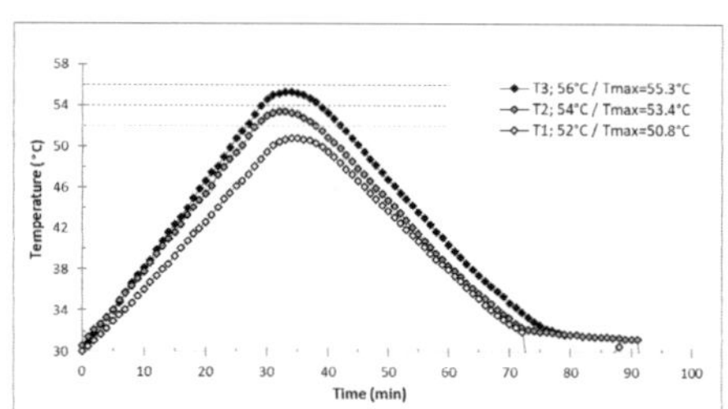

Campden BRI
food and drink innovation

Pasteurisation validation - prediction

Microorganism	Temp. (°C)	Beer	Calculated Campden BRI trial - various z-values		Calculated Del Vecchio trial - z=6.94	
			PU_{tot}	L_{TProc}	PU_{tot}	L_{TProc}
Acetobacter pasteurianus (z_{ale} =5.17, z_{stout}=6.71)	52	Ale	0.2	2.3	0.7	7.9
		Stout	0.6	4.5	0.7	5.1
	54	Ale	0.6	6.9	1.6	17.6
		Stout	1.4	10.2	1.6	11.3
	56	Ale	1.5	16.2	3.0	32.9
		Stout	2.8	19.6	3.0	21.2
Lactobacillus brevis (z_{ale} =9.48, z_{stout}=8.68)	52	Ale	2.0	10.0	0.7	3.6
		Stout	1.5	10.2	0.7	4.8
	54	Ale	3.5	17.5	1.6	7.9
		Stout	2.9	19.0	1.6	10.6
	56	Ale	5.5	27.5	3.0	14.8
		Stout	4.7	31.1	3.0	19.7
Lactobacillus hilgardii (z_{ale} =5.97, z_{stout}=6.08)	52	Ale	0.4	5.7	0.7	10.2
		Stout	0.4	5.3	0.7	8.9
	54	Ale	1.0	14.4	1.6	22.6
		Stout	1.1	13.4	1.6	19.8
	56	Ale	2.1	29.9	3.0	42.3
		Stout	2.2	27.4	3.0	37.0

Campden BRI
food and drink innovation

Pasteurisation validation - results

Microorganism	Temp. (°C)	Beer	Calculated Campden BRI trial - various z-values		Calculated Del Vecchio trial - z=6.94		Achieved Campden BRI validation
			PU_{tot}	L_{TProc}	PU_{tot}	L_{TProc}	L_{TProc}
Acetobacter pasteurianus (z_{ale} =5.17, z_{stout}=6.71)	52	Ale	0.2	2.3	0.7	7.9	4.8
		Stout	0.6	4.5	0.7	5.1	7.5
	54	Ale	0.6	6.9	1.6	17.6	>8.7
		Stout	1.4	10.2	1.6	11.3	>8.7
	56	Ale	1.5	16.2	3.0	32.9	>8.7
		Stout	2.8	19.6	3.0	21.2	>8.7
Lactobacillus brevis (z_{ale} =9.48, z_{stout}=8.68)	52	Ale	2.0	10.0	0.7	3.6	>8.8
		Stout	1.5	10.2	0.7	4.8	>8.7
	54	Ale	3.5	17.5	1.6	7.9	>8.7
		Stout	2.9	19.0	1.6	10.6	>8.6
	56	Ale	5.5	27.5	3.0	14.8	>8.7
		Stout	4.7	31.1	3.0	19.7	>8.6
Lactobacillus hilgardii (z_{ale} =5.97, z_{stout}=6.08)	52	Ale	0.4	5.7	0.7	10.2	>8.8
		Stout	0.4	5.3	0.7	8.9	>8.7
	54	Ale	1.0	14.4	1.6	22.6	>8.7
		Stout	1.1	13.4	1.6	19.8	>8.6
	56	Ale	2.1	29.9	3.0	42.3	>8.7
		Stout	2.2	27.4	3.0	37.0	>8.6

Campden BRI
food and drink innovation

Conclusions

- Three the most heat-resistant microorganisms able to grow in the beers were *L. brevis* and *L. hilgardii* and *A. pasteurianus*
- The thermo tolerance of these bacteria were similar
- z-values of the three most heat resistant bacteria were either lower (5.17 °C) or higher (9.48°C) then z-value (6.94°C) reported by Del Vecchio *et al.*
- Using Del Vecchio's z-value under- or over estimated the lethality of the process
- Process lethality was calculated correctly when z-values determined in this study were used
- Only 1.6 PU was needed to eliminate 6 log of spoilage bacteria
- Further studies, conducted on a broad range of organisms (including yeast ascospores) and other beer styles, are still required

Campden BRI
food and drink innovation

Future work – yeast ascospores

Ascospore producing yeast tested		D_{60} (McIlvaine buffer; pH=4)
Saccharomyces cerevisiae	ATCC 9080	15.2
Pichia manshurica	Wine isolate	5.4
Saccharomycodes ludwigii	Wine isolate	4.2
Saccharomyces cerevisiae	Kombucha isolate	3.4
Saccharomyces spp	Brewery isolate	3.3
Saccharomyces spp	Brewery isolate	2.4
Pichia farinosa	Beer isolate	2.4

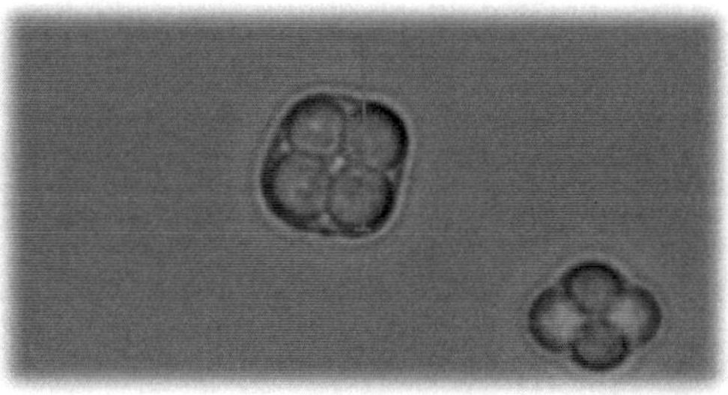

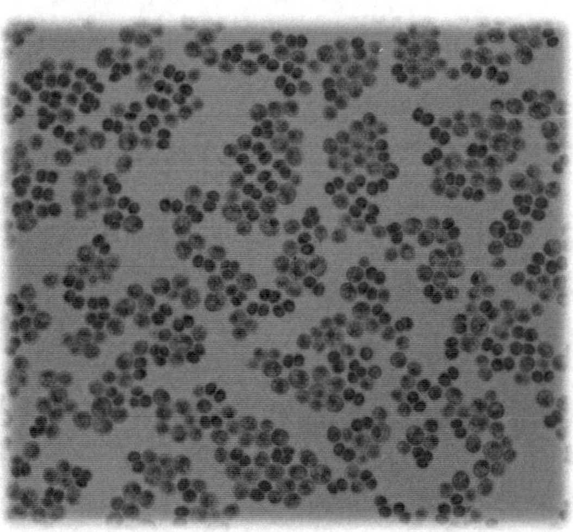

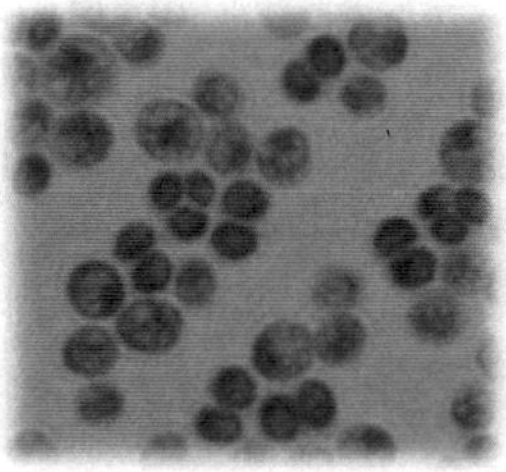

Campden BRI
food and drink innovation

Research article

Received: 12 February 2018 Revised: 10 June 2018 Accepted: 11 June 2018 Published online in Wiley Online Library

(wileyonlinelibrary.com) DOI 10.1002/jib.520

Challenging the assumptions around the pasteurisation requirements of beer spoilage bacteria

Grzegorz Rachon,* Christopher J. Rice, Karin Pawlowsky and Christopher P. Raleigh

Current recommendations for beer pasteurisation are based on the study in 1951 by Del Vecchio and coworkers. In this work, 14 beer spoilage bacteria were screened for their ability to grow or survive in ale and stout together with the determination of their thermo tolerance at 60°C. Using a capillary tube method, the *D*-value (decimal reduction time) and *z*-value (temperature resistance coefficient) of the three thermo tolerant bacteria (*Acetobacter pasteurianus*, *Lactobacillus brevis* and *Lactobacillus hilgardii*) were determined. Validation of pasteurisation at a range of pasteurisation units (PU) in packaged product were performed in a tunnel pasteuriser. This study showed that eight of the 14 microorganisms were able to grow in both beer styles, whilst different thermo tolerances were observed amongst the spoilage bacteria. Effective pasteurisation of the selected microorganisms was achieved at significantly lower PU values than those recommended by the European Brewery Convention *Manual of Good Practice*. In package pasteurisation conducted at 1.6 PU resulted in greater than an 8-log reduction in viable cell numbers, resulting in 'commercial sterility'. Although this study demonstrated that successful pasteurisation was achieved for vegetative cells at significantly lower PU values than those recommended, further studies are required to demonstrate the optimal level of pasteurisation for spore forming bacteria and for yeast.

Keywords: pasteurisation; spoilage bacteria; validation; capillary tubes; heat inactivation

BREF – Brewers' Research & Education Fund
(British Beer & Pub Association)

Thank you

Campden BRI
food and drink innovation

L19 - TAKING HOP AROMA APART: SUPERCRITICAL CO2 FRACTIONATED OILS OFFERS NEW FLAVOUR POSSIBILITIES FOR BREWERS.

Christina Dietz[1]; David Cook[1]; Ray Marriott[2]; Colin Wilson[2]; Rebecca Ford[1]

1 - University of Nottingham, School of Biosciences, Division of Food Sciences, Sutton Bonington Campus, Loughborough, UK; 2 - Totally Natural Solutions Ltd, Unit 3a Arnold Business Park, Branbridges Road, East Peckham, Kent, UK

Abstract

Advanced, clean label hop products obtained by sustainable extraction are required to comply with future legislation and to manage fluctuations in hop oil content and composition caused by climate change. Hop oil fractions with unique sensory characteristics can be produced by sequential extraction using organic solvents. However, little research is available regarding the sensory characterisation of hop oil fractions. Two studies were designed to measure olfactory, gustatory and trigeminal-type differences between supercritical (sc)CO2 fractions and sub-fractions.

Magnum hop oil was fractionated into five fractions of different polarities. The total oil and fractions were evaluated in 4% EtOH/H2O by a trained sensory panel (n=10) using a newly developed attribute lexicon with a modified quantitative descriptive analysis (QDA) approach. Significant differences were found between all samples for multiple olfactory, gustatory, and trigeminal sensations. In addition, the chemical profiles of the samples were analysed using gas chromatography-mass spectrometry (GCMS). Interestingly, Principal Component Analysis of sensory and analytical data revealed that particularly the polar mono- and sesquiterpene alcohol sub-fractions are responsible for cross-modal interactions eliciting both aroma and/or taste and trigeminal sensations.

In order to identify compounds driving these cross-modal interactions, selected fractions were further divided into smaller characterised sub-fractions and individual molecules. These were applied in a lager beer base (4.5% v/v) and profiled analytically using GC-MS and sensorially by a trained panel using QDA. The outcomes of each study indicate that matrix-, concentration-, and perceptual interaction effects determine sensory impressions of the sub-fractions.

Short biography

Christina is a Sensory and Brewing Science PhD researcher at the University of Nottingham (UK) holding a BSc degree in Food Management from Weihenstephan-Triesdorf University of Applied Sciences (Germany) and a MSc degree in Food Technology from Wageningen University (The Netherlands). Since February 2017, Christina is working on her PhD project, which focuses on multimodal stimuli (gustatory/olfactory/trigeminal) in hops that contribute to different sensory characteristics in beer. Her PhD project is sponsored by Totally Natural Solutions Ltd (UK).

L20 - EXPLORING HOP DERIVED CONTRIBUTORS TO BEER BITTERNESS USING DATA-INDEPENDENT ACQUISITION (DIA)

Nils Rettberg[1]; Laura Knoke[1]; Julia Hildebrandt[1]; Sarah Thörner[1]; Jörg Maxminer[1]

1 - VLB Berlin - Research Institute for Beer and Beverage Analysis

Abstract

Untargeted metabolite profiling (UMP) has become a relevant method to understand molecular dynamics of complex processes such as beer production. UMP targets a non-discriminative comparison of samples in order to identify significant (unknown) metabolites that relate to product quality. UMP methods commonly do not use any sample cleanup steps and use generic LC-methods combined with high resolution mass spectrometry.

Beer quality control is based on multiple well established quantitative methods and it seems unlikely that UMP substitutes those in near future. However, additional information for quantification can be retrospectively extracted from UMP data. Therefore, it seemed reasonable to evaluate, if and to which extend, UMP and quantification of beer key quality attributes can be combined in a single LC run.

The aim of the current study was to evaluate if quantitation of beer bitter acids, representing a complex and relevant group of non-volatile beer metabolites, can be integrated into an UPLC-Q-TOF based UMP. In order to identify and quantify the full spectrum of major and minor beer bitter compounds we employed a DIA approach on beer samples varying in their bitterness characteristics. An accurate mass database for bitter acids ionized by ESI (-) and fragmented MS^e by was compiled, calibration was established by either using hop extracts or bitter acid DCHA-salts. The data collected by UPLC-Q-TOF MS is discussed in comparison with results of conventional beer bitter acid analysis (EBC 9.8 and EBC 9.47) as well as sensory evaluation (quantitative descriptive analysis).

Short biography

Nils (born 1983) is a trained brewer and maltster holding a diploma in biotechnology with focus on brewing science from TU Berlin (TUB). After accumulating sufficient skills in brewing and drinking beer, he developed a deep interest in analyzing molecules that make this excellent beverage either taste terribly good or horribly stale. Therefore, after graduating in 2011, Nils worked as research associate at TU Berlin (Chair of Bioanalytics) and VLB Berlin (Research Institute for Special Analyses). In 2014 he successfully defended his doctoral thesis on "Comprehensive analysis of hop secondary metabolites". He is currently in charge of the VLB Research Institute for Beer and Beverage Analysis. This Institute comprises an accredited lab for beer and beverage analysis, as well a group of scientists dedicated to versatile research projects in the field of hops and beer.

L21 - DIMETHYL SULFIDE IN BEER: THE POTENTIAL ROLE OF HOPS

Elise Salanouve[1]; Stéphane Delpech[1]; Clément Viel[1]; Scott R. Lafontaine[1]; Thomas H. Shellhammer[2]; Laurent Dagan[1]

1 - NYSEOS, ZA Parc 2000, 53 rue Claude François, 34080 Montpellier, France; 2 - Department of Food Science and Technology, Oregon State University, 100 Wiegand Hall, Corvallis, Oregon 97331, United States

Abstract

Dimethyl sulfide (DMS) is typically considered an off-flavor in beer, with a cooked-vegetable aroma, at a low flavor threshold of 30 g/L. DMS has a significant effect on the flavor of beer and at low levels may contribute to beer aroma complexity.[1] Brewers tend to control DMS amount by focusing on its precursors and their decomposition during the brewing process. The conventional wisdom is that malted barley is the principal source of DMS precursors since DMS mainly results from the thermal decomposition of S-methyl methionine (SMM) in malt and from the reduction of DMSO during fermentation.[2] Focusing on this key parameter, we have hypothesized that hops contain SMM since it has been reported in the literature that SMM plays a major role in sulfur transport in plants.[3,4] A screening of American commercial IPA's revealed high concentrations of DMS and its precursors. Based on these results, experiments were conducted probing the role of hops on the release of DMS after bottling. Our results demonstrated that hops can contribute to DMS in beer, and dry-hopped beers contain significant concentrations of DMS precursors that may be released upon storage depending on hopping rate, storage time and temperature. Our work then highlights the necessity to better analyze and understand the role of hops as potential sources of DMS in beer.

(1) Bamforth, C. *et al. J. Am. Soc. Brew. Chem.* **2014**

(2) Baldus, M. *et al. Brew. Sci.* **2018**

(3) Bourgis, F. *et al. The Plant Cell* **1999**

(4) Trossat, C. *et al. Plant Physiology* **1996**

Short biography

Dr. Elise Salanouve obtained her PhD in Molecular chemistry at Université Pierre et Marie Curie in Paris in 2014. She worked in a biotech start-up as a R&D scientist focusing on agronomy and bio-sourced chemistry. Since 2018, she has joined NYSEOS, as a R&D project manager, a laboratory specialized on aroma compounds analysis in the fields of enology and brewing. She is in charge of R&D programs that support actively winemakers and brewers to better understand and control their aroma beverages.

BEECH WOOD SAWDUST AS A SUSTAINABLE FILTER AID FOR PRE-COAT FILTRATION

Max Panglisch, Thomas Kunz, Frank-Jürgen Methner

Institute of Food Technology and Food Chemistry
Chair of Brewing Science
TU Berlin

Aim of the study

- Pre-coat filtration with kieselguhr is state of the art in beer production
  - finite raw material with high waste disposal costs
  - improper handling can cause health problems
  - negative influence on oxidative stability of beer by iron entry

Can beech wood sawdust apply as a sustainable filter aid for beer filtration?

Laboratory, pre-coating and filtration trials were carried out to:

- evaluate handling during filtration, technological problems
- examine influences of key beer parameters and oxidative beer stability
- verify flavor impacts on filtrated beer

All results are compared to reference values generated with kieselguhr as filter aid

Introduction of beech wood sawdust

Technische Universität Berlin

wood chips

wood shavings

wood sawdust

European beech
(Fagus sylvatica)

distributed beech wood sawdust batches:

particle size: **<0,065mm**

particle size: **0,08-0,125mm**

particle size: **<0,125mm**

particle size: **>0,125mm**

laboratory trials →

Which of those batches has suitable filtration characteristics?

Which one is applicable for further pre-coat and filtration trials?

37. EBC Antwerp 2019 | Beech wood sawdust as a sustainable filter aid for pre-coat filtration | Panglisch
Slide 3

Laboratory trials

microscope evaluation

Technische Universität Berlin

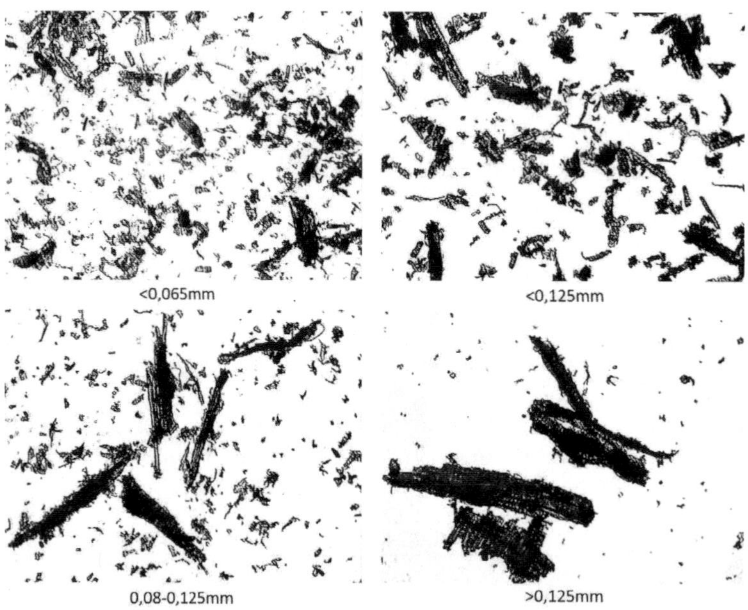

37. EBC Antwerp 2019 | Beech wood sawdust as a sustainable filter aid for pre-coat filtration | Panglisch
Slide 4

Laboratory trials
sedimentation characteristics

Technische Universität Berlin

1.

(<0,065mm; <0,125mm; 0,08-0,125mm; >0,125mm)

2.

(<0,065mm; <0,125mm; 0,08-0,125mm; >0,125mm)

3.

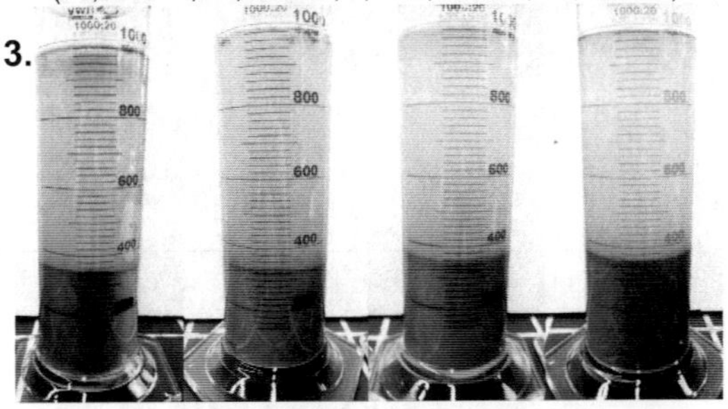

(<0,065mm; <0,125mm; 0,08-0,125mm; >0,125mm)

1. sedimentation after 5 minutes
2. sedimentation after 25 minutes
3. sedimentation after 24 hours

Effects:
- large particles sediment faster
- Coloring effect
- high turbidity

37. EBC Antwerp 2019 | Beech wood sawdust as a sustainable filter aid for pre-coat filtration | Panglisch
Slide 5

Laboratory trials
filtration tests with "Filtercheck" acc. to " Raible Filtertest"

Technische Universität Berlin

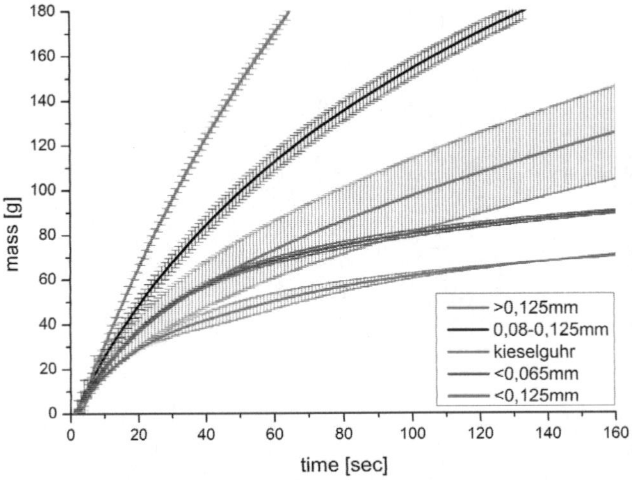

analysis	unit	kieselguhr	<0,125mm	<0,065mm	0,08-0,125mm	>0,125mm
haze 90° /25° (20° C)	[EBC]	1,78 / 1,40	2,80 / 1,65	4,10 / 2,20	32,8 / 24,4	>99,9

37. EBC Antwerp 2019 | Beech wood sawdust as a sustainable filter aid for pre-coat filtration | Panglisch
Slide 6

Treatment of beech wood sawdust

procedure

Technische Universität Berlin

Input of several washable substances causing turbidity, increasing the color and affecting the flavor

➡ **beech wood sawdust needs to be treated before usage for filtration needs**

- 5 consecutively performed short rinsing steps with cold water
- gentle drying process with hot air (70 ° C)
- mechanical crushing of the chunks

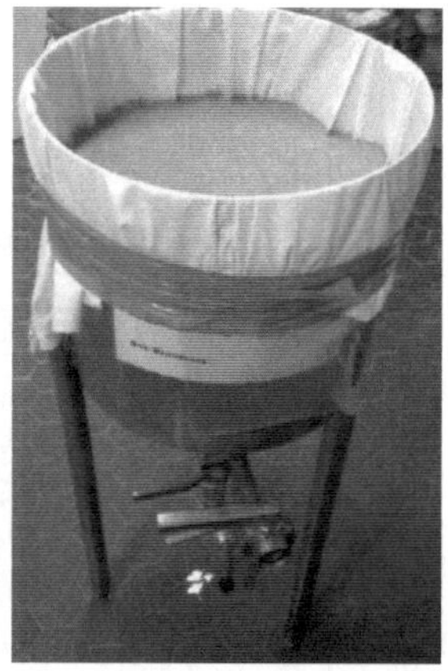

Treatment of beech wood sawdust

iron content of rinsing water samples

Technische Universität Berlin

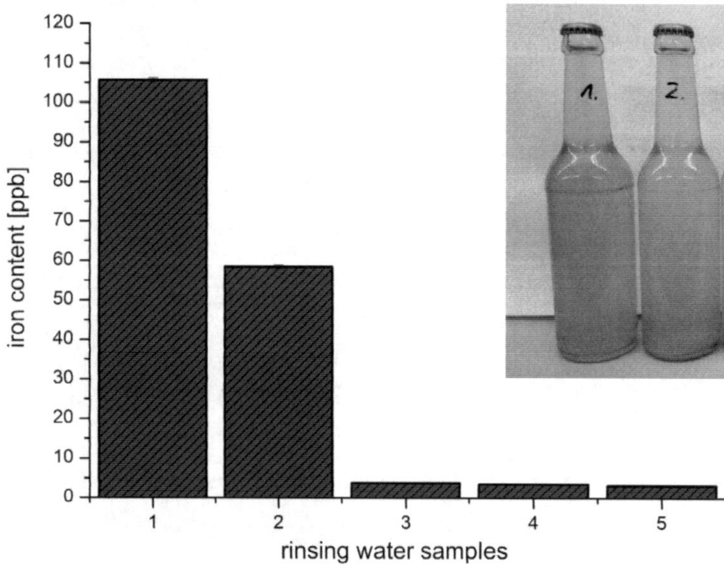

Treatment of beech wood sawdust

polyphenols / anthocyanogenes

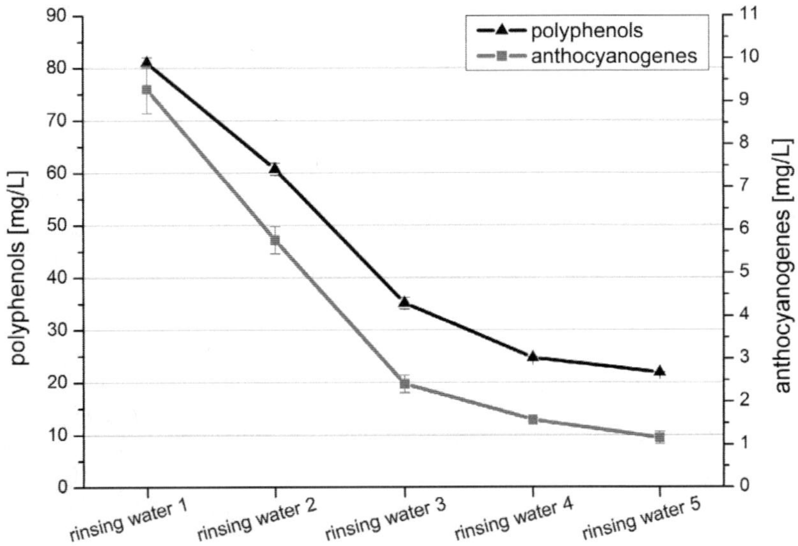

Filtrox filtration pilot plant

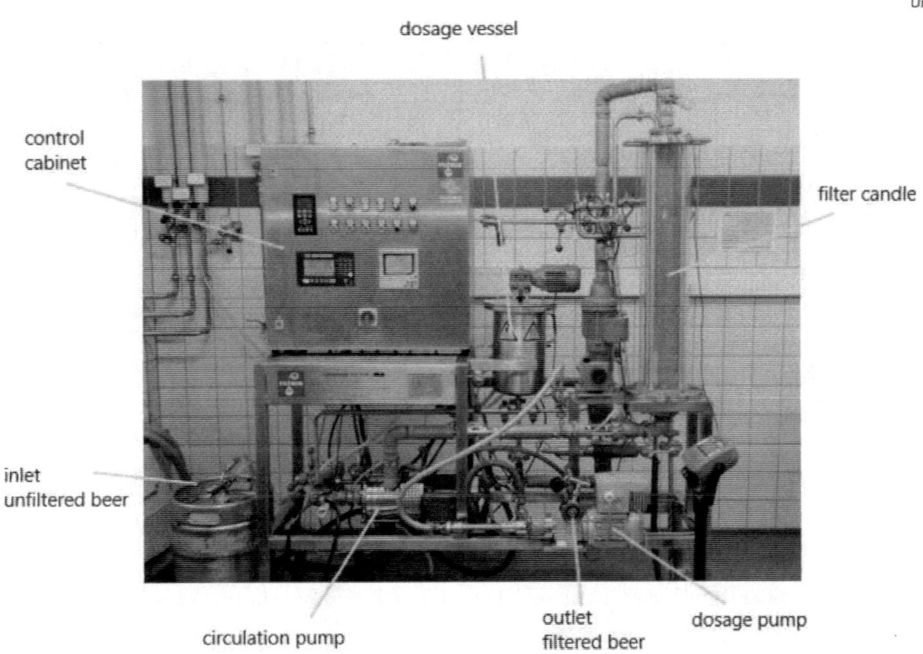

Filtration filtration pilot plant

experimental setup filtration trials

Technische Universität Berlin

- result of several pre-coating trials is that sawdust with particle size of <0,065mm has best technological characteristics

- focus in this lecture is the differentiation of untreated and treated sawdust

  1. **<0,065mm**
  2. **<0,065mm (treated sawdust)**

- further trials:
  - comparison of different particle size distributions
  - examination of higher pre-coat and dosage amount during filtration
  - comparison of different unfiltrates (pale/dark; normal/high gravity)
  - combination of sawdust with additives

Filtrox filtration pilot plant

handling; filtration performance

Technische Universität Berlin

start of the pre-coating

filter cartridge coated

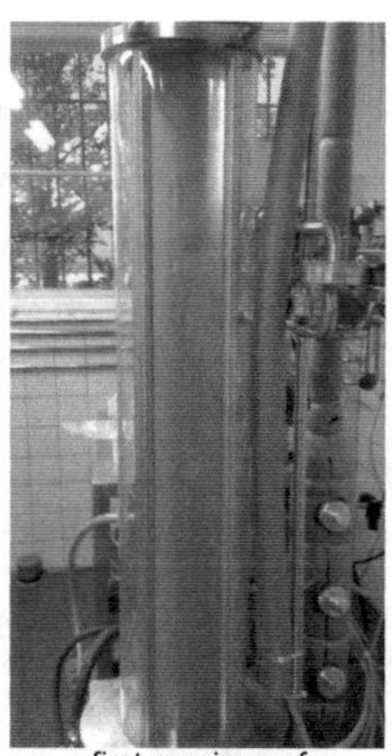

first runnings of unfiltered beer

filtration

end of filtration, increased filter cartridge

- satisfying homogenous and stable filter cake is built, robust against pressure surges and drops
- Pressure difference after pre-coating is between 0,05-0,2 bar
- increase of pressure difference between filter inlet and outlet is relatively slow compared to kieselguhr
  - filter cake with sawdust is not as compressible as with kieselguhr
    - influence on filter life

Filtrox filtration pilot plant
Filtration performance

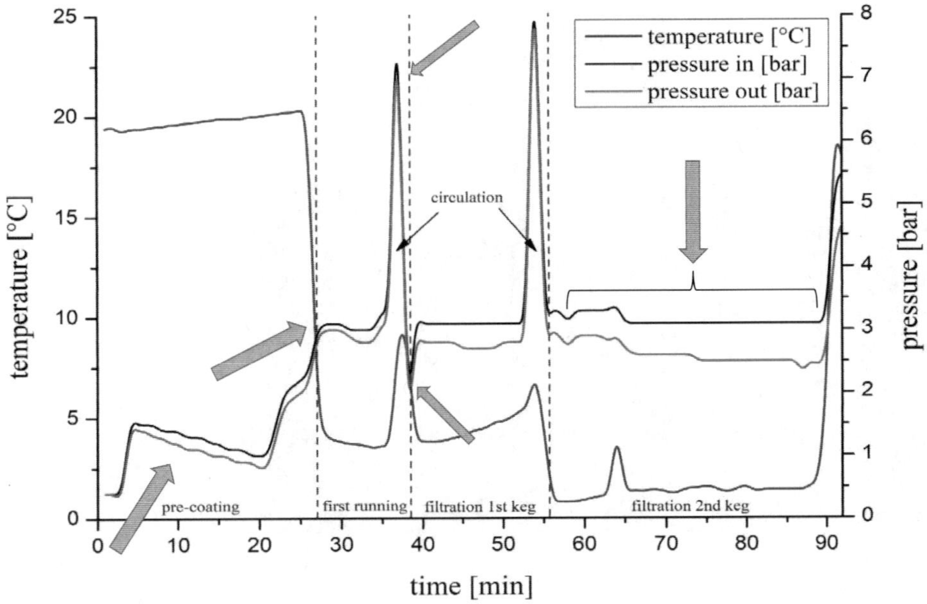

37. EBC Antwerp 2019 | Beech wood sawdust as a sustainable filter aid for pre-coat filtration | Panglisch
Slide 13

Filtrox filtration pilot plant
results without large deviations

analysis	unit	unfiltrate	reference kieselguhr	<0,065mm sawdust	<0,065mm sawdust treated
original gravity	[%]	12.03	11.77	11.58	11.57
pH-value	[-]	4.26	4.252	4.28	4.29
viscosity	[mPa*s]	1.562	1.562	1.554	1.542
bitterness	[IBU]	28.6	28.4	29	28
total nitrogen (12%)	[mg/l]	805	782	776	780
ß-glucans	[mg/l]	155	153	157	157
sulphur dioxide	[mg/l]	8.8	8.4	8	7.6
photometric iodine	[dE]	0.45	0.2	0.21	0.23

37. EBC Antwerp 2019 | Beech wood sawdust as a sustainable filter aid for pre-coat filtration | Panglisch
Slide 14

Filtrox filtration pilot plant
beer color

Technische Universität Berlin

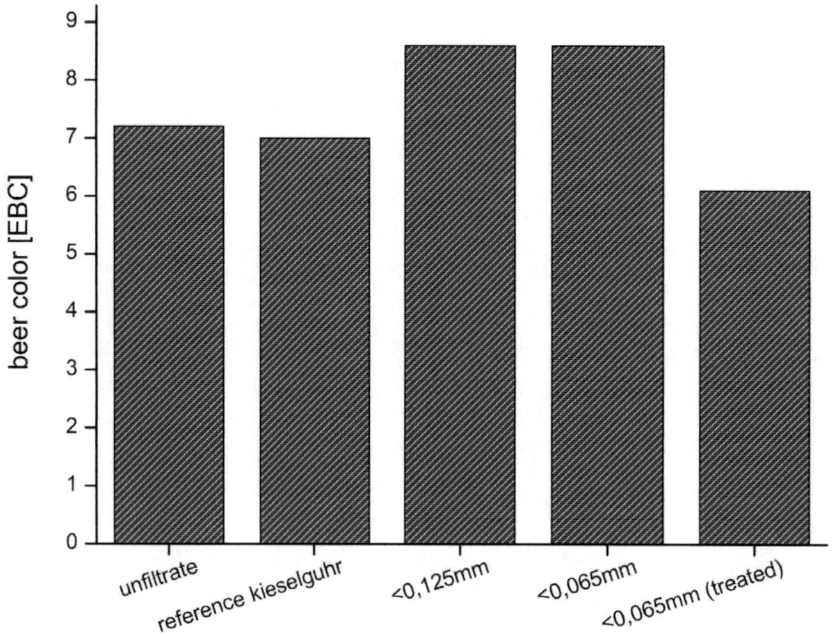

37. EBC Antwerp 2019 | Beech wood sawdust as a sustainable filter aid for pre-coat filtration | Panglisch
Slide 15

Filtrox filtration pilot plant
Polyphenols / Anthocyanogenes

Technische Universität Berlin

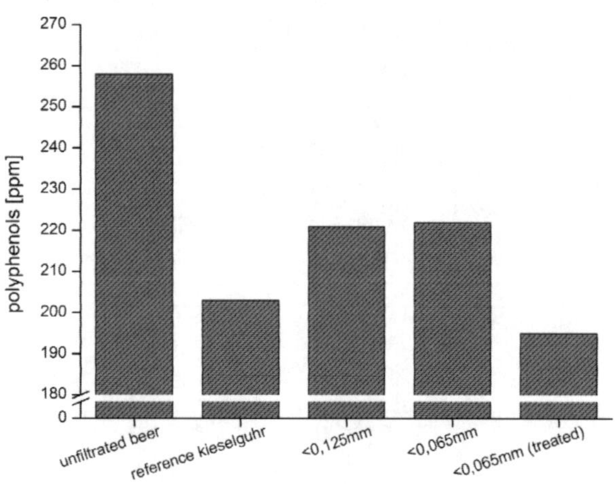

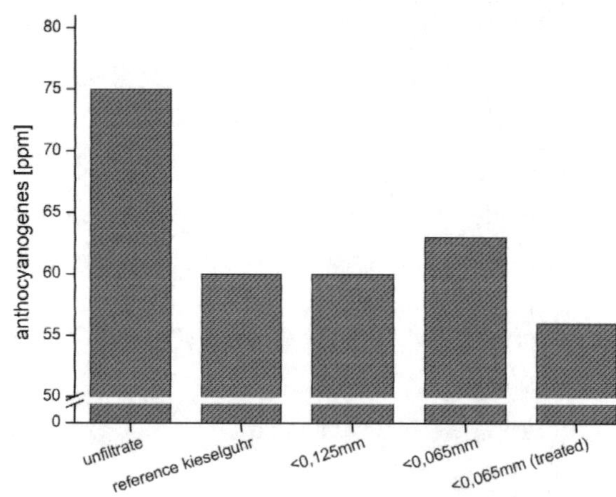

37. EBC Antwerp 2019 | Beech wood sawdust as a sustainable filter aid for pre-coat filtration | Panglisch
Slide 16

Filtrox filtration pilot plant
haze

Technische Universität Berlin

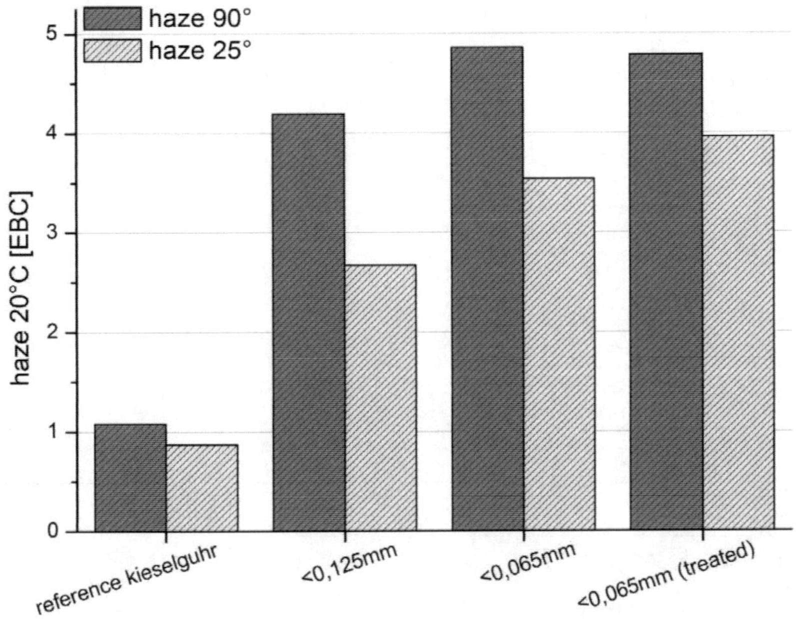

37. EBC Antwerp 2019 | Beech wood sawdust as a sustainable filter aid for pre-coat filtration | Panglisch
Slide 17

Filtrox filtration pilot plant
iron

Technische Universität Berlin

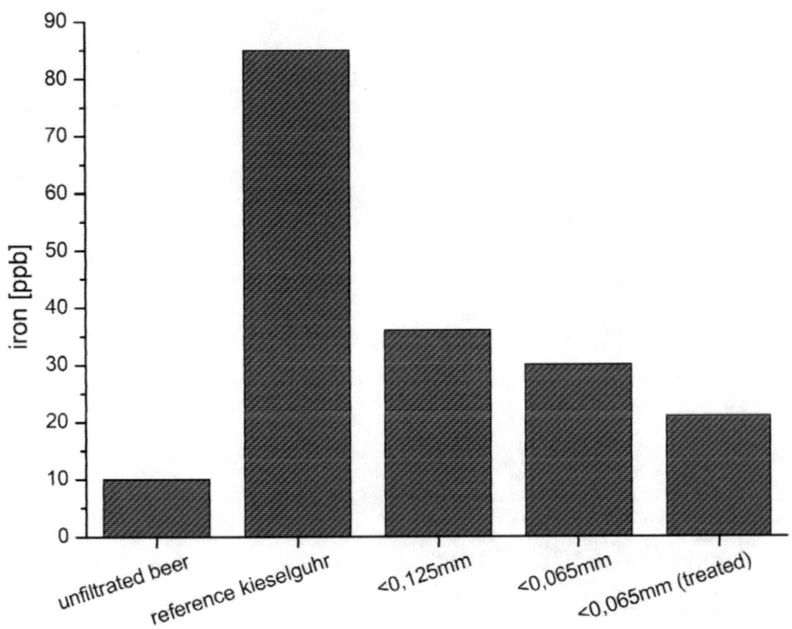

37. EBC Antwerp 2019 | Beech wood sawdust as a sustainable filter aid for pre-coat filtration | Panglisch
Slide 18

Filtrox filtration pilot plant

Endogenous Oxidative Potential (EAP), T500

Technische Universität Berlin

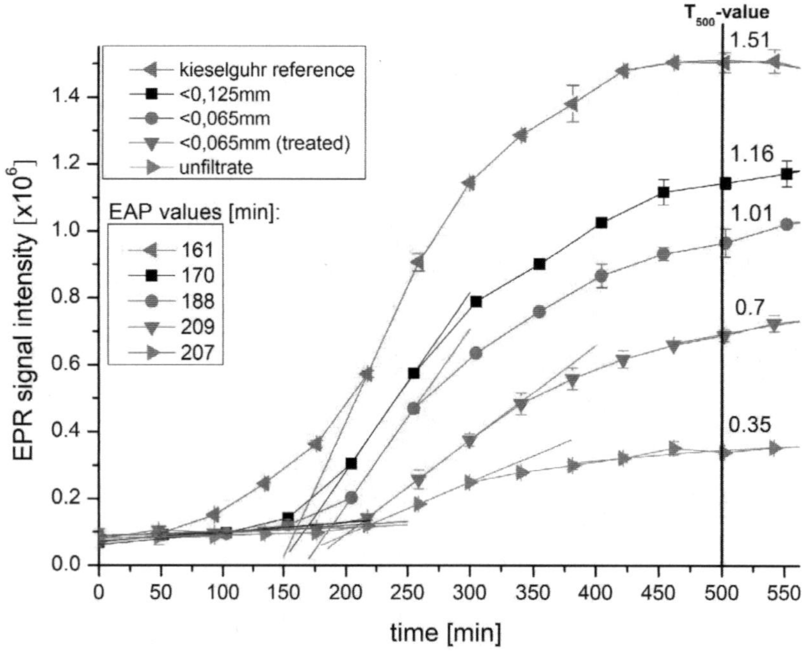

Sensory Analysis

results triangle-test with significance levels

Technische Universität Berlin

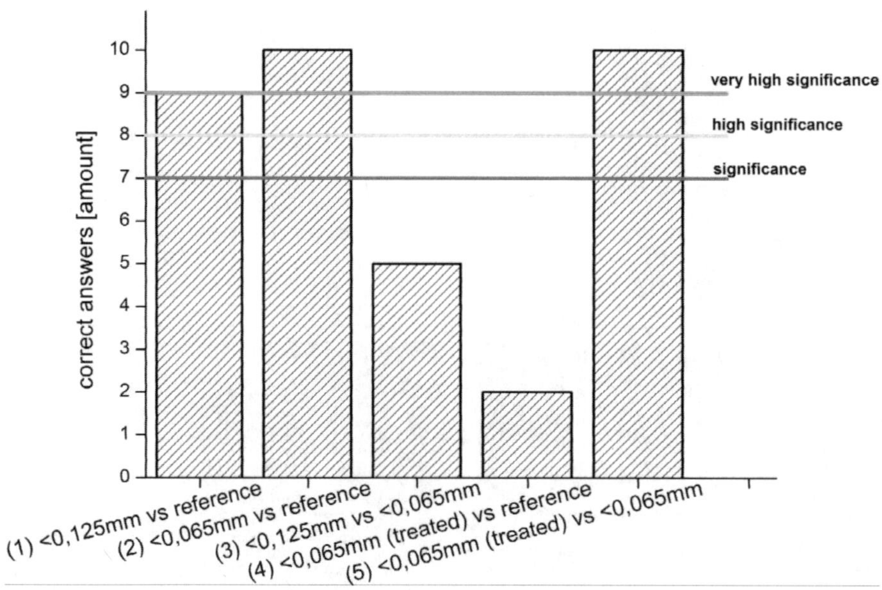

Descriptive sensory analysis

tasters' smell and taste attributes

Technische Universität Berlin

	kieselguhr	untreated sawdust	treated sawdust
smell	aromatic, floral, fresh	smelly, bread crust, sweet, wooden, dully	aromatic, neutral, fresh, fruity, clear
taste	palatable, full-bodied, pleasant bitterness, balanced, mild	astringent, wooden, scratchy bitterness, sweet, cardboard, phenolic	herb, balanced, clear, pleasant bitterness, rounded

Conclusion

Technische Universität Berlin

Pro

handling:

- homogenous pre-coating and filtration with a candle pre-coat filter is possible;
- regular dosage during filtration is easy because of a fast homogenization of the sawdust particles with water;

analysis:

- improved antioxidative stability because of a significantly lower iron content in the pre-treated sawdust samples;
- no influence on important parameters as original gravity, pH-value, bitterness, ß-glucans, nitrogen and sulphur dioxide content;

sensory analysis:

- pre-treated beech wood sawdust has no negative effects on beer flavor;

Contra

haze:

- filtrate is slightly opaque - the haze value is above the sensing threshold (around 4 EBC);

analysis :

- during treatment of beech wood sawdust there is parallel rinsing out of polyphenols and anthocyanogenes;

sensory analysis :

- the filtrated beers with untreated beech wood sawdust as filter aid are sensorial strongly contaminated (a wooden flavor covers nearly all other aroma impressions).

Potential improvements

Technische Universität Berlin

pre-treatment process:

- optimization of rinsing steps regarding use of water, water temperature, ratio water : sawdust, addition of acids (lower pH) or complexing agents
  - maximum output of iron
  - minimum output of polyphenols

- mixing of untreated and treated sawdust in order to use positive wooden aromas
  - could be suitable for special beer types / interesting for craft brewing scene
    - syringaldehyde, vanillin, eugenol, guaiacol etc.
  - simulation of storage in wooden barrels

Potential improvements

Technische Universität Berlin

filtration process:

- combinations of different particle sizes as it is practice using kieselguhr ensure higher selectivity of the filter cake and therefore better haze values of filtrates

- optimization of pre-coating flow velocity and pressure for beech wood sawdust

- there is adequate potential left regarding a reduction of haze values under sensing threshold
  - usage of pre-clarified beer e.g. by centrifuge
  - usage of stabilization aids like gallotannins
  - combination of sawdust and cellulose fibers

analysis	sawdust	sawdust treated	sawdust treated, preclarified	sawdust + cellulose fibers
original gravity	11.58	11.57	11.07	11.31
color	8.5	6.1	6.7	7.7
haze 90/25 (20° C)	4.86/3.54	4.79/3.96	**2.34/3.31**	**0.91/0.93**

Acknowledgements

Technische Universität Berlin

Prof. F.-J. Methner, T. Kunz
Staff
Brewing and Beverage Technology
TU Berlin, Germany

SMOKE IT ALL
EASY SMOKING IN THE GRILL
Genuine Smoking Dust

P. N. Petersen, O. Vistisen
Smoke it All A/S, Denmark

Thank you for your Attention!

Technische Universität Berlin

I am now delighted to answer your questions

and later talk with you while having a delicious Belgium beer. Cheers!

The role of wooden barrels in lambic beer production

Dr. ir. Jonas De Roos

37th European Brewery Convention
Antwerp, Belgium
June 2-6, 2019

VRIJE UNIVERSITEIT BRUSSEL

Research Group of Industrial Microbiology and Food Biotechnology
Prof. Dr. ir. Luc De Vuyst
Prof. Dr. ir. Frédéric Leroy
Prof. Dr. Stefan Weckx

Introduction

Aims

Role of wooden barrels

Conclusions

Introduction

Belgian beers

Lambic beer

Lambic beer fermentation and maturation

Wooden barrels

Belgian beers

Bottom fermentation

Lager beers

Top fermentation

Ales

Saccharomyces bayanus
Saccharomyces pastorianus
(*Saccharomyces cerevisiae* x *Saccharomyces eubayanus*)

Saccharomyces cerevisiae

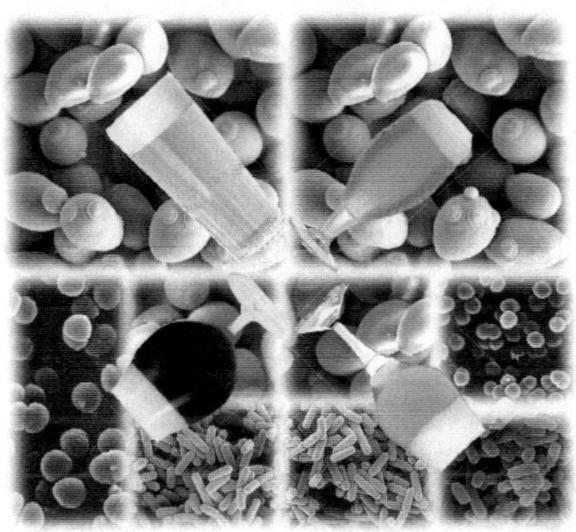

In-house mixed starter culture of yeasts and lactic acid bacteria

Autochthonous non-starter microorganisms of both yeasts and bacteria

Mixed fermentation

Red, red-brown, and old-brown acidic ales

Spontaneous fermentation

Acidic beers

Lambic beer

Local, traditional, sour-tasting beer

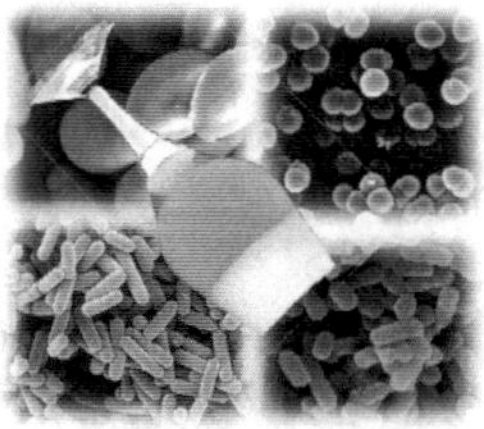

Ten breweries in the Senne valley (Brussels)

Two additional breweries in South West Flanders

Belgian-style acidic ales:
e.g., American coolship ales

Verachtert & Iserentant, 1995; De Keersmaecker *et al.*, 1996; Bokulich *et al.*, 2012
Pothakos *et al.*, 2016; Spitaels *et al.*, 2015, 2017; De Roos and De Vuyst, 2019 JSFA

Lambic beer

The basis for

Kriek

Whole sour cherries

Framboise

Raspberries

Faro

Rock sugar

Gueuze

Mixture of young lambic
(6 months to 1 year)
and old lambic
(1+ years)

Pothakos *et al.*, 2016; Spitaels *et al.*, 2015, 2017; De Roos and De Vuyst, 2019 JSFA

Lambic beer

Spontaneous fermentation

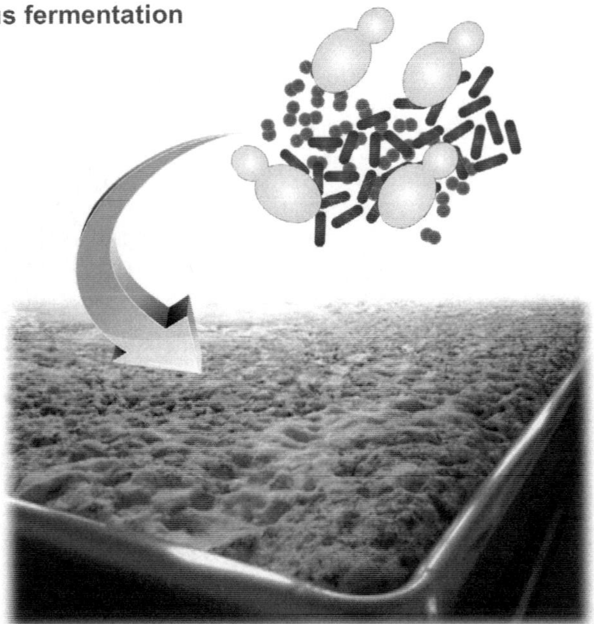

7 © IMDO-VUB — Pothakos *et al.*, 2016; Spitaels *et al.*, 2015, 2017; De Roos and De Vuyst, 2019 JSFA

Lambic beer

Fermentation and maturation in wooden barrels for up to three years

Oak

Chestnut

8 © IMDO-VUB — Pothakos *et al.*, 2016; Spitaels *et al.*, 2015, 2017; De Roos and De Vuyst, 2019 JSFA

Lambic beer fermentation and maturation

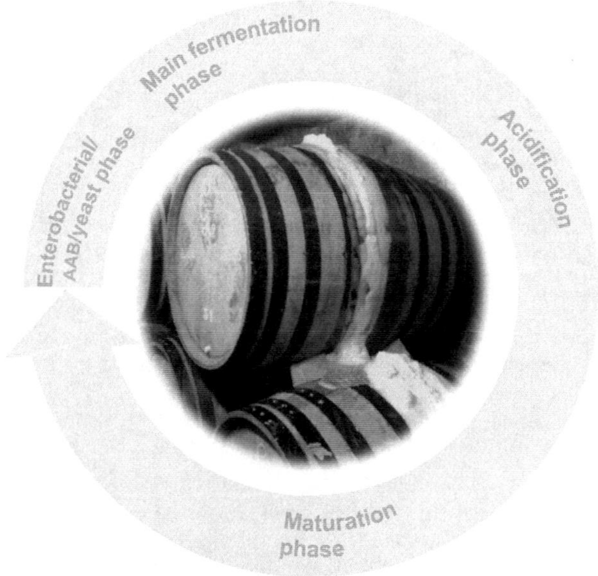

 De Roos and De Vuyst, 2019 JSFA; De Roos *et al.*, 2018 AEM; De Roos *et al.*, 2018 FiM

Lambic beer fermentation and maturation

Enterobacteria
Acetic acid bacteria (AAB)
(Oxidative) yeasts

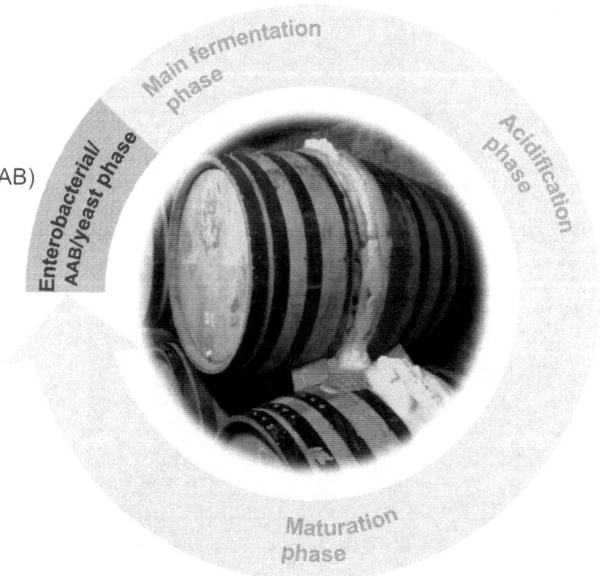

 De Roos and De Vuyst, 2019 JSFA; De Roos *et al.*, 2018 AEM; De Roos *et al.*, 2018 FiM

Lambic beer fermentation and maturation

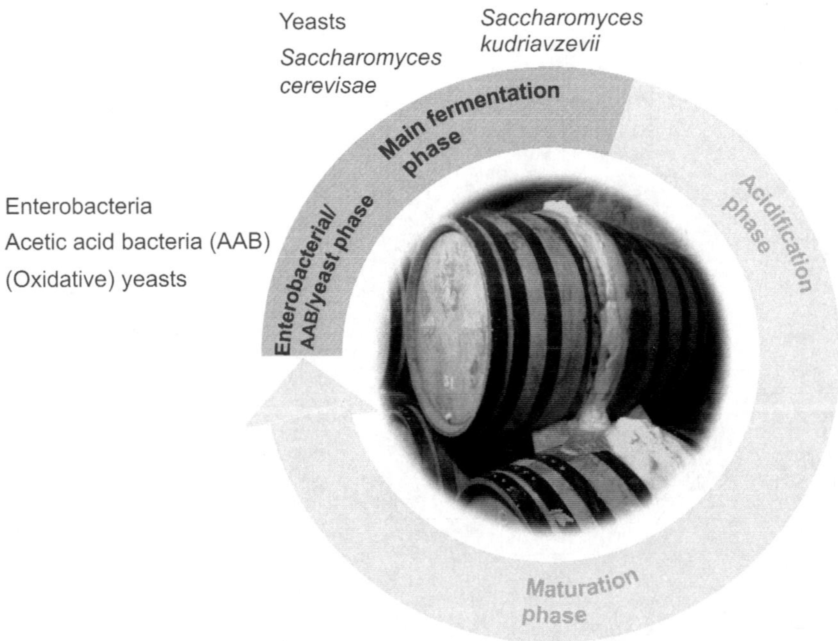

11 © IMDO-VUB De Roos and De Vuyst, 2019 JSFA; De Roos *et al.*, 2018 AEM; De Roos *et al.*, 2018 FiM

Lambic beer fermentation and maturation

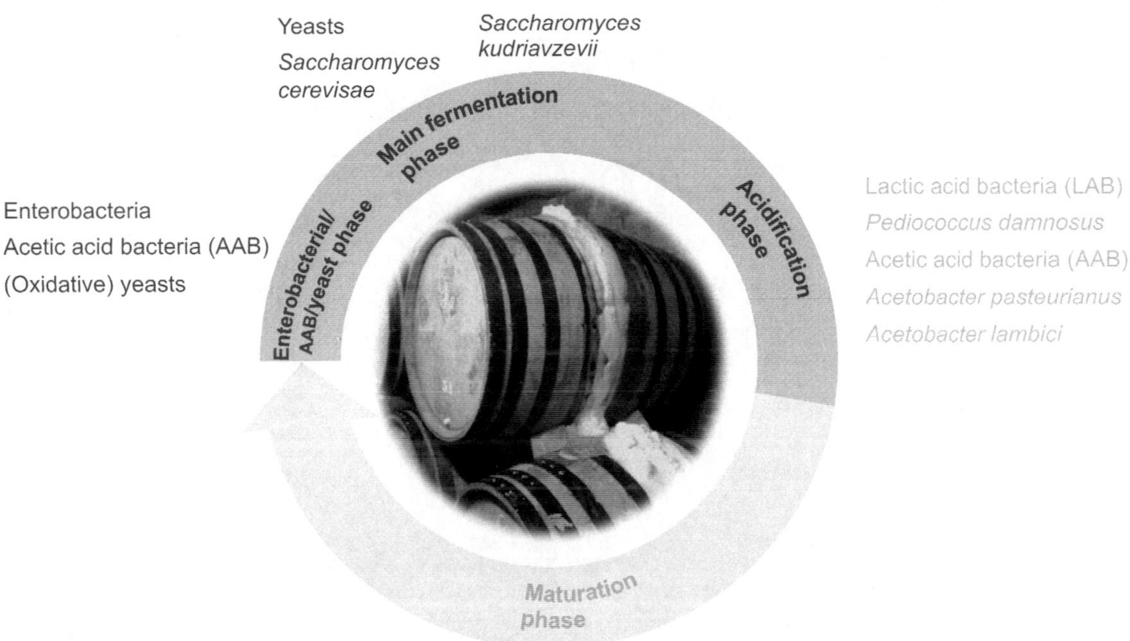

12 © IMDO-VUB De Roos and De Vuyst, 2019 JSFA; De Roos *et al.*, 2018 AEM; De Roos *et al.*, 2018 FiM

Lambic beer fermentation and maturation

Yeasts
Saccharomyces cerevisae

Saccharomyces kudriavzevii

Enterobacteria
Acetic acid bacteria (AAB)
(Oxidative) yeasts

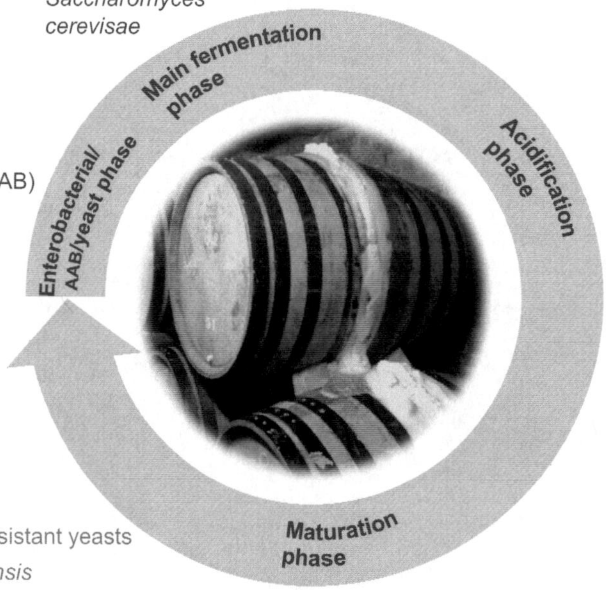

Lactic acid bacteria (LAB)
Pediococcus damnosus
Acetic acid bacteria (AAB)
Acetobacter pasteurianus
Acetobacter lambici

Cycloheximide-resistant yeasts
Dekkera bruxellensis
Brettanomyces custersianus

De Roos and De Vuyst, 2019 JSFA; De Roos *et al.*, 2018 AEM; De Roos *et al.*, 2018 FiM

The use of wooden barrels in fermentation processes

Del Campo *et al.*, 2003; Guzzon *et al.*, 2011; White *et al.*, 2017
Spitaels *et al.*, 2017; De Roos and De Vuyst, 2019 JSFA

The use of wooden barrels in fermentation processes

Used for:

Oxygen ingress

Flavor formation

Limited number of batches

Del Campo *et al.*, 2003; Guzzon *et al.*, 2011; White *et al.*, 2017

15 © IMDO-VUB

Spitaels *et al.*, 2017; De Roos and De Vuyst, 2019 JSFA

The use of wooden barrels in fermentation processes

Used for:

Oxygen ingress

Unlimited number of batches

Used for:

Oxygen ingress

Flavor formation

Limited number of batches

Del Campo *et al.*, 2003; Guzzon *et al.*, 2011; White *et al.*, 2017

16 © IMDO-VUB

Spitaels *et al.*, 2017; De Roos and De Vuyst, 2019 JSFA

The use of wooden barrels in fermentation processes

Oelofse *et al.*, 2008; Guzzon *et al.*, 2011; Curtin *et al.*, 2012; Steensels *et al.*, 2015

De Roos and De Vuyst, 2019 JSFA

The use of wooden barrels in fermentation processes

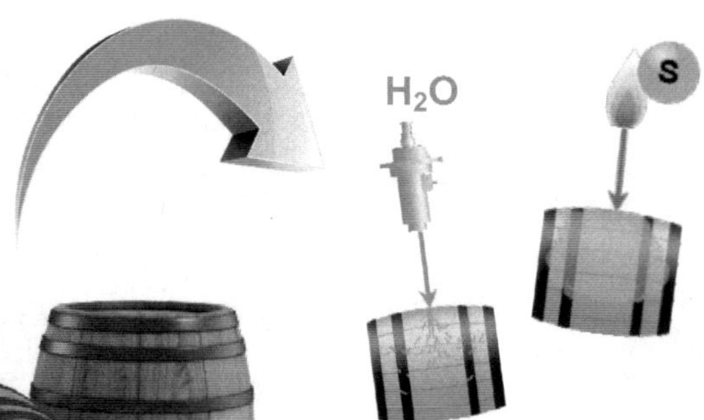

Difficult to sanitize:

Porosity

Resistance

Brettanomyces

Oelofse *et al.*, 2008; Guzzon *et al.*, 2011; Curtin *et al.*, 2012; Steensels *et al.*, 2015

De Roos and De Vuyst, 2019 JSFA

The use of wooden barrels in fermentation processes

Difficult to sanitize:

Porosity

Resistance

Brettanomyces

Resident microbiota:

Isolated from the inside of wooden (lambic) barrels

Very limited detection culture-dependently

Role in lambic beer production?

Oelofse *et al.*, 2008; Guzzon *et al.*, 2011; Curtin *et al.*, 2012; Steensels *et al.*, 2015

Spitaels *et al.*, 2015; De Roos and De Vuyst, 2019 JSFA

Aims

Aims

Assessment of the role of the wooden barrels during lambic beer production

- Fermentation and maturation
- Inoculation source

The role of wooden barrels

The role of wooden barrels

Adapted aseptic sampling

Adapted aseptic sampling

Started spring 2015

Oak casks
Horizontally positioned

Started spring 2015

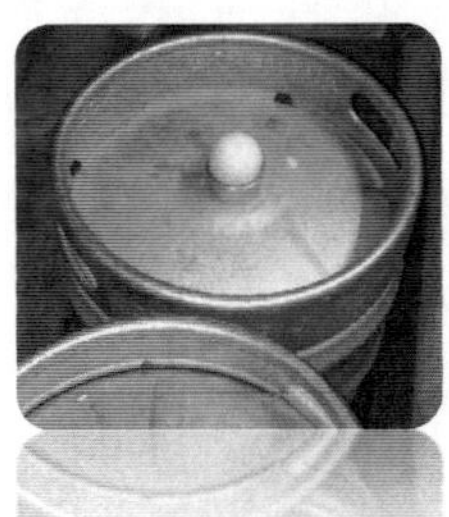

Stainless steel vats
Vertically positioned

Adapted aseptic sampling

Started spring 2015

Oak casks
Horizontally positioned

Started spring 2015

Stainless steel vats
Vertically positioned

Adapted aseptic sampling

Started spring 2015

Oak casks
Horizontally positioned

Started spring 2015

Stainless steel vats
Vertically positioned

↓

0 h, 1 h, 24 h, 3 d, 1 w, 2 w, 3 w, 6 w, 3 m, 6 m, 9 m, 13 m, 18 m, 24 m, 30 m

Adapted aseptic sampling

Started spring 2015

Oak casks
Horizontally positioned

Started spring 2015

Stainless steel vats
Vertically positioned

0 h, 1 h, 24 h, 3 d, 1 w, 2 w, 3 w, 6 w, 3 m, 6 m, 9 m, 13 m, 18 m, 24 m, 30 m

Microbiological analysis

Metabolite target analysis

The role of wooden barrels

Adapted aseptic sampling

Microbial distribution during production

Microbial community dynamics: culture-dependent

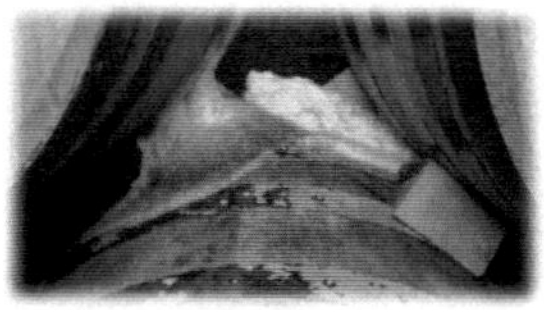

 De Roos *et al.*, 2018 AEM; De Roos *et al.*, 2018 FiM

Microbial community dynamics: culture-dependent

- Selective plating and incubation on different agar media

Agar Media	mDMS	mMRS	YPG	YPGc
Amphotericin B (ppm)	5	5		
Cycloheximide (ppm)	200	200		50
Chloramphenicol (ppm)			200	200
Temperature (°C)	30	20	30	30
Time (days)	7	10	7	7
Incubation	Aerobic	Anaerobic	Aerobic	Aerobic

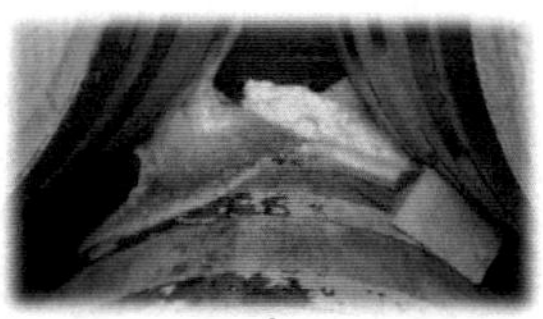

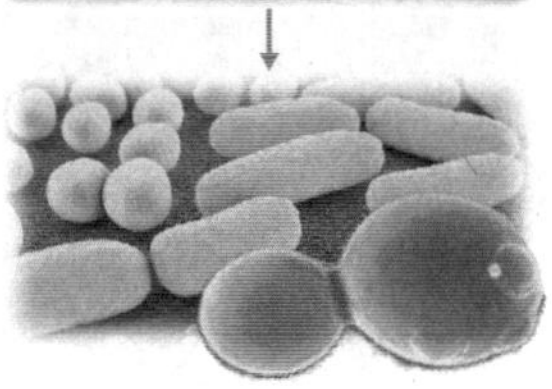

 De Roos *et al.*, 2018 AEM; De Roos *et al.*, 2018 FiM

Microbial community dynamics: culture-dependent

- Selective plating and incubation on different agar media

Agar Media	mDMS	mMRS	YPG	YPGc
Amphotericin B (ppm)	5	5		
Cycloheximide (ppm)	200	200		50
Chloramphenicol (ppm)			200	200
Temperature (°C)	30	20	30	30
Time (days)	7	10	7	7
Incubation	Aerobic	Anaerobic	Aerobic	Aerobic

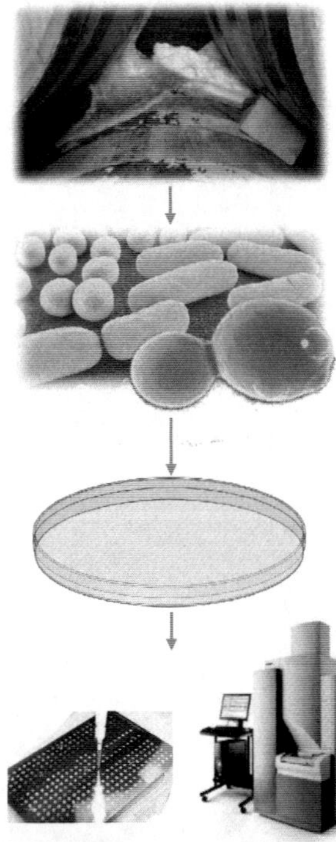

- Identification of colonies
  - Picking of a relevant number of colonies
  - MALDI-TOF MS
  - Comparative gene sequence analysis
    - 16S rRNA gene/*dnaK* gene (bacteria)
    - Internal Transcribed Spacer (ITS) region (fungi)

  De Roos *et al.*, 2018 AEM; De Roos *et al.*, 2018 FiM

Microbial community dynamics: culture-dependent

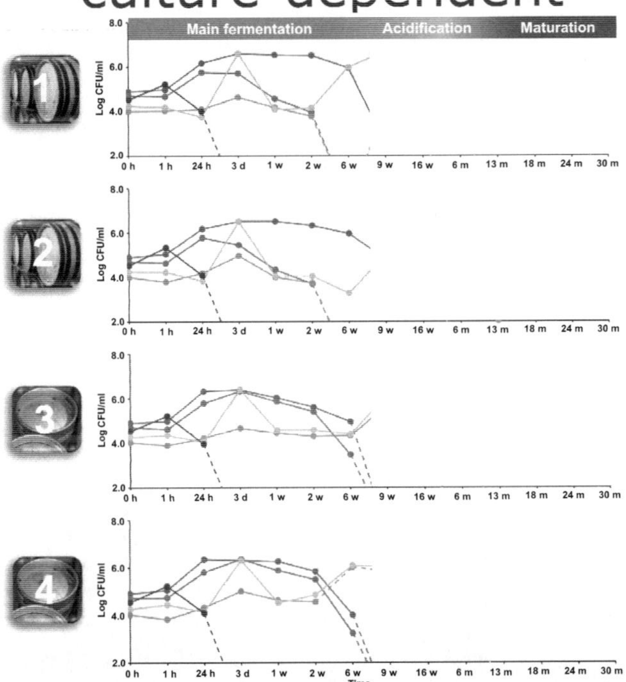

Yeasts

Cycloheximide-resistant Yeasts

LAB

AAB

Microbial community dynamics: culture-dependent

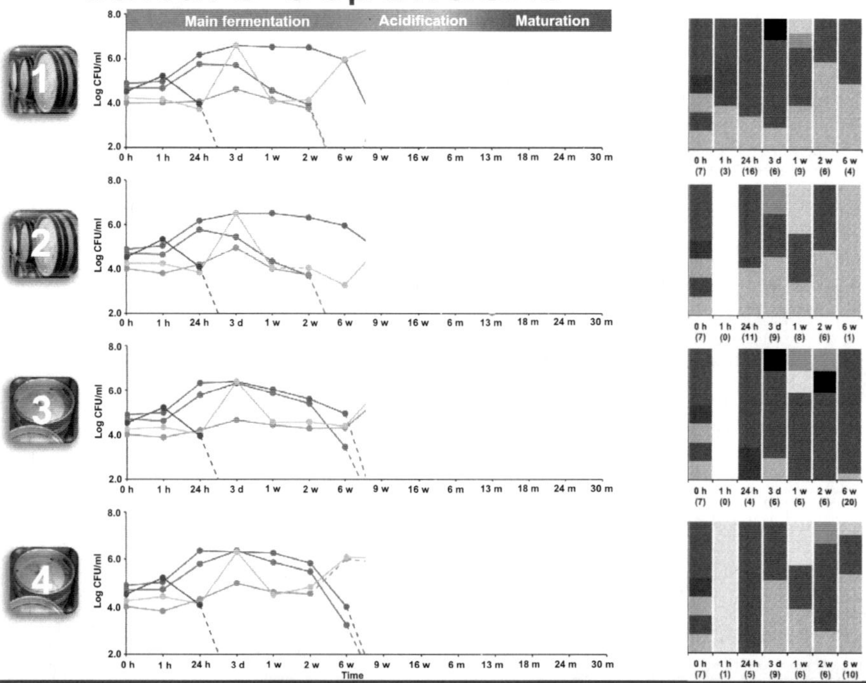

33 © IMDO-VUB

Microbial community dynamics: culture-dependent

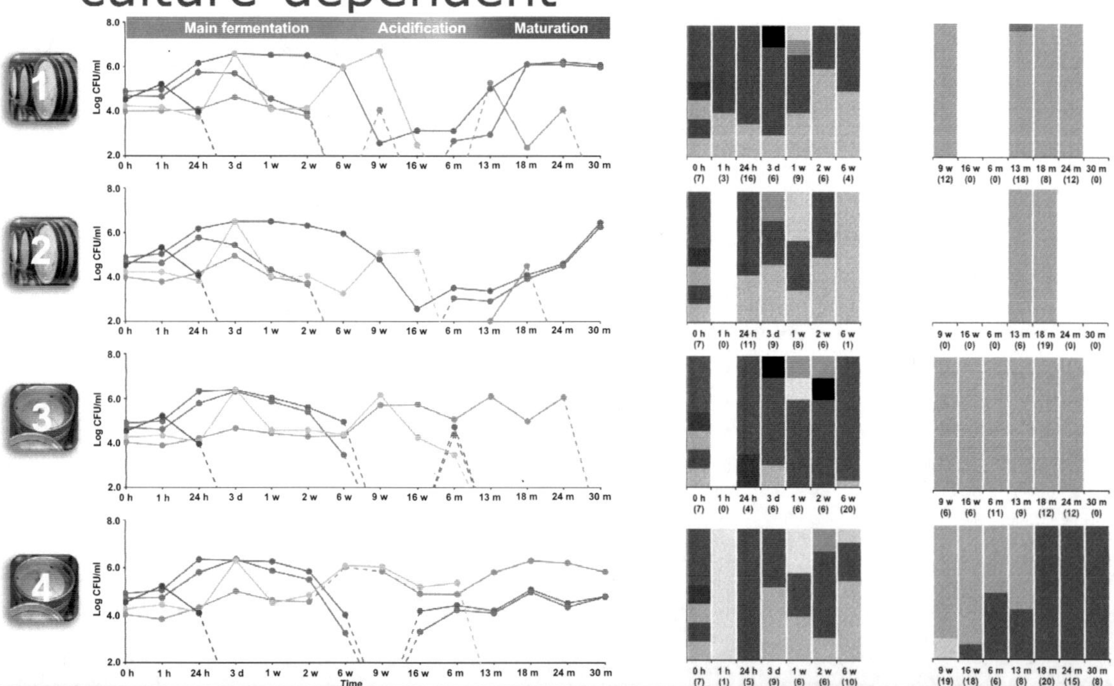

34 © IMDO-VUB

Microbial community dynamics: culture-dependent

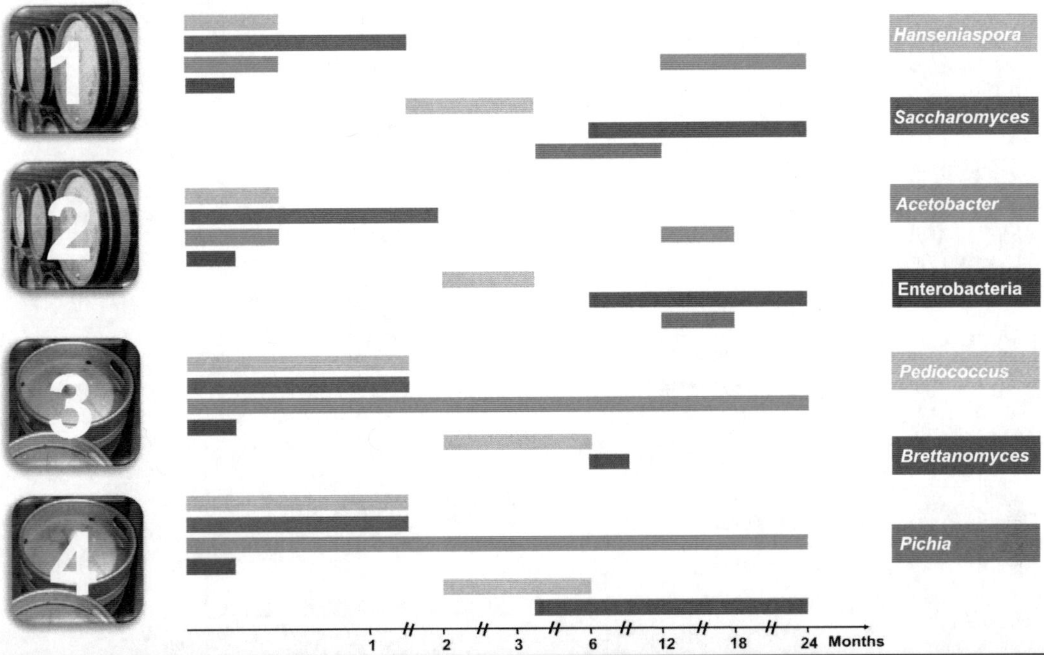

35 © IMDO-VUB

Microbial community dynamics: culture-dependent

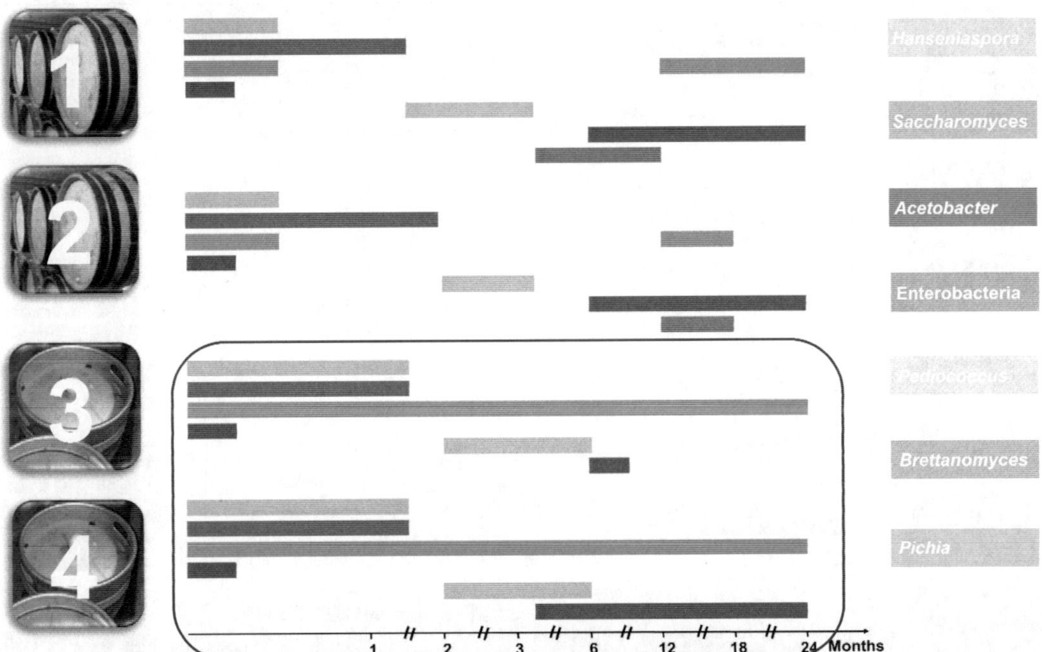

36 © IMDO-VUB

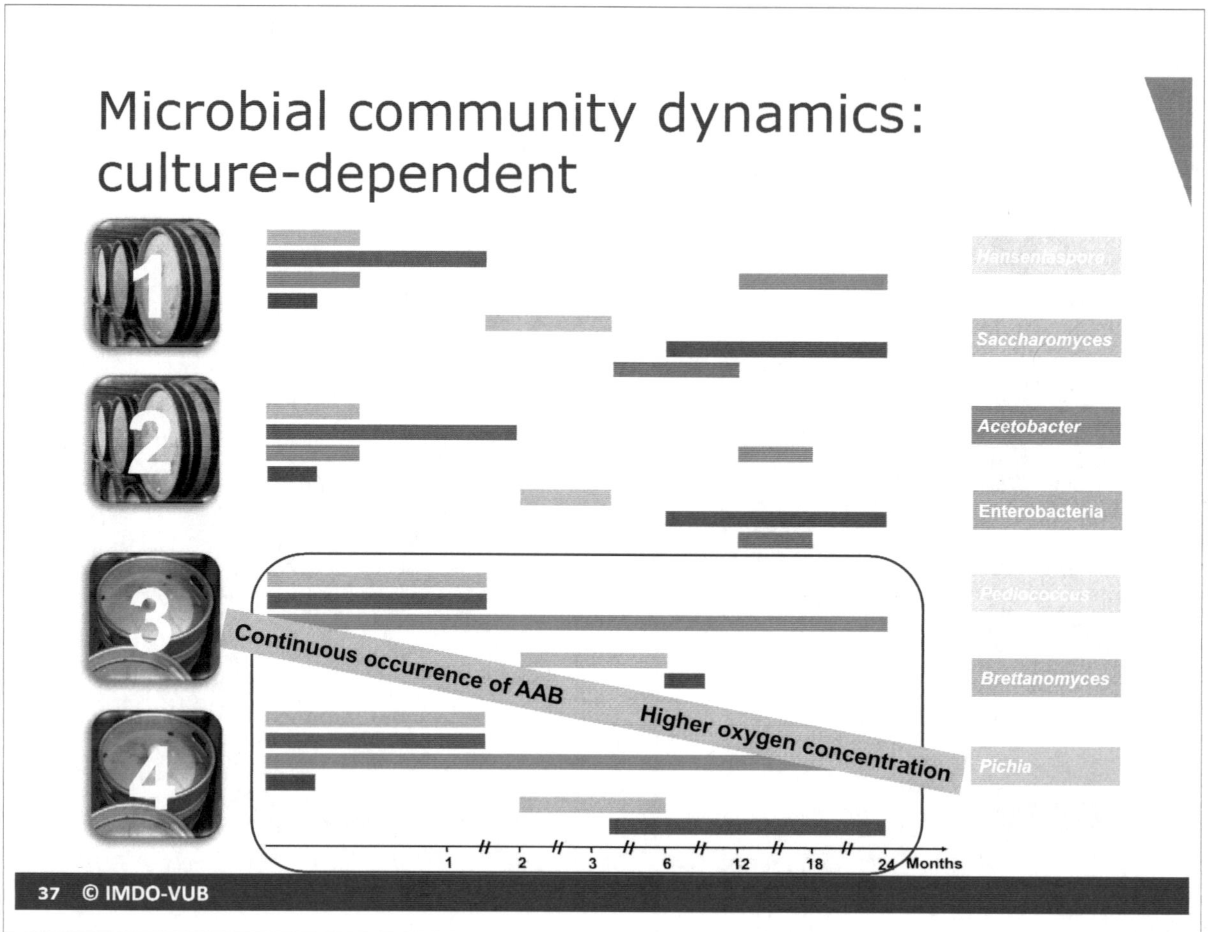
Microbial community dynamics: culture-dependent
1
2
3
4
Hanseniaspora
Saccharomyces
Acetobacter
Enterobacteria
Pediococcus
Brettanomyces
Pichia
Continuous occurrence of AAB
Higher oxygen concentration
1
2
3
6
12
18
24
Months
37 © IMDO-VUB

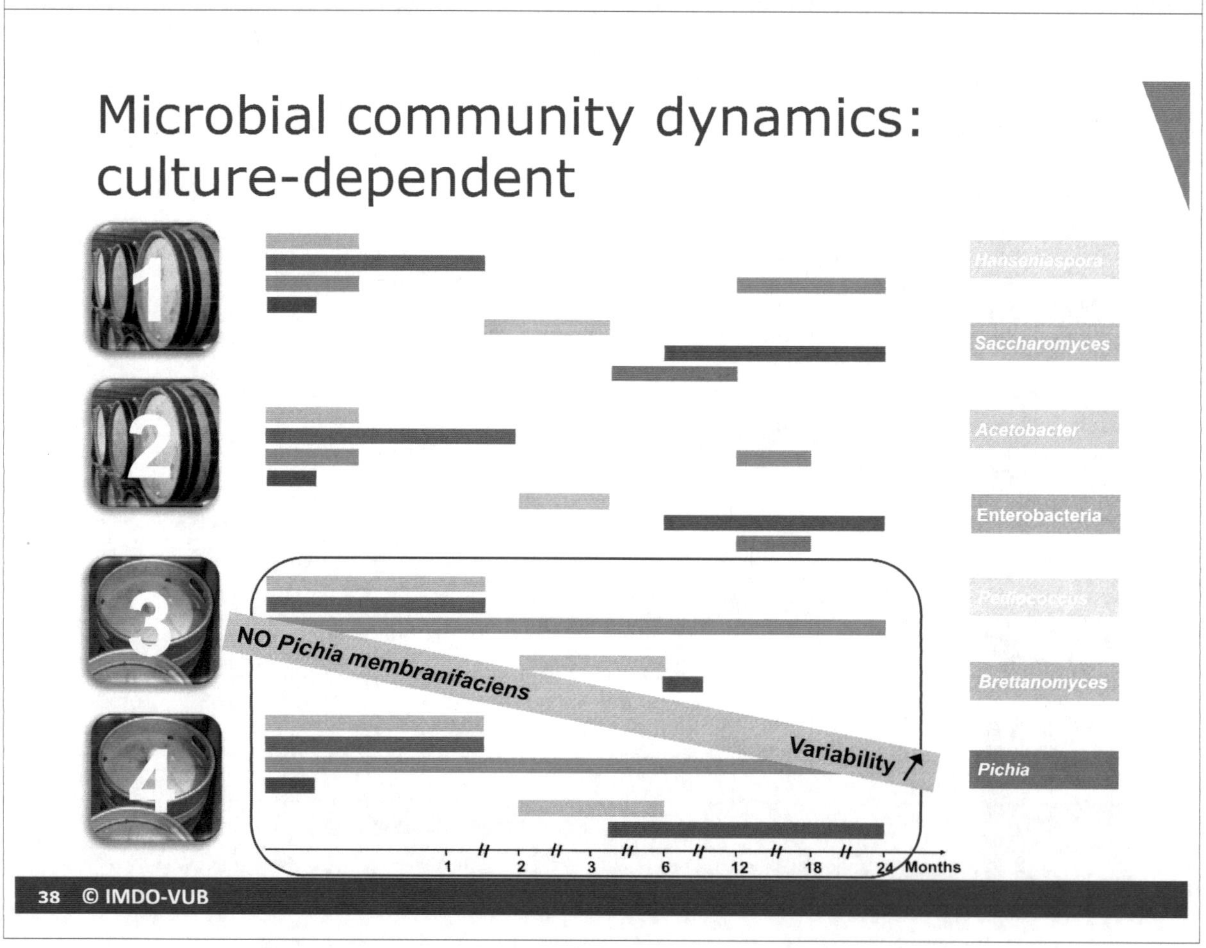
Microbial community dynamics: culture-dependent
1
2
3
4
Hanseniaspora
Saccharomyces
Acetobacter
Enterobacteria
Pediococcus
Brettanomyces
Pichia
NO Pichia membranifaciens
Variability ↗
1
2
3
6
12
18
24
Months
38 © IMDO-VUB

The role of wooden barrels

Adapted aseptic sampling

Microbial distribution during production

Substrate and metabolite distribution during production

Substrate and metabolite analysis

Equipment	HPLC-RI/UV	UPLC-MS/MS	GC-FID
Calibration	External standard	External standard	Internal standard
Compounds	Ethanol Acetic acid Lactic acid	Malic acid	Acetoin Ethyl acetate Ethyl lactate

 De Roos *et al.*, 2018 FiM

Substrate and metabolite distribution

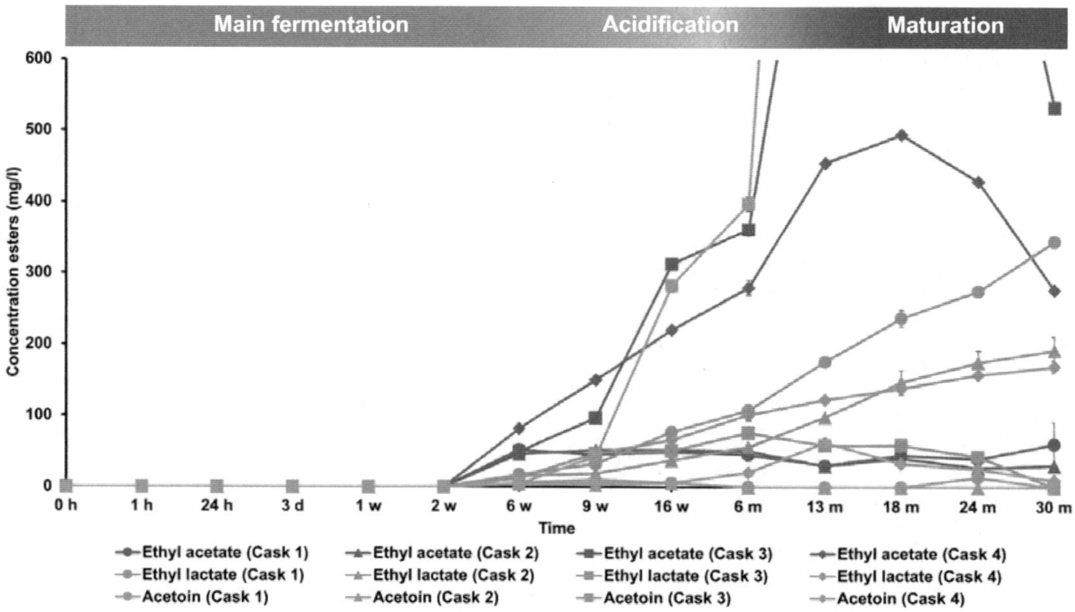

Substrate and metabolite distribution

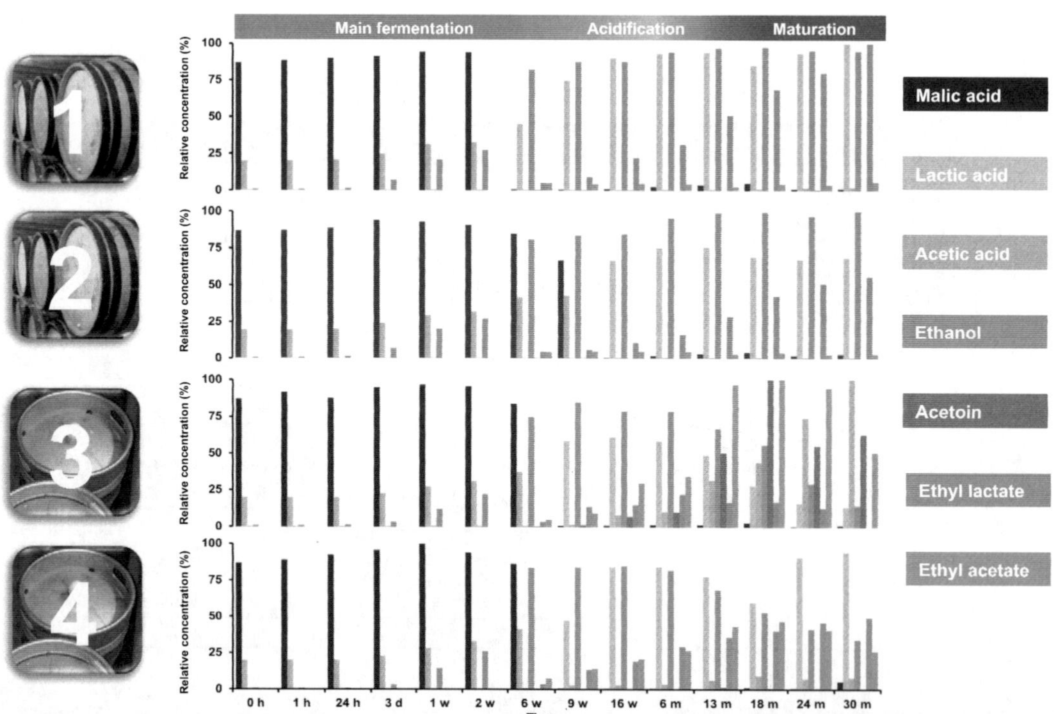

Substrate and metabolite distribution

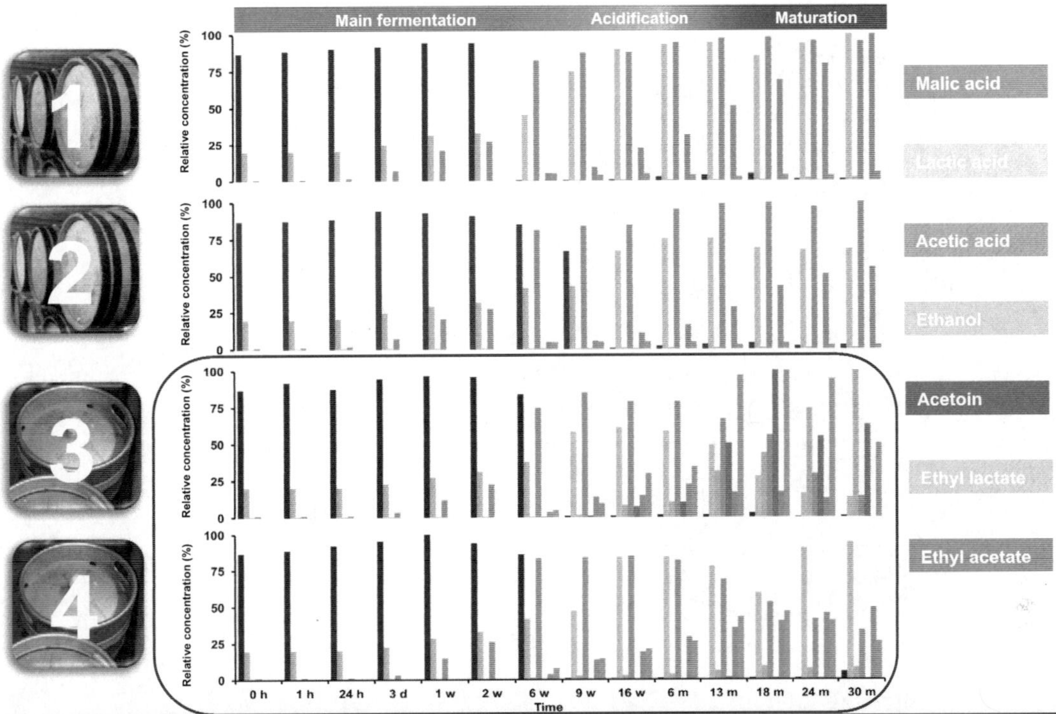

Substrate and metabolite distribution

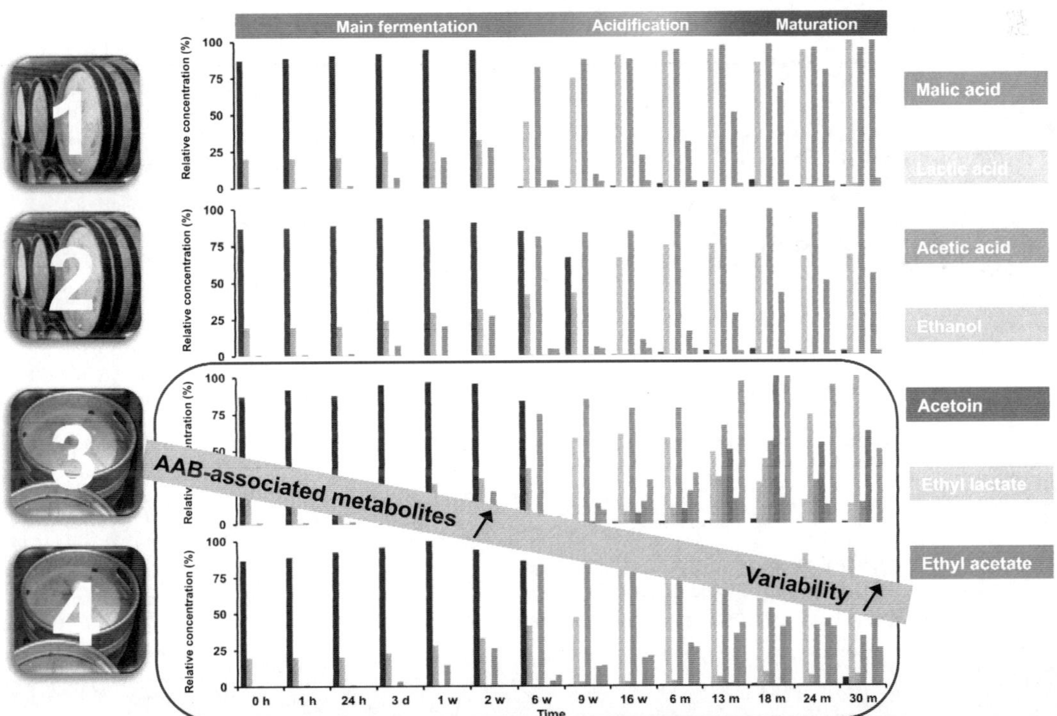

The role of wooden barrels

Adapted aseptic sampling

Microbial distribution during production

Substrate and metabolite distribution during production

Barrel swabbing of interior surfaces

45 © IMDO-VUB

Barrel swabbing of interior surfaces

Casks

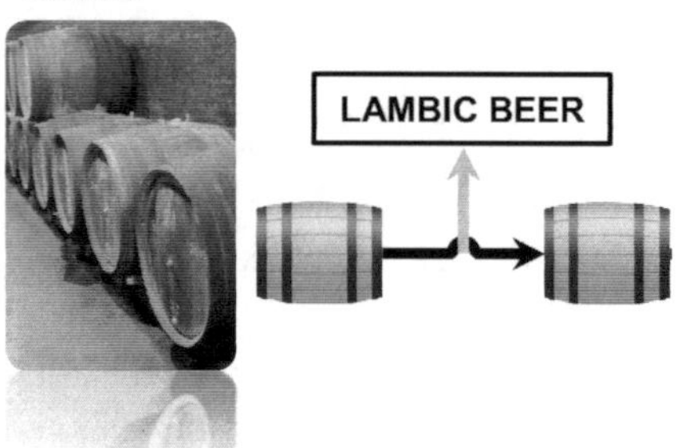

Foeders

46 © IMDO-VUB De Roos *et al.*, 2018 AEM

Barrel swabbing of interior surfaces

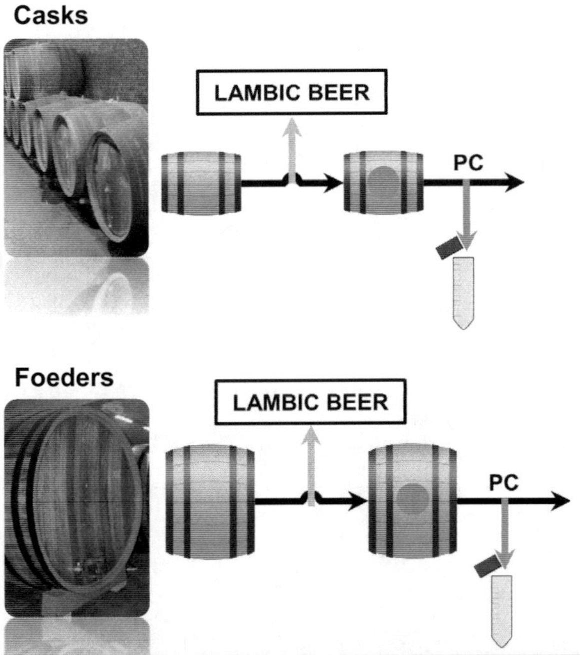

 De Roos *et al.*, 2018 AEM

Barrel swabbing of interior surfaces

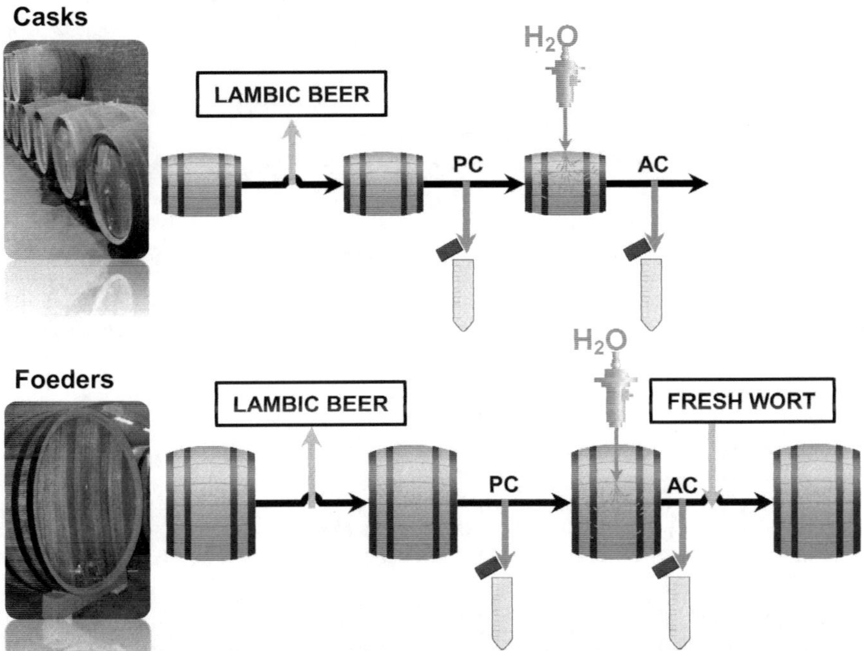

 De Roos *et al.*, 2018 AEM

Barrel swabbing of interior surfaces

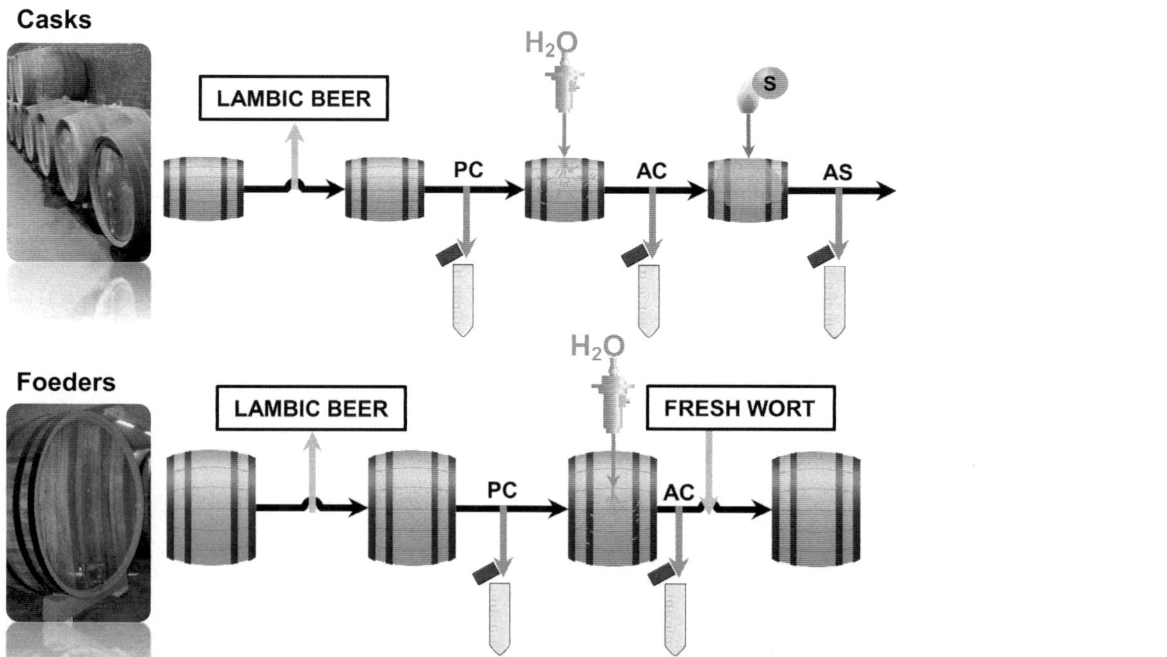

49 © IMDO-VUB De Roos *et al.*, 2018 AEM

Barrel swabbing of interior surfaces

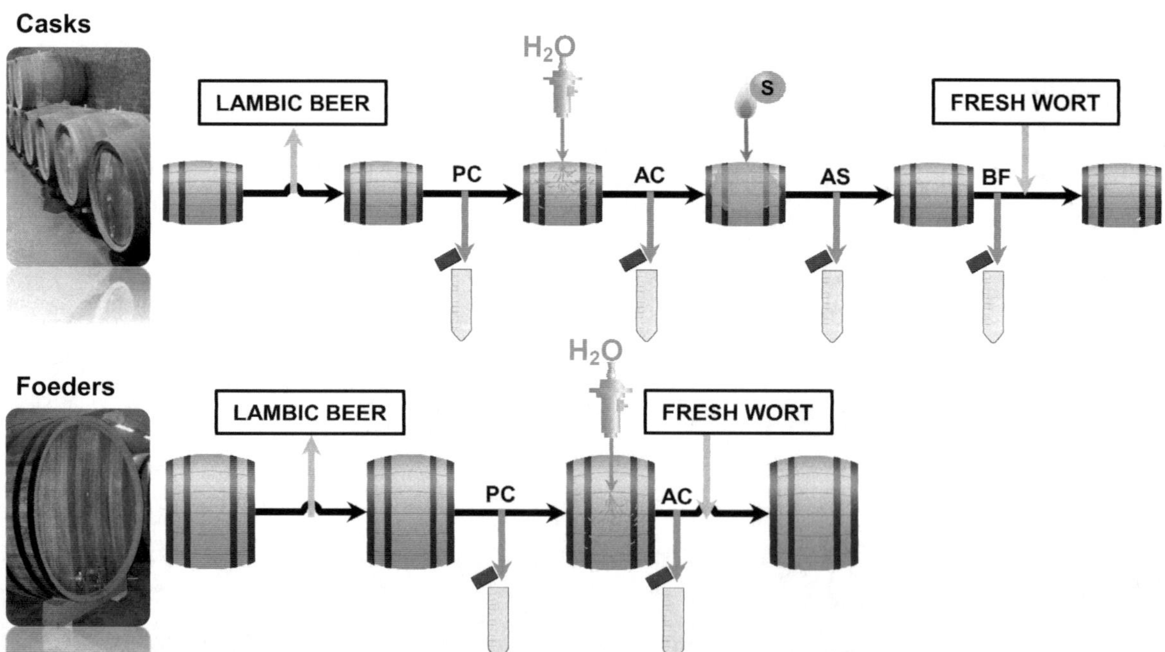

50 © IMDO-VUB De Roos *et al.*, 2018 AEM

The role of wooden barrels

Adapted aseptic sampling

Microbial distribution during production

Substrate and metabolite distribution during production

Barrel swabbing of interior surfaces

Microbial composition of the interior surfaces

Amplicon sequencing

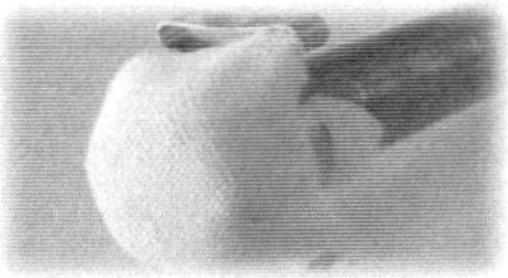

De Roos *et al.*, 2018 AEM

Amplicon sequencing

- DNA extraction on cell pellets and PCR
  - Enzymatic, chemical, and mechanical treatments for cell lysis
  - Column purification
  - PCR of the bacterial V4 (16S rRNA) and fungal ITS1 (rRNA) region

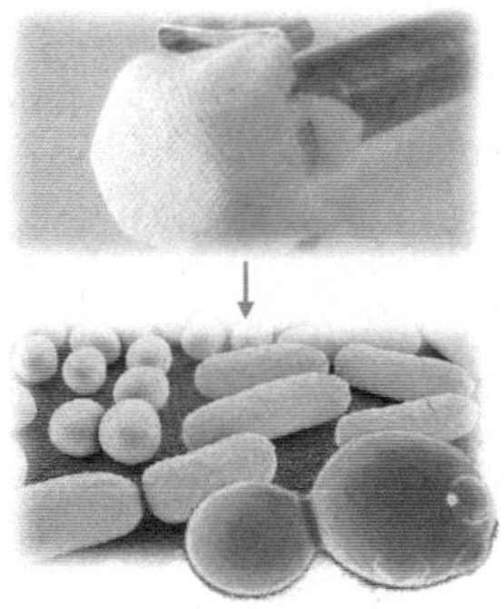

De Roos *et al.*, 2018 AEM

Amplicon sequencing

- DNA extraction on cell pellets and PCR
  - Enzymatic, chemical, and mechanical treatments for cell lysis
  - Column purification
  - PCR of the bacterial V4 (16S rRNA) and fungal ITS1 (rRNA) region

- Sequencing and data processing
  - Taxonomical analysis
    - mothur and DADA2 software
  - Statistics
    - Alpha-diversity
      - Diversity and evenness
    - Beta-diversity
      - PCoA analysis
      - PERMANOVA
      - ANCOM

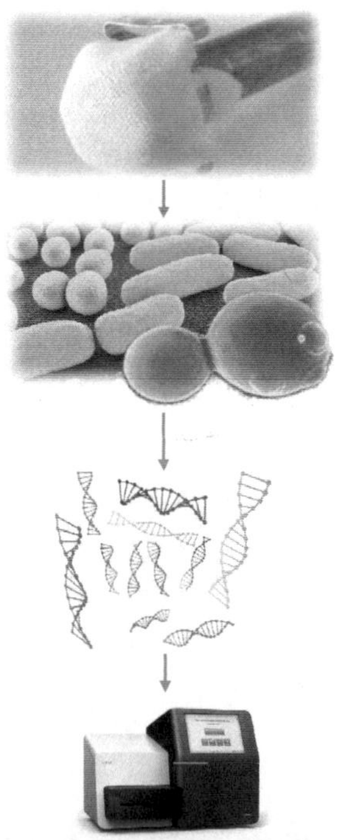

De Roos *et al.*, 2018 AEM

Amplicon sequencing: bacteria

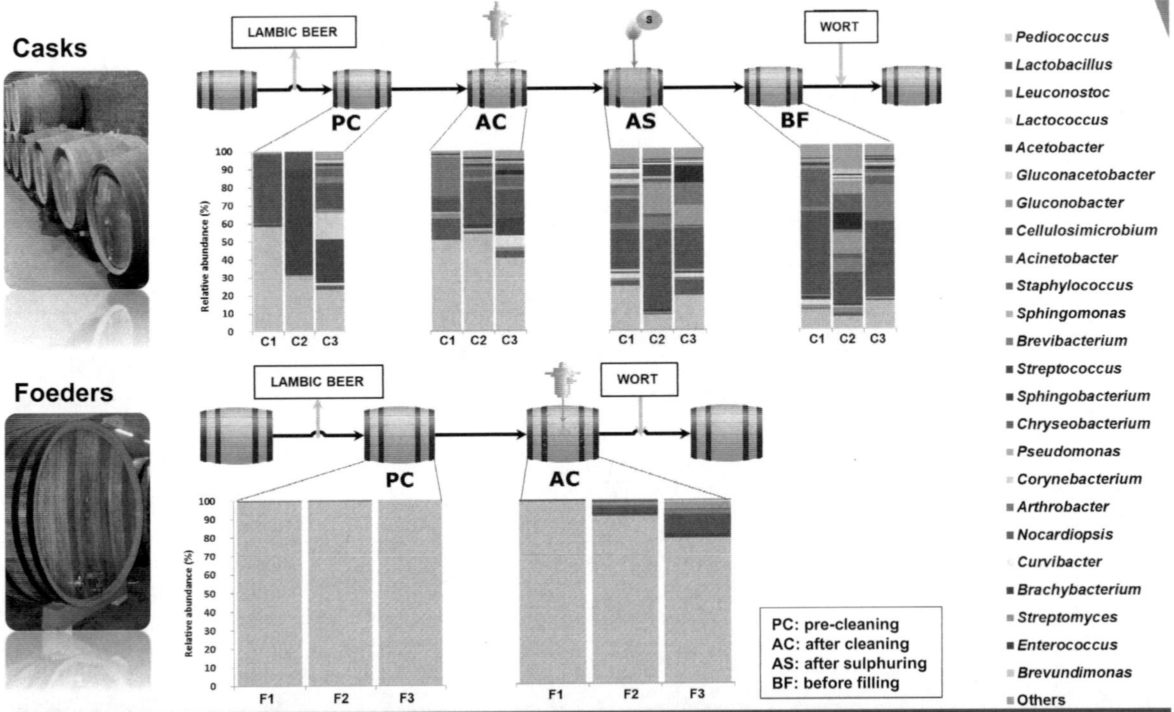

55 © IMDO-VUB

De Roos *et al.*, 2018 AEM

Amplicon sequencing: bacteria

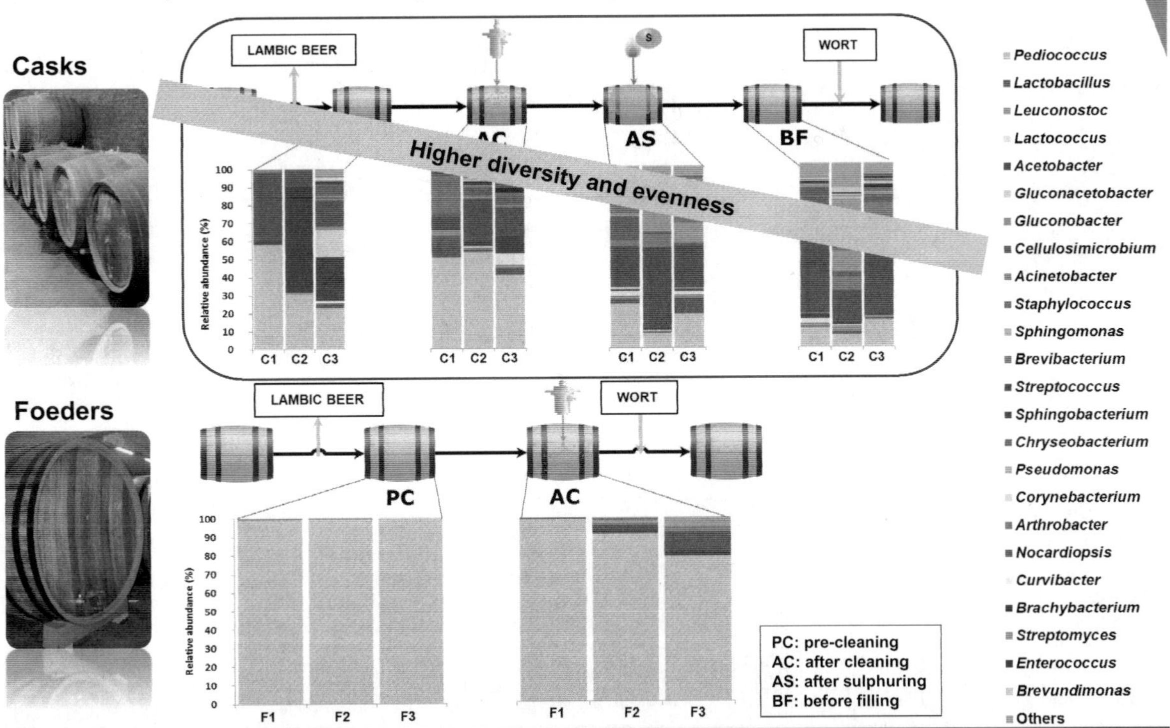

56 © IMDO-VUB

De Roos *et al.*, 2018 AEM

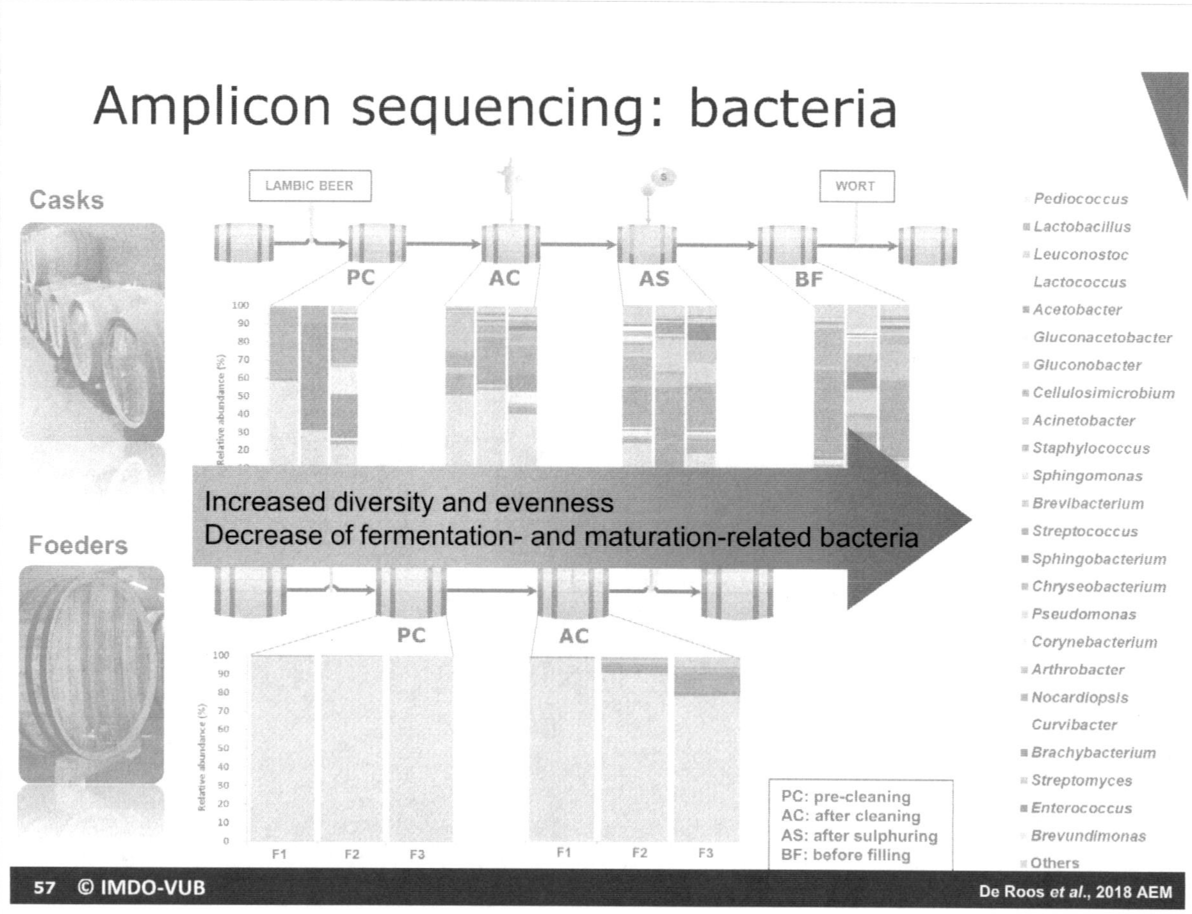
Amplicon sequencing: bacteria
Casks
LAMBIC BEER
WORT
PC
AC
AS
BF
Relative abundance (%)
Increased diversity and evenness
Decrease of fermentation- and maturation-related bacteria
Foeders
PC
AC
F1
F2
F3
F1
F2
F3
PC: pre-cleaning
AC: after cleaning
AS: after sulphuring
BF: before filling
Pediococcus
Lactobacillus
Leuconostoc
Lactococcus
Acetobacter
Gluconacetobacter
Gluconobacter
Cellulosimicrobium
Acinetobacter
Staphylococcus
Sphingomonas
Brevibacterium
Streptococcus
Sphingobacterium
Chryseobacterium
Pseudomonas
Corynebacterium
Arthrobacter
Nocardiopsis
Curvibacter
Brachybacterium
Streptomyces
Enterococcus
Brevundimonas
Others
57 © IMDO-VUB
De Roos et al., 2018 AEM

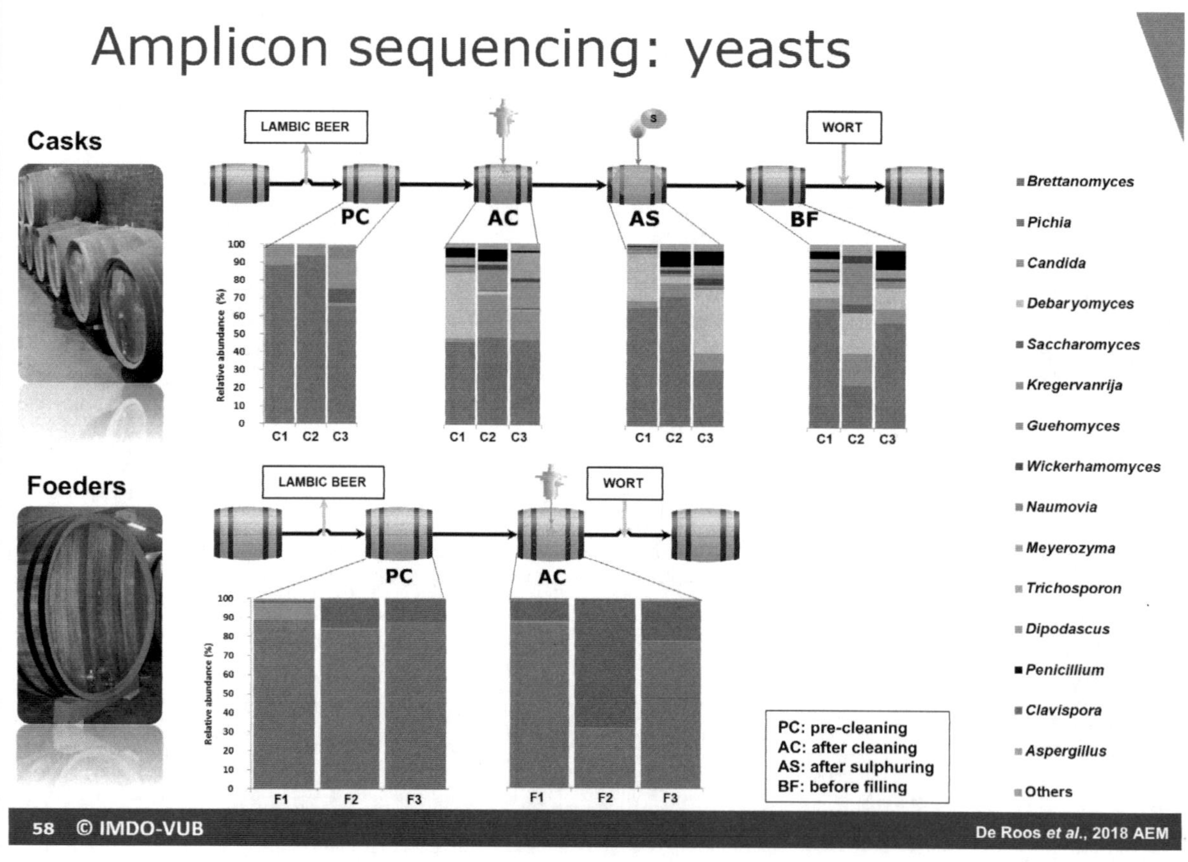
Amplicon sequencing: yeasts
Casks
LAMBIC BEER
WORT
PC
AC
AS
BF
Relative abundance (%)
C1 C2 C3
C1 C2 C3
C1 C2 C3
C1 C2 C3
Foeders
LAMBIC BEER
WORT
PC
AC
F1
F2
F3
F1
F2
F3
PC: pre-cleaning
AC: after cleaning
AS: after sulphuring
BF: before filling
Brettanomyces
Pichia
Candida
Debaryomyces
Saccharomyces
Kregervanrija
Guehomyces
Wickerhamomyces
Naumovia
Meyerozyma
Trichosporon
Dipodascus
Penicillium
Clavispora
Aspergillus
Others
58 © IMDO-VUB
De Roos et al., 2018 AEM

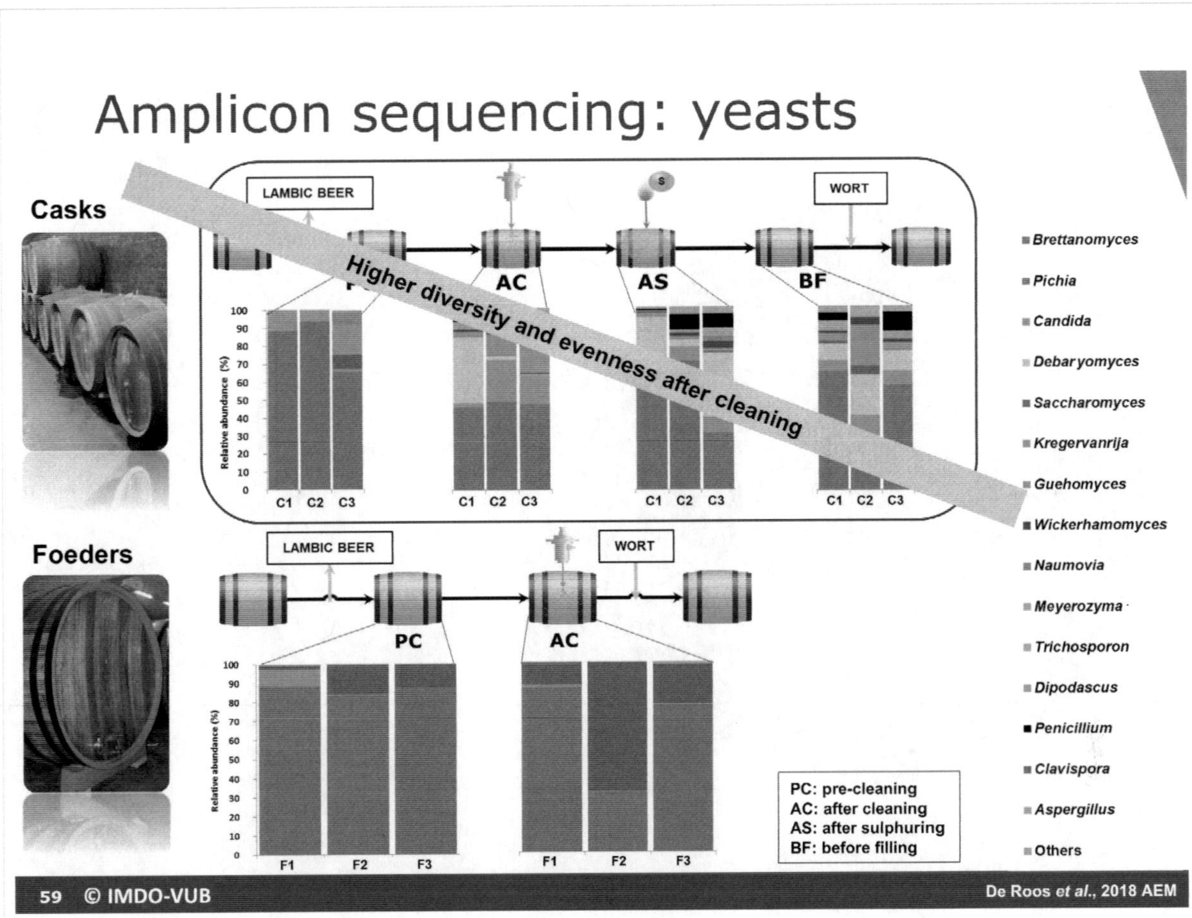
Amplicon sequencing: yeasts
Casks
LAMBIC BEER
WORT
AC
AS
BF
Higher diversity and evenness after cleaning
Relative abundance (%)
C1 C2 C3
Foeders
LAMBIC BEER
WORT
PC
AC
F1 F2 F3
PC: pre-cleaning
AC: after cleaning
AS: after sulphuring
BF: before filling
Brettanomyces
Pichia
Candida
Debaryomyces
Saccharomyces
Kregervanrija
Guehomyces
Wickerhamomyces
Naumovia
Meyerozyma
Trichosporon
Dipodascus
Penicillium
Clavispora
Aspergillus
Others
59 © IMDO-VUB
De Roos et al., 2018 AEM

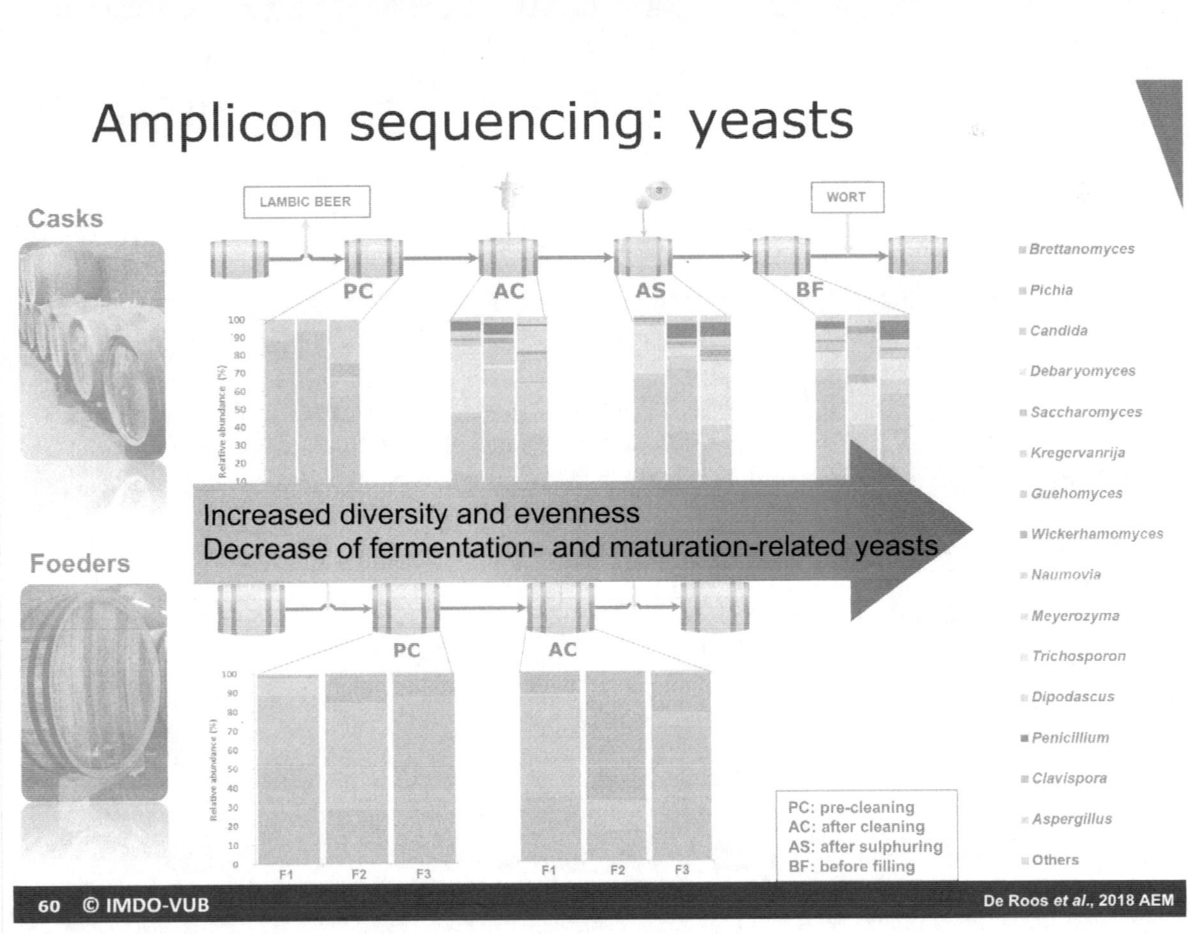
Amplicon sequencing: yeasts
Casks
LAMBIC BEER
WORT
PC
AC
AS
BF
Relative abundance (%)
Increased diversity and evenness
Decrease of fermentation- and maturation-related yeasts
Foeders
PC
AC
F1 F2 F3
PC: pre-cleaning
AC: after cleaning
AS: after sulphuring
BF: before filling
Brettanomyces
Pichia
Candida
Debaryomyces
Saccharomyces
Kregervanrija
Guehomyces
Wickerhamomyces
Naumovia
Meyerozyma
Trichosporon
Dipodascus
Penicillium
Clavispora
Aspergillus
Others
60 © IMDO-VUB
De Roos et al., 2018 AEM

Conclusions

Conclusions

- Additional inoculation source, besides air
  - Fermentation- and maturation-related microorganisms were present at the end of cleaning
- Steering of the production process
  - Installment of a microaerobic environment during fermentation and maturation

Acknowledgements

Prof. Dr. ir. Luc De Vuyst
Prof. Dr. Stefan Weckx
Prof. Dr. ir. Frederic Leroy
IMDO team

Prof. Dr. Peter Vandamme
Dr. Maarten Aerts
Mrs. Margo Cnockaert

Brewery Oud Beersel
Mr. Gert Christiaens

....

VSC Vlaams Supercomputer Centrum

UNIVERSITEIT GENT

VUB VRIJE UNIVERSITEIT BRUSSEL
SRP, IRP, IOF projects

HERCULES STICHTING

KMO-PORTEFEUILLE
ERKEND DIENSTVERLENER

How aging in tropical wood barrels may create an identity for Brazilian beer

Giovanni Casagrande Silvello;

Bortoletto, A. M.; Corrêa, A. C.; de Castro, M. C.; Alcarde, A. R.

University of São Paulo, USP-ESALQ, Piracicaba, São Paulo, Brazil
giovanni.silvello@usp.br

INTRODUCTION

- Worldwide beer culture:
  - Belgium;
  - Germany;
  - UK;
  - USA.
- Brazilian entrance in craft beer culture based on classical culture;

INTRODUCTION

LTQB

ESALQ

37th CONGRESS EBC ANTWERP EUROPEAN BREWERY CONVENTION 02–06 June 2019

- Brazilian culture increasing maturity:
  - Brazilian yeast strain selection (Furlan, 2016);
  - Catharina Sour (BJCP, 2018) beer style containing native fruits;
- Increasing sources for creating a national identity;
  - Beer cultural aspect;
  - New products.

INTRODUCTION

LTQB

ESALQ

37th CONGRESS EBC ANTWERP EUROPEAN BREWERY CONVENTION 02–06 June 2019

- Huge biodiversity;
- Extensive biomes.

- Dense Ombrophilous Forest
- Open Ombrophilous Forest
- Mixed Ombrophilous Forest
- Semi-deciduous Seasonal Forest
- Deciduous Seasonal Forest
- Campinarana heath forest
- Savannah
- Steppe Savannah
- Steppe
- Pioneer Formations
- Ecotone
- Ecological refuge
- Water

Source of data: IBGE (2006).

INTRODUCTION

Safe? (Toxicity)

Wood species	Major composition (%)				Density (g·cm^{-3})
	Cellulose	Lignin	Extractives	Minerals	
Oak (*Quercus sp*)	65,2	25,9	10,5	0,5	0,626
'Amburana' (*Amburana cearensis*)	59,1	27,4	17,9	0,6	0,557
'Cabreúva' (*Myrocarpus frondosus*)	63,6	26,1	10,5	0,5	0,815

Available? (Sustainability)

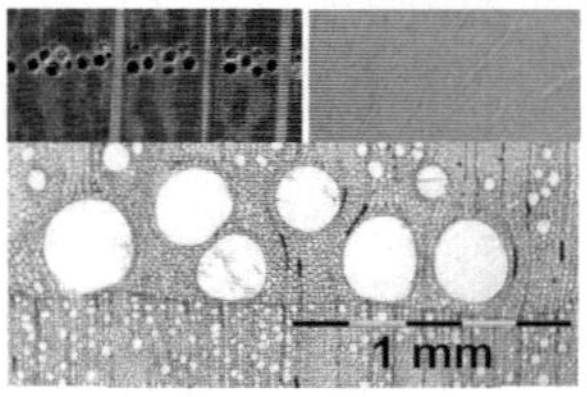

Quercus alba
(American Oak)

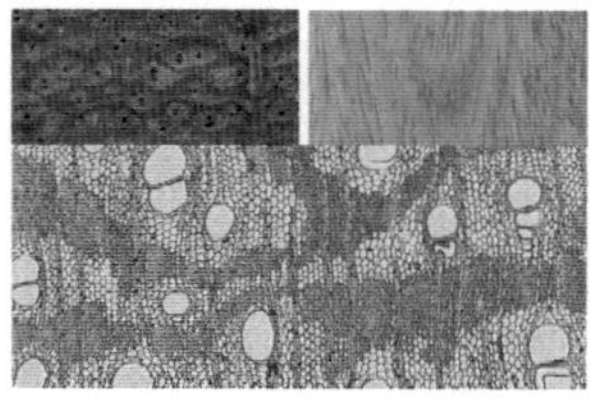

Amburana cearensis
('Amburana')

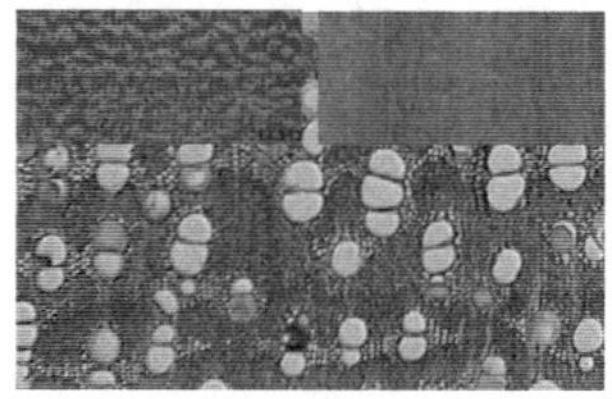

Myrocarpus frondosus
('Cabreúva')

Source: Delta Intkey and Tropical Timber

INTRODUCTION

- Deforestation vs. Reforestation

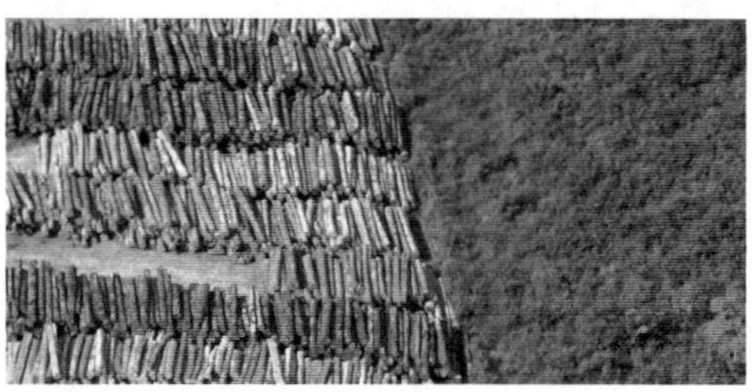

1,675 views | Mar 14, 2019, 02:43pm

How A Cachaça Distillery In Brazil Is Protecting The Future Of An Entire Industry

Claudia Alarcón Contributor
Dining & Drinking

INTRODUCTION

- Brazilian cooperage origins:
- *Cachaça* market in 18th century – commercial path between RJ and MG;
- Mules carrying wooden vessels:
  - Increasing demand for wooden casks;
  - Natural selection of woods (tacit knowledge);
  - Attractive taste and safe for consumers (anti-microbial properties).

INTRODUCTION

- Brazilian cooperage industry.

Pictures: rsnoalambique.blogspot.com

MATERIAL & METHODS

- Beer production:
  - 86.0 % Pilsner malt;
  - 6.9 % Special B;
  - 1.7 % Melanoiden;
  - 5.4 % Cane sugar.
  - 1.57 g/ L Target hop (10 % alpha);
- Wort: **<u>24.5 °Plato</u>**, pH 5.3, 33.5 EBU and 59.8 EBC;
- Fermentation carried at 23.5 °C (until 12.3 °Plato) → → 20 °C (until 2.3 °Plato) → *lagering* 4 weeks at 7 °C;
  - Yeast strain US-05 (Fermentis).

MATERIAL & METHODS

- Beer production:
  - Wood barrels pre-treated and filled;
    - Wood species:
      - *Amburana cearensis* (Fr. Allem.) A.C. Smith – 'Amburana';
      - *Myrocarpus frondosus* Allemão – 'Cabreúva';
      - *Quercus alba* L. – 'American Oak'.
  - Barrels conditioned in an aging chamber:
    - Temperature 17.0 ± 0.3 °C
    - Humidity 81 ± 3 %);

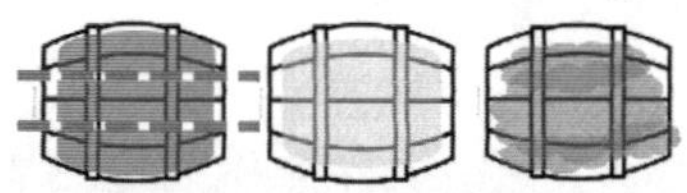

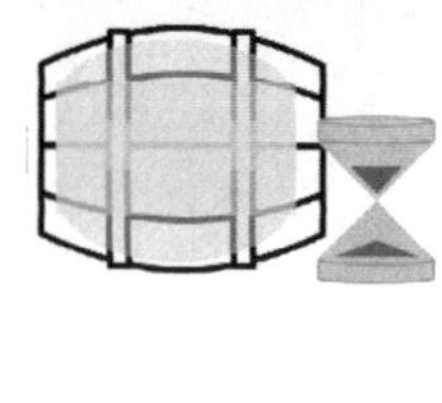

MATERIAL & METHODS

- Sensory analysis (trained panel – 7 members):

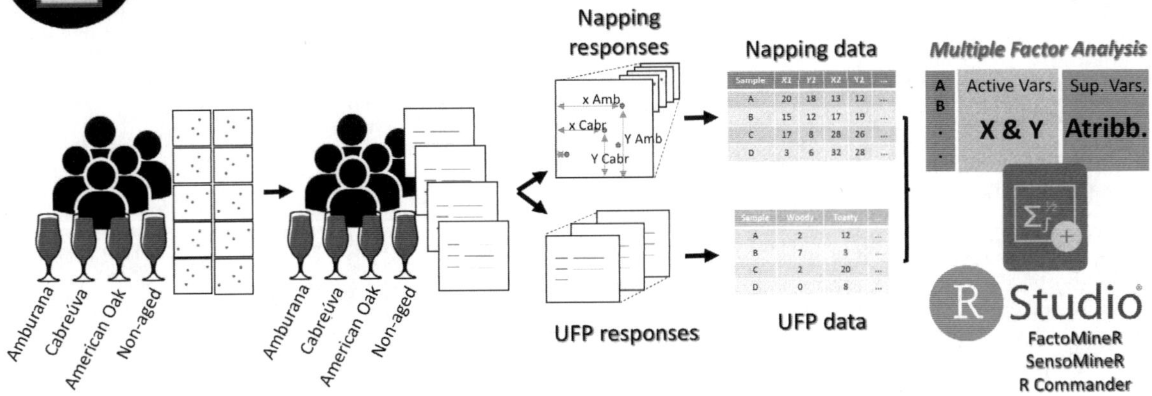

RESULTS & PERSPECTIVES

- Napping:
  - Plot of samples;
- UFP:
  - Plot of most relevant attributes;
  - Mode of attributes per sample = 9;
  - Min. = 6, Max. = 20;
- Differences among wood aged and non-aged (dim. 1);

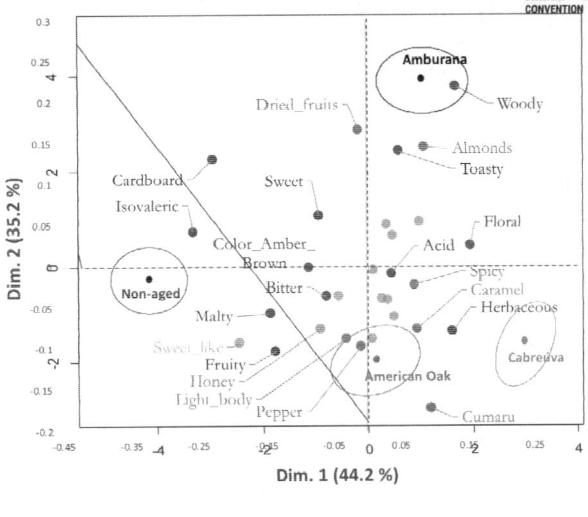

RESULTS & PERSPECTIVES

- Differentiation of 'Amburana' (dim. 2).
- Differentiation of American Oak (dim.3).

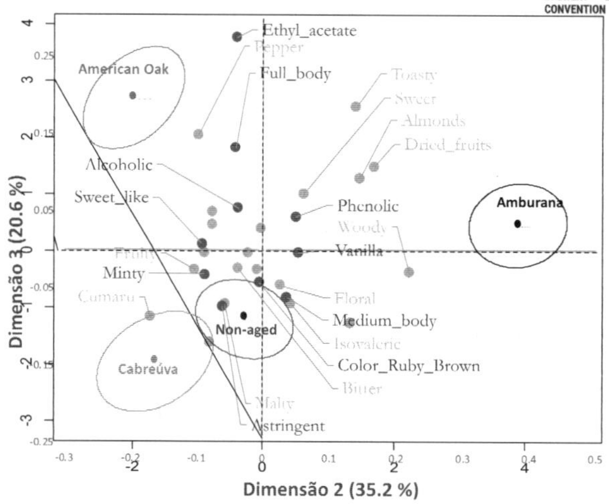

RESULTS & PERSPECTIVES

Treatment	Attributes (UFP)		Flavor major class[1]	Flavor major class[2]
'Amburana'	**Sweet** **Phenolic** Dried Fruits Toasty	**Woody** **Vanilla** Almonds	Sweet, Resinous, Nutty, Burnt, Phenolic, Fruity	Chemical, Fruity, Spicy, Vegetal, Heat induced
'Cabreúva'	**Astringent** **Herbaceous** **Bitter** Minty	**Spicy** **Floral** Cumaru Ruby Brown Color	Floral, Spicy, Grassy, Mouth-feel and taste impact	Spicy, Vegetal, Floral, Taste and Texture impact
American Oak	Caramel Honey **Ethyl acetate**	Alcoholic **Pepper**	Alcoholic, Ester, Caramel, Sweet, Spicy	Heat induced, Spicy, Biological, Chemical
Non-aged Beer	**Malty** Fruity Sweet-like Acid	**Isovaleric** **Cardboard** Amber Brown Color	Malty, Fruity, Sweet, Fatty acid, Stale, Taste impact	Fruity, Chemical, Biological, Heat induced

[1]Beer Flavor Wheel developed by Meilgaard, Dalgliesh and Clapperton (1979).
[2]Beer Aroma Wheel developed by Schmelzle (2009).

RESULTS & PERSPECTIVES

Wood species	Flavor Compounds (wood extracts)	Description (data basis)
Quercus alba (American Oak)	**Vanillin**	**Vanilla**
	γ-lactone	**Coconut**
	5-Hydroximethylfurfural	Caramel
Quercus sessile, Quercus robur, Quercus petrae (European Oak)	**Syringaldehyde**	**Pepper**
	Coniferaldehyde	Cinnamon
	Sinapaldehyde	**Spicy**
	Eugenol	Clove
Myrocarpus frondosus (Cabreúva)	Rutin	Antioxidant
	Alkaloids	**Bitter taste**
	β-pinene	Piney, Resinous
	β-copaene	**Spicy, Woody**
	Germacrene B	**Earthy, Spicy, Woody**
	Spathulenol	**Green odor**
Amburana cearensis (Amburana)	Butein (chalcone)	Antioxidant
	2-ethylhexanoic acid	Mild odor
	Dodecanol (Lauryl alcohol)	**Fatty, Floral**
	Dihydrocoumarin	**Hay, Herbal , Coconut**
	Coumarin	**Sweet-aromatic, Vanilla , Bitter**

FINAL CONSIDERATIONS

- Tropical wood-barrels aging: unique flavours and valuable products (special product category);
- Availability of raw-materials and creative craft industry;
- Consolidation of a Brazilian brewing culture:
  - Technology maturity;
  - Innovative capacity;
  - Internal organization (implementation).

FINAL CONSIDERATIONS

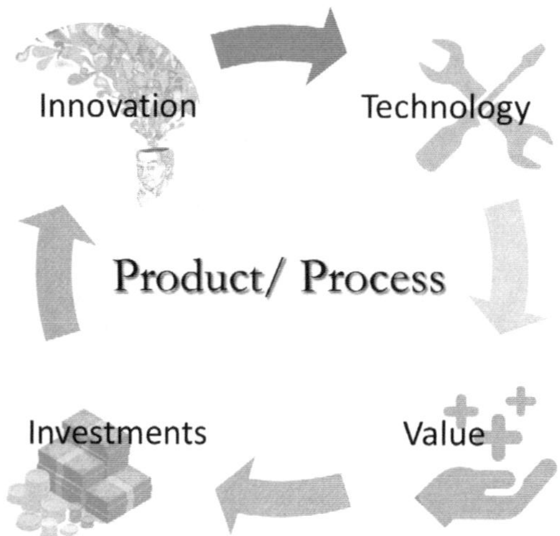

L25 - MALTING BARLEY AGRONOMY & QUALITY : EVOLUTION OF THE RAW MATERIAL QUALITIES OTHER 20 YEARS IN FRANCE

Marc Schmitt[1]

1 – IFBM-Qualtech, Nancy, France

Abstract

In France, the registration of a new barley variety is very structured. Three main organisations take part in the process : the **CTPS** (Permanent Technical Committee of Breeding) which is involved in the registration on the French catalogue and on the malting barley variety list, the **CBMO** (Beer Malt Barley Committee) which is involved in the testing for registration on the Malteurs and Brasseurs de France preferred variety list and **IFBM** which is the technical tool for tests (micromalting, pilot malting and brewing, analyses on barley, malts, worts and beers).

Such a system is aimed at selecting new barley varieties according to Breeder, Maltster and Brewer needs.

The presentation will describe and discuss the evolution of the quality other the last 20 years for the 2 row spring, the 2 row & 6 row winter barleys. The lecture will be documented with the data of barley registration and the pilot tests made for the CBMO.

Short biography

Marc SCHMITT is Scientific Director at IFBM, the French Institute of Malting and Brewing. He joined the Institute at the beginning of 2002. His first job at IFBM was Malting & Brewing Laboratory Manager. He became R&D manager in 2008 and Deputy Scientific Director in 2014. His works concern mainly raw materials. Since 2008, he is in charge of the works of CBMO (barley – malt – beer committee) in France. He has made several presentations for congresses et technical meetings.

He has been lecturer in biochemistry and biosciences engineering at Nancy University. He completed his PhD in food biochemistry which was awarded by Nancy University (1996).

From 1994 to 2002, he managed the Central laboratory of the French Cheese Institute.

Member of the Steering Committee of EBC Analysis Committee, he is chairman of the "Barley – Malt" subcommittee..

Member of the EBC "Brewing Science Group"

Chairman of the Malting Barley Ring test of BIPEA.

European Brewers Convent

04.06.2019
Dipl. –Brmst. Christoph Remmelberger
Technical sales – Malting Plants

Revival of an old method of malting
-
The SCHULZ MALTING SYSTEM

source: Der bayerische Bierbrauer from the year 1929

2

Historic Germination drums

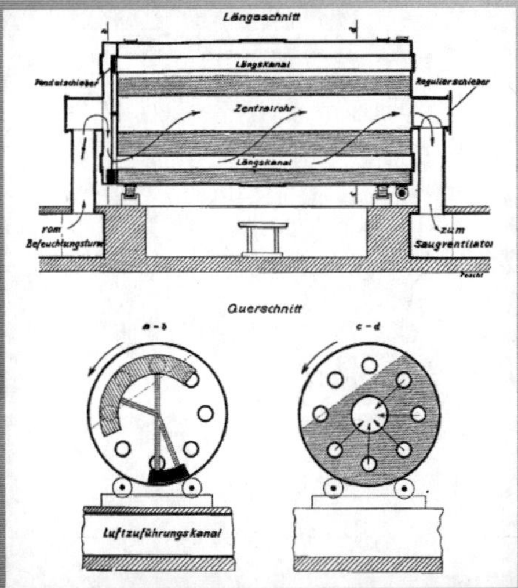

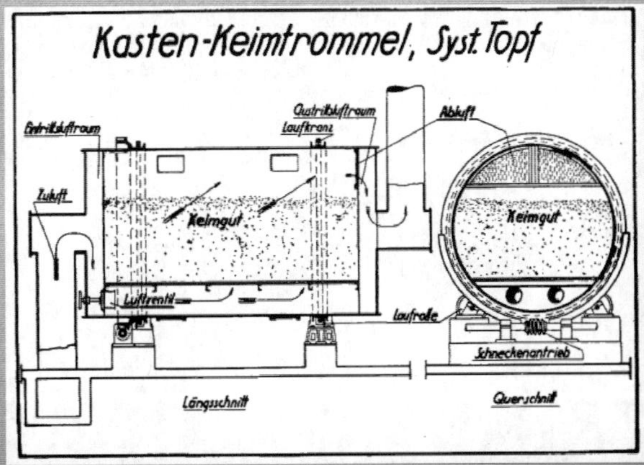

source: Die Technologie der Malzbereitung; Ludwig Narziss

Regionalization

Past:

Opposite was happening in the malting industry:

- Concentration of production took place in the last decades
- Malt production concentrated on a few places and plants got bigger
- Malt as raw material for breweries/bakeries is not available locally

Today:

Regional aspects and thinking enjoys a far distribution

Today still connected:

- Brewery and agriculture/ farming
- Brewery and malthouse
- Agriculture and malthouse
- Beer and Grain linked to local production

→ Possibilities for regionalization exists

SCHULZ Malting System prototype

SCHULZ Malting System

SCHULZ Malting System

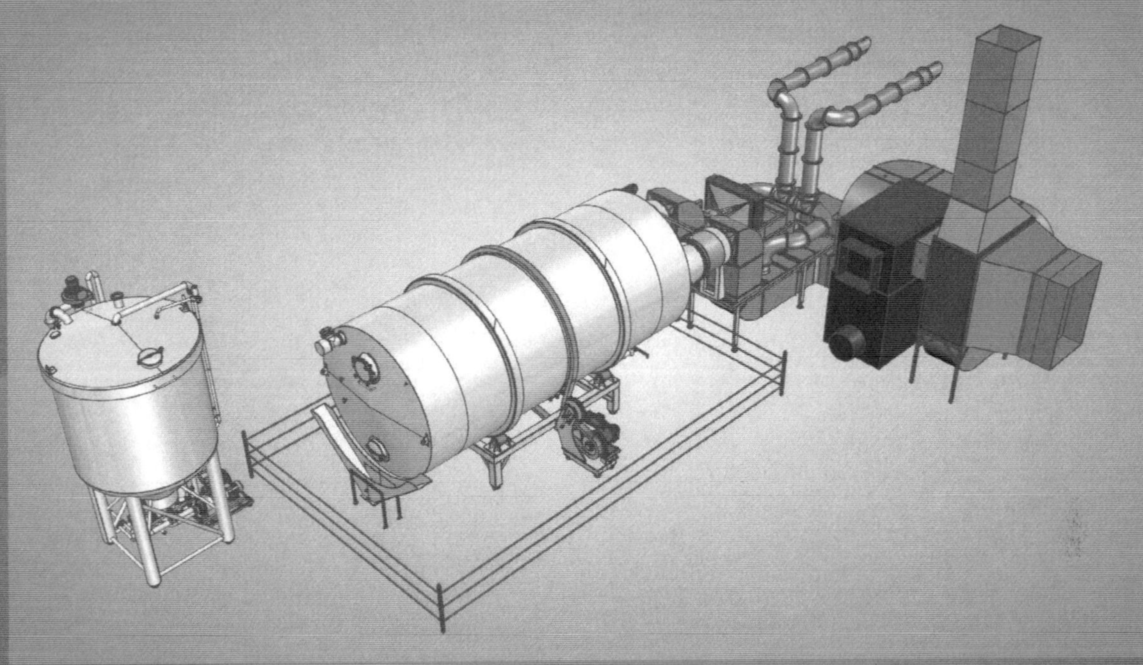

Characteristics + Thoughts during the development

- small Batch sizes: 2; 5; 10 and 25 tons of barley
- High level automation – to operate next to running brewery
- Small effort and short time for cleaning
- Homogeneous malt (gentle and ideal mixing)
- Space-saving (Germination and kilning in one vessel)
- Installation in existing buildings possible
- Easy unit add-on option with additional germination-kilning combo drums when using existing components
- Unique appearance and design

Steeping vessel

- Conical funnel steeping vessel
- Automated removal of floating barley
- Mixing and washing by re-circulation
- 2 possibilities of pneumatic mixing and ventilation
- CO2 suction
- Excellent hygienic conditions
- Closed vessel with hood (Dustfree Filling)
- Automatic cleaning with spray balls

Steeping vessel

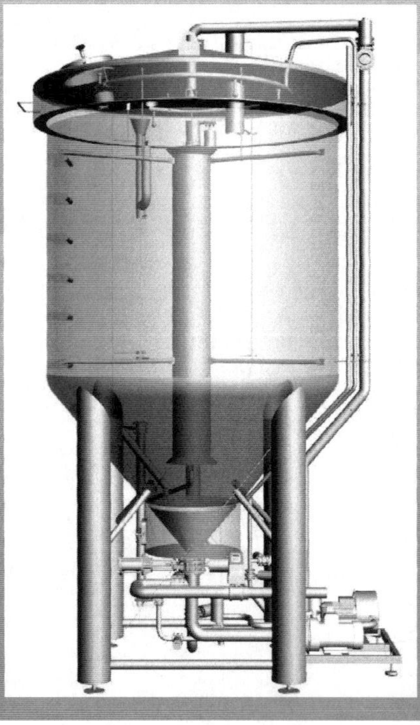

Germination-Kilning-Drum

Germination: Aeration and cooling

- Automatic docking on and off of air channels
- Exact regulation of fresh and circulating air volume according to parameters of operator
- Intelligent energy-saving mode for using the cold outside air in winter
- Air humidification system

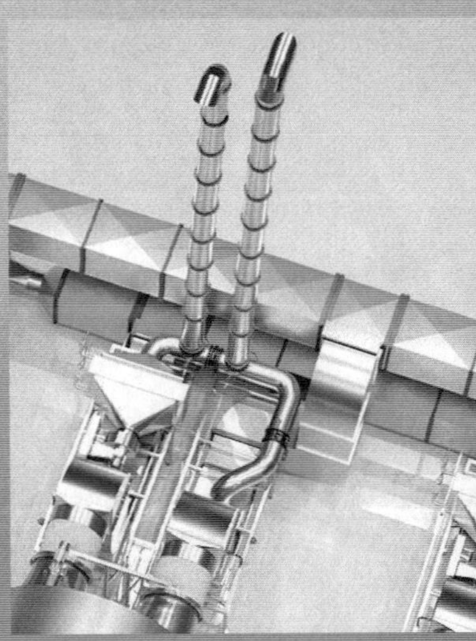

Germination: Rotation and mixing

- Automatic mixing by rotations of the drum
- Consistent flat surface of green malt
- Spraying possible while rotation → Best dispersion of water
- Transport of green malt not necessary (gentle treatment)

Withering and kilning

- Indirect air heater with gas-powered burner (natural or liquid gas), steam or hot water
- Exact regulation of fresh and circulating air volume according to parameters of operator
- Kilning temperatures up to 130 °C (266°F)

Automatic system

SCHULZ

➢ Clearly structured visualization of components, measurements and control elements

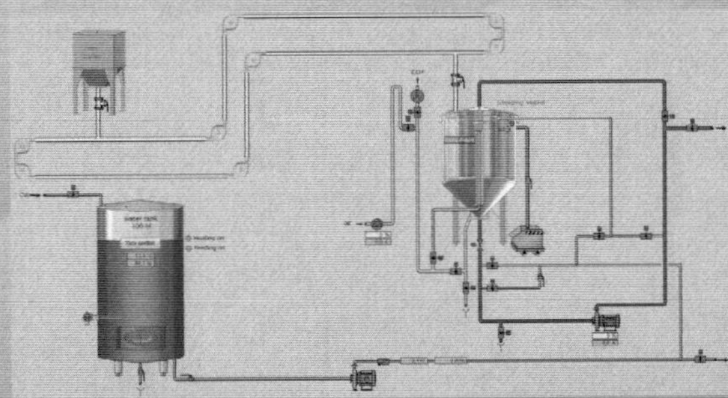

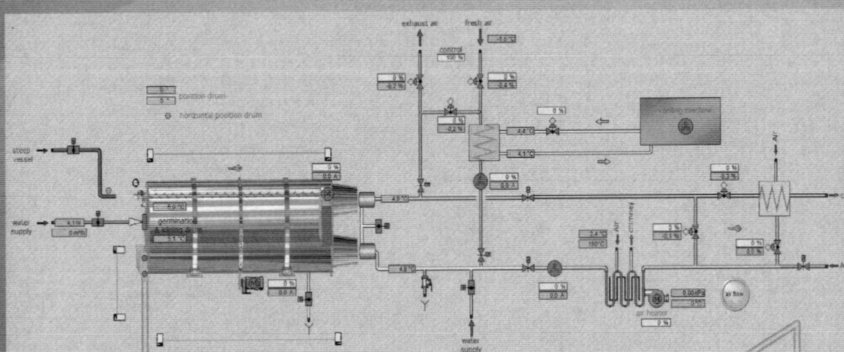

Automatic system

SCHULZ

➢ High level for automation cuts down on work hours (around 1h per day)
➢ Simple operation -No programming knowledge needed
➢ Easy and flexible creation of process recipes
➢ Recording of all measurements

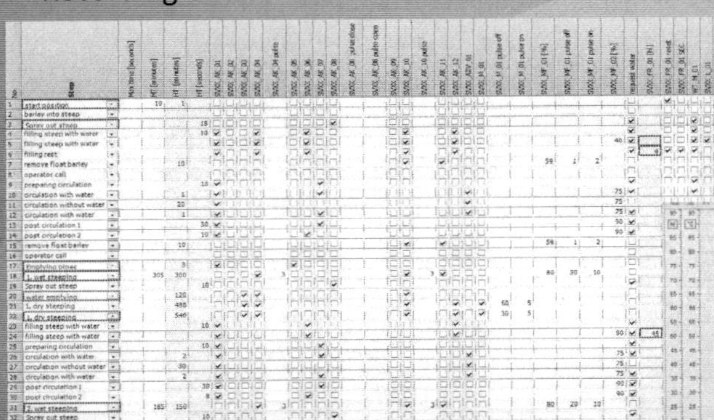

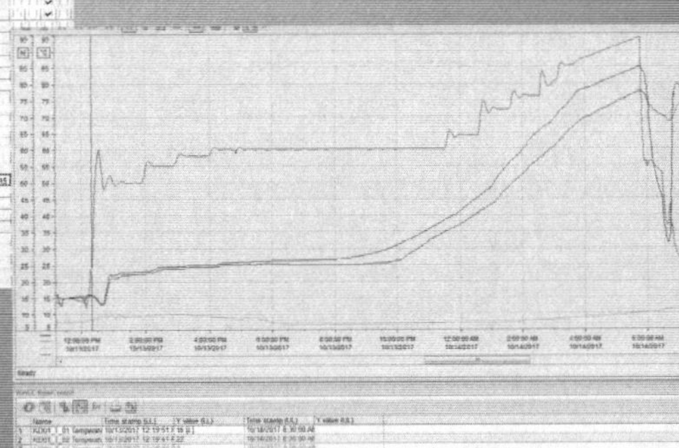

Expansion

- Easy expansion up to 3 Germination-Kilning-Drums
- Possibility to produce 3 batches in one week
- Use of existing equipment for all 3 drums:
  - ➢ Steeping vessel
  - ➢ Kilning fan
  - ➢ Air heater
  - ➢ Glass-tube heat exchanger

Annual capacity

	QUANTITY OF GERMINATION/KILNING DRUM			
UNIT SIZE	1	2	3	
2 to barley/ batch	100 to 80	200 to 160 to	300 to 240 to	barley finished malt
5 to barley/ batch	250 to 200 to	500 to 400 to	750 to 600 to	barley finished malt
10 to barley/ batch	500 to 400 to	1000 to 800 to	1500 to 1200 to	barley finished malt
25 to barley/ batch	1250 to 1000 to	2500 to 2000 to	3750 to 3000 to	barley finished malt

*calculated with 50 weeks of production in one year

High flexibility

- Allows for experimental processing of alternative grains (barley, wheat, rye,...) for a maximum product variety
- Flexible batch sizes of 25-100 % of the max. load possible
- Production of a wide range of malt types (130°C; closed system), also suitable for the baking industry
- Allows for alternative grain processing for different industries

Chances

Malting drums open doors to new and enriching possibilities!

Production of unique malt types!

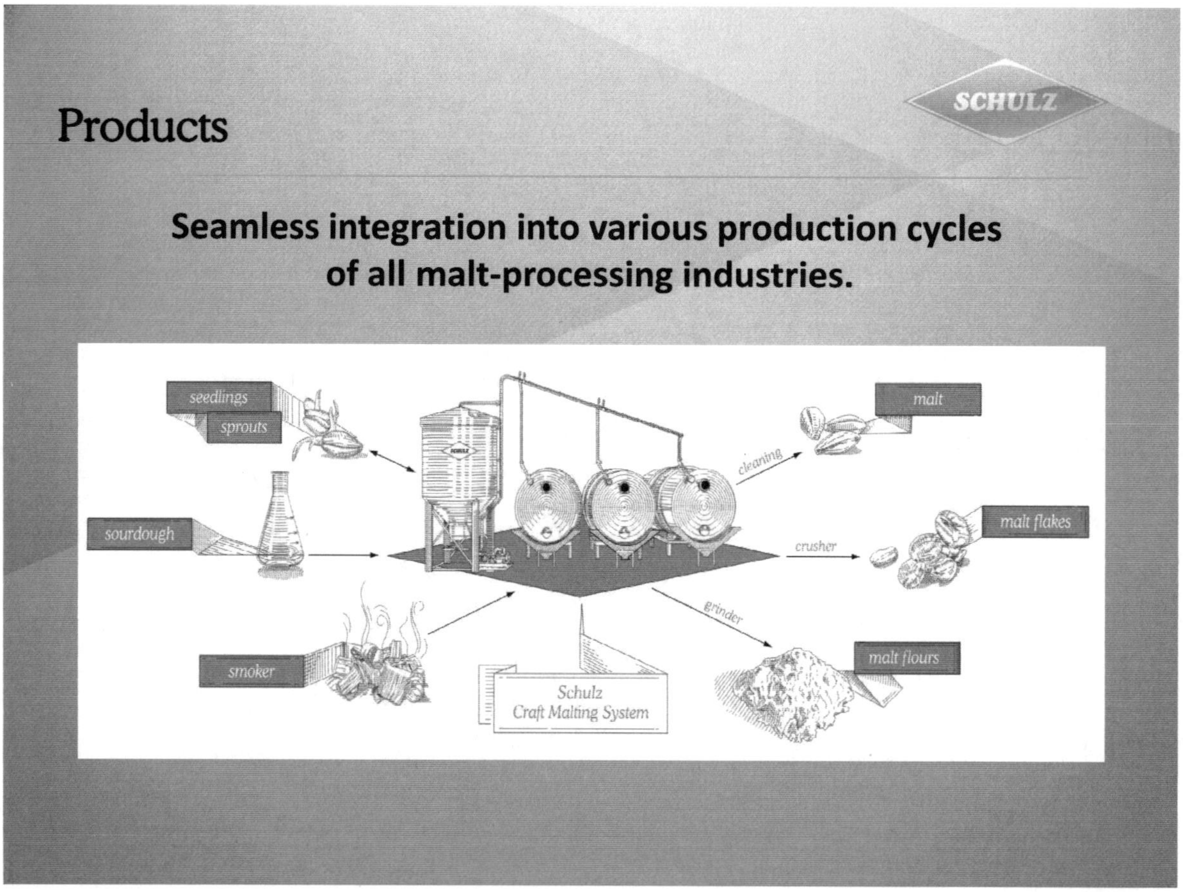
Products
SCHULZ
Seamless integration into various production cycles
of all malt-processing industries.
seedlings
sprouts
sourdough
smoker
Schulz
Craft Malting System
cleaning
crusher
grinder
malt
malt flakes
malt flours

Business segments
SCHULZ
Maltster
Brewer
Miller
Farmer
Distiller
Baker

Pictures & business models

MASTRI BIRRAI UMBRI

SCHULZ

1 x 10 to SCHULZ malting plant before expansion
MASTRI BIRRAI UMBRI COOP. ARL. (Italy)

Pictures & business models

MASTRI BIRRAI UMBRI

SCHULZ

2 x 10 to SCHULZ malting plant after expansion
MASTRI BIRRAI UMBRI COOP. ARL. (Italy)

Pictures & business models

MASTRI BIRRAI UMBRI

SCHULZ

2 x 10 to SCHULZ malting plant and 70 hl brew house
MASTRI BIRRAI UMBRI COOP. ARL. (Italy)

Pictures & business models

CERCLE DES AGRICULTEURS GENÈVE

SCHULZ

1 x 5 to SCHULZ malting plant
Cercle des Agriculteurs de Genève (Switzerland)

Pictures & business models

1 x 10 to SCHULZ malting plant (expandable)
Root Shoot Malting (USA)

Pictures & business models

1 x 5 to SCHULZ malting plant
Warbro Kvarn AB (Sweden)

Pictures & business models

6 x 10 to SCHULZ malting plant
1886 Malt House (USA)

under construction:............

- 2to malting plant without expansion:

  Research and Development malting facility

- 2to malting plant for Japanese Whisky Distillery:

  including expansion possibility
  including smoker to produce smoked malt

SCHULZ
Thank you for your attention.
www.schulz-craftmalting.de
www.kaspar-schulz.de
www.facebook.com/Schulz1677
www.instagram.com/kasparschulz/
KASPAR SCHULZ
Brauereimaschinenfabrik &
Apparatebauanstalt GmbH
Kaspar-Schulz-Str.1
96052 Bamberg, Germany

L27 - FIRST TIME QUANTIFICATION OF CYSTEINYLATED ALDEHYDES IN MALT AND BREWERY SAMPLES

Paula Bustillo Trueba[1]; Barbara Jaskula-Goiris[1]; Mark Sanders[3]; Jean-Paul Vincken[3]; Jessika De Clippeleer[2]; Erik Van Der Eycken[1]; Joshep De Brabanter[1]; Luc De Cooman[1]; Gert De Rouck[1]; Guido Aerts[1]

1 - KU Leuven; 2 - Ghent University, Belgium; 3 - Wageningen University & Research, The Netherlands

Abstract

Beer flavour is not stable during storage. The mechanisms of formation of aldehydes, which are considered as a major cause of off-flavours during beer aging, remain uncertain. Increase in aldehydes by *de novo* formation or the release from bound-state precursors have been suggested as the cause of aldehyde appearance. In a previous study, we proposed the bound form between cysteine and aldehydes, also called cysteine-bound aldehydes or 2-substituted 1,3-thiazolidine-4-carboxylic acids, as potential sources of the increasing aldehyde levels found during beer aging. In this study, cysteine-bound aldehydes of selected marker aldehydes were synthesized and used as reference compounds. Implementation of the methodology for the detection and quantification of these compounds was done by Liquid Chromatography-Mass Spectrometry (LC-MS). Malt samples and samples collected at different points during the brewing process (from the onset of mashing until the fresh beer, as well as during beer aging) were analysed for their free and cysteine-bound aldehydes content using HS-SPME-GC-MS (Headspace-Solid Phase Microextraction-Gas Chromatography-Mass Spectrometry) and LC-MS (Liquid Chromatography-Mass Spectrometry), respectively. In malt, we were able to quantify all selected aldehyde markers, both free and cysteine-bound. Furthermore, quantification of free aldehydes was possible in all brewing samples, whereas cysteine-bound aldehydes detection and quantification was dependent on the nature of the 2-substitution in the thiazolidine ring. In conclusion, even though cysteine-bound aldehydes were not identified unambiguously in the final beer, our experimental data support the general concept of bound-state aldehydes as potential contributors to increasing aldehyde levels and, therefore, to beer flavour deterioration.

Short biography

Paula Bustillo Trueba. Obtained degree in Chemistry by the University of Salamanca (Spain) in 2015. At the moment PhD student at the University of Leuven (KU Leuven, Belgium) in the cluster for Bioengineering Technology (CBeT), Laboratory of Enzyme, Fermentation and Brewing Technology (EFBT). The research topic is "Unravelling formation and/or release of beer staling aldehydes".

L28 - QUANTITATIVE MONITORING OF BEER STALING ALDEHYDES DURING THE MALTING PROCESS

Weronika Filipowska[1]; Maciej Ditrych[1]; Barbara Jaskula-Goiris[1]; Guido Aerts[1]; Gert De Rouck[1]; Luc De Cooman[1]

1 - KU Leuven, Bioengineering Technology TC, Ghent Technology Campus

Abstract

With time, many (bio)chemical transformations take place in packaged beer, resulting in the appearance of off-flavours – in particular the so-called "staling aldehydes". These volatiles can be formed via *e.g.* Maillard reactions, fatty acids oxidation, Strecker degradation of amino acids or they can be released from their bound-state forms (*e.g.* bound to bisulfite and/or cysteine). To date, the origin of beer staling aldehydes and their precursors still remains unknown. Malt, being the major raw material used in the brewing process, contains relatively high concentrations of beer staling aldehydes.

In this study, a novel HS-SPME-GC-MS methodology was used to evaluate the behaviour of aldehydes during the industrial-scale malting process. The critical factors influencing the formation of aldehydes and their precursors were investigated. Additionally, various malting treatments (*e.g.* sulphuring during drying, different kilning temperatures) were compared and levels of aldehydes were correlated with standard quality parameters of malt (colour, pH, moisture, free amino nitrogen *etc.*).

The results showed that the malting process considerably impacts the amount of aldehydes introduced into the brewing process. Heat-load plays the key role in the aldehyde formation/release. Aeration and barley variety also have an impact. In turn, sulfuring seems to be less important. The presentation on the one hand discusses the relevance of the malting technology on malt quality and beer flavour stability, and, on the other hand, it introduces analytical methodology allowing to monitor the levels of staling aldehydes at the very early stage of the brewing process starting with the selection of the raw materials.

Short biography

Biotechnologist, graduated BSc in 2015 and MSc in 2016 at Wroclaw University of Environmental and Life Sciences, Poland. Industrial and research intern in multiple countries (Poland, Iceland, Spain) working on the projects of biofuel production, yeast cell stress and sustainable production in the food industry.

Currently Marie Sklodowska-Curie Fellow within European Joint Doctorate in Food Science (EJDFoodSCI) project. PhD student at the Laboratory of Enzyme, Fermentation and Brewing Technology at KU Leuven, Belgium. PhD student at Bioenergy & Brewing Science Department at University of Nottingham, United Kingdom. Interested in relation between malting process and beer flavour instability.

KU LEUVEN

Non-Starch Carbohydrates from unmalted cereals as mouthfeel contributors in NABLAB and Light beers

N.A. Langenaeken[1], N. Coenen[2], D.P. De Schutter[3], C. Clasen[2], C.M. Courtin[1]

[1]Laboratory of Food Chemistry and Biochemistry and Leuven Food Science and Nutrition Research Centre (LFoRCe), KU Leuven, Kasteelpark Arenberg 20, 3001 Heverlee, Belgium

[2]Soft Matter, Rheology and Technology Section, KU Leuven, Celestijnenlaan 200F, Heverlee, Belgium

[3]Fund Baillet-Latour, Brouwerijplein 1, 3000 Leuven, Belgium

NABLAB's are on the rise

Sales of non-alcoholic beer surge

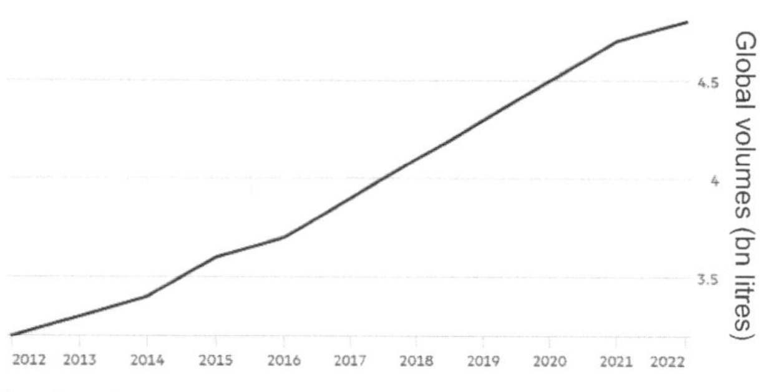

Source: Euromonitor
© FT

Sales growth 2012-2017 (%)

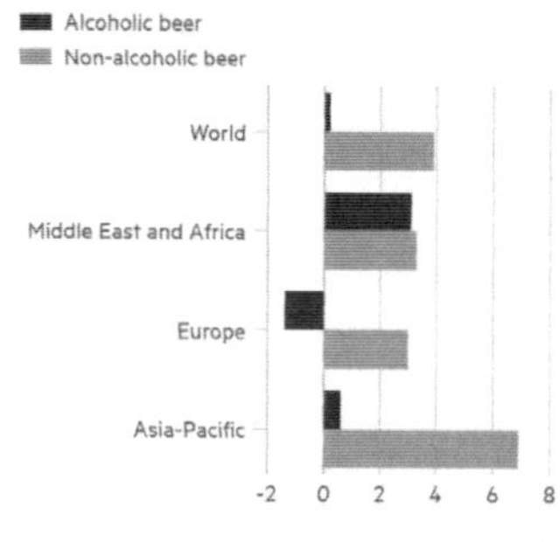

Source: Euromonitor
© FT

NABLAB in the market

Why some people **oppose** to drinking NABLAB ➡

- Aroma
- Texture/ Mouthfeel
- Psychology

Mouthfeel is complex

Carbonation (0 – 1 sec)
- Sting
- Bubble size
- Foam volume
- Total CO_2

Fullness (1 – 2 sec)
- Viscosity
- Density

Afterfeel (10 – 60 sec)
- Stickiness
- Oily mouthcoat
- Astringency

Mouthfeel is complex

Human level

- Saliva production
- Amylase activity
- pH
- Food interference
- State of mind
- …

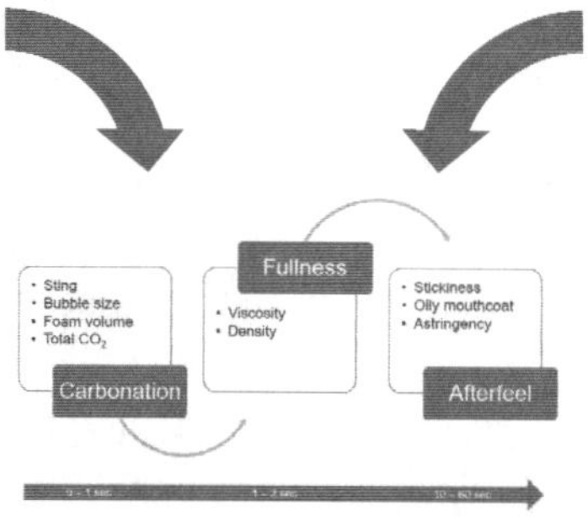

Beer level

- Alcohol } ↘ NABLAB
- Dextrin } ↘ Light
- Protein
- β-glucan
- Glycerol
- Ions
- Polyphenols
- CO_2
- …

Arabinoxylan?

Use of arabinoxylan (AX) to improve mouthfeel?

- Hypothesis: Can we improve mouthfeel of NABLAB's or light beers by increasing their arabinoxylan content?
- Advantages:
  - AX is present in malt and adjuncts
  - AX is known to contribute to viscosity
  - (AX is healthy)
- Threats:
  - AX can block wort and beer filtration
- Opportunities:
  - Use of AX in non-alcoholic beers

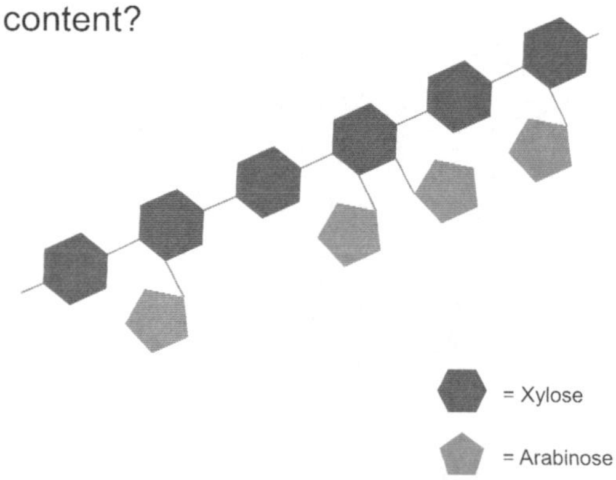

Outline

- Proof of concept
- Selection of raw material
- Optimize application in brewing process

7 Laboratory of Food Chemistry and Biochemistry KU LEUVEN

Proof of concept | Approach

- Use of purified AX (Megazyme, 23 cSt, medium viscosity)
    - 323 kDa (~1800 avDP)
- 4 samples:

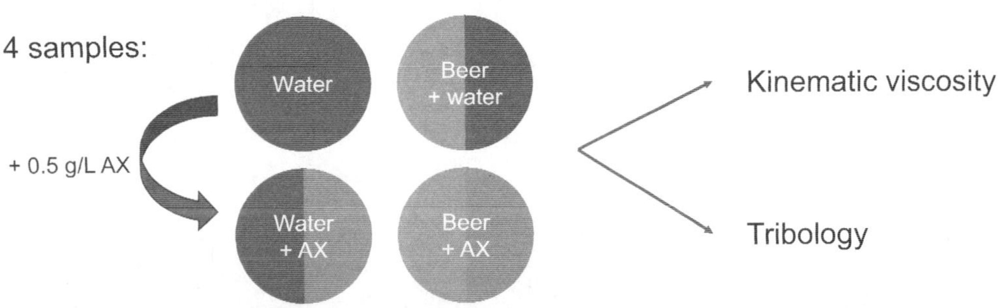

8 Laboratory of Food Chemistry and Biochemistry KU LEUVEN

Intermezzo | Tribology

- **Tribology** is the science and engineering of interacting surfaces in relative motion. It includes the study and application of the principles of friction, lubrication and wear.

- Traditional tribology: machines, gears, wheels,…
- Biotribology: human hip or knee joints,…
- Last decade: increasing interest as mouthfeel assessment for food and beverages!

9 Wikipedia

Laboratory of Food Chemistry and Biochemistry KU LEUVEN

Intermezzo | Tribology

- Ball-on-three-plates set-up:
  - Friction/lubrication will affect torque, resulting in a measurement of force.
  - This force is divided by the applied normal force, resulting in the friction factor.
  - The lower the friction factor, the better the lubrication properties

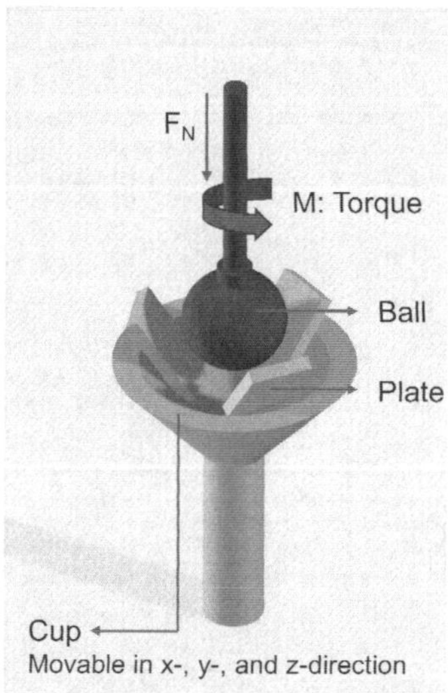

10 Adapted from Anton-Paar

Laboratory of Food Chemistry and Biochemistry KU LEUVEN

Intermezzo | Stribeck curves

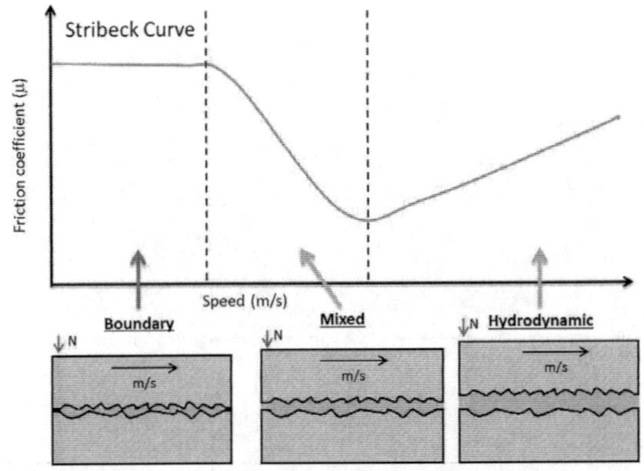

- Boundary region is dependent on material properties
- Mixed region is of most interest for mouthfeel
- Hydrodynamic region is more related to viscosity (and non-existent in our set-up)

Proof of concept | Results

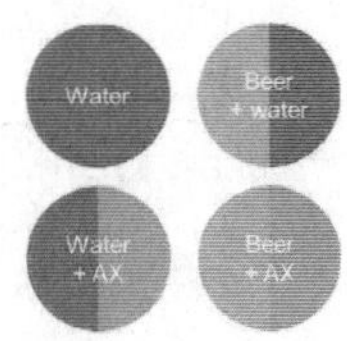

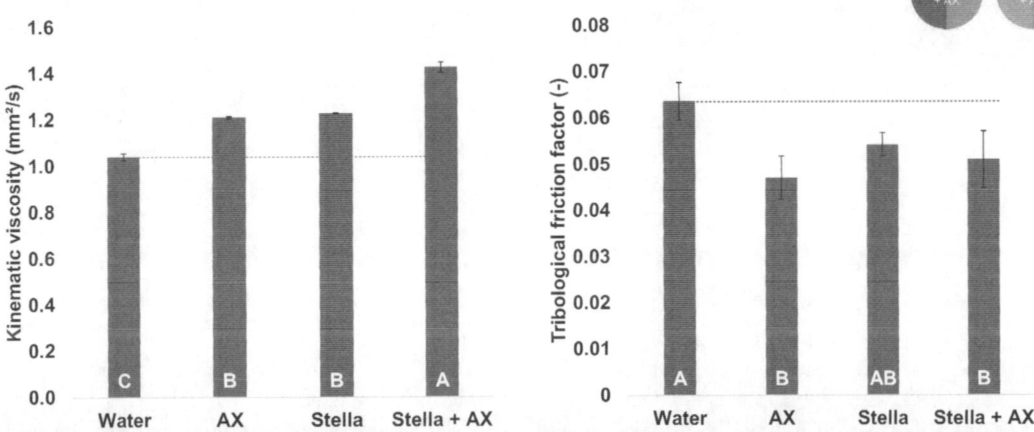

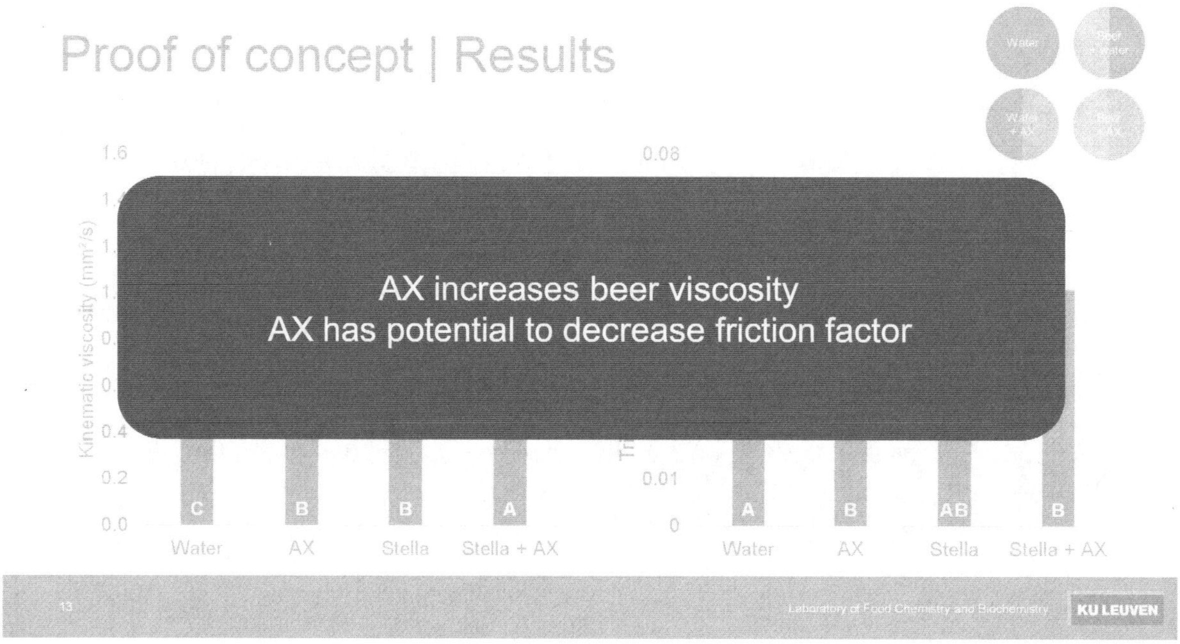
Proof of concept | Results
Water
Beer + water
Water + AX
Beer + AX
1.6
0.4
0.2
0.0
Kinematic viscosity (mm²/s)
C
B
B
A
Water
AX
Stella
Stella + AX
0.06
0.01
0
A
B
AB
B
Water
AX
Stella
Stella + AX
AX increases beer viscosity
AX has potential to decrease friction factor
13
Laboratory of Food Chemistry and Biochemistry
KU LEUVEN

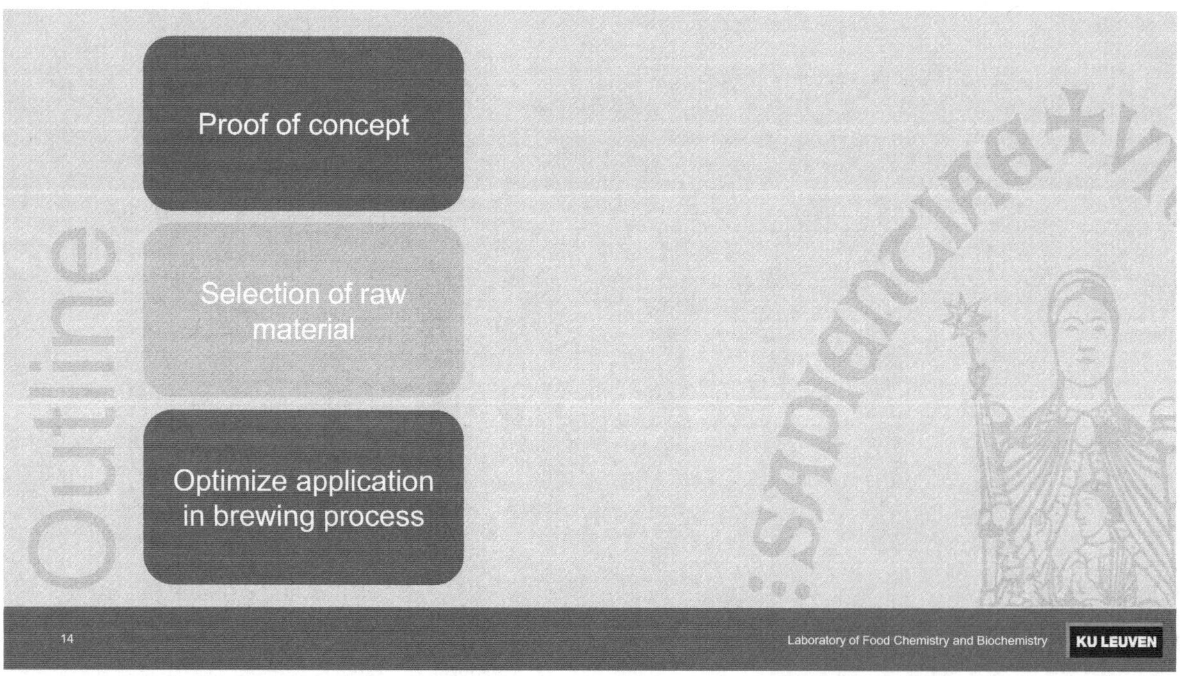
Outline
Proof of concept
Selection of raw material
Optimize application in brewing process
14
Laboratory of Food Chemistry and Biochemistry
KU LEUVEN

Raw material selection | Approach

- Goal: to enhance mouthfeel in beers by use of adjuncts that contain non-starch carbohydrates.

- Samples:

	β-glucan (%dm)	AX (%dm)
100% barley malt (= control)	0.5	6.5
80% barley malt + 20% barley	3.5	6.5
80% barley malt + 20% rye	1.5	8.8
80% barley malt + 20% oat flakes	3.5	2.4

15 Laboratory of Food Chemistry and Biochemistry KU LEUVEN

Raw material selection | Approach

- Mashing → Lochner device (300 mL)
  - 45 °C – 30 min
  - 62 °C – 30 min
  - 72 °C – 30 min
  - 78 °C – 10 min
- Wort filtration → Paper filter
- Boiling → Autoclave
- Fermentation → Schott bottles

Carbohydrate analyses:
- Dextrin
- AX
- β-Glucan

Mouthfeel estimators:
- Density
- Kinematic viscosity
- Tribological friction factor

16 Laboratory of Food Chemistry and Biochemistry KU LEUVEN

Raw material selection | Results

	100% Malt	20% Barley	20% Rye	20% Oats
Original extract (% w/w)	11.24 ± 0.01	11.13 ± 0.00	11.11 ± 0.08	11.25 ± 0.00
App. extract (% w/w)	0.76 ± 0.03	0.93 ± 0.01	1.05 ± 0.06	0.88 ± 0.00
ABV (% v/v)	5.52 ± 0.01	5.37 ± 0.01	5.29 ± 0.01	5.46 ± 0.00
pH	3.96 ± 0.00	4.02 ± 0.00	4.00 ± 0.00	3.94 ± 0.01
Color Value (EBC)	15.1 ± 0.0	13.2 ± 0.0	13.6 ± 0.1	12.6 ± 0.1
RDF (%)	76.4 ± 0.2	75.2 ± 0.1	74.3 ± 0.3	75.6 ± 0.0

*Beers brewed in duplicate

- Similar ABV & pH
- Decrease in color value for adjunct beers

Raw material selection | Beer composition

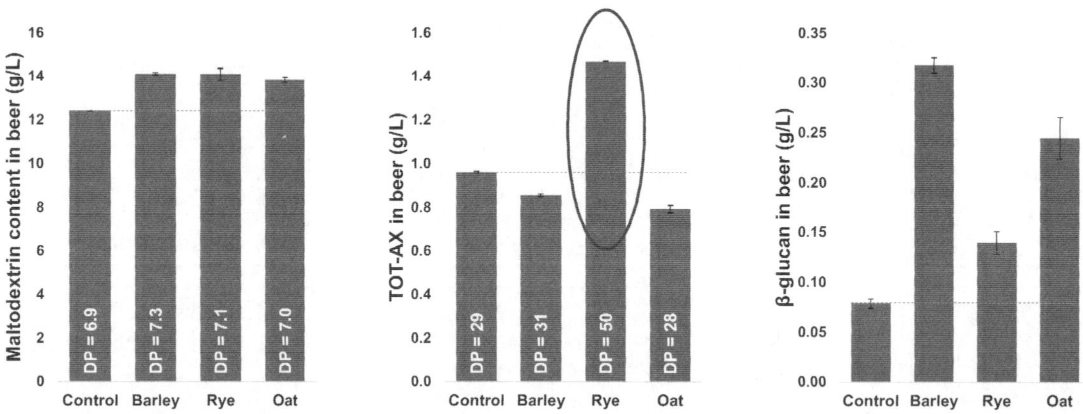

Raw material selection | Sensory estimators

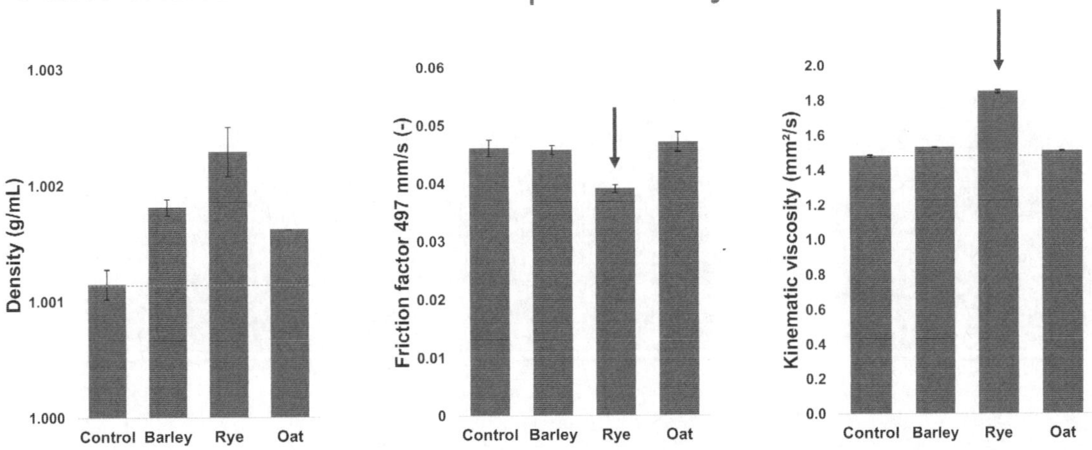

Raw material selection | PCA

- AX correlates with:
  - Kinematic viscosity
  - Friction factor (inversely)
- Not correlated:
  - β-glucan
  - Maltodextrin

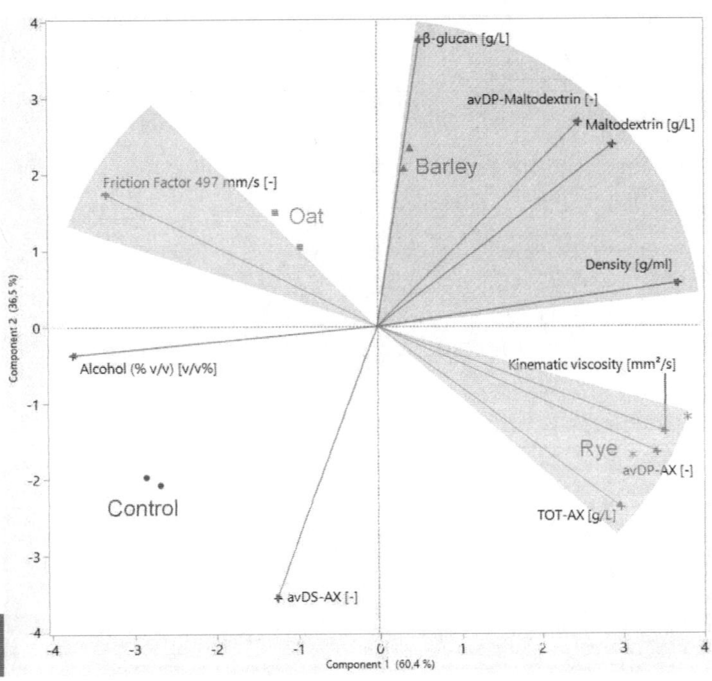

Raw material selection | PCA

- AX correlates with:

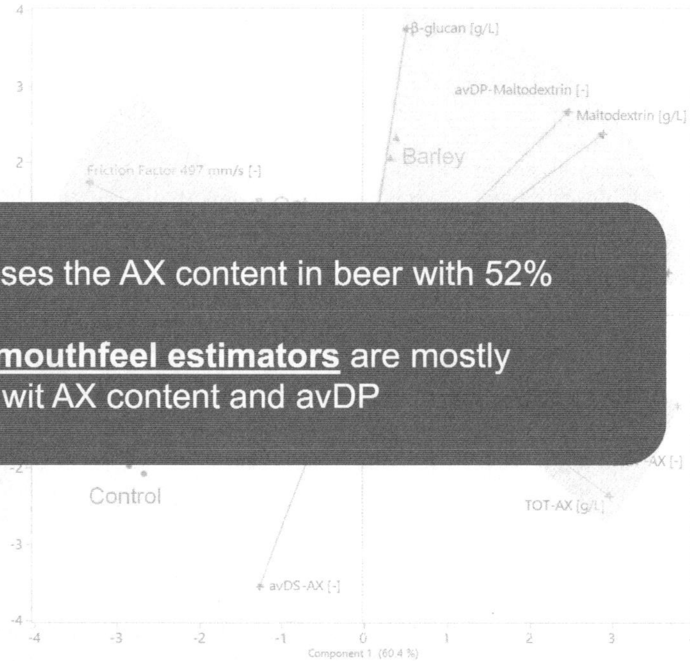

<u>Unmalted Rye</u> increases the AX content in beer with 52%

Within this set-up, **<u>mouthfeel estimators</u>** are mostly correlated wit AX content and avDP

21

Outline

- Proof of concept
- Selection of raw material
- Optimize application in brewing process

22 Laboratory of Food Chemistry and Biochemistry KU LEUVEN

Optimization rye beer | Approach

- <u>Goal</u>: to improve AX solubilization in rye beers

- <u>Samples</u>:
  - 100% barley malt — Disk milled at 2mm
  - 80% barley malt + 20% rye — Ball milled (smaller particles)
  - 80% barley malt + 20% rye — Disk milled at 2 mm (larger particles)
  - 80% barley malt + 20% rye — + Xylanase

Optimization rye beer | Particle size analysis

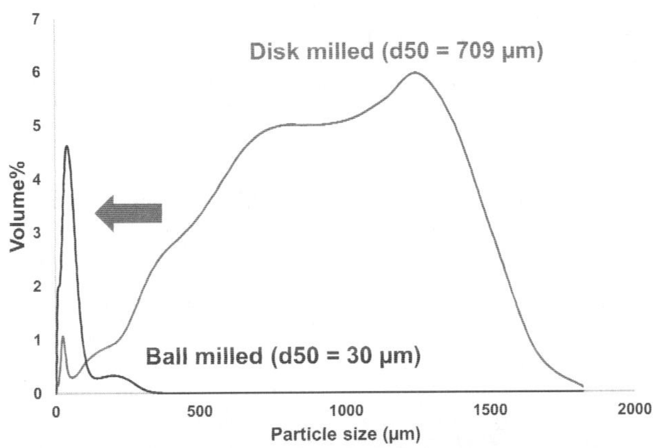

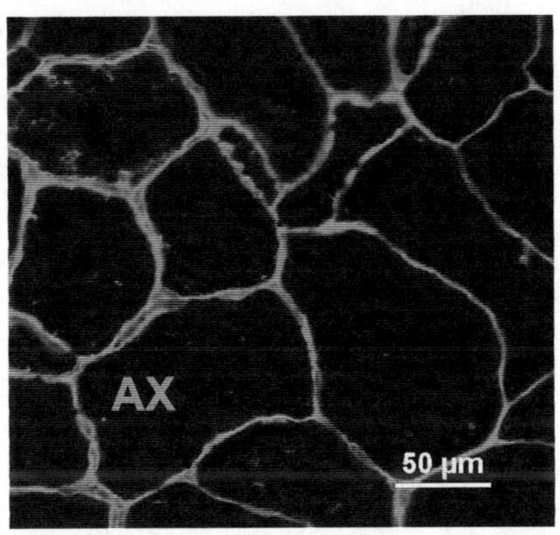

Optimization of rye beer | Results

	Control	Ball milled rye	Disk milled rye	Rye + Xylanase
p (original extract) (% w/w)	10.91 ± 0.01	10.80 ± 0.05	10.81 ± 0.05	10.92 ± 0.01
Ea (app. extract) (% w/w)	0.62 ± 0.00	0.97 ± 0.04	0.94 ± 0.01	1.04 ± 0.00
Alcohol (% v/v)	5.41 ± 0.01	5.17 ± 0.05	5.19 ± 0.03	5.20 ± 0.01
pH	4.1 ± 0.0	4.1 ± 0.0	4.1 ± 0.0	4.1 ± 0.0
Color Value (EBC)	12.9 ± 0.0	13.8 ± 0.2	13.5 ± 0.0	13.8 ± 0.1
RDF (%)	77.3 ± 0.0	74.7 ± 0.3	74.9 ± 0.0	74.2 ± 0.0

*Beers brewed in duplicate

- Slightly higher alcohol content for control beer
- Lower RDF for rye beers

25 Laboratory of Food Chemistry and Biochemistry KU LEUVEN

Optimization of rye beer | Beer composition

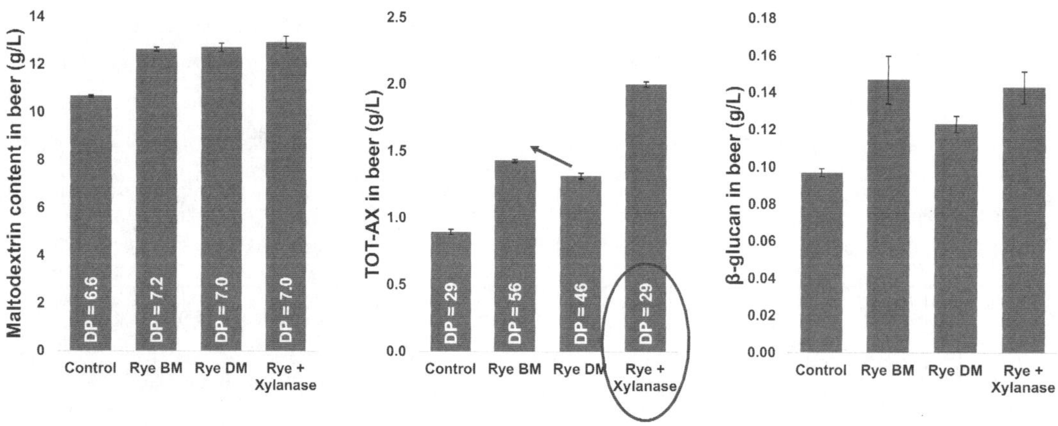

26 Laboratory of Food Chemistry and Biochemistry KU LEUVEN

Optimization of rye beer | Mouthfeel estimators

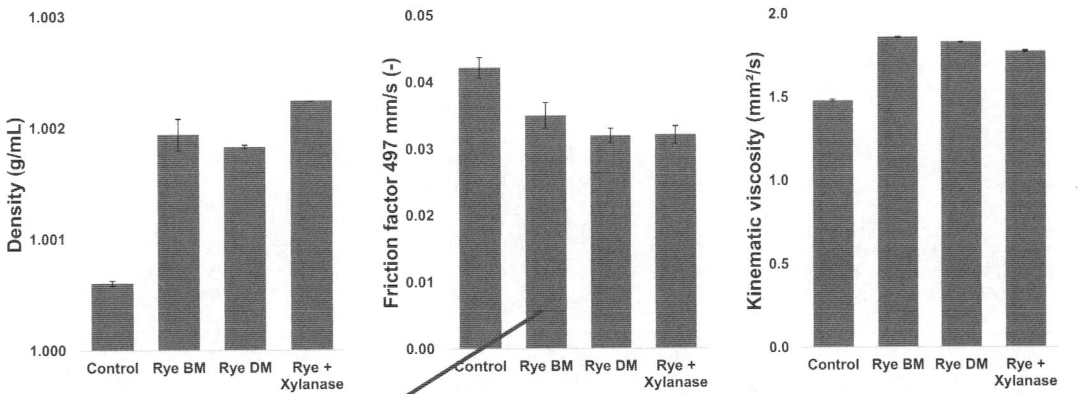

24% of rye AX solubilized avDP = 164 → (~50% of beer AX)
8% of malt AX solubilized avDP = 29 → (~50% of beer AX)

Optimization of rye beer | Mouthfeel estimators

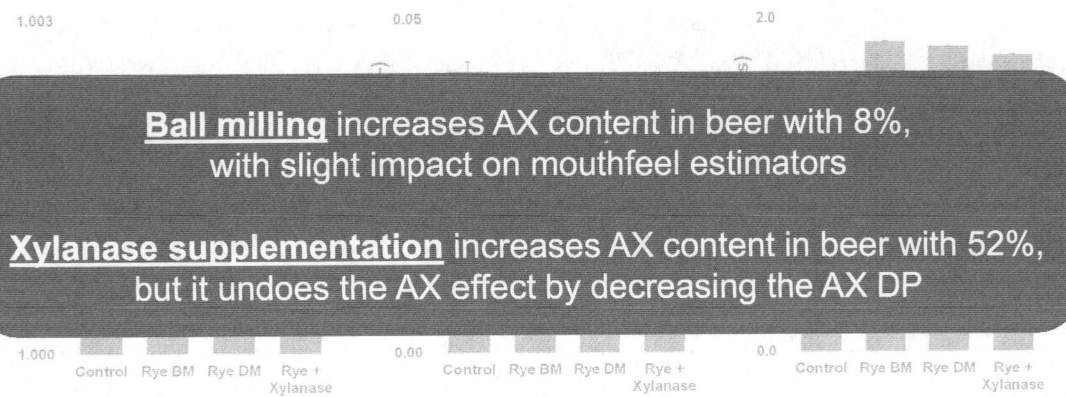

Ball milling increases AX content in beer with 8%, with slight impact on mouthfeel estimators

Xylanase supplementation increases AX content in beer with 52%, but it undoes the AX effect by decreasing the AX DP

Wrap-up

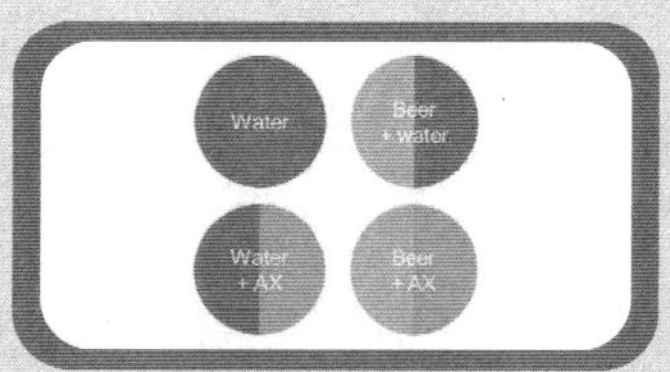

→ AX has potential to increase beer viscosity and lubrication

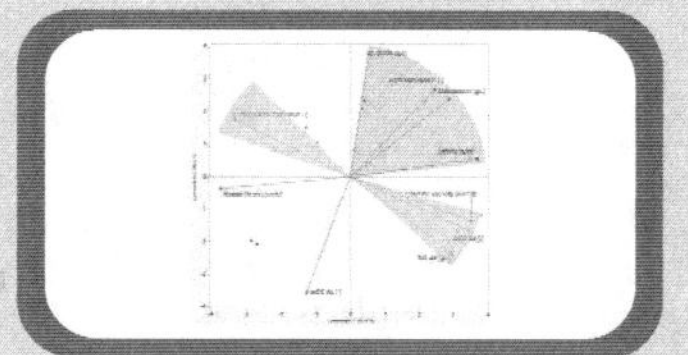

→ Unmalted rye is most suitable to increase AX content in beer

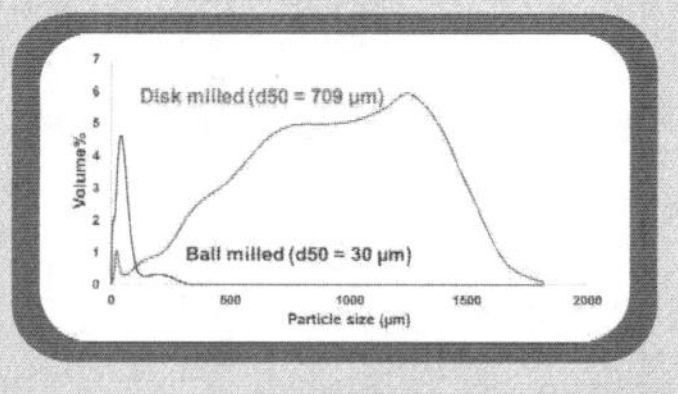

→ Fine milling or enzyme supplementation can be used to improve AX solubilization, but the impact is limited

Thanks to…

All colleagues from
Lab of Food Chemistry and Biochemistry

Collaborations
Prof. Clasen (KUL) – Tribology

Fund Baillet-Latour

Questions?

School of Food and Nutritional Sciences
University College Cork
Ireland

University College Cork, Ireland
Coláiste na hOllscoile Corcaigh

Chance and Challenge: Non-*Saccharomyces* yeasts in non-alcoholic and low alcohol beer brewing

Konstantin Bellut & Elke K Arendt

School of Food and Nutritional Sciences, University College Cork, Ireland

37th CONGRESS EBC ANTWERP EUROPEAN BREWERY CONVENTION 02–06 June 2019

37th Congress of the European Brewery Convention
Antwerp, Belgium
4 June, 2019

Hot topic: alcohol-free beer

University College Cork, Ireland
Coláiste na hOllscoile Corcaigh

Alcohol-free Heineken 0.0 lands in the US
10-Jan-2019 By Beth Newhart

Carlsberg: 'We are seeing a significant increase in alcohol-free'
07-Feb-2019 By Rachel Arthur

AB InBev: 8% of beer volumes now come from no and low alcohol
04-Mar-2019 By Rachel Arthur
Approximately 8% of AB InBev's global beer volumes now come from no and low alcohol beers, as the world's largest brewer works towards its goal of 20% by 2025. "We're committed to leading innovation in this space," it says.

AB INBEV, HEINEKEN, CARLSBERG
How are the biggest brewers embracing the growth of no and low alcohol beers?
19-Mar-2019 By Rachel Arthur
The world's largest brewers have been ramping up in alcohol beer: proof that the category is no longer a se central part of the beer market moving forward.

Craft beer trends: 'Huge growth' in lower alcohol brews and beer in cans
02-Apr-2019 By Rachel Arthur
The UK craft beer market is seeing 'huge growth' in cans and no & low alcohol beers, according to online craft beer distribution marketplace EeBriaTrade.

04/06/2019 https://www.beveragedaily.com/Beer 2

NABLAB Market growth

UCC
University College Cork, Ireland
Coláiste na hOllscoile Corcaigh

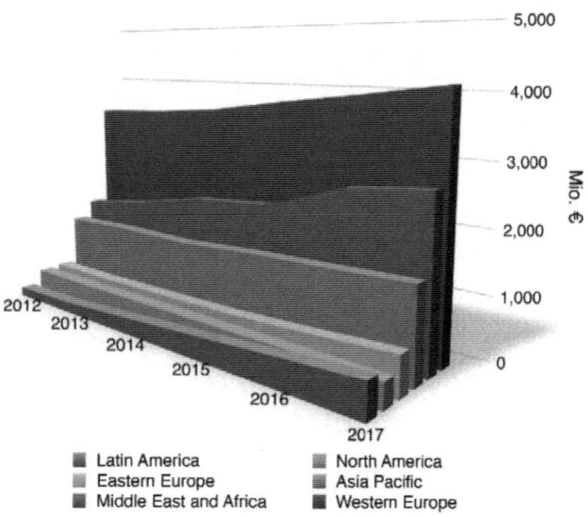

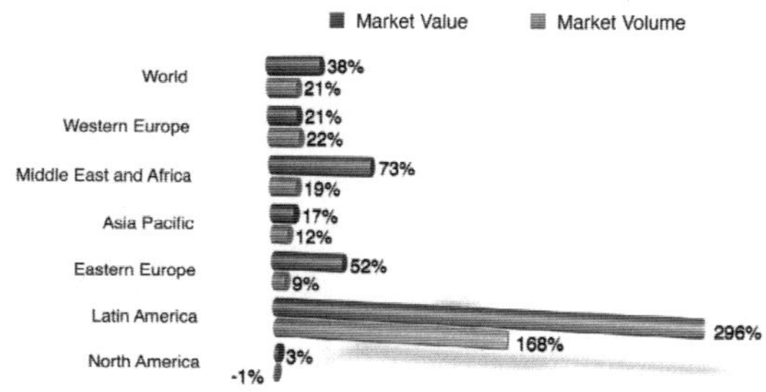

Every **3rd** beer launched in the MENA region in 2016 was non-alcoholic.

04/06/2019 Bellut K, Arendt EK (2019) Chance and Challenge: Non-*Saccharomyces* yeasts in nonalcoholic and low alcohol beer brewing: A Review. J Am Soc Brew Chem 77:77–91. 3

NAB market: Western Europe

UCC
University College Cork, Ireland
Coláiste na hOllscoile Corcaigh

The Western European NAB market held **41%** of the worlds market volume in 2017.

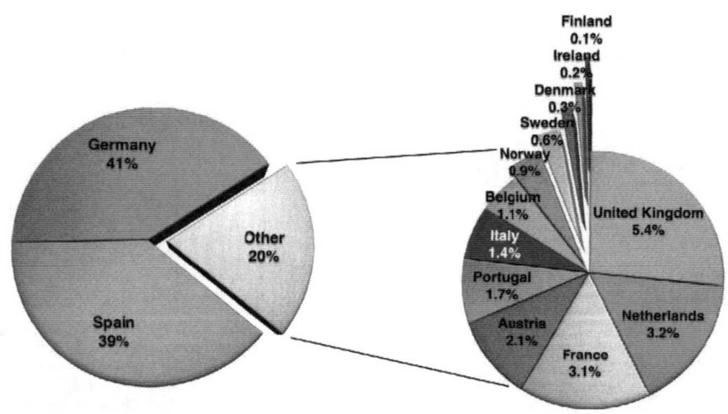

There are **400** brands of NAB on the German market.

04/06/2019 Bellut K, Arendt EK (2019) Chance and Challenge: Non-*Saccharomyces* yeasts in nonalcoholic and low alcohol beer brewing: A Review. J Am Soc Brew Chem 77:77–91. 4

NAB market: North America

High ABV as a way to differentiate from milder main-stream lager beer.

Every **2nd** beer launch in 2015 had a high ABV of 6.6% and more.

Compared to only **1 in 50** with a low ABV of 3.5% and less.

New surveys say that **54%** of American Millennials are actively trying to reduce alcohol consumption.

➔ NAB Boom yet to come?

Legislation to escalate NAB production

After the introduction of a zero-tolerance drunk driving law in Colombia,

NAB launches increased from **6%** to **16%**.

After the sales ban on beer in small shops and kiosks in Indonesia,

NAB launches escalated from **4%** to **33%**.

Take Home Message

TAKE HOME MESSAGE

- Large global NAB market growth
- Market growth expected to continue
- Western Europe and MENA the biggest regions
- North American NAB market growth yet to come?
- New legislation is escalating NAB launches

04/06/2019 7

Production methods

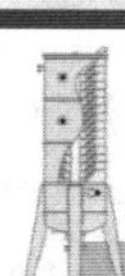

DEALCOHOLIZATION

Physical

- **Thermal**
  - Rectification
  - Evaporation
  - Spinning Cone Column
- **Membrane**
  - Dialysis
  - Reverse Osmosis
  - Osmotic Distillation

LIMITED FERMENTATION

Biological

- **Standard Brewery Equipment**
  - Arrested/limited fermentation
  - Changed mashing
  - Special yeast
- **Special Equipment**
  - Continuous Fermentation

04/06/2019

Brányik T, Silva DP, Baszczyňski M, et al (2012) A review of methods of low alcohol and alcohol-free beer production. J Food Eng 108:493–506.

8

The taste problem

UCC University College Cork, Ireland Coláiste na hOllscoile Corcaigh

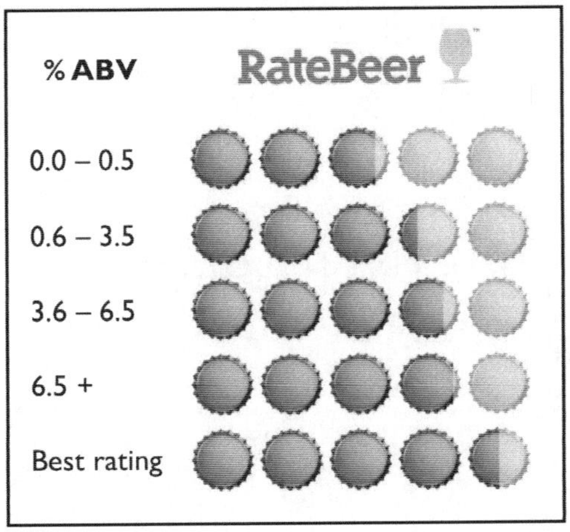

How do I expect NAB to taste like?

How do I want NAB to taste like?

Does NAB have to taste just like beer?

04/06/2019

Bellut K, Arendt EK (2019) Chance and Challenge: Non-*Saccharomyces* yeasts in nonalcoholic and low alcohol beer brewing: A Review. J Am Soc Brew Chem 77:77–91.

9

Dealcoholization vs. Limited fermentation

UCC University College Cork, Ireland Coláiste na hOllscoile Corcaigh

Dealcoholized:

- Sour & bitter taste
- Cabbage-like aroma
- Mouthcoating texture

Limited fermentation:

- Sweet taste
- Malty, honey-like aroma or
- Hop aroma

→ Clear correlation between production method and taste

→ Taster preference for sweet and slightly fruity NAB

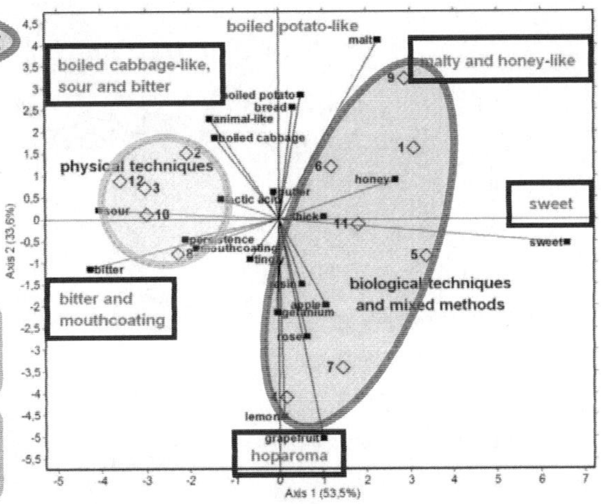

04/06/2019

Schmelzle A, Lindemann B, Methner F-J (2013) Sensory descriptive analysis and consumer acceptance of non-alcoholic beer. BrewingScience 66:144–153.

10

Impact of labeling

UCC
University College Cork, Ireland
Coláiste na hOllscoile Corcaigh

Content: **BEER**

Expectation = Experience

➔ Fulfilled expectations

Decreased emotional responses

Content: **NAB**

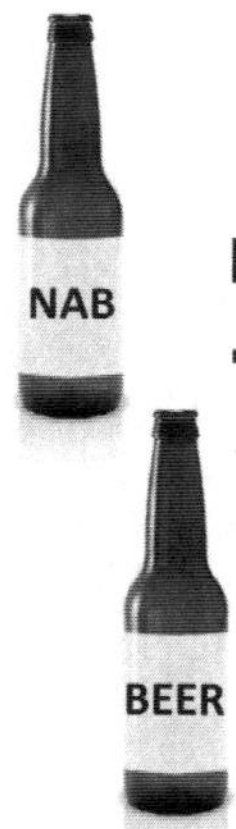

Expectation > Experience

➔ Unfulfilled expectations

Increased liking

More fulfilled

04/06/2019 Silva AP, Jager G, Voss HP, et al (2017) What's in a name? The effect of congruent and incongruent product names on liking and emotions when consuming beer or non-alcoholic beer in a bar. Food Qual Prefer 55:58–66. 11

Consumer conceptualization

UCC
University College Cork, Ireland
Coláiste na hOllscoile Corcaigh

Functional Emotional

- NAB as substitute for beer and soft drinks
- Healthier alternative
- Sometimes described as "fake beverage"

➢ High level of comparison with regular beer

➢ Unfulfilled expectations

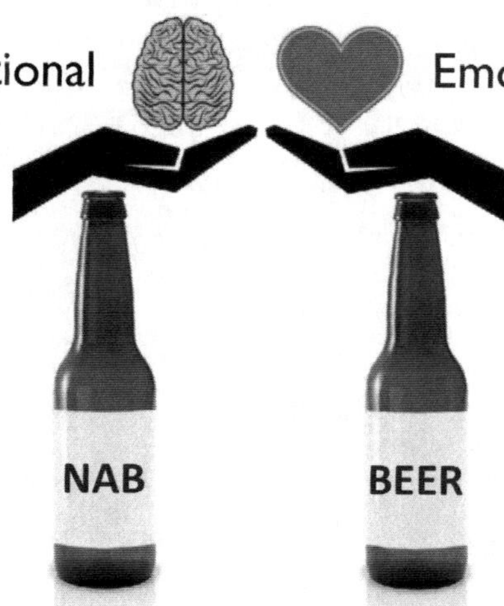

NAB should be treated as a beverage in its own right

➔ Prevent disappointment by the consumer

➔ Avoid direct conceptual comparisons with beer

04/06/2019 Silva AP, Jager G, van Bommel R, et al (2016) Functional or emotional? How Dutch and Portuguese conceptualise beer, wine and non-alcoholic beer consumption. Food Qual Prefer 49:54–65. 12

Take Home Message

University College Cork, Ireland
Coláiste na hOllscoile Corcaigh

Cont...

BEER

TAKE HOME MESSAGE

- Clear taste difference between the production methods
  - -> Dealcoholized: sour & bitter / Limited ferm.: sweet & worty
- Consumer: Unfulfilled expectations during NAB consumption & negative connotations with the term "NAB"

-> Suggestions to treat NAB as a beverage on its own right and to avoid comparison with regular beer

Non-*Saccharomyces* in NAB brewing

University College Cork, Ireland
Coláiste na hOllscoile Corcaigh

Making use of the limited fermentative capacity of non-*Saccharomyces* yeasts.

7°P wort

Brewers' yeast:	Non-*Sacch.* yeast:
• RDF 70%	• RDF 15%
• 2.6% ABV	• 0.5% ABV

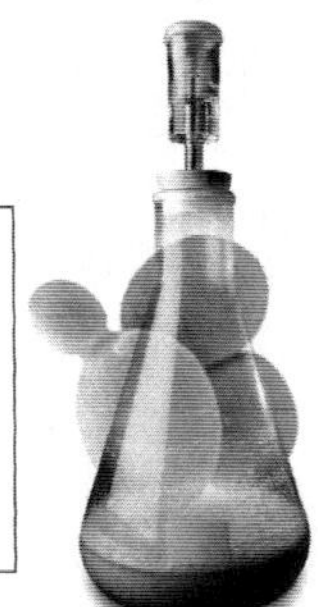

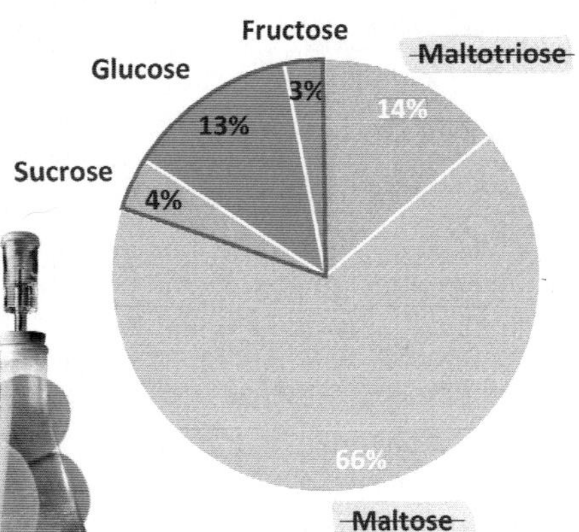

Wort sugars

Non-*Saccharomyces* in NAB brewing

UCC
University College Cork, Ireland
Coláiste na hOllscoile Corcaigh

- Known to have a big impact on the flavor of fermented products
  - Wine & other fermented alcoholic beverages
- Have the potential to introduce novel untypical flavors into NAB brewing
  - Mostly fruity flavors due to high ester production

Candida spp.
Cyberlindnera spp.
Hanseniaspora spp.
Mrakia gelida
Pichia spp.
Saccharomycodes ludwigii
Torulaspora delbrueckii
Zygosaccharomyces spp.

Saccharomycodes ludwigii

UCC
University College Cork, Ireland
Coláiste na hOllscoile Corcaigh

- First mentioned in 1928 (Glaubitz & Haehn)
- Most extensively studied non-*Sacch.* yeast for NAB brewing
- Most recently: Use as a control strain in comparison with other non-*Saccharomyces* spp.

Narziß et al. (1992) 0.68% ABV in 11.5°P
- Increased diacetyl values; worty taste

Meier-Dörnberg & Hutzler (2014) 0.5% ABV in 7°P; Wheat Beer style
- Typical wheat beer flavors missing

De Francesco et al. (2018) 1.23% ABV in 12°P
- Cereal-like & malty

Bellut et al. (2018) 0.5% ABV in 6.6°P
- Sweet taste, worty & bread-like flavors

Low aroma production → worty & cereal-like flavors stay dominant

Candida spp.

UCC
University College Cork, Ireland
Coláiste na hOllscoile Corcaigh

- Abundantly distributed in nature
- 60+ species in food: kefir, sourdough, coffee, soy sauce, …
- Principal human pathogenic yeasts
- More and more species considered as emergent pathogens

Estela-Escalante et al. (2016, 2017) 1.50% ABV in 12.5°P
- *C. zemplinina*, isolated from overripe grapes
- No sensory study conducted

Li et al. (2011; patent) 0.47% ABV in 9.0°P
- *Scheffersomyces shehatae* comb. nov. (2012)
- High ester concentration
- Lack of typical sweet & worty off-taste

Low potential due to strong safety concerns

04/06/2019 17

Cyberlindnera spp.

UCC
University College Cork, Ireland
Coláiste na hOllscoile Corcaigh

- Formerly *Williopsis* spp.
- Known for high production of acetate esters (ethyl, isoamyl and 2-phenylethyl acetate)

Van Rijswijck et al. (2017) 0.6% ABV in 12.0°P
- *C. fabianii* screening
- High ester ratio

Liu & Quek (2016) 1.7% ABV in 13.8°P
- *C. mrakii* NCYC 500
- High concentrations of esters, particularly isoamyl acetate
- ➔ Extra-fruity low-alcohol beer

High potential to create very fruity NAB, masking worty off-flavors

04/06/2019 18

Pichia spp.

UCC
University College Cork, Ireland
Coláiste na hOllscoile Corcaigh

- Large yeast genus, found in natural habitats: soil, freshwater, insects, plants and fruits
- Food: wine fermentations, brines, cheese

COMMERCIALIZED

Van Rijswijck et al. (2017) 0.5-0.8% ABV in 12.0°P
- *P. kudriavzevii* screening
- High ester ratio (slightly higher compared to *C. fabianii* strains)

Saerens & Swiegers (2016; patent) 0.2-0.7% ABV in 8.3°P
- *P. kluyveri*
- High concentrations of esters, particularly isoamyl and phenylethyl acetate
- Reported beer-like flavor

High potential and commercially available

04/06/2019 19

Torulaspora delbrueckii

UCC
University College Cork, Ireland
Coláiste na hOllscoile Corcaigh

- Found in a large variety of habitats such as fruits, malt and soil
- Use in the wine industry
  ➔ fruity flavor
- First commercially sold non-*Saccharomyes* yeast in winemaking

MALTOSE UTILIZATION STRAIN-DEPENDENT

Michel et al. (2016) 0.8-0.9% ABV in 12.0°P
- Pear-like or citrus fruit-like character

Canonico et al. (2016) 2.6% ABV in 12.3°P
- Fruity, citric character

Tataridis et al. (2016) 2.3% ABV in 12.2°P
- Estery, fruity; full-bodied

Bellut et al. (2018) 0.5% ABV in 6.6°P
- low fruity character, wort-like, bread-like

Potential highly strain-dependent

04/06/2019 20

Zygosaccharomyces spp.

UCC
University College Cork, Ireland
Coláiste na hOllscoile Corcaigh

- Genus comprises several of the most challenging and notorious spoilage yeasts
- Can grow at high osmotic pressure, low pH, high ethanol concentrations

POOR FLAVOR

Mortazavian et al. (2014) 0.4% ABV in 6.0°P
- *Z. rouxii*
- Low acceptance during sensory

De Francesco et al. (2015) 0.9% ABV in 12.0°P
- *Z. rouxii*
- Diacetyl values above flavor threshold

Bellut et al. (2018) 0.5% ABV in 6.6°P
- *Z. bailii, Z. kombuchaensis*
- Diacetyl character, honey-like or grassy, fruity

Low potential due to poor flavor

04/06/2019 21

Take Home Message

UCC
University College Cork, Ireland
Coláiste na hOllscoile Corcaigh

TAKE HOME MESSAGE

- Maltose-negative yeasts bear a great potential for NAB brewing
- Potential to create fruity NAB with a novel taste
- But: it has to be the right species & strain
- Research still mostly only in early stages
-> The search continues

Stay for next presentation
-> learn about *Mrakia gelida*

TASTE AND AROMA

04/06/2019 22

Lachancea fermentati for NABLAB brewing

UCC
University College Cork, Ireland
Coláiste na hOllscoile Corcaigh

- *Lachancea* genus possesses the for yeast uncommon metabolic trait to produce **lactic acid** during alcoholic fermentation
- *Lachancea thermotolerans*
  - reducing pH and enhancing total acidity in wine fermentations
  - produce sour beers in single culture fermentations

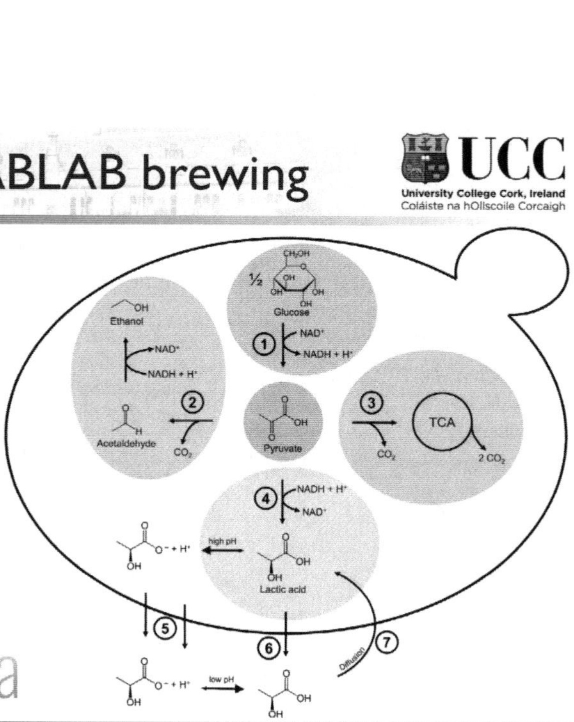

lachancea

Bellut K, Michel M, Hutzler M, et al Investigation into the application of *Lachancea fermentati* strain KBI 12.1 in low alcohol beer brewing. J Am Soc Brew Chem *in press*

Characterization of strain KBI 12.1

UCC
University College Cork, Ireland
Coláiste na hOllscoile Corcaigh

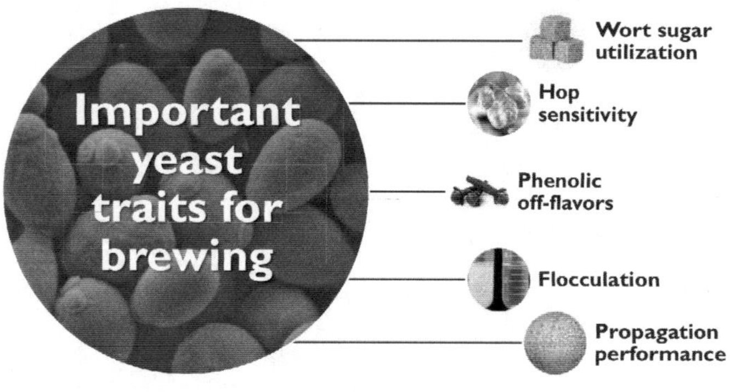

✓ No maltotriose utilization

✓ Marginal hop sensitivity

✓ No phenolic off-flavors

✓ "Medium" flocculation

✓ Good propagation performance

Bellut K, Michel M, Hutzler M, et al Investigation into the application of *Lachancea fermentati* strain KBI 12.1 in low alcohol beer brewing. J Am Soc Brew Chem *in press*

Fermentation performance

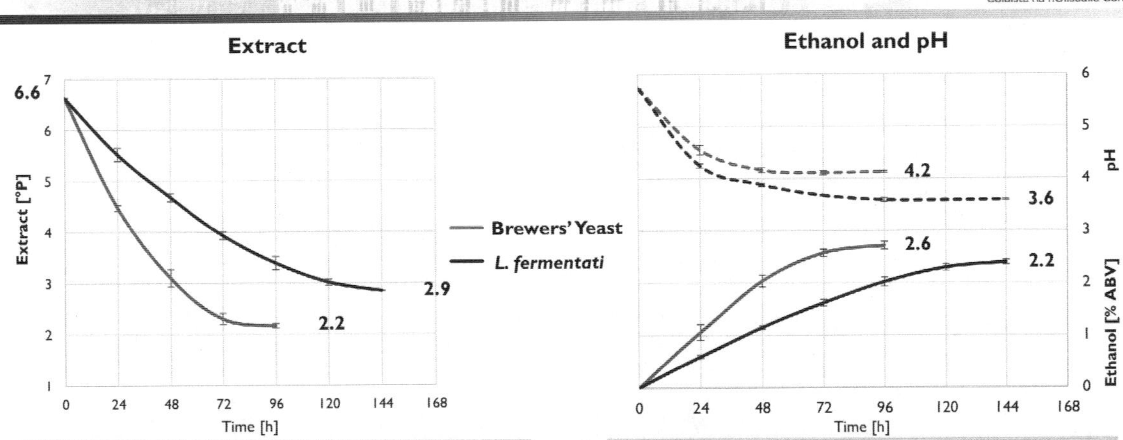

- Higher residual extract due to missing maltotriose utilization

- Substantial pH drop due to lactic acid production
- Lower ethanol production

04/06/2019 Bellut K, Michel M, Hutzler M, et al Investigation into the application of *Lachancea fermentati* strain KBI 12.1 in low alcohol beer brewing. J Am Soc Brew Chem *in press* 25

Beer analyses

	WLP001		KBI 12.1
	S. cerevisiae		*L. fermentati*
Ethanol [%vol.]	2.6	↓	2.2
Final gravity [%real]	2.1	↑	2.9
Fermentation time [h]	96	↑	144
Final pH	4.2	↓	3.6
FAN [mg/L]	29	↑	75
Σ Higher alcohols [mg/L]	82	↓	65
Σ Esters [mg/L]	4.3	↑	13.2
Diacetyl [mg/L]	0.04	▬	0.04

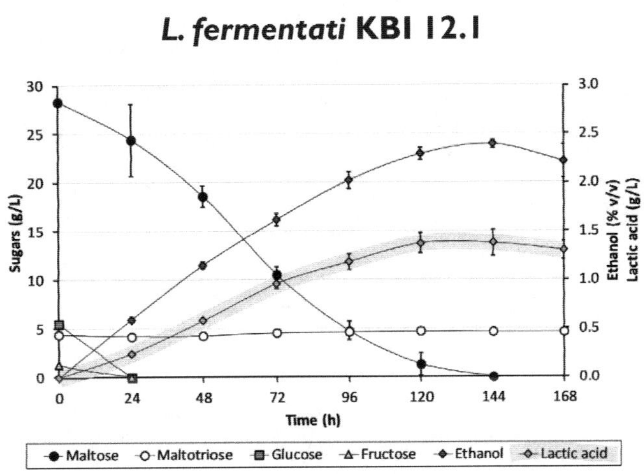

04/06/2019 Bellut K, Michel M, Hutzler M, et al Investigation into the application of *Lachancea fermentati* strain KBI 12.1 in low alcohol beer brewing. J Am Soc Brew Chem *in press* 26

Take Home Message

University College Cork, Ireland
Coláiste na hOllscoile Corcaigh

TAKE HOME MESSAGE

- Lachancea fermentati strain KBI 12.1 is a promising yeast for the production of NABLAB
- No fermentation of maltotriose -> - 15% fermentable extract
- Lactic acid production can counteract residual wort sweetness
- Significant ester production, POF negative, medium flocculation, diacetyl production below threshold, flavor perceived as fruity

04/06/2019 27

Back to the taste problem

University College Cork, Ireland
Coláiste na hOllscoile Corcaigh

How do I expect NAB to taste like?
Not like beer...
How do I want NAB to taste like?
Good!
Does NAB have to taste just like beer?
No!

- Although dealcoholization methods are improving it is physically impossible to exactly copy the taste of regular beer
- NAB should taste ~~like beer~~ good
- There are 75+ beer styles out there. We should broaden our horizon in terms of the taste of NAB
- Non-*Saccharomyces* yeasts have a great potential to create new and unique NAB styles

04/06/2019 28

Conclusion: Chance & Challenge

University College Cork, Ireland
Coláiste na hOllscoile Corcaigh

Chance

- For breweries to create unique NAB with novel flavor characteristics
- For small and craft breweries: no additional equipment required

Challenge

- For researchers & brewers to find the right strain and investigate its suitability
- For the industry to implement and up-scale
- To overcome the low consumer acceptance of NAB (Marketing issue?)

Thank you for your attention

University College Cork, Ireland
Coláiste na hOllscoile Corcaigh

Questions?

JOURNAL OF THE AMERICAN SOCIETY OF BREWING CHEMISTS
2019, VOL. 77, NO. 2, 77–91
https://doi.org/10.1080/03610470.2019.1569452

ASBC Taylor & Francis

REVIEW ARTICLE

Check for updates

Chance and Challenge: Non-*Saccharomyces* Yeasts in Nonalcoholic and Low Alcohol Beer Brewing – A Review

Konstantin Bellut[a] and Elke K. Arendt[a,b]

[a]School of Food and Nutritional Sciences, University College Cork, College Road, Cork, Ireland; [b]APC Microbiome Ireland, University College Cork, College Road, Cork, Ireland

Bellut K, Arendt EK (2019) Chance and Challenge: Non-*Saccharomyces* yeasts in nonalcoholic and low alcohol beer brewing: A Review. J Am Soc Brew Chem 77:77–91. https://doi.org/10.1080/03610470.2019.1569452

University College Cork, Ireland
Coláiste na hOllscoile Corcaigh

37th CONGRESS EBC ANTWERP
EUROPEAN BREWERY CONVENTION 02–06 June 2019

Special thanks and acknowledgements go to:

Prof. Elke Arendt

And the whole Cereal & Beverage Science Group at UCC

Fund Fonds
Fonds Baillet Latour

An innovative production of low alcohol beer using a *Mrakia gelida*

Valeria Sileoni[1], Giovanni De Francesco[1];Ombretta Marconi[1]; Ciro Sannino[2]; Giuseppe Perretti[3]

1 - Italian Brewing Research Centre, University of Perugia;

2 - Department of Agriculture, Food and Environmental Science, Industrial Yeasts Collection DBVPG, University of Perugia;

3 - Department of Agriculture, Food and Environmental Science, University of Perugia

Italian Brewing Research Centre

University of Perugia

EBC Congress 2-6 June 2019, Antwerp

V. Sileoni

STATE OF THE ART

Due to the increasing consumer demand, the production of low alcoholic (0.6 – 3.5 % ABV) and non alcoholic (ABV ≤ 0.5 %) beer is the new goal of the present brewing producers.

The growing popularity is mainly due to the increasing consciousness about well-being, so diet, health but also drink-drive awareness, religion and lower taxes

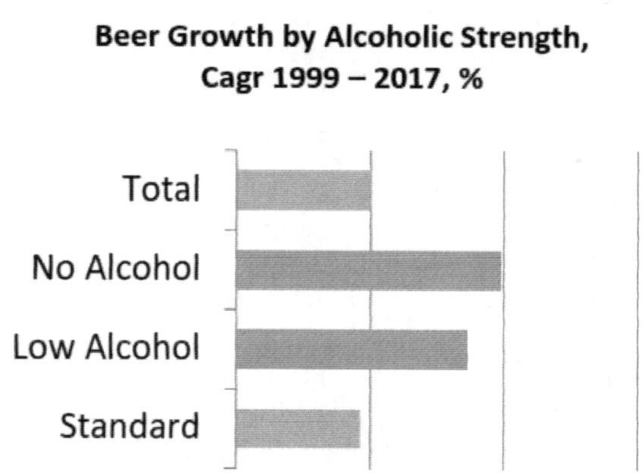

GlobalData.

2-6 June 2019, Antwerp

01 An innovative production of low alcohol beer using a *Mrakia gelida* V. Sileoni

STATE OF THE ART

Percentage increase in market value and market size of non alcoholic beer (≤ 0.5% ABV) in the 5-year period from 2012 to 2017 (Euromonitor International).

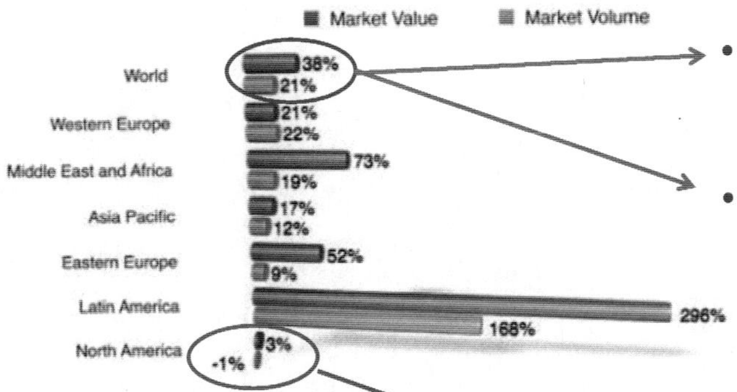

- Market Value + 38% (from 7.1 to 9.9 billion Euros)
- Market volume + 21% (from 31.9 to 38.7 Mio. hl)

Craft brewers

2-6 June 2019, Antwerp

02 An innovative production of low alcohol beer using a *Mrakia gelida* V. Sileoni

STATE OF THE ART

Regional market share of non alcoholic beer (≤ 0.5% ABV) in 2017 (Euromonitor International).

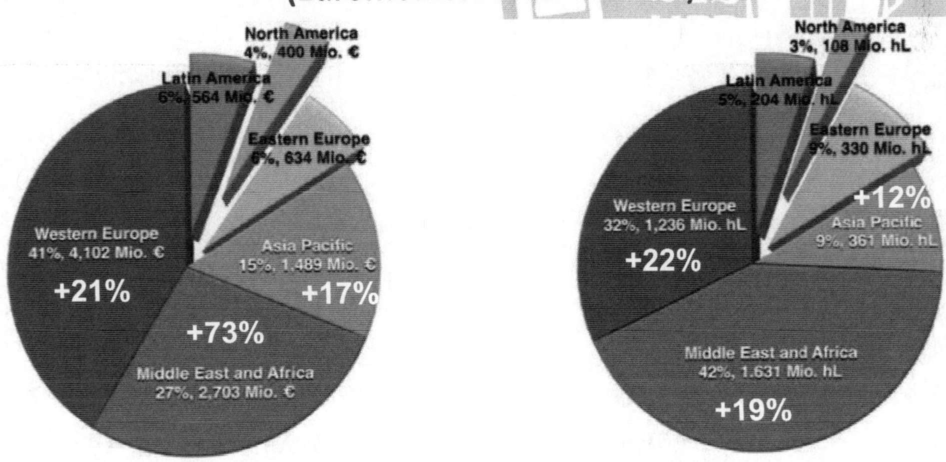

- Increment due to health and wellness trends, especially for young people.
- Asiatic consumers, in general, prefer less strong beer in terms of ABV, compared to the global market.
- High percentage of young people in Middle East and Africa (40 %).
- High percentage of Muslims in Middle East and Africa.

2-6 June 2019, Antwerp

03 An innovative production of low alcohol beer using a *Mrakia gelida* V. Sileoni

STATE OF THE ART

Low alcoholic (0.6 – 3.5 % ABV) and non alcoholic beer (≤ 0.5% ABV) main production methods

Biological:

Continuous fermentation (Special equipment)

Arrested/limited fermentation

Changed mashing

Use of non-conventional yeasts (NCYs)

Physical:

Vacuum rectification (thermal)

Falling film evaporation (thermal)

Dialysis (membrane)

Reverse osmosis (membrane)

Osmotic distillation (membrane)

Branyik T., 2012 (ICT). A review of methods of low alcohol and alcohol-free beer production. Journal of Food Engineering 108 493-506.

STATE OF THE ART

Increasing interest in non-conventional Yeast for low alcoholic (0.6 – 3.5 % ABV) and non alcoholic beer (≤ 0.5% ABV) production

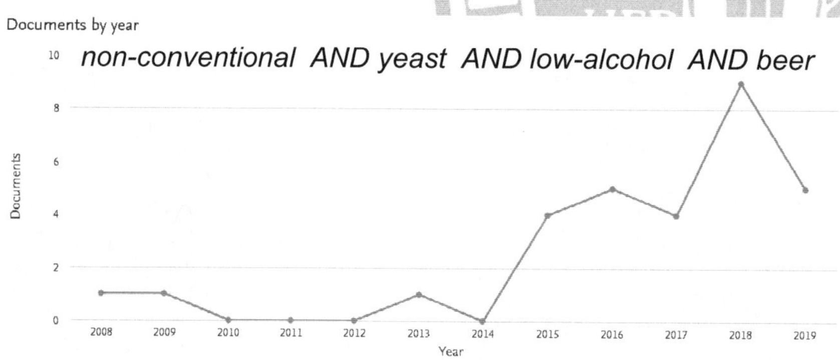

*source: Scopus

- possibility to use a **standard brewery equipment** (avoiding substantial investment into the equipment necessary for dealcoholization).
- opportunity for **small and craft breweries**.
- chance to tap into **unconventional atypical flavors, such as fruity notes** (i.e. apricot, litchi, pear, citrus fruit), which are not the usual common flavors in beer, with a consequent advantage in terms of consumer acceptance.

STATE OF THE ART

On the other hand, non-conventional Yeast:

- **Sterile and careful handling** of the yeast is very important to avoid contamination and **pasteurization** becomes essential due to the residual sugars in the finished product.
- The high residual extract (mostly maltose) in the finished products leads to a **sweet taste** for beer produced with NCYs.
- When brewing with **adjuncts**, it has to be considered if the NCY is able to ferment the sugar(s).
- Finally, the NCYs **safety** with regards to human consumption should be considered ("Qualified Presumption of Safety" (QPS; European Food Safety Authority) or "Generally Recognized As Safe" (GRAS; American Food and Drug Administration) status).

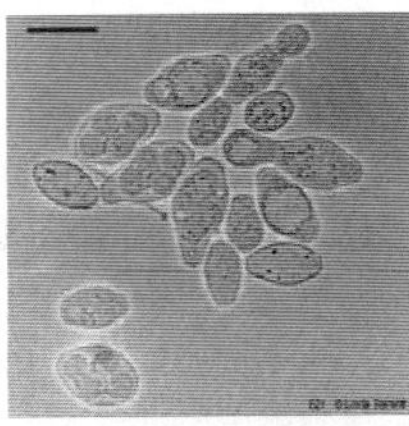
Saccharomycodes ludwigii

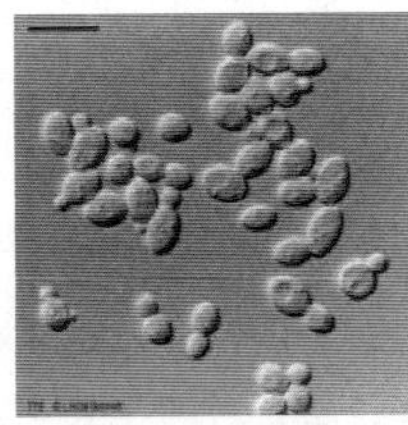
Zygosaccharomyces rouxii

2-6 June 2019, Antwerp

06 An innovative production of low alcohol beer using a *Mrakia gelida* V. Sileoni

STATE OF THE ART

What is meant with «non-conventional Yeast»

- The so-called nonconventional yeasts are **non-*Saccharomyces* yeasts with limited abilities to ferment wort sugars** for the production of low-alcohol and nonalcoholic beers in single culture fermentation.
- The non-*Saccharomyces* species ***Saccharomycodes ludwigii*** has been applied commercially for this purpose for many years and is the most popular species with regards to the number of studies conducted in the past.
- Anyway, the beers produced by *S. ludwigii* resulted in a **predominantly poor sensorial evaluation**, so investigations into new non-Saccharomyces species are justified.

Alcohol Reduction
• *Saccharomycodes ludwigii:* low fermentation activity in maltose and maltotriose
• *Torulaspora delbrueckii:* inability to ferment maltose and maltotriose
• *Pichia kluyveri:* limited ability to ferment glucose
• *Zigosaccharomyces rouxii:* total or partial inability to ferment maltose

2-6 June 2019, Antwerp

07 An innovative production of low alcohol beer using a *Mrakia gelida* V. Sileoni

2-6 June 2019, Antwerp

Aim of the project

Enhance the yeasts biodiversity for low alcoholic/non alcoholic beer.

↓

Exploit the ability of Basidiomycota (considered low- or even non-fermenting yeast, because sensible to ethanol) to ferment different source of sugars.

↓

In particular, little literature described the ability of the some strains of the genus Mrakia to produce (and partially tolerate) ethanol[1,2]. (This genus includes psychrophilic species mainly isolated from cold environments[3]).

↓

Study the possibility of use Mrakia strains to produce low alcoholic/non alcoholic beer FIRST TIME REAL APPLICATION (pilot plant brewery)

1. *Thomas-Hall et al., Extremophiles (2010) 14:47–59*
2. *Tsuji et al., Mycoscience (2018) 59:54-58*
3. *Turchetti et al., FEMS Microbiol Ecol (2008) 63:73–83*

2-6 June 2019, Antwerp

MATERIALS AND METHODS

Interdisciplinary work on NCYs

dsa3

Italian Brewing Research Center

Industrial Yeasts Collection

2-6 June 2019, Antwerp

10

S AND METHODS

DBVPG number	Specie	Maltose fermentation at 10°C		
		15 days	16 days	19 days
4974	Mrakia blollopis	-	-	-
4844	Mrakia frigida	-	-	-
4977	Mrakia gelida	-	-	-
4983	Mrakia gelida	-	-	-
4995	Mrakia gelida	-	-	-
5106	Mrakia gelida	-	-	+
5202	Mrakia gelida	-	-	-
5218	Mrakia gelida	-	-	-
5219	Mrakia gelida	-	-	-
5306	Mrakia gelida	-	-	-
5308	Mrakia gelida	-	+	
5952	Mrakia gelida	-	-	+
5953	Mrakia gelida	+		
10473	Mrakia cf. gelida	-	-	-
5928	Mrakia cf. gelida	w	+	
5947	Mrakia cf. gelida	-	-	-
10084	Mrakia cf. gelida	-	-	-
5980	Mrakia psychrophila	-	-	-
5981	Mrakia psychrophila	-	-	-
4728	Mrakia robertii	-	-	-
4761	Mrakia robertii	-	-	-
4774	Mrakia robertii	-	-	-
4782	Mrakia robertii	-	-	-
4801	Mrakia robertii	-	-	-
5309	Mrakia robertii	-	-	-
5358	Mrakia robertii	-	-	-
5384	Mrakia robertii	-	-	-
5385	Mrakia robertii	-	-	-
5968	Mrakia robertii	-	-	-
10007	Mrakia robertii	-	-	-
10046	Mrakia robertii	-	-	-

Test for the ability to ferment maltose

unable to grow at T > 17°C

Symbols:
- not fermented;
+ fermented;
w, weak.

DBVPG
Industrial Yeasts Collection

. Sileoni

2-6 June 2019, Antwerp

MATERIALS AND METHODS

Lab scale microfermentation screening for volatile compounds

- 5 mL of suspensions of 10^6 cells per mL were used to inoculate flask containing 250 mL of standard wort (all pilsner malt, 12°P, pH 5.4, color 8 EBC).
- The flask were incubated on a rotary shaker under aerobic conditions for 6 days at 10°C (150 rpm), until 10^7 cells.
- 10 mL of this fermenting broth was used to inoculate 250 mL of wort, incubated on a rotary shaker under anaerobic conditions for 6 days at 10°C (150 rpm) for fermentation.

	Mrakia gelida strains				
	DBVPG 5106	DBVPG 5308	DBVPG 5928	DBVPG 5952	DBVPG 5953
Total alcohols (mg/L)	25.9±0.51	25.4±0.07	26.2±0.11	31.7±0.12	32.0±0.37
Total esters (mg/L)	0.26±0.01	0.15±0.01	0.15±0.01	0.89±0.08	0.14±0.01
Total aldehydes (µg/L)	7868±582	24871±4138	13405±897	26135±360	8720±241
Diacetyl (µg/L)	95.6±11.95	329.5±0.21	221.9±1.38	74.7±2.92	286.0±0.96
Ethanol (% v/v)	0.66±0.01	0.54±0.03	0.62±0.03	0.66±0.02	0.69±0.04

11 An innovative production of low alcohol beer using a *Mrakia gelida* — V. Sileoni

MATERIALS AND METHODS

Wort Production

The samples were produced in the pilot plant of CERB, Italian Brewing Research Centre, University of Perugia (Spadoni , Orvieto, Italy) with a production capacity of 110 liters.

The following raw materials were used:

- 25 kg of Pale Malt (Durst-Malz, Bruchsal-Heidelsheim, Bruchsal, Germany), crushed in a malt mill (two-rolls distance 1 mm) + 75 L of heated water (1:3).
- 100 g of hop (Hallertau Magnum hop, 25 IBU, 13% α-acids, Barth-Haas, Nurnberg, Germany).

2-6 June 2019, Antwerp

MATERIALS AND METHODS

Wort Production

80 L of sparging water pH 5.2

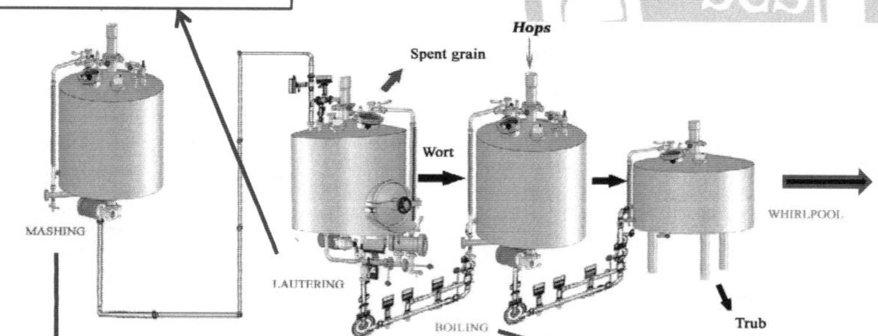

1 hour

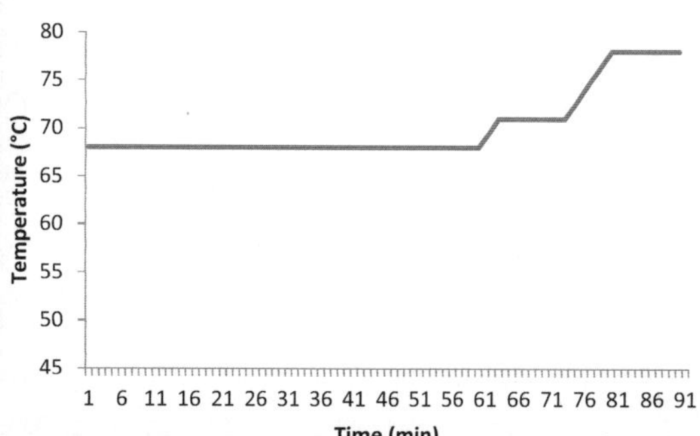

+ 15 mL of lactic acid
+ 17 g of $CaCl_2$
↓
pH 5.5

2-6 June 2019, Antwerp

MATERIALS AND METHODS

Fermentation

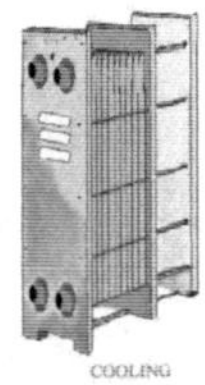

Cooled wort aerated with sterile air

Inoculated with 4% of fermenting broths (10^7 CFU/mL) of:

- The strain *Mrakia gelida* DBVPG 5952.
- The commercial strain WSL17 *Saccharomycodes ludwigii* (Hefebank Weihenstephan, Hallertau Deutschland, Germany) (control).

25 L fermenter

23°C for *S. ludwigii* 10°C for *M. gelida*

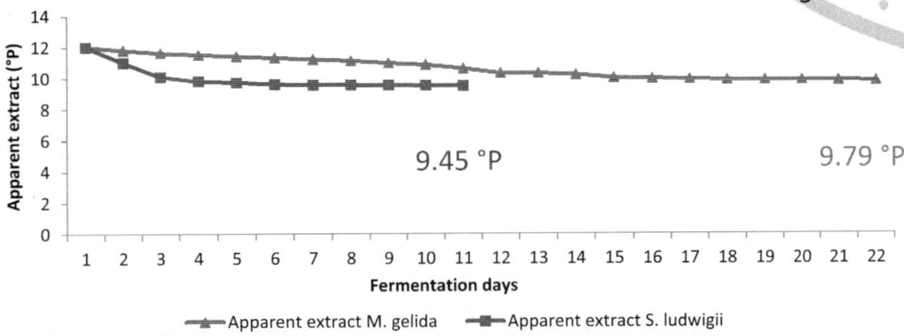

2-6 June 2019, Antwerp

MATERIALS AND METHODS

Maturation and bottle conditioning

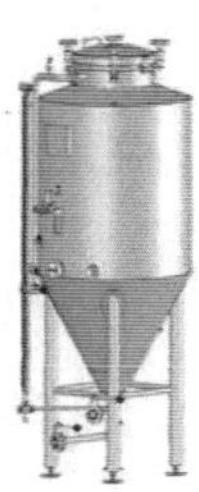

Maturation 10 days 2°C

dextrose
5 g/L

15 days at
- 20 °C for *S. ludwigii*
- 10 °C, for *M. gelida*

Sensory analysis: trained panel (12 people aged between 25 and 55 years) through description analysis according to Analytica-EBC method 13.10. Scores from 0 (absent) to 9 (extremely strong).

Chemical analysis

BOTTLE CONDITIONING

1. natural carbonation (low cost),
2. Richer flavour profile (additional production of esters and higher alcohols),
3. beer stabilization (reduced sensitivity to oxidations).

2-6 June 2019, Antwerp

RESULTS

Wort analysis

Wort Quality parameters	
Original extract (°P)	12.0 ± 0.1
Colour (EBC-U)	8.0 ± 0.8
Total nitrogen (mg/L)	903 ± 68
FAN (mg/L)	160 ± 7
Viscosity (mPa·s)	1.76 ± 0.02
Haze (EBC-U)	47.5 ± 1.7
pH	5.44 ± 0.04

°P: Plato degrees, EBC-U: European Brewery Convention units, FAN: Free Amino Nitrogen. n=2 technological repetitions. Mean ± standard deviation.

Sugars profile

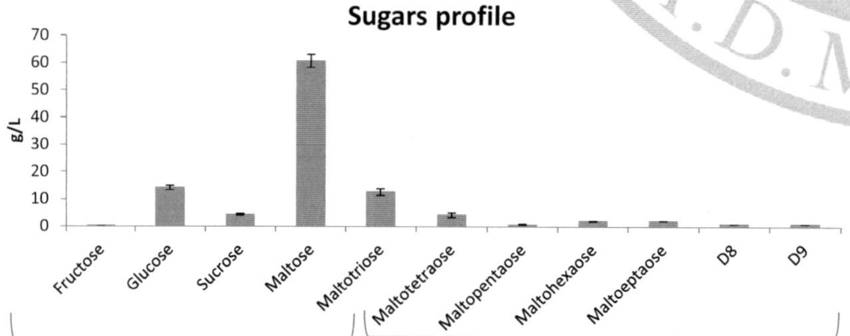

Fermentable sugars 91.96 g/L

Dextrins 11.02 g/L

2-6 June 2019, Antwerp

RESULTS

Beer analysis – standard quality attributes

Beer Quality attributes	Fermented *M. gelida*	Fermented *S. ludwigii*	Refermented *M. gelida*	Refermented *S. ludwigii*
Original extract (°P)	11.5 a	11.5 a	11.5A	11.4 A
Apparent extract (°P)	9.83 a	9.42 a	9.36 A*	9.33 A*
Real extract (°P)	9.75 a	9.53 a	9.43 A	9.26 A
Alcohol (%v/v)	1.16 a	1.23 a	1.40 A*	1.32 A*
ADF (%)	17.9 a	20.5 a	22.0 A	20.3 A
Colour (EBC-U)	11.7 a	12.0 a	11.6 A	12.1 A
pH	4.75 a	4.60 b	4.71 A	4.60 B
Total nitrogen (mg/L)	821a	777 a	812 A	789 A
Haze (EBC-U)	2.20 a	1.40 a	1.55 A	2.40 A

°P: Plato degrees, ADF: Apparent degree of fermentation, EBC-U: European Brewery Convention units. n=2 technological repetitions. Mean ± standard deviation.

Values in the same row followed by a different letter are statistically different. Lowercase letters for fermented samples and capital letters for refermented samples. * indicates differences between fermented and refermented samples ($p<0.05$).

2-6 June 2019, Antwerp

RESULTS

Beer analysis – sugars profile

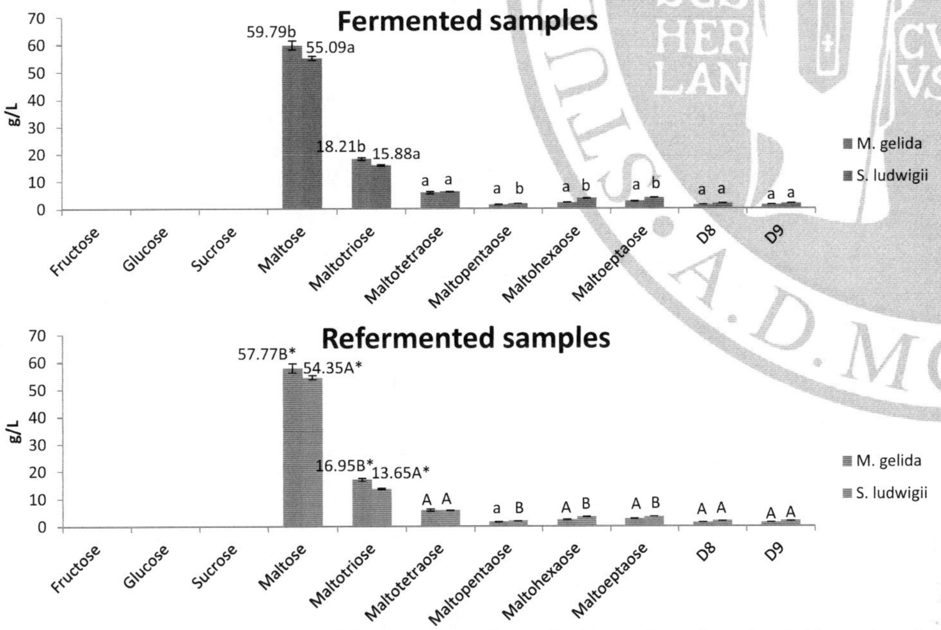

2-6 June 2019, Antwerp

Bars with a different letter are statistically different. Lowercase letters for fermented samples and capital letters for refermented samples. * indicates differences between fermented and refermented samples (p<0.05).

RESULTS

Beer analysis – volatile compounds – higher alcohols

Volatile compounds	Perception Threshold#	Fermented M. gelida	Fermented S. ludwigii	Refermented M. gelida	Refermented S. ludwigii
1-Propanol (mg/L)	800	5.66 b	2.61 a	6.41 B*	2.81 A*
Isobutanol (mg/L)	200	9.79 a	13.0 b	10.5 A	13.7 B
3-Methyl-1-butanol (mg/L)	70	5.96 a	14.3 b	6.63 A	14.4 B
2-Methyl-1-butanol (mg/L)	65	1.44 a	5.50 b	1.65 A	5.59 B
2-Phenylethanol (mg/L)	125	2.65 a	6.83 b	2.65 A	6.89 B

°P: Plato degrees, ADF: Apparent degree of fermentation, EBC-U: European Brewery Convention units. n=2 technological repetitions. Mean ± standard deviation.
Values in the same row followed by a different letter are statistically different. Lowercase letters for fermented samples and capital letters for refermented samples. * indicates differences between fermented and refermented samples (p<0.05).

2-6 June 2019, Antwerp

RESULTS

Beer analysis – volatile compounds – esters

Volatile compounds	Perception Threshold#	Fermented M. gelida	Fermented S. ludwigii	Refermented M. gelida	Refermented S. ludwigii
Ethyl acetate (mg/L)	25-30	0.61 a	9.31 b	3.52 A*	14.91 B*
Isoamyl acetate (mg/L)	1.6	n.d.	0.027	n.d.	0.029
Ethyl hexanoate (mg/L)	0.23	0.009 a	0.011 a	0.010 A	0.012 A
Ethyl octanoate (mg/L)	0.9	0.006 a	0.009 a	0.006 A	0.009 A

°P: Plato degrees, ADF: Apparent degree of fermentation, EBC-U: European Brewery Convention units. n=2 technological repetitions. Mean ± standard deviation.
Values in the same row followed by a different letter are statistically different. Lowercase letters for fermented samples and capital letters for refermented samples. * indicates differences between fermented and refermented samples ($p<0.05$).

2-6 June 2019, Antwerp

RESULTS

Beer analysis – volatile compounds – carbonyl compounds

Volatile compounds	Perception Threshold#	Fermented M. gelida	Fermented S. ludwigii	Refermented M. gelida	Refermented S. ludwigii
Acetaldehyde (μg/L)	25000	2301 a	918 b	2526 A	1743 A*
2-Methylbutanal (μg/L)	1250	1.30 a	1.84 a	0.59 A*	4.58 B*
3-Methylbutanal (μg/L)	600	7.24 a	6.22 a	2.37 A*	11.5 B*
Hexanal (μg/L)	600	1.02 a	0.68 b	0.59 A*	0.79 A
Furfural (μg/L)	150000	7.33 a	10.28 b	6.88 A	5.63 A*
Methional (μg/L)	250	7.78 a	5.60 b	6.50 A	9.54 B*
Phenylacetaldehyde (μg/L)	1600	9.19 a	10.30 a	7.95 A	17.64 B*
2,3-Butanedione (diacetyl) (μg/L)	100-400	7.75 a	8.05 a	5.04 A	5.20 A

°P: Plato degrees, ADF: Apparent degree of fermentation, EBC-U: European Brewery Convention units. n=2 technological repetitions. Mean ± standard deviation.
Values in the same row followed by a different letter are statistically different. Lowercase letters for fermented samples and capital letters for refermented samples. * indicates differences between fermented and refermented samples ($p<0.05$).

2-6 June 2019, Antwerp

RESULTS

Beer analysis – sensory profile – aroma

Visual aspect	M. gelida	S. ludwigii
color	yellow	yellow
head foam	fine	poor
clearance	good	good

Fruity descriptors:
- M. gelida: apricot, grape and litchi
- S. ludwigii: apricot

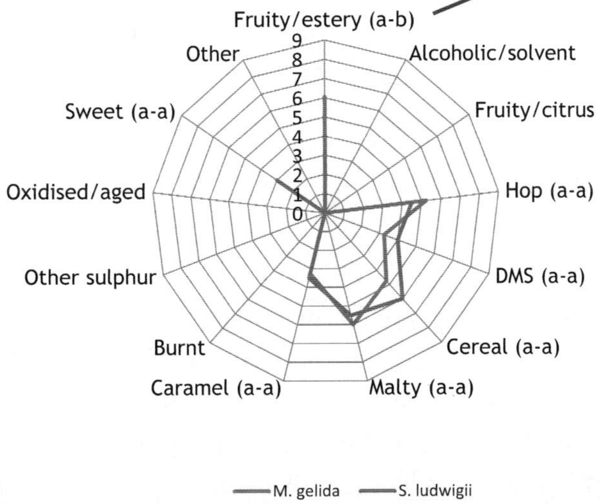

2-6 June 2019, Antwerp

RESULTS

Beer analysis – sensory profile – taste

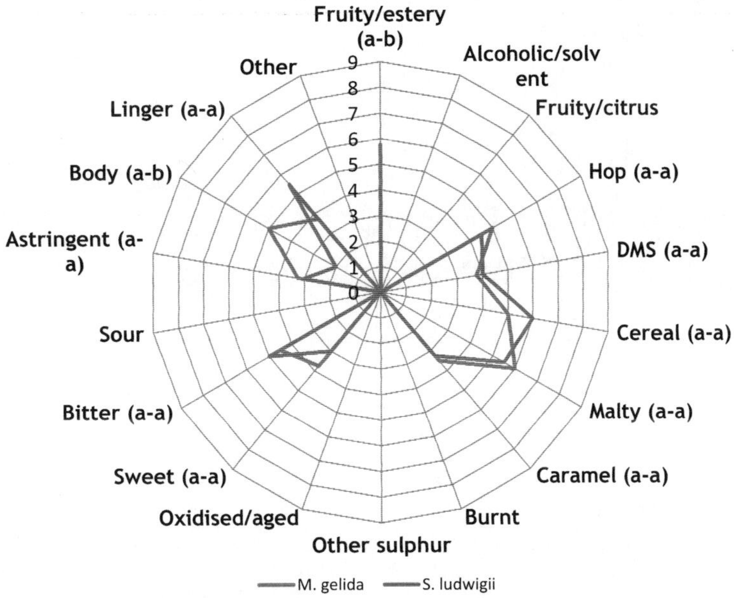

2-6 June 2019, Antwerp

Conclusions

- *M. gelida* could represent a good candidate to be used as starter in brewing low alcoholic and non alcoholic beer.
- The quality attributes, volatile compounds and sensory profile of beer produced by *M. gelida* were comparable to the ones obtained by commercial yeasts for low alcoholic and non alcoholic beer.
- *M. gelida* can also be a new entry in GRAS ("Generally Recognized As Safe") list considering also that the experiments performed by Thomas-Hall et al. (2010)*, showed that no abnormalities were observed in any of the rats fed with beer produced using *Mrakia* strains and all organs were found to be healthy;
- Additionally, due to its psychrophilic aptitude, *Mrakia* strains are not able to grow at human body temperature, and so they can't be considered a pathogen for human and animals.

**Thomas-Hall et al., Extremophiles (2010) 14:47–59*

2-6 June 2019, Antwerp

24 An innovative production of low alcohol beer using a *Mrakia gelida* V. Sileoni

June 4, 2019
37th EBC Congress - Antwerp

Temporal Dominance of Sensations Paired with Dynamic Wanting:

A New Method of Sensory Evaluation for Beer Drinkability

Takahiro Wakihira[1]; Seiko Miyashita[1]; Minoru Kobayashi[1]; Kazuhiko Uemura[1]; Pascal Schlich[2]

1 - Asahi Breweries, Ltd., Research Laboratories for Alcohol Beverages, Moriya, Japan;
2 - Centre des Sciences du Goût et de l'Alimentation, AgroSup Dijon, CNRS, INRA, Univ. Bourgogne Franche-Comté, F-21000 Dijon, France

---

Background

What is "Drinkability"?

German: *weiter trinken*
German–English translation: continue drinking

Issues surrounding "Drinkability"

- No clear definition
- No established method to asses it
- No clear understanding of the factors affecting it

Objective and Goal

Objective

To develop a new method for evaluating beer drinkability

Goal

To clarify the factors affecting drinkability

Defining "Drinkability"

Agreed concept

A highly drinkable beer makes us say,

Defining "Drinkability"

One more, please!

Repetition of "I want to have another sip"

<u>We defined drinkability as:</u>

The power to evoke "Wanting to drink continuously."

Method: Temporal Dominance of Sensations

Grasp <u>changes</u> in sensory perceptions while drinking

Schlich et al. (2003)

Temporal Dominance of Sensations (TDS)

Select the most dominant attribute over time

START

BITTER	MALTY
SOUR	HOPPY
SWEET	ESTERY
STIMULATION	OTHER

STOP

Visualize changes in the dominance rate

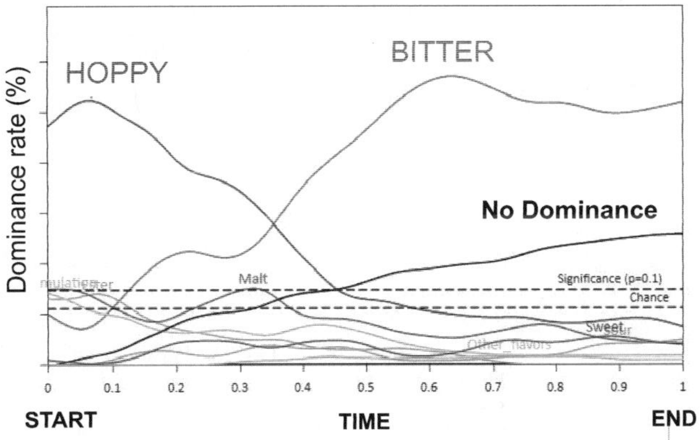

TimeSens

Test Flow: Multi-sip Drinkability Test

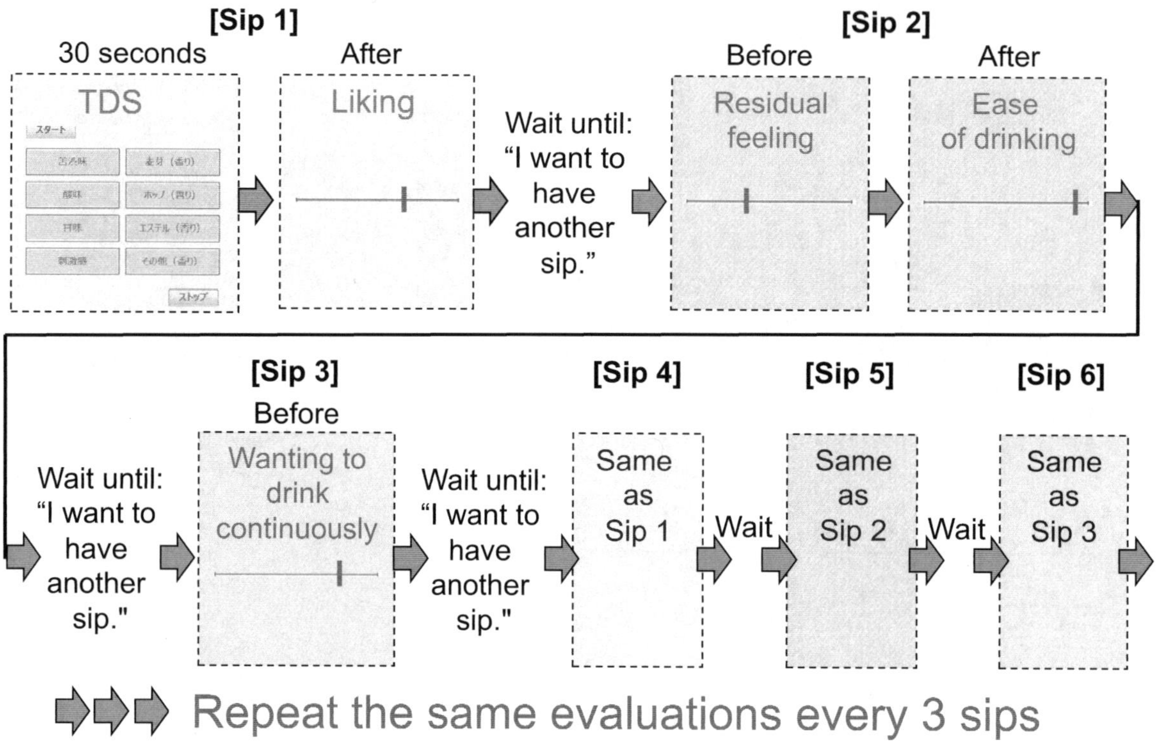

Overview of Test 1 and Test 2

[Test 1]

- Expert beer panel (N=11) evaluated 4 products, 2 repetitions each, for a total of 8 evaluations per panelist.

*Expert beer panel = Employees of the R&D center who are trained in tasting and evaluating beer.

[Test 2]

- Naïve panel (N=44) evaluated 3 products, no repetitions, for a total of 3 evaluations per panelist.

*Naïve panel = Employees of the R&D center who are not trained, but drink beer once a week or more.

These tests were conducted in an *ad libitum* setting for drinking amount and drinking pace.
(maximum: 500 mL for Test 1, 350 mL for Test 2)

Test 1 Results

Test 1 Results: Wanting

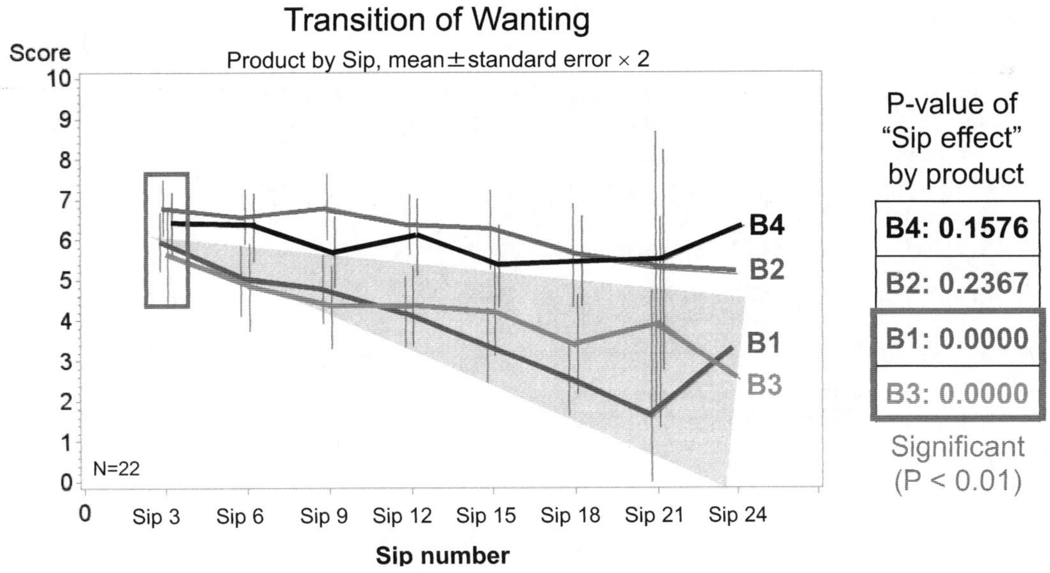

Product differences in "Wanting" increased while drinking.

Test 1 Results: Ease of Drinking

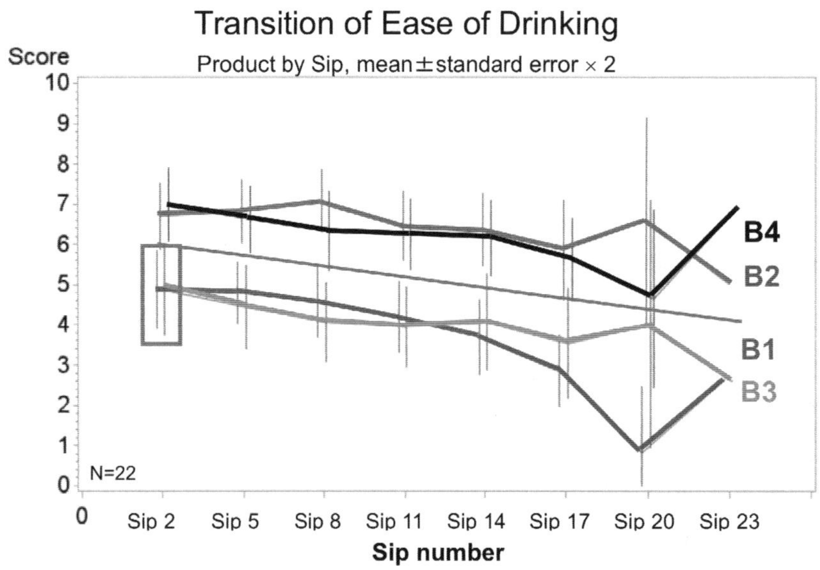

Scores for B1 and B3 were lower from the beginning.

"Ease of drinking" at each sip could lead to "wanting to drink continuously (= Drinkability)."

Separate study presented in ***Poster #XX.***

Test 1 Results: Flavor Characteristics

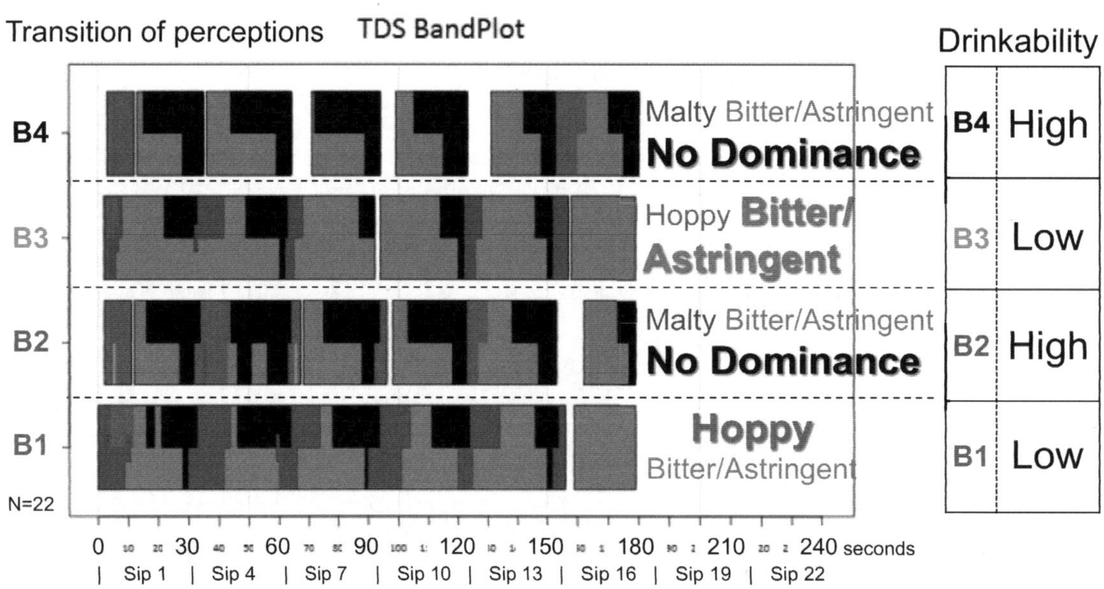

The greater the sensory load produced, the less one wants it. <u>**Fewer build-up effects could lead to high drinkability.**</u>

Test 2 Results

Test 2 Results: Wanting

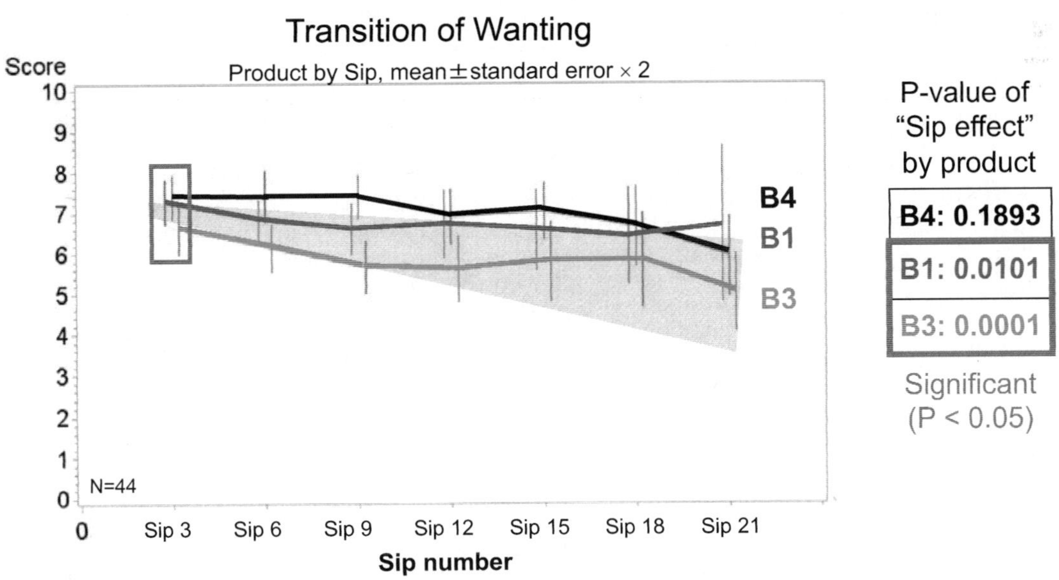

As with Test 1, product differences in "Wanting" increased while drinking. [Reproducibility confirmed]

Test 2 Results: Flavor Characteristics B4 vs. B3

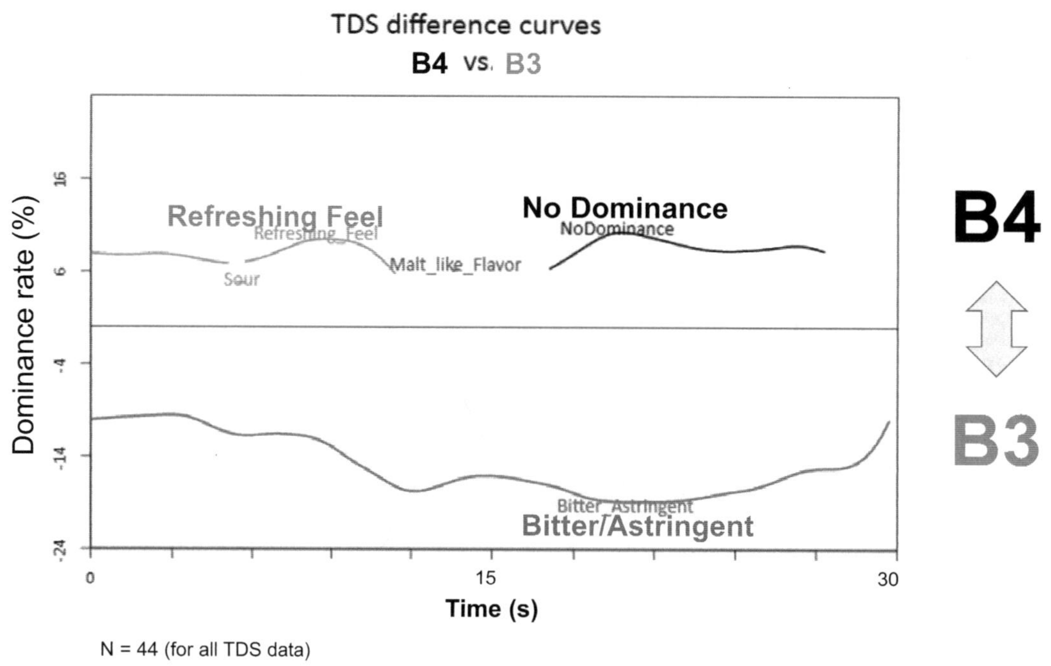

Test 2 Results: Flavor Characteristics B4 vs. B1

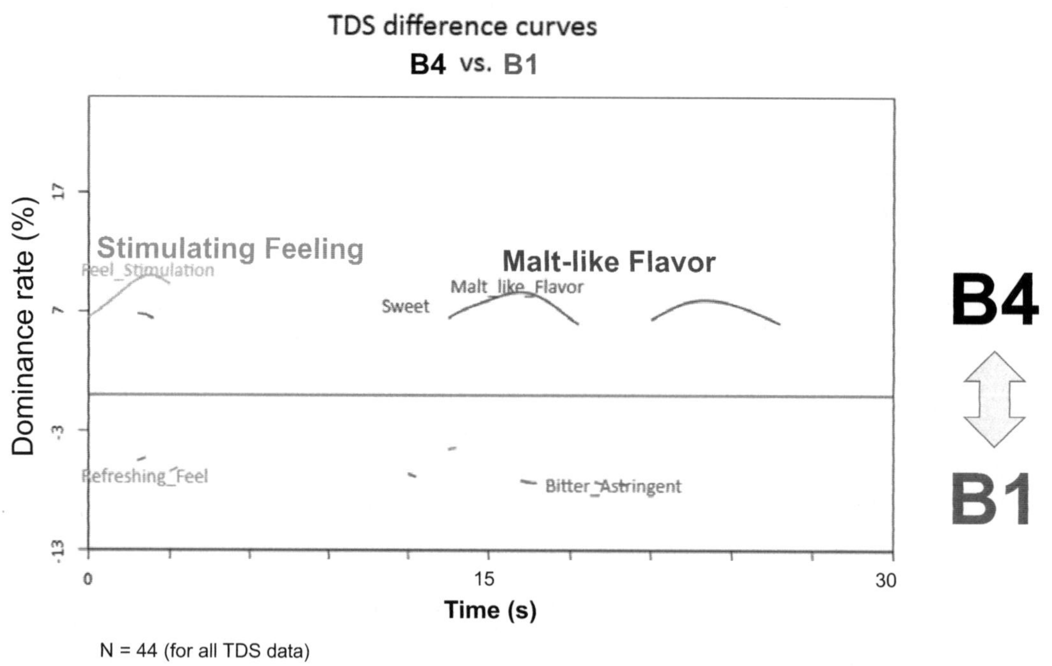

Test 2 Results: Wanting and Perceptions B4 vs. B1

Linking wanting segmentation to temporal perception

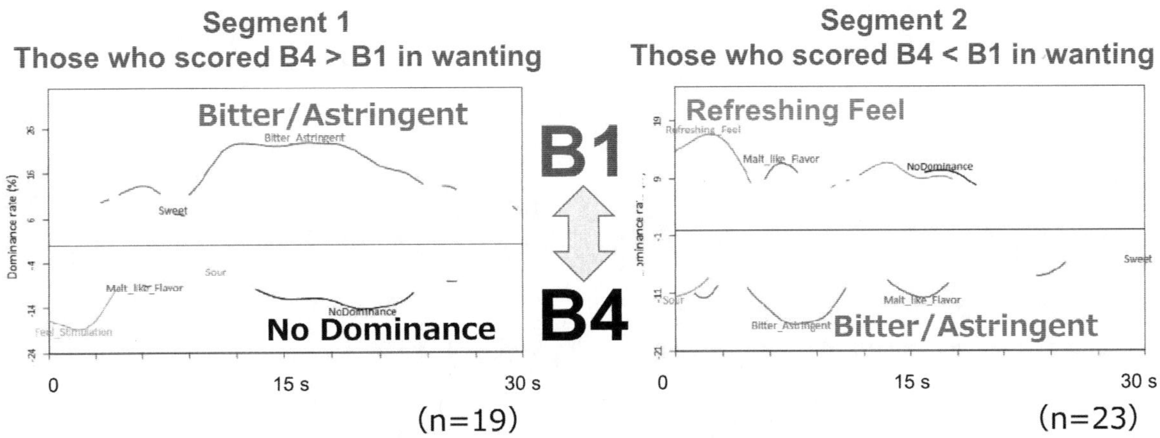

Perceptional differences were observed between those who wanted B4 and those who wanted B1.

Test 2 Results: Duration of Bitter/Astringent

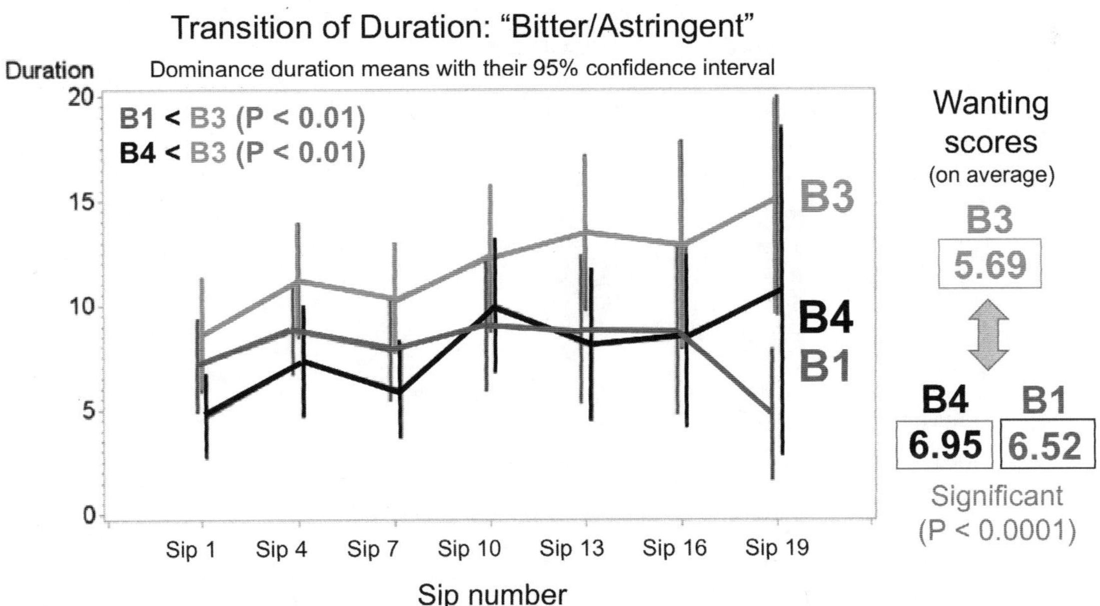

Build-up of Bitter/Astringent could lead to low drinkability.

Wanting and Volume of Beer Consumed

Test 1 Results: Wanting and Volume of Beer Consumed

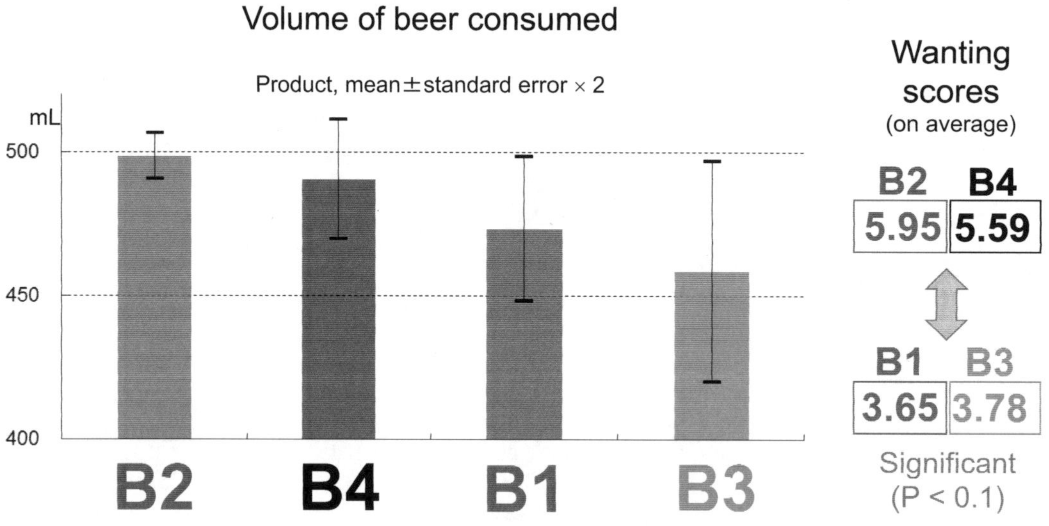

B2/B4 tended to be consumed more than B1/B3, but not at a statistically significant level.

Test 2 Results: Wanting and Volume of Beer Consumed

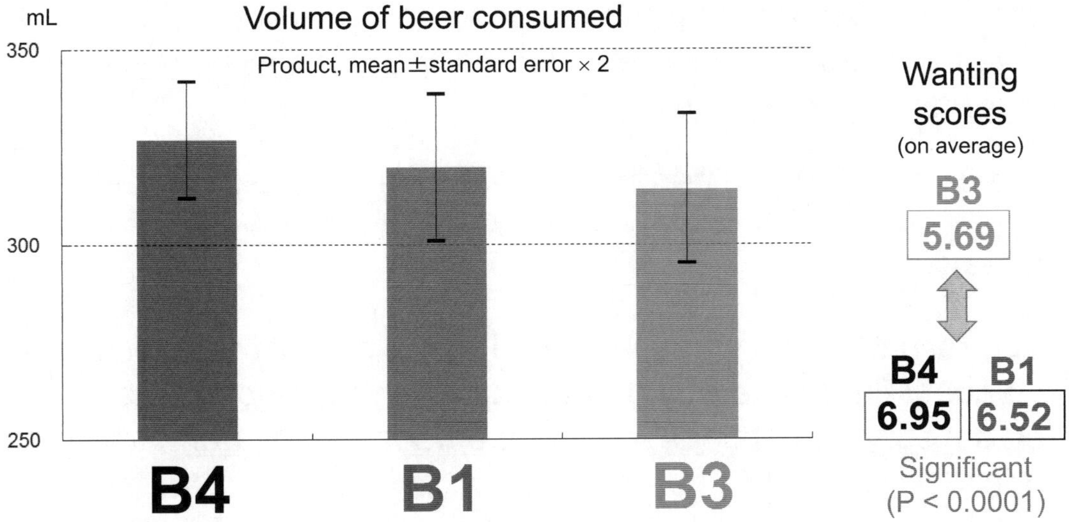

No significant differences in consumed volume were observed, but the trend in consumed volume was in line with the wanting scores.

Summary

Summary

- ✓ We developed a new method based on TDS paired with dynamic wanting, which enabled us to discriminate beers in terms of drinkability.
- ✓ We found product differences in wanting: although almost nonexistent at the beginning, these gradually became larger.
- ✓ The results of this study indicated that a product with less standout flavor, thereby producing fewer build-up effects on sensory perceptions, could be a beer with higher drinkability.

  (Further investigations will be needed for other factors.)
- ✓ The results also indicated that "ease of drinking" at each sip could lead to beer drinkability.

  (To be presented further in ***Poster #XX***.)

Future Prospects

- ✓ Apply this method to general consumers on a larger scale.
- ✓ Apply this method with designed prototypes to clarify further the factors affecting drinkability.
- ✓ Develop new methods more relevant to drinking contexts in real life, such as "with food".
- ✓ Seek methods to enable a better understanding of consumers in terms of both state of mind and behavior while drinking.

Vision of the Sensory Team
"Connect consumers with Asahi through sensory perceptions."

Thank you for your attention!

L33 - UNDERSTANDING AROMATIC STABILITY IN DRY-HOPPED BEER

Margaux Huismann[1]; Fraser Gormley[2]; Dzeti Dzait[2]; Alex Speers[3]; Dawn Maskell[1]

1 - International Centre for Brewing & Distilling; 2 - BrewDog; 3 - Canadian Institute of Fermentation Technology

Abstract

Considerable research has been conducted in an effort to understand hop aroma and how it influences dry-hopped beer. A large component of flavour/ aroma in hops are terpenes and terpenoids. However, little is understood about the extraction of terpenes and terpenoids during dry hopping and their overall stability in beer. It was hypothesised that as hop dose and ethanol content increased, the concentration of terpenes/ terpenoids extracted would be greater. A two-factor study examining the effect of hop dose rate and ethanol content was designed to test the hypothesis. The study found that terpene and terpenoid extraction did not linearly increase with higher dose rates and ethanol content, but that solubility might be dependent on the polarity of the specific terpenes and terpenoids. With this in mind, a second study was undertaken to assess the stability of terpenes and terpenoids in packaged beer. As sensory panels have suggested that 'fresh' hop aromas decrease over time, it was hypothesised that terpene/ terpenoid concentration declines over time. A dry-hopped ale was aged at 4°C and 20°C over the course of 120 days with bi-weekly sensory and GC/MS-SPME examination. The results of this experiment found some terpene/ terpenoid compounds to decrease over time (e.g., myrcene) while others were observed to increase after 40-60 days of bottle storage (e.g., *trans*-geraniol). These results may be due to yeast mediated biotransformation or changes in other analytes in bottle. Increasing our understanding of dry-hop aroma stability is crucial to beer quality.

Short biography

Margaux Huismann received a B.Sc. in Microbiology from the University of Wisconsin-La Crosse in 2014. Over the course of 2014-2015, she completed her M.Sc. in Brewing and Distilling at Heriot-Watt University at the International Centre for Brewing and Distilling (ICBD) in Edinburgh. During her Master's, she was employed at Stewart Brewing and Barney's Beer and worked with a team of ICBD Master's students to create Edinburgh Gin's Seaside Gin. She began her Ph.D in 2015 and was a young member committee representative of the Scottish Section of the Institute of Brewing and Distilling. Her Ph.D. research focuses on 'Understanding the Physical-Chemical Stability of Dry-Hopped Beer', and is jointly funded by Interface Food & Drink, the Institute of Brewing & Distilling and BrewDog.

BARTH-HAAS GROUP

Influence of whirlpool temperature on hop aroma intensity in ale beers

Marius Hartmann
Barth-Haas Group
University of Applied Science Weihenstephan-Triesdorf

Opener

BARTH-HAAS GROUP

- How do science and brewing passion go together?

BARTH-HAAS GROUP

Agenda

1. Influence of whirlpool temperature
2. Theory of late hopping and hop aroma
3. Set up of whirlpool trials
4. Chill down matters

The Craft Beer Movement

BARTH-HAAS GROUP

The brewing industry has changed in the past decade

- More interest in hop aroma forwarded beers
- Challenges with brewing methods and procedures
- Challenges with beer quality and stability
- Rethinking techniques for special beers

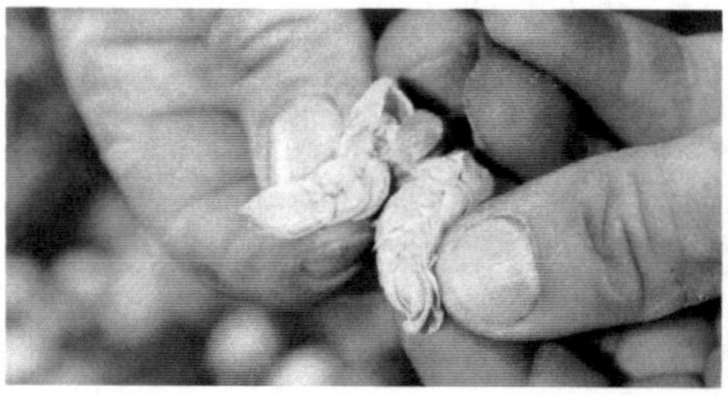

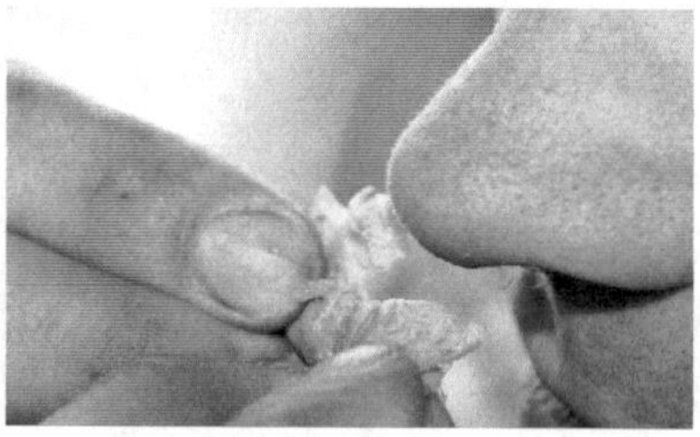

The goal of the project

BARTH-HAASGROUP

Does a lower temperature of the wort in the whirlpool improve hop aroma in a hop forwarded Ale?

- Intensity & Improvement
- What kind of aroma compounds are responsible for a possible change of flavour?
- Are there differences in isomerization?

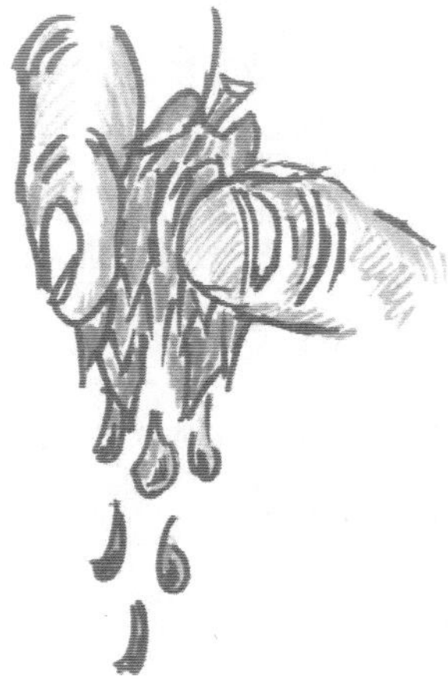

5

BARTH-HAASGROUP

Agenda

1. Influence of whirlpool temperature
2. Theory of late hopping and hop aroma
3. Set up of whirlpool trials
4. Chill down matters

6

Whirlpool hopping – some facts

BARTH-HAASGROUP

Proven benefits of wort temperature reduction in the whirlpool:

- Reduced transformation from DMS-P to free DMS
- Reduced thermal stress (TBI-value)
- Reduced isomerization
- Increased intensity of hop aroma
- Increased hot break
- Decreased concentration of aging compounds

(Dissertation Kaltner, 2001)

Late hopping – some facts

BARTH-HAASGROUP

Whirlpool effect is affected by:

1. Shearing force (cavitation, pump design)
2. Wort remaining too long in Whirlpool
3. Pump speed
4. Quality of the raw materials and mashing procedure
5. Size of the hot break material
6. Built in components
7. Wrong dimensioning of the geometric (1:3)
8. Hop addition with pellets is limited in the whirlpool (300 g/hl)
9. >300 g/hl hop pellets „hop strainer" is beneficial for late hopping
10. **Lower temperature**

Late hopping

BARTH-HAAS GROUP

- ➢ Late hopping in the whirlpool

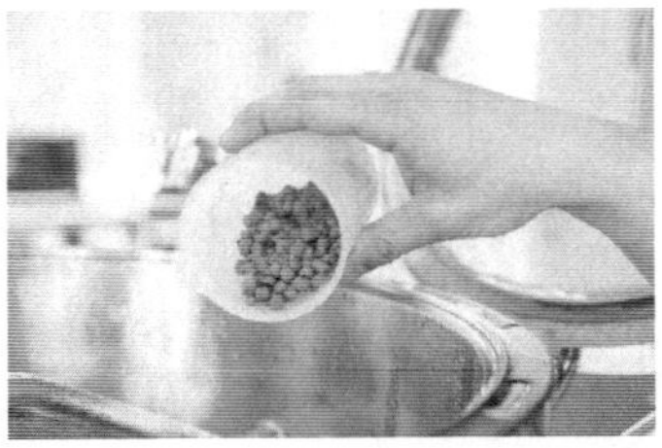

- ➢ Late hoping in a hopstrainer
  (HopBack: BrauKon)

Process Solutions

BARTH-HAAS GROUP

Pre Cooling Systems 1 & 2

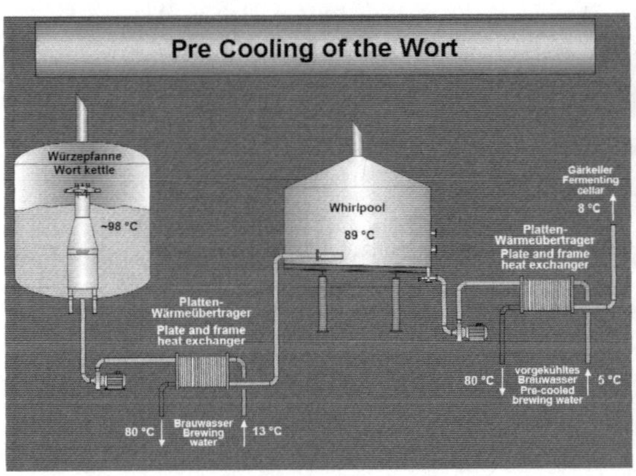

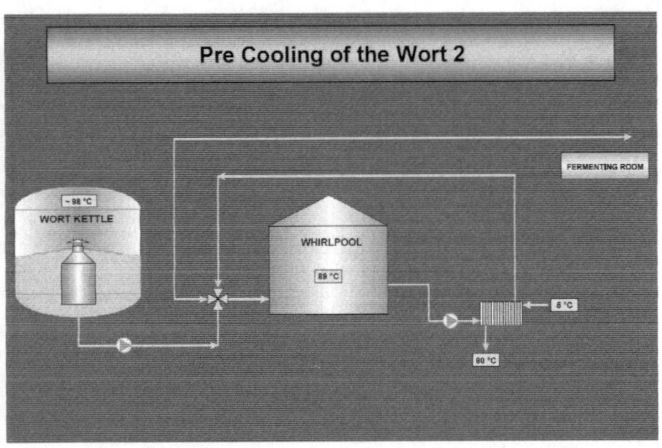

GEA

Process solutions

BARTH-HAAS GROUP

Pre Cooling System 3

- Continuous system:

Heat exchanger:

→ Warm up wort before boiling 75 °C to 81 °C

→ Chill wort after boiling from 99 °C to 92 °C

SCHULZ *X–LOOP*

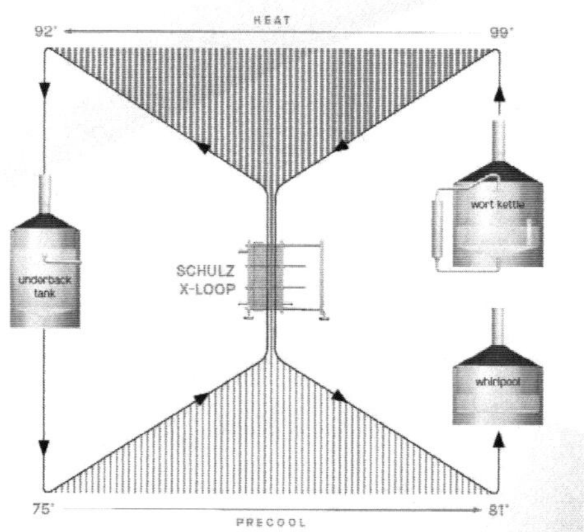

Pre Cooling System 4

- Brew high gravity beer and cold water addition in line during knockout wort in whirlpool

Sensory relevant hop aroma compounds in beer

BARTH-HAAS GROUP

Sensory description	Substance	Conc. In hopped beers	Threshold in beer
Black currant	4 MMP	10-200 ppb	10-50 ppb
Black currant, grapefruit	3 MH/3MHA	10-80 ppb	55 ppb
Resinous	Myrcene	5-100 ppm	10-1000 ppm
Floral, citrussy, lavender	Linalool	10-200 ppm	8-80 ppm
Floral, rose	Geraniol	5-100 ppm	4-40 ppm
Citrussy, lime, Lychee	Beta-Citronellol	5-50 ppm	8 ppm
Estery, apple	ethyl 2-methyl-butanoate	1-10 ppm	1,2 ppm
Fruity, estery	ethyl 4-methyl-pentanoate	1-3 ppm	1 ppm
Fruity, herbal	Cis rose-oxid	n. d.	5-50ppb
Green, metallic	(z)-1,5-octadien-3-on	n. d.	0,0034 ppm
Hoppy, pineapple	(e,z)1,3,5-Undecatriene	n. d.	no data
Cedarwood	Caryophylla-3,8-dien-(13)-dien-5-beta-ol	n. d.	no data
Cheesy/onion/garlic	Various sulfur compounds	1-10 ppm	1-500 ppm

- Synergetic, additive & masking effects

BARTH-HAAS GROUP

Agenda

1. Influence of whirlpool temperature
2. Theory of late hopping and hop aroma
3. Set up of whirlpool trials
4. Chill down matters

Whirlpool Trials

BARTH-HAAS GROUP

Set up of trials

- 8 Pale Ale Beers identical recipe
- only difference is the whirlpool temperature

Beer recipe

- A light single hop (US-Mosaic) Pale Ale (80 l):
- 12 ° Plato (80 % Pale Ale Malt, 17 % Munich Malt, 3 % Sour Malt)
- fermented with Ale Yeast (US-05) 18 °C in a UNI-Tank

BARTH-HAAS GROUP

The Brewhouse

4-Vessel brewhouse (Kaspar-Schulz)

Automated system for hop dosing

Specialties

Wort chilling during whirlpool rest

Lautertun as hop-back

„SchoKolino“ – gentle boiling system

Fermentation/Lagering

Open/ closed fermenters

Uni- Tanks (chilled by glycol chilling system)

Dry-hopping system – „Rocket“

 15

Heat exchanger (finned tube)

BARTH-HAAS GROUP

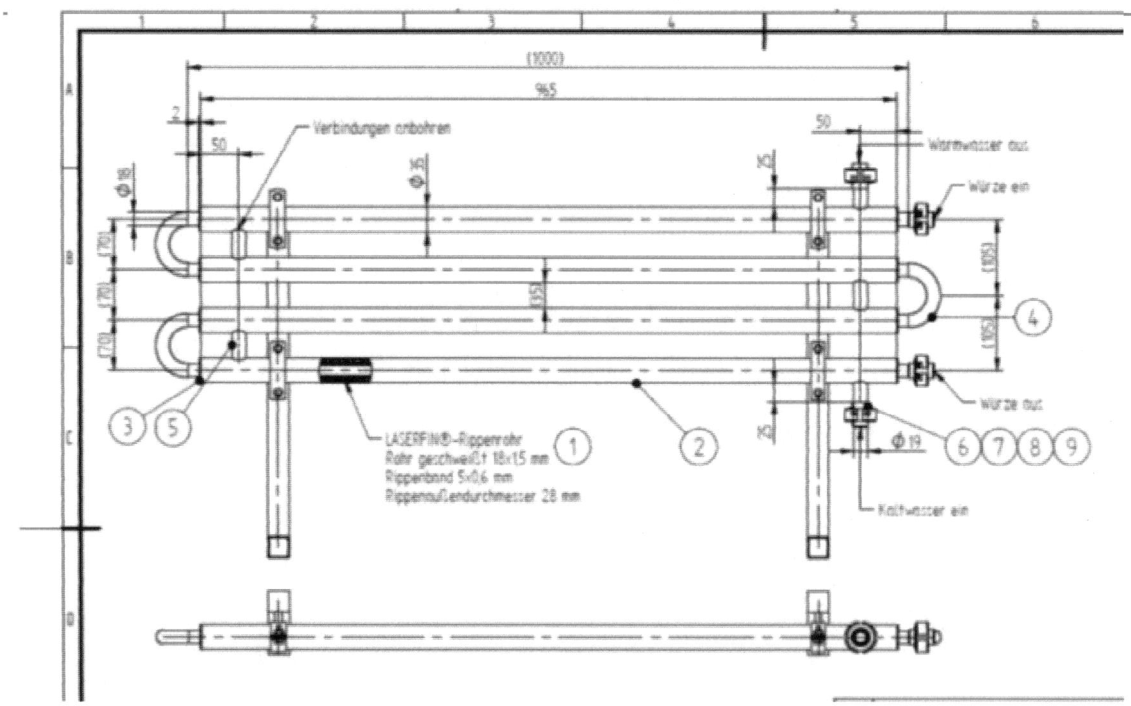

 16

Whirlpool Trials

Hops addition (Pellets)

- Beginning of boiling: 35 g US Mosaic (12.7 % Alpha Acids) for 18 BE
- Boiling: 70 min, External boiler 104°C
- Late hopping: 117 g US Mosaic (1.70 ml Oil / 100 g hops) = 2.5 ml Oil/100 l

Main parameters

- Pre-chilling of the wort before whirlpool:
- Late hopping at 93 °C, 83 °C and 73 °C
- Whirlpool rest 20 min
- Knock out time: 1 min 30 sec
- Open the cold water valve

Agenda

1. Influence of whirlpool temperature
2. Theory of late hopping and hop aroma
3. Set up of whirlpool trials
4. Chill down matters

Results

BARTH-HAASGROUP

Name	p (original extract) (%w/w)	Alcohol (% v/v)	Ea (app. Extract) (%w/w)	pH	CO2 (g/l)	O2 (mg/l)
Beer 1 (73°C)	11,95	5,03	2,66	4,53	3,29	0,01
Beer 2 (83°C)	12,1	5,18	2,66	4,51	3,36	0,02
Beer 3 (93°C)	12,1	5,15	2,75	4,6	3,23	0,04

Tests: Sensory / Analytical

BARTH-HAASGROUP

- **Sensory Tests (trained panel)**
- **12 taster**

1. Barth-Haas Hopsessed Tasting Scheme (descriptive)
2. Triangle Test (difference)
3. DLG- Scheme (descriptive)

Hopsessed Scheme

BARTH-HAAS GROUP

BARTH-HAAS GROUP UNIFORM TASTING SCHEME

CATEGORY	SPECIFIC ATTRIBUTES	INTENSITY (1 - 10)
FLORAL	Elderflower, camomile blossom, lily of the valley, jasmine, apple blossom, rose, geranium, carnation, lily, lilac, lavender	
CITRUS	Grapefruit, orange, lemon, lime, bergamot, lemon grass, ginger, tangerine	
SWEET FRUITS	Banana, watermelon, honeydew melon, peach, apricot, passion fruit, lychee, dried fruit, plum, pineapple, cherry, kiwi, mango, guava	
GREEN FRUITS	Pear, apple, quince, gooseberry, white wine grape	
RED BERRIES	Cassis, blueberry, raspberry, blackberry, strawberry, red currant, black currant, wild strawberry, cranberry	
CREAM CARAMEL	Butter, chocolate, yoghurt, honey, cream, caramel, toffee, coffee, tonka bean, vanilla	
WOODY AROMATIC	Tobacco, cognac, barrique, leather, woodruff, incense, myrrh, resin, cedar, pine, earth	
MENTHOL	Mint, balm, camphor, menthol, wine yeast	
HERBAL	Lovage, marjoram, tarragon, dill, parsley, basil, fennel, coriander, rosemary, thyme, green tea, black tea, mate tea, sage	
SPICY	Pepper, chilli, curry, juniper, aniseed, liquorice, fennel seeds, clove, cinnamon, gingerbread, coriander seeds, nutmeg	
GREEN GRASSY	Green grassy, tomato leaves, green pepper, fresh cut grass, hay, nettle, cucumber	
VEGETAL	Celery root, celery stock, leek, onion, artichoke, garlic, wild garlic	

FOR BEER	RATING
AROMA INTENSITY (1-10)	
AROMA QUALITY (1-10)	
BITTER INTENSITY (1-10)	
BITTER QUALITY (1-10)	
HARMONY (1-10)	
BODY AND MOUTHFEEL (1-10)	
PREFERENCE	
ESTIMATED BITTERNESS (IBU)	

BARTH-HAAS GROUP

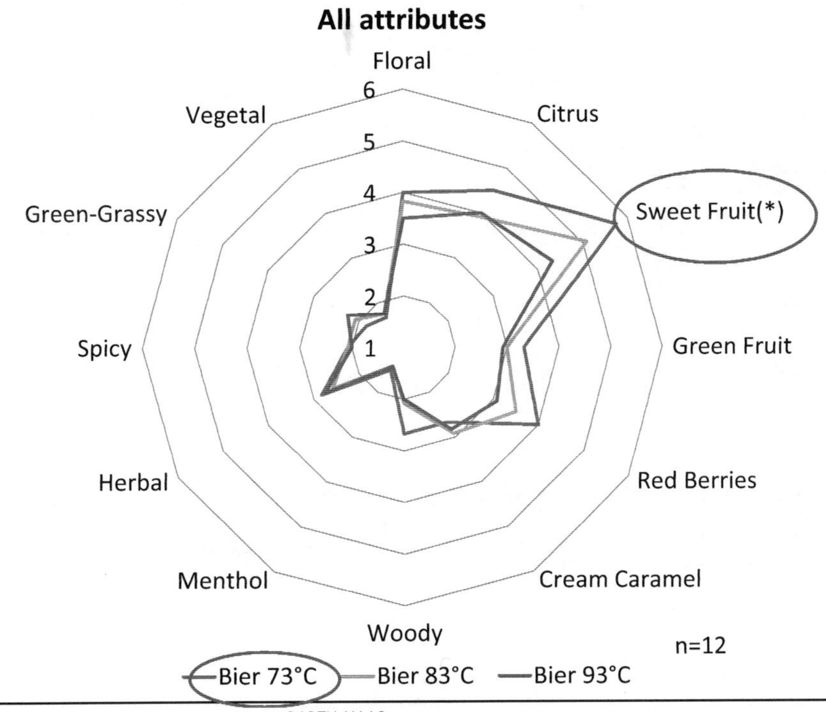

Specific descriptors

BARTH-HAAS GROUP

➢ **1) Beer (73 °C** during late hopping):

Mild, fruity, complex, sweet fruits, pear, currant

➢ **2) Beer (83 °C** during late hopping)

Fruity and blanaced, not as fruity as beer (1)

➢ **3) Beer (93 °C** during late hopping)

Not as fruity as the other beers, little unpleasent bitterness

Sweet Fruits

BARTH-HAAS GROUP

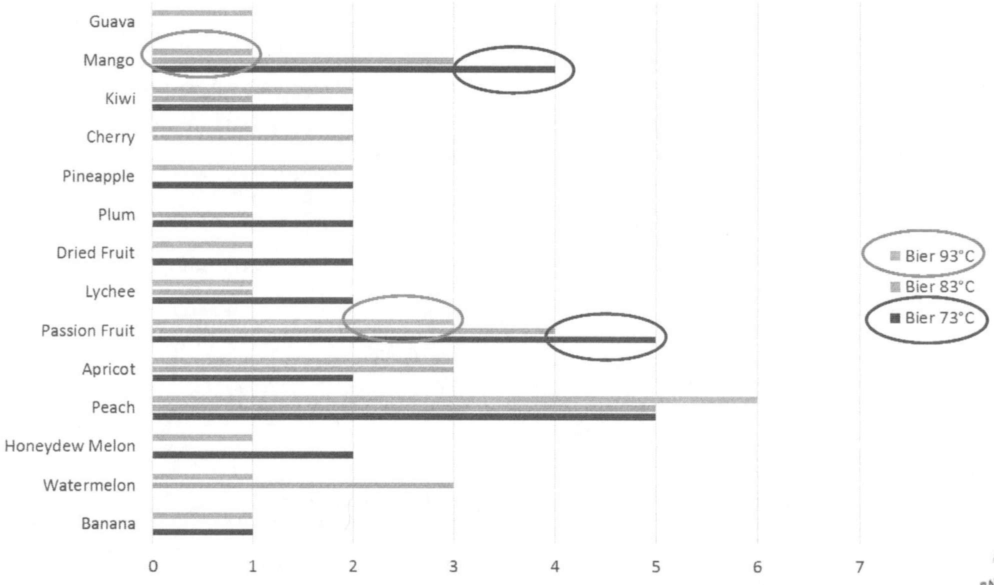

n=12

All attributes

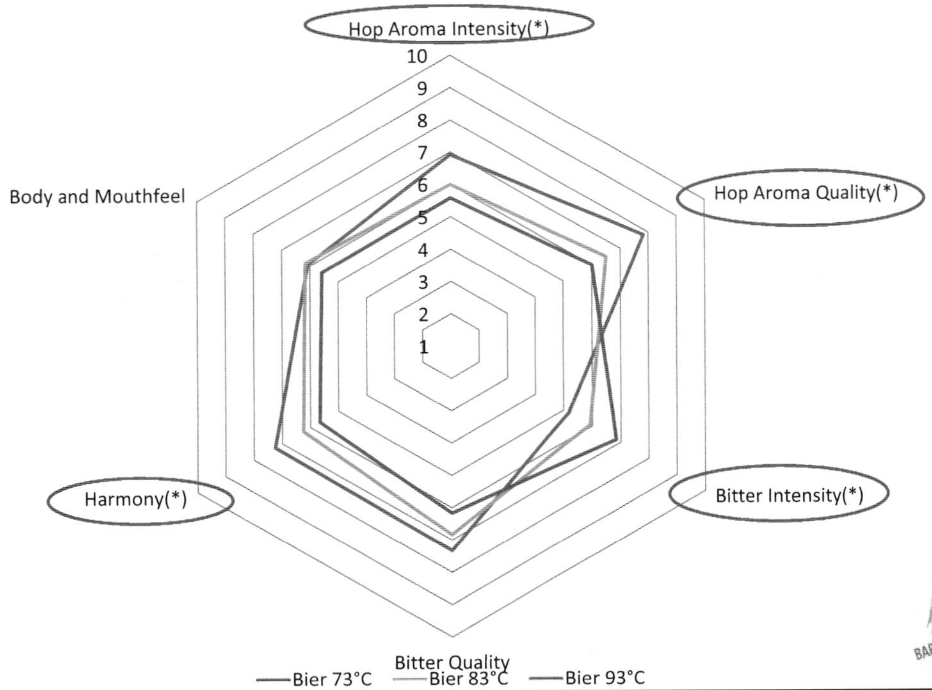

25

Rank all at last opportunity:

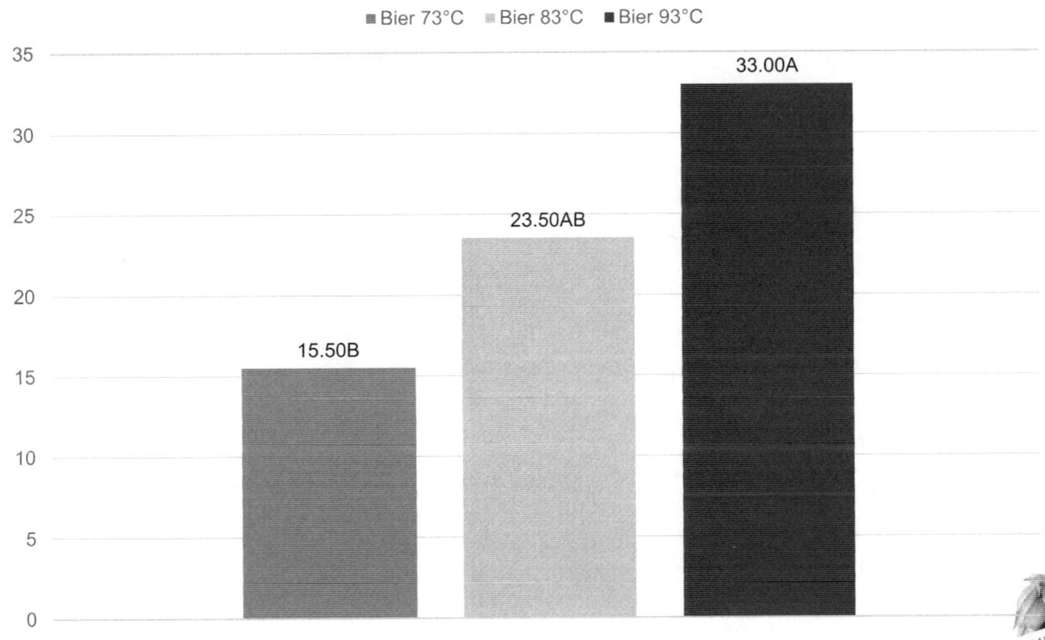

26

Results triangle tests

BARTH-HAASGROUP

Result of the comparison between beer 1 (73 °C) and beer 2 (83 °C):

- 7 of 10 taster identified the different sample→ significant difference with a level of significance of α = 5 %

Reslult of the comparioson between beer 2 (83 °C) and beer 3 (93 °C):

- 4 of 10 taster identified the different sample→ a significant difference could not be proved with a level of significance of α = 5 %

Thiols and precursors

BARTH-HAASGROUP

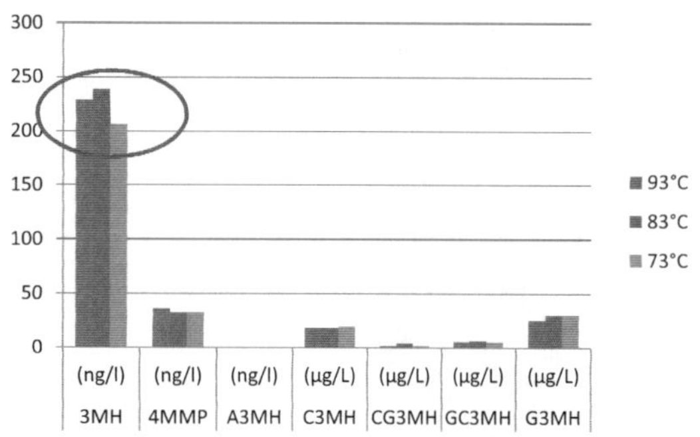

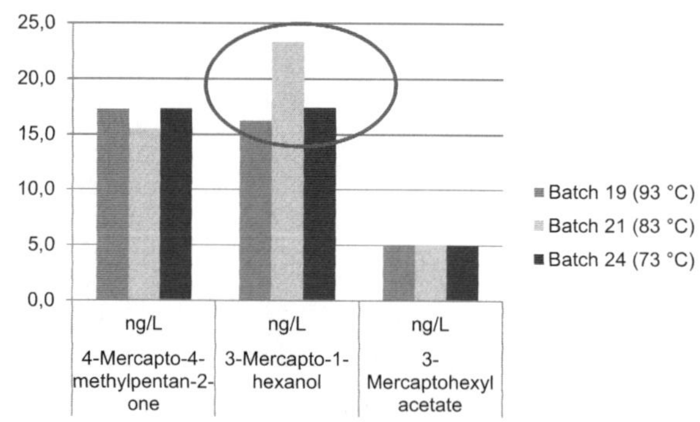

BARTH-HAAS GROUP

Other hop aroma compounds

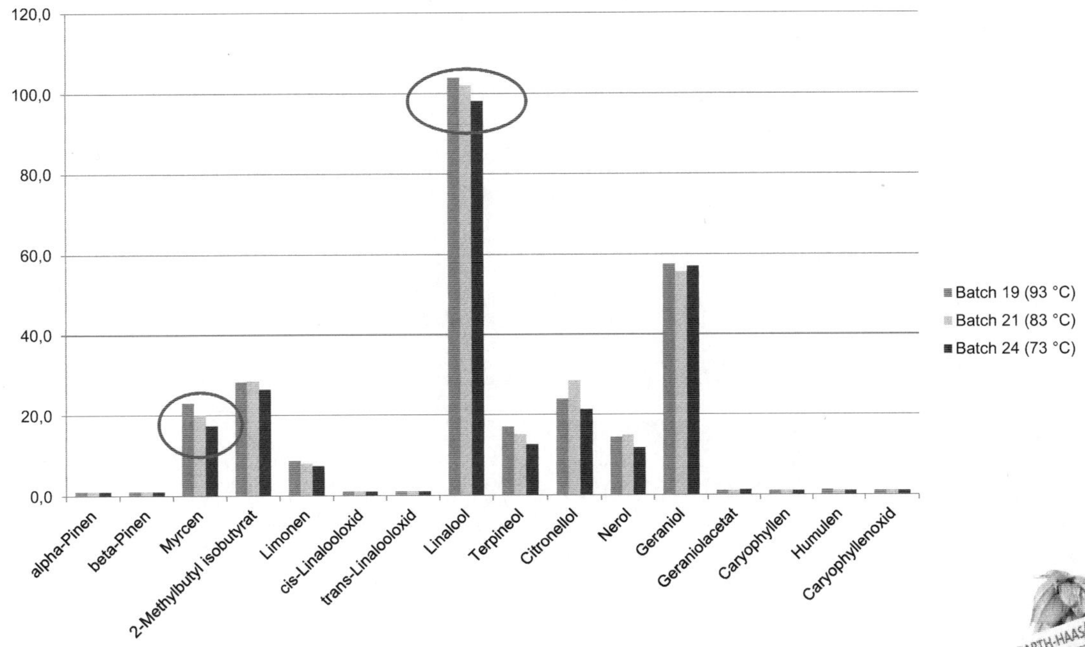

BARTH-HAAS GROUP

Analytical and sensory results of the bitterness

➢ **Analytical results:**

Beer 1 (73°C) = 25 BU

Beer 2 (83°C) = 28 BU

Beer 3 (93°C) = 30 BU

➢ **Sensory results:**

Beer 1 (73°C) = 27 BU

Beer 2 (83°C) = 28 BU

Beer 3 (93°C) = 31 BU

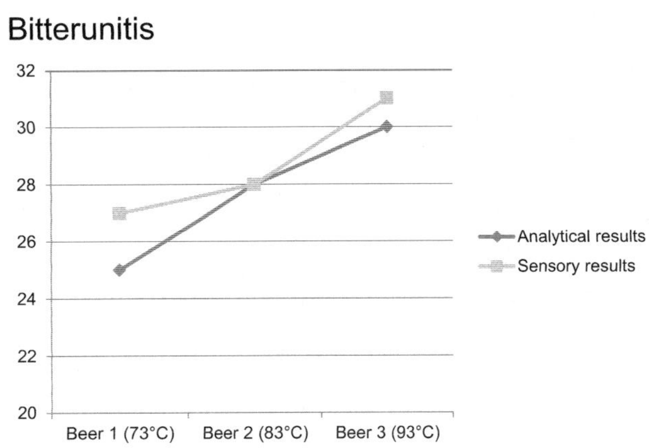

BARTH-HAAS GROUP

Isomerization during „late hopping“

- EBC-Bitterunits ($NATECO_2$)
- The first hop addition produced **18 BU**
- Hop addition in the whirlpool:
- 117,65g US-Mosaic (12,8 % Alpha Acids) per 80 Liter

1. Bei 73°C → **+ 7 BU (25 BU)** → ca. 4 %
2. Bei 83°C → **+10 BU (28 BU)** → ca. 5 %
3. Bei 93°C → **+12 BU (30 BU)** → ca. 6 %

BARTH-HAAS GROUP

Conclusion

- Same recipe & brewing technique but different beer taste
- Significant difference between beers (73 and 83 °C)
- Improvement of the beer taste for all parameters with lower temperatures
- No correlation between analysed aroma compounds and sensory evaluation
- Increased sensory bitterness corresponds with measured IBU
- Contrary to prevailing literature linalool compounds do not necessarily correlate with aroma intensity

© 2018 BARTH-HAAS – All rights reserved. Proprietary and confidential.

Future studies:

BARTH-HAAS GROUP

- Interaction of hop aroma compounds (e.g. Esters with thiols)
- Saturation of aroma compounds at high oil concentrations?
- Behaviour of different hop products (enriched pellets, hop extract)

33

At the Barth-Haas Campus

BARTH-HAAS GROUP

E-Mail: marius.hartmann@johbarth.de
Marius Hartmann
Brewmaster of the Concept Brewery Nürnberg

34

European Brewing Convention

37th Congress – 02-06 June 2019

How to Optimize the Utilization of Hop Cysteine and Glutathione S-Conjugates in Late and Dry Hopping : Focus on Dual Hops and Saaz

Cécile Chenot

Sonia Collin

Croix du Sud, 2 L7.05.07 B-1348 Louvain-la-Neuve, Belgium

cecile.chenot@uclouvain.be

June, 4th 2019, Antwerp

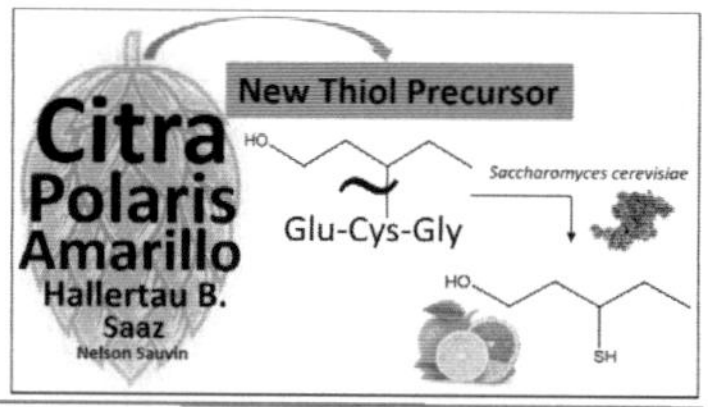

UCL UNIVERSITÉ CATHOLIQUE DE LOUVAIN

Introduction

Hops & Beer

⇒ **Hop in the brewing process, since the 8th century AD**

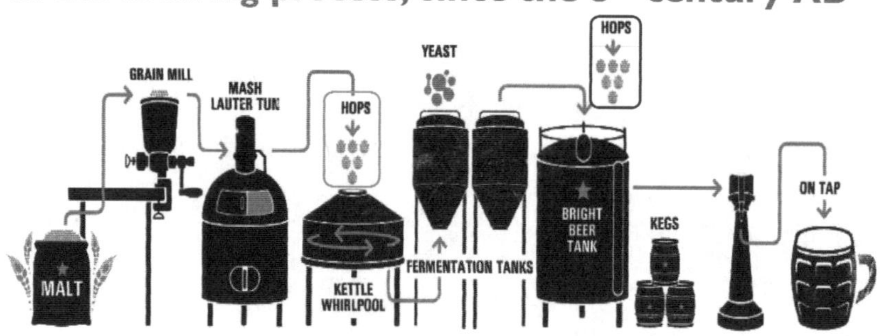

- Bitterness : iso-alpha acids, ...
- Flavors : essential oils (0.5-3 % w/w)
- Foam
- Antibacterial
- Reducing power: polyphenols

Few active flavors in beer (late/dry hopping)

- Acids, ketones, esters (hexanoic acid, β-damascenone, ethyl cinnamate, γ-nonalactone)
- Terpenoids (humuladienone, linalool ..)
- Thioesters
- Polysulfides (DMTS)
- Thiols (MBT)
- **Polyfunctional thiols**

Bitter, Aromatic & Dual hops

UCL – EARTH & LIFE INSTITUTE -1-

Lermusieau, Bulens & Collin, J. Agric. Food Chem., 49 (2001) 3867-3874

Lermusieau & Collin, J. ASBC, 61 (2003) 109-113

Introduction

Polyfunctional thiols

⇒ Distance between SH and OH/=O/Ace : 3 carbons
(except for MBT, 2SEol/A, 6SHol and 4S4MPol)

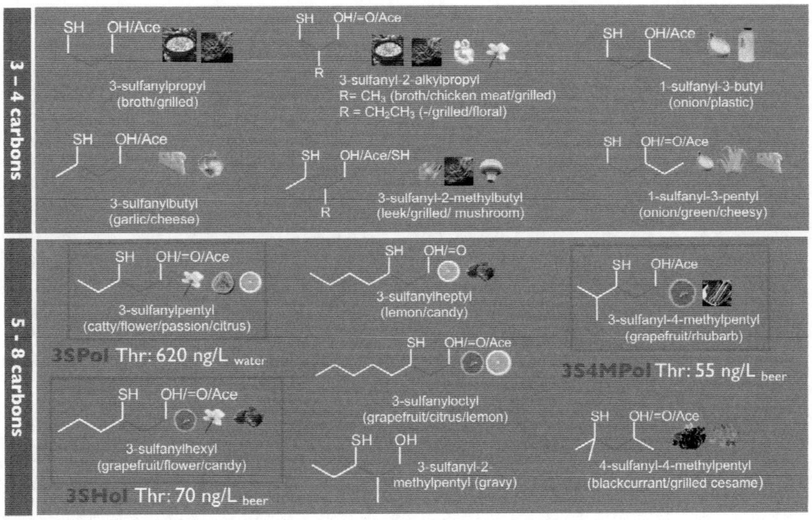

⇒ **Strong organoleptic** impact due to very low thresholds

Vermeulen, Lejeune, Tran & Collin, J. Agric. Food Chem., 54 (2006) 5061-5068
Gros, Nizet & Collin, J. Agric. Food Chem., 59 (2011) 8853-8865
Kankolongo, Gros, Nizet & Collin, J. Agric. Food Chem., 63 (2015) 3022-3030

UCL – EARTH & LIFE INSTITUTE —— **-2-** ——

Introduction

Free polyfunctional thiols in hops

⇒ 41 Free thiols found in hops and beers

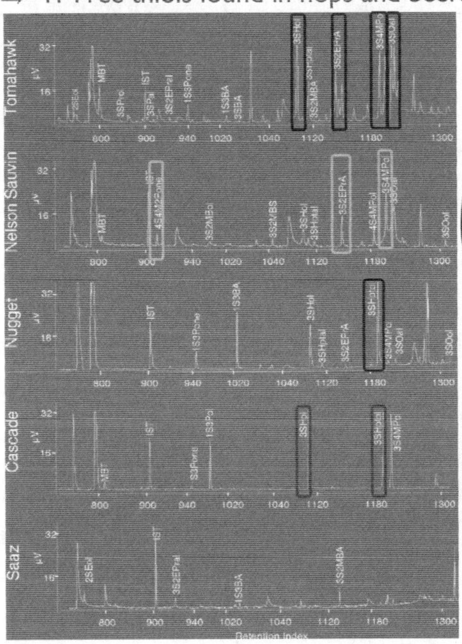

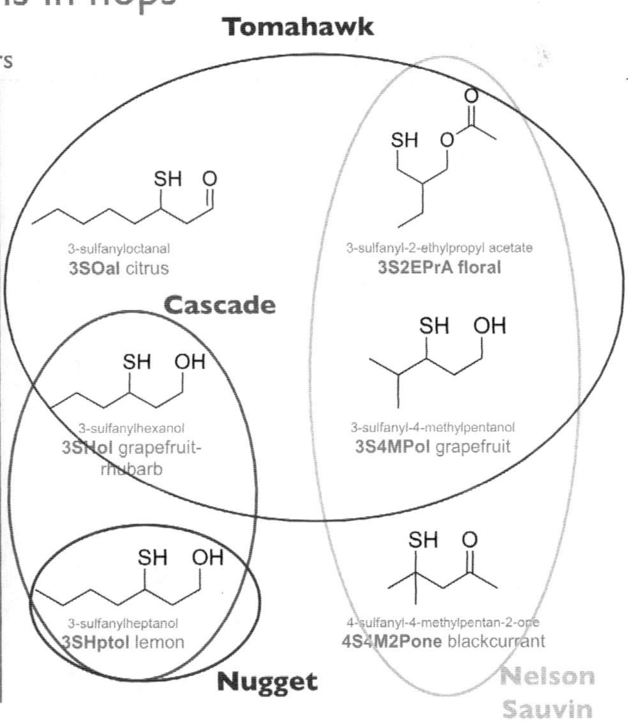

UCL – EARTH & LIFE INSTITUTE —— **-3-**

Gros, Nizet & Collin, J. Agric. Food Chem., 59 (2011) 8853-8865

Introduction

Free and bounded polyfunctional thiols in hops

⇒ Evolution through boiling and fermentation

Suspicion of thiol precursors

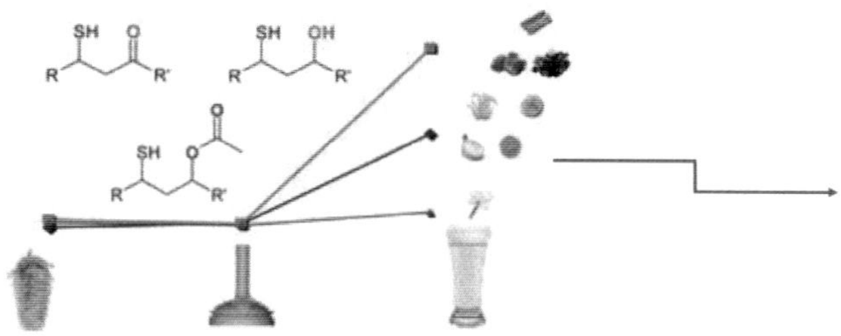

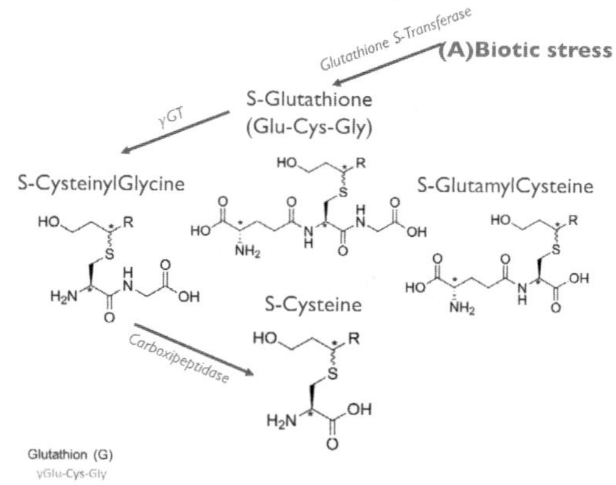

⇒ S-conjugates already found in other plants

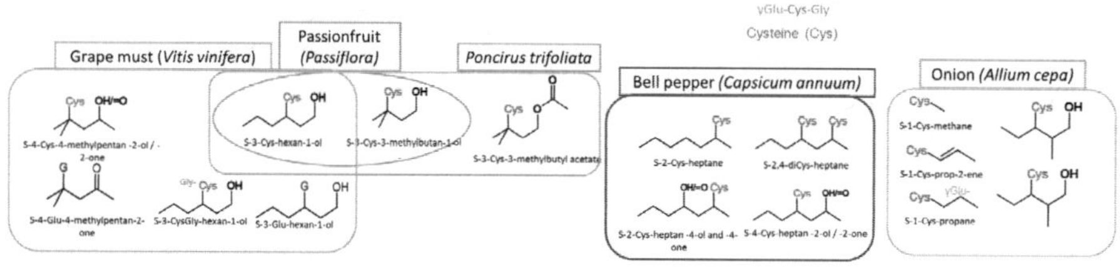

Gros *et al.* 2011; Gros *et al.* 2012; Tominaga *et al.*, 1998; Peyrot des Gachons *et al.*, 2002; Federizzi *et al.*, 2009; Capone *et al.*, 2011, Starkenmann *et al.*, 2007, Tominaga *et al.*, 2000, Starkenmann et *al.*, 2011a, 2011b ; 2011c

UCL – EARTH & LIFE INSTITUTE -4-

Introduction

First evidence of cysteine S-conjugates in hop

2. Direct method

1. Indirect method

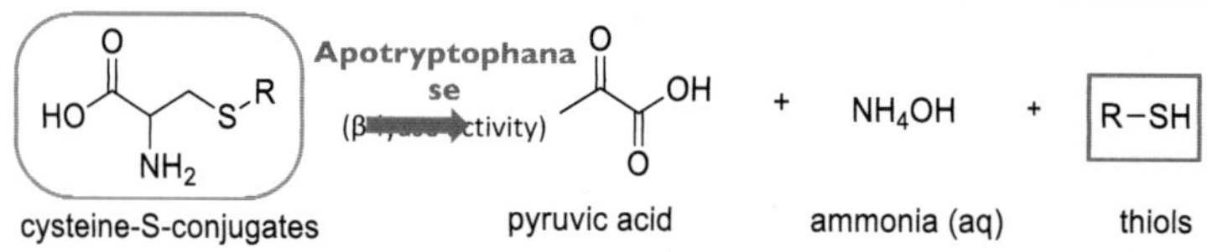

Chemical synthesis of standards

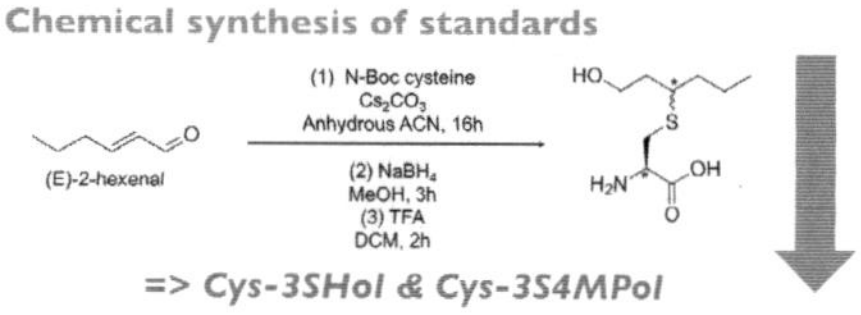

=> *Cys-3SHol & Cys-3S4MPol*

Liquid chromatography (HPLC-ESI(+)-MS/MS)

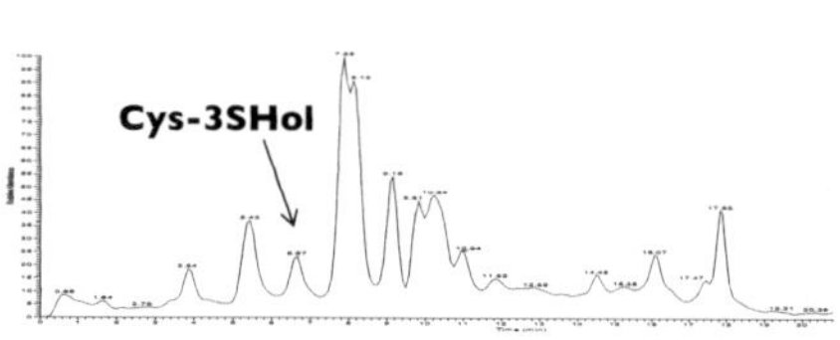

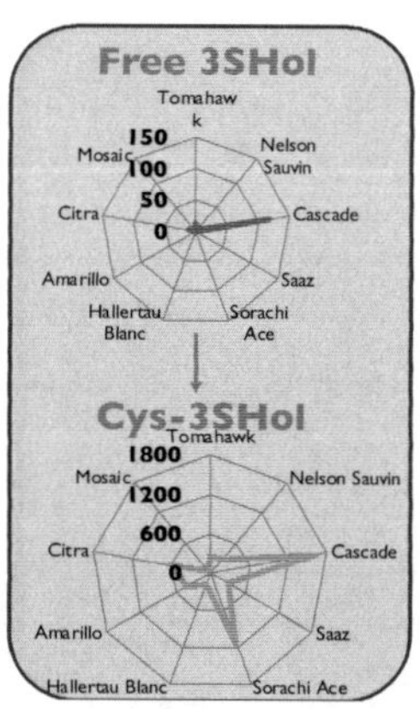

Gas chromatography (GC-PFPD)

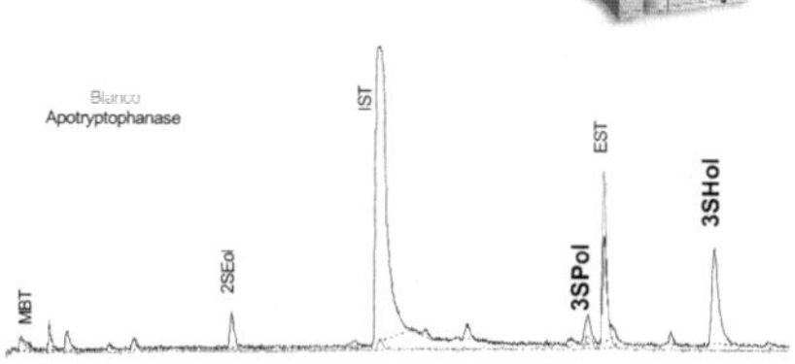

Gros, Peeters & Collin, J. Agric. Food Chem., 60 (2012) 7805-7816
Gros, Tran & Collin, J. Inst. Brew., 119 (2013) 221-227
Kankolongo, Gros, Nizet & Collin, J. Agric. Food Chem., 63 (2015) 3022-3030

UCL – EARTH & LIFE INSTITUTE -5-

First evidence of glutathione S-conjugates in hops

Direct method

Chemical synthesis of standards

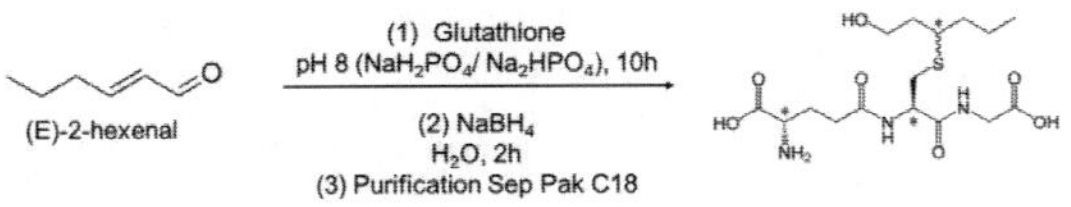

=> ***G-3SHol & G-3S4MPol***

Liquid chromatography (HPLC-ESI(+)-MS/MS)

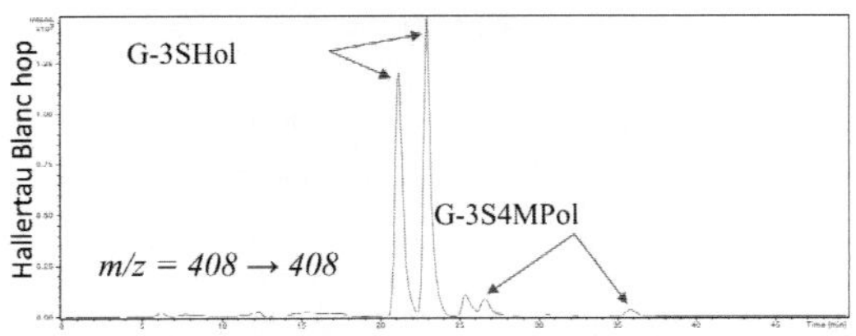

3SHol

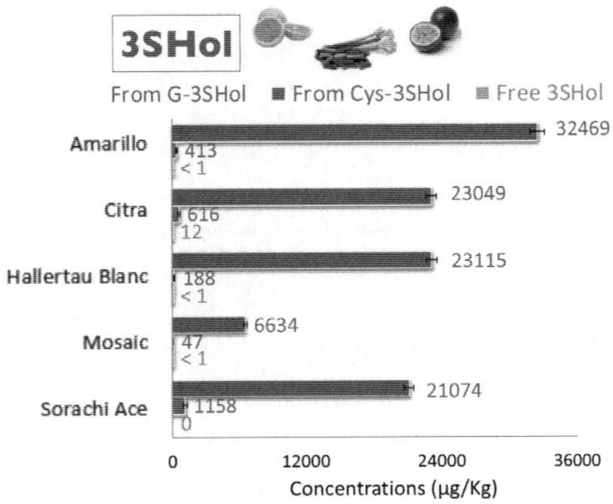

3S4MPol

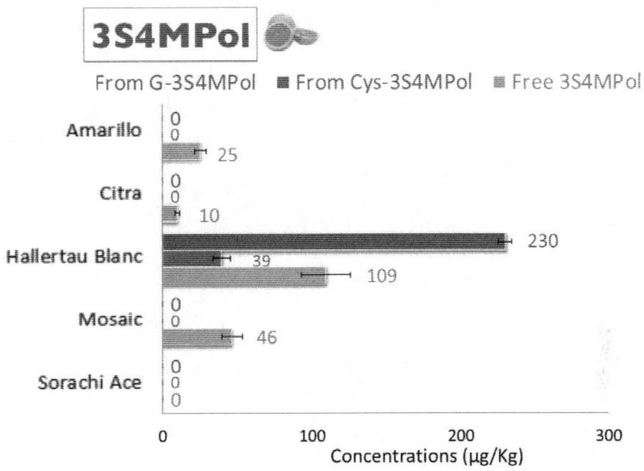

Kankolongo, Decourrière, Alonso, Bodart, Robiette, Collin. J. Agric. Food Chem., 2016, *64*, 8572–8582
Decourrière, Chenot, Ferreira, Collin. 15th Weurman Flavour Research Symposium

Optimization of S-conjugates extraction

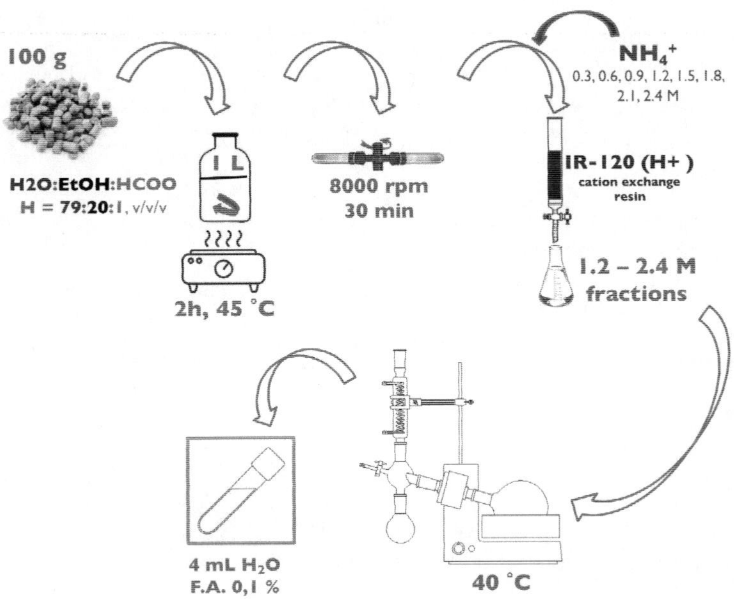

Kankolongo, Decourrière, Alonso, Bodart, Robiette, Collin. J. Agric. Food Chem., 2016, *64*, 8572–8582
Decourrière, Chenot, Ferreira, Collin. 15th Weurman Flavour Research Symposium

Optimized analysis

HPLC analyses

- ➢ Comparison of HPLC columns for S-conjugates resolution

Astec® Cyclobond® I 2000 RSP column 25cmx4.6mm, 5µm | **Hypersil GOLD™ aQ column 10cmx2.1mm, 3µm**

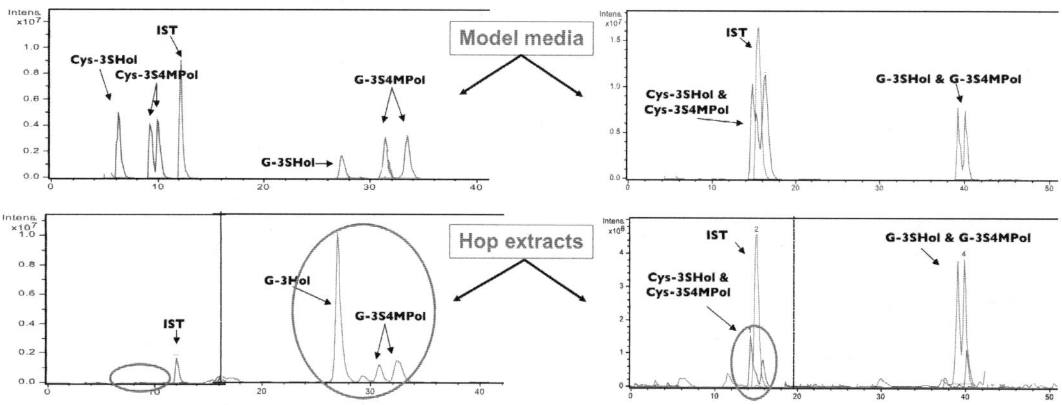

(+) Dictinction between 3SHol & 3S4MPol precursors

(-) Disparition of peaks for Cys-conjugates in hop extracts ! (coelution with a.a. in much higher amount)

(-) No distinction between 3SHol & 3S4MPol precursors

(+) No disparition of peaks for Cys-conjugates in hop extract !

New S-conjugates

Synthesis of Cys- and G-3SPol (3-sulfanyl-pentan-1-ol)

- ➢ **3SPol** *(in µg/Kg)*

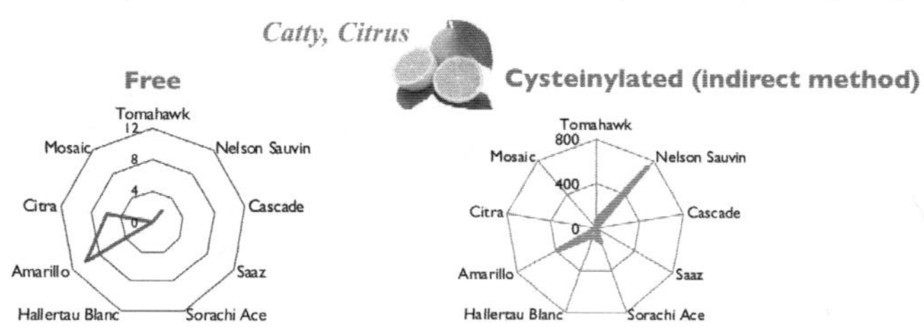

- ➢ **New syntheses of Cys-3SPol & G-3SPol**

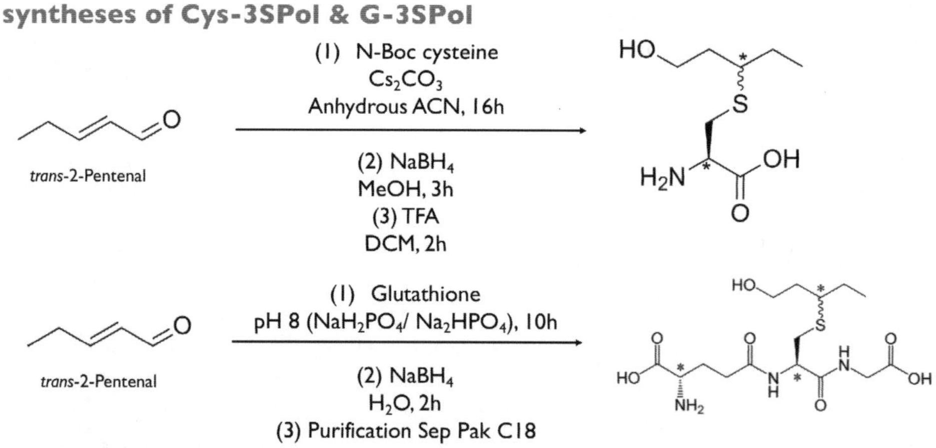

Chenot, Robiette, Collin. J. Agric. Food Chem., 2019, 67, 4002–4010

Synthesis of CysGly-3SPol

CysGly-precursors obtained from G-precursors via enzymatic reaction with γGT

- (1) 5 mL ultra pure water adjusted at pH 7 + 200 ppm G-3SPol = BLANCO
- (2) 4 mL ultra pure water adjusted at pH 7 + 1 mg γGT (24 U) in 1 mL + 200 ppm G-3SPol

] Stir for 30 min at 37° C

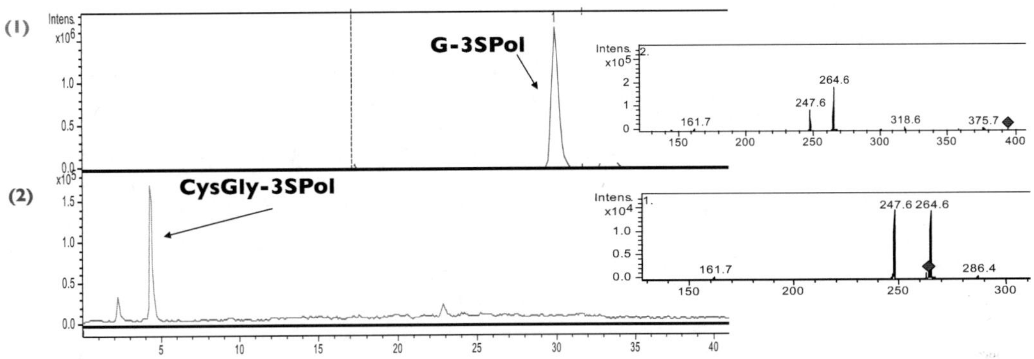

Chenot, Robiette, Collin. J. Agric. Food Chem., 2019, 67, 4002–4010

Standard analyses

- HRMS

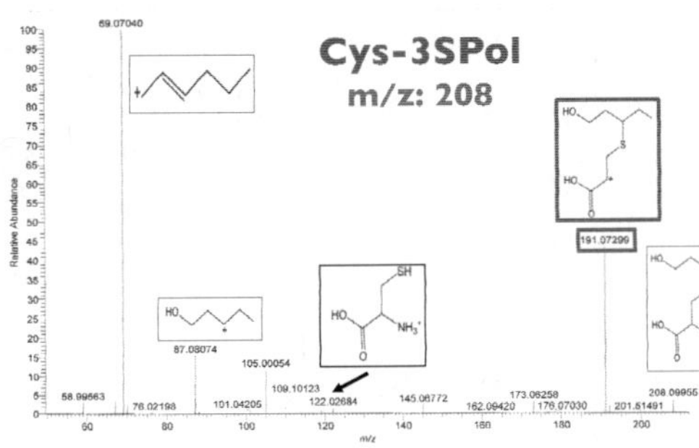

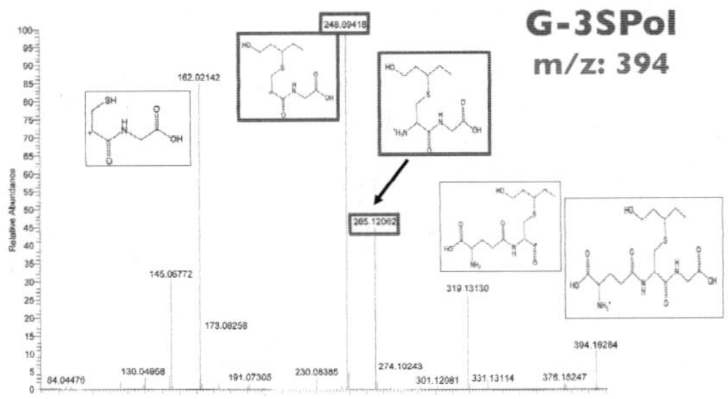

- HPLC-ESI(+)-MRM of synthesized standards (Hypersil Gold aQ column)

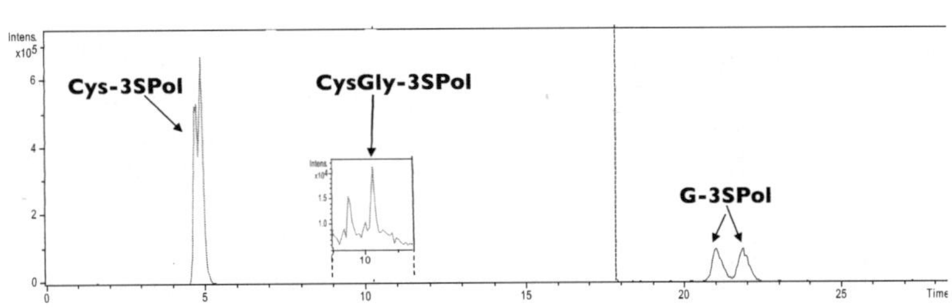

Hop analyses

New S-conjugates

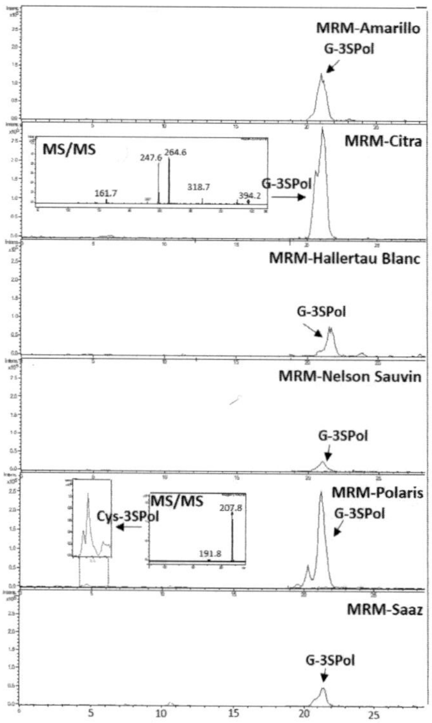

3SPol Potential

(in ppb of free thiol equivalents)

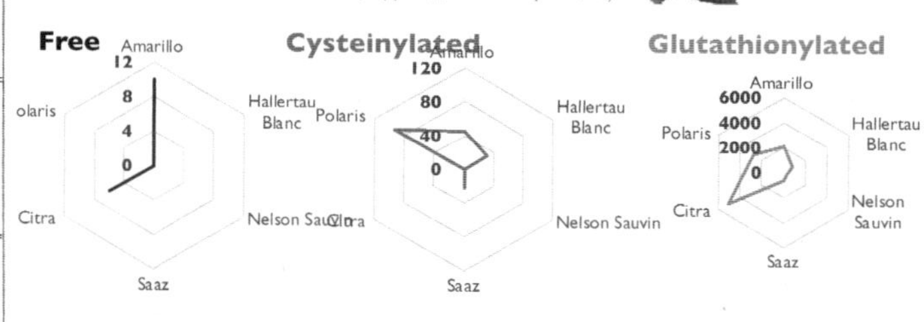

⇒ 3SPol precursors: ubiquitous to all the studied varieties, including Saaz

⇒ No relationship between amount of
Free (max: Amarillo)
Cysteinylated (max: Polaris)
Glutathionylated (max: Citra)

⇒ Glutathionylated form = HUGE potential vs free and cysteinylated

UCL – EARTH & LIFE INSTITUTE -12-

Chenot, Robiette, Collin. J. Agric. Food Chem., 2019, 67, 4002–4010

Production pathways

New S-conjugates

LOX: Lipoxygenase
HL: Hydroperoxide Lyase
EI: Enal Isomerase
GST: Glutathione S-Transferase
ADH: Alcohol Deshydrogenase

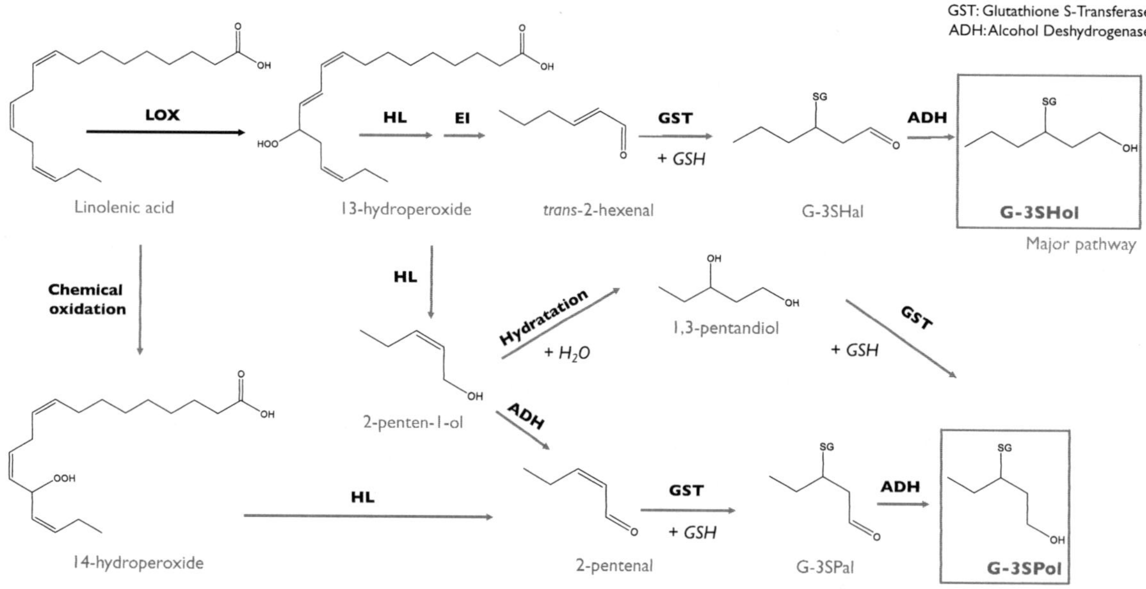

UCL – EARTH & LIFE INSTITUTE -13-

S-conjugates Summary

mg/Kg of bound forms in hop	Amarillo (USA, 2015)	Citra (USA, 2017)	Hallertau Blanc (Germany, 2015)	Nelson Sauvin (New Zealand, 2018)	Polaris (Germany, 2017)	Saaz (Czech Republic, 2017)
Cys-3SPol	d	d	d	d	0.162[a]	d
G-3SPol	7.5[bc]	18.1[a]	3.0[cd]	1.4[d]	9.8[b]	2.5[cd]
Cys-3SHol + Cys-3S4MPol	2.1[b]	0.3[d]	1.3[c]	0.2[d]	4.9[a]	0.4[d]
G-3SHol	101.0[a]	91.0[a]	77.1[ab]	20.1[c]	118.2[a]	95.7[a]
G-3S4MPol	nd	nd	0.3[b]	d	3.6[a]	d

nd: not detected, d: <50 μg kg^{-1}

⇒ **Citra**: the richest in term of G-3SPol

⇒ **Polaris**: the richest in term of Cys-adducts

⇒ & the richest in term of G-3S4MPol (and G-3SHol)

⇒ **Nelson Sauvin**: the poorest in term of thiol adducts

⇒ **Saaz**: few sulfides, polysulfides, free thiols and Cys-adducts

⇒ but high G-potential (more delicate flavors)

Real « precursors » ? Possible release by yeast ?

Chenot, Robiette, Collin. J. Agric. Food Chem., 2019, 67, 4002–4010

Thiols release

Yeast activity on Cys-adducts

⇒ First evidenced in bottle refermentation

⇒ **Release of 19 polyfunctional thiols after 3 weeks**

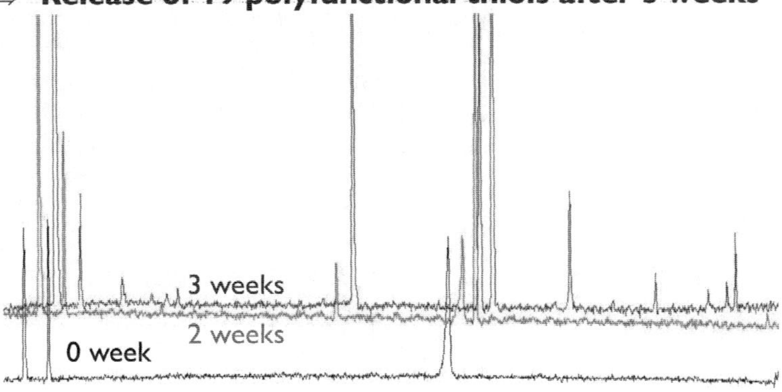

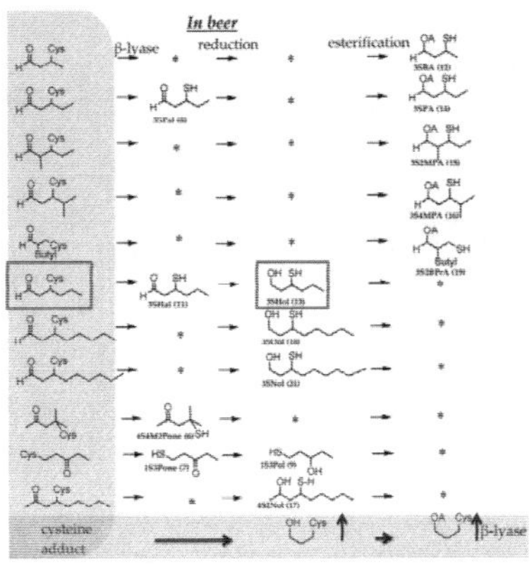

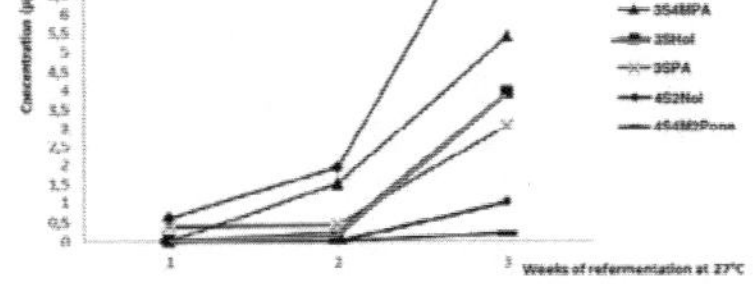

Fig. 3. Evolution of 2SEA, 3S4MPA, 3SHol, 3SPA, 4S2Nol, and 4S4M2Pone concentrations through bottle refermentation.

Nizet, Gros, Peeters, Chaumont, Robiette Collin. J. Am. Soc. Brew. Chem. 71(1):15-22, 2013

Fermentation trials & thiols extraction

Thiols release

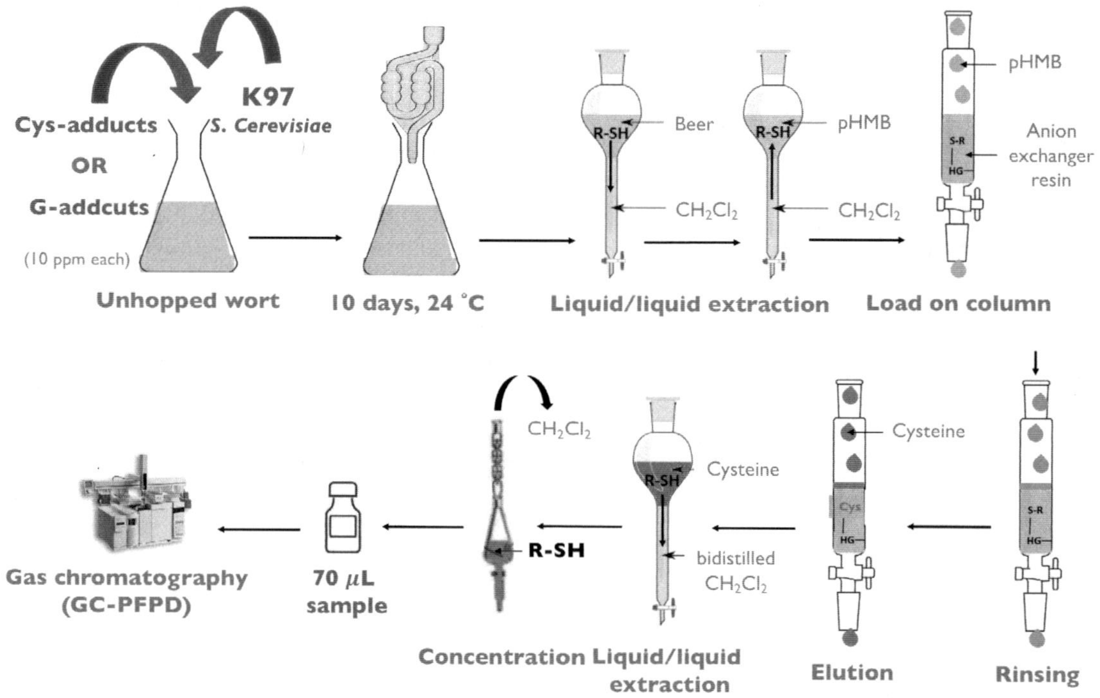

Thiols quantitation from Cys-adducts release

Thiols release

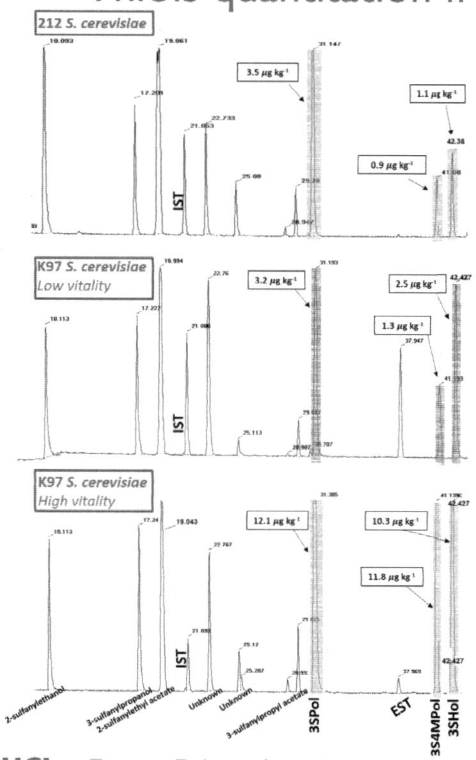

Molar conversions:

% Released from	212	K97 *low*	K97 *High*
3SPol	0.05 %	0.05 %	**0.12 %**
3S4MPol	0.03 %	0.02 %	**0.16 %**
3SHol	0.02 %	0.04 %	**0.13 %**

⇒ 2-sulfanylethanol, 2-sulfanylethyl acetate, 3-sulfanylpropanol and 3-sulfanylpropyl acetate: issued from cysteine and homocysteine **yeast Ehrlich pathways**

⇒ **0.02-0.16 %** molar conversion for Cys-adducts

⇒ **Enhanced release** by brewing yeast **K97** with high vitality *(Fermentis)*

Chenot, Robiette, Collin. J. Agric. Food Chem., 2019, 67, 4002–4010

Thiols release

Thiols quantitation from G-adducts release

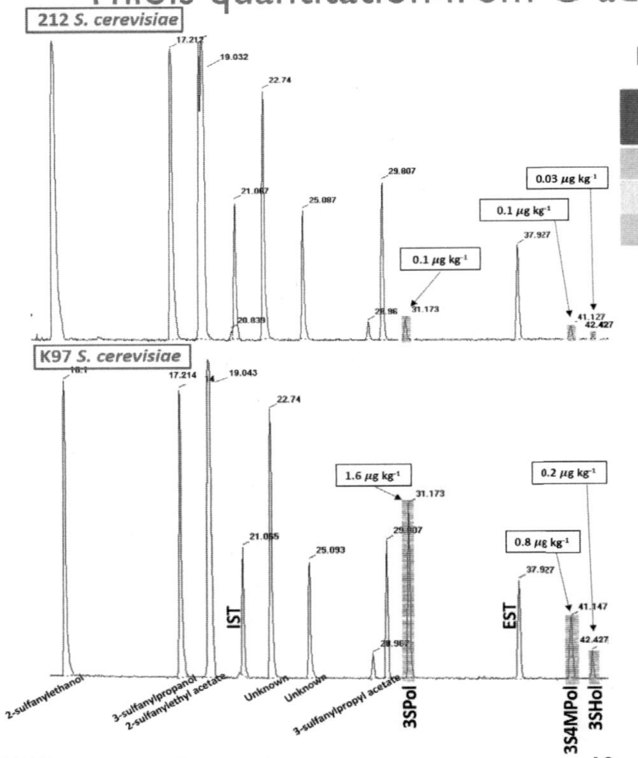

Molar conversions:

% Released from	212	K97
3SPol	0.005 %	**0.05 %**
3S4MPol	0.004 %	**0.03 %**
3SHol	0.001 %	**0.01 %**

⇒ **First evidence** of **strong thiol release** from G-3SPol, G-3SHol & G-3S4MPol by brewing yeast **K97** *(Fermentis)* vs a commonly used 212 *S. cerevisiae*

⇒ Up to **0.05%** for G-3SPol

⇒ **Lower** activity (0.01%) on G-3SHol

Chenot, Robiette, Collin. J. Agric. Food Chem., 2019, 67, 4002–4010

Conclusions & perspectives about New S-conjugates

- **First evidence of 3-sulfanyl-pentan-1-ol cysteine & glutathione adducts in hops**
  - G-3SPol ubiquitous to all the studied samples (Citra ++)
  - G-3SHol > G-3SPol in all the studied samples (Polaris ++)
  - G-adducts >>>> Cys adducts > Free thiols (especially for Saaz)
- **Investigation of new bound forms and new hop varieties**

Conclusions & perspectives about thiol release from S-conjugates

- **Free thiols released from Cys-adducts by yeast through primary fermentation** (0.1%) **& bottle refermentation** (3 weeks)

- **Confirmation of the "precursor" status of G-adducts**
  - Release of thiols from glutathione adducts by some *S. cerevisiae* strains
  - K97 (Fermentis) emerged as a super producer (0.05%)
  - Preferential activity for some G-thiol structures (3SPol > 3SHol)

- **Understanding the release mechanisms**
  - Effect of fermentation parameters (T°, yeast strains, yeast vitality, number of cells, dry/liquid yeasts, …)
  - Impact of the thiol structure

UCL – Earth & Life Institute -20-

Acknowledgments

- **Financial support:**

Fermentis
LESAFFRE FOR BEVERAGES

- **Awards and grants:**

BARTH-HAAS GROUP GIRACT ARAEBUL ABInBev

- **Hop suppliers:**

YAKIMA CHIEF -HOPS- Hopsteiner

UCL – Earth & Life Institute -21-

EARTH & LIFE INSTITUTE

Thank you for your attention

Questions ?

UCL UNIVERSITÉ CATHOLIQUE DE LOUVAIN

L36 - FUNCTIONALITY OF HOP PROANTHOCYANIDINS IN BREWING: AN ALTERNATIVE TO TANNIC ACID?

Arthur Gadon[1]; David Cook[1]

1 - University of Nottingham, UK

Abstract

The polyphenolic fraction of hops contains proanthocyanidin polymers, also known as tannins, which can bind to metal ions and haze active proteins, thus yielding potential benefits for beer flavour and colloidal stability. Our group has demonstrated that polyphenol-rich extracts of hops, and of spent hops resulting from the CO_2 extraction of alpha, have fining activity when added post-fermentation. Activity is attributed to the proanthocyanidins (flavan-3-ol polymers) crosslinking yeast cells and haze active proteins, to form flocs which settle out and fine the resulting beer. Fining activity is related to the size of aggregates formed. In recent work we applied analytical ultracentrifugation (AUC) to characterise the molecular weight ranges of polyphenols extracted from different hop varieties. This has demonstrated the presence of high molecular weight polymers within these extracts with molecular weights of up to 90 kDa. Fining activity of extracts has been shown to correlate with the presence of such high molecular weight tannins, with Saaz hops consistently found to out-perform other varieties in this regard. The functionality of proanthocyanidin rich hop extracts has parallels with the known activities of tannic acids. Logically, researchers have also investigated the addition of polyphenolic fractions of hops in the brewhouse as a part of the overall strategy of stabilising beers. This paper will review current knowledge of the functionality of hop proanthocyanidins in brewing and raises the possibility that these co-products of hop processing might provide sustainable and clean label solutions as a part of overall fining strategies in the brewery.

Short biography

David Cook is AB InBev Professor in Brewing Science at the University of Nottingham, UK and has more than 20 years of experience conducting research and teaching relating to brewing, analytical food chemistry and flavour technology.

David is Course Director for the University of Nottingham's innovative e-learning based Masters courses in Brewing Science, which are designed to be studied part-time as Continuing Professional Development courses for brewers and allied professionals.

The Cook group specialises in applied research and interacts extensively with the malting and brewing sectors. Current research activity focuses on malting science and technology, beer flavour formation and stability and the reduction of primary energy usage in malting and brewing. Sustainable bioprocessing of raw materials and co-products is a core focus of the group.

Pall Corporation

Filtration. Separation. Solution.SM

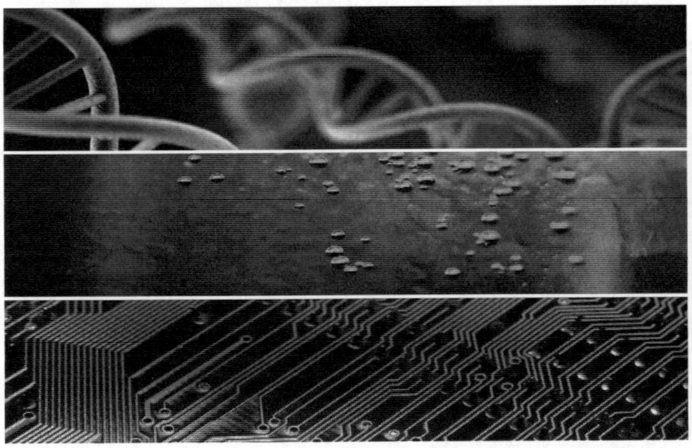

Microplastics in Beer

Facts about Filtration

Ursula Brendel-Thimmel
Pall Corporation

Content

- Plastics in the environment versus microplastics in food and beverages
- Definition of microplastics
- Toxicology of microplastics
- Key literature on microplastics in food and beverages
- Facts on filtration
  - Can particles be found in filtered product?
  - Do filters add particles?
  - Which filters remove particles?
- Summary and Outlook

Plastics in the Environment - Pollution of Oceans and Rivers

ZEIT ONLINE

Wegwerfplastik

Plastik ist außer Kontrolle

In der EU ist Wegwerfplastik künftig nicht mehr erlaubt. Das Verbot mag spät kommen und unzureichend sein, doch es ist der einzig richtige Beschluss.

ZEIT ONLINE. 27.3.2019.

Microplastics in Beverages and Beer

- „Plastik in Bier- das Reinheitsgebot war einmal“ [1] (Plastics in beer)
- „Mikroplastik in Mineralwasser und Bier” [2] (Microplastics in mineral water and beer)
- „Bei einem NDR-Test war kein einziges Getränk frei von Rückständen“ [3] (None of the tested beverages free from residues)

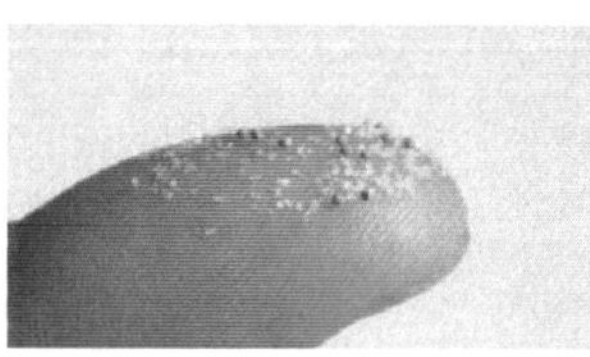

Definition Microplastics

- **Definition by EFSA** (European Food Safety Authority) [4]:

  "For this statement, they are defined as a heterogeneous mixture of differently shaped materials referred to as fragments, fibres, spheroids, granules, pellets, flakes or beads, in the range of **0.1 - 5,000 µm.** A distinction can be made between primary and secondary microplastics. Primary microplastics are plastics that were originally manufactured to be that size while secondary microplastics originate from fragmentation of larger items, e.g. plastic debris."

- **Bundesinstitut für Risikobewertung** (BfR) [5]

  Die Größenangaben für Mikroplastik sind in der Literatur nicht einheitlich definiert und schwanken meist zwischen **0,001 mm bis kleiner als 5 mm**. (varying definitions in the literature between **0.001 – 5 mm (1 – 5,000 µm)**)

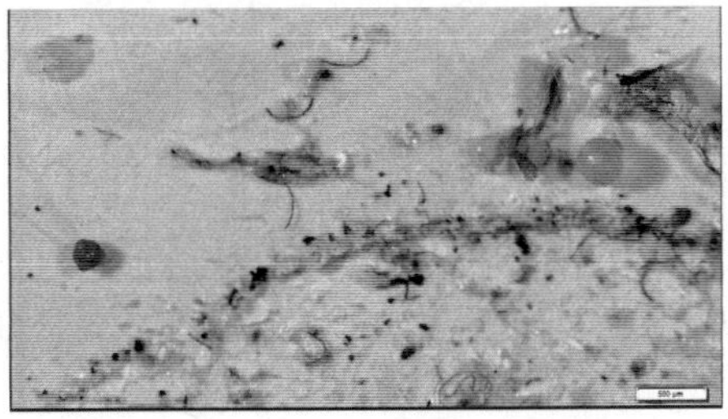

PALL Pall Corporation

Toxicology of Microplastics

- **Statement of the EFSA regarding toxicology from 2016** [6]:
- Hazards can origin from
  - particles themselves (form, surface)
  - additives (extractables)
  - adhering contaminants
  - microbial contaminations
- Lack of experimental data, toxicological risk stays unclear at this time
- No limits for daily intake so far

PALL Pall Corporation

Key Literature

Liebezeit, G. and E. Liebezeit[7]

Synthetic particles as contaminations in German beers. Food additives & Contaminants: Part A,2014,31(9):p. 1574-8

Schymanski, Darena et al.[8]

Analysis of microplastics in water by micro-Raman spectroscopy: release of plastic partciles from different packaging into mineral water, Water Research 129 (2018) 154-162

Oßmann et al.[9]

Small-sized microplastics and pigmented particles in bottled mineral water,

Water Research 141 (2018) 307-316

7

PALL Pall Corporation

Microplastics in Beverages

- Dr. Nils Rettberg´s presentation at the VLB Tagung 2019
  Microplastics Analysis - a review of current workflows and challenges[10]

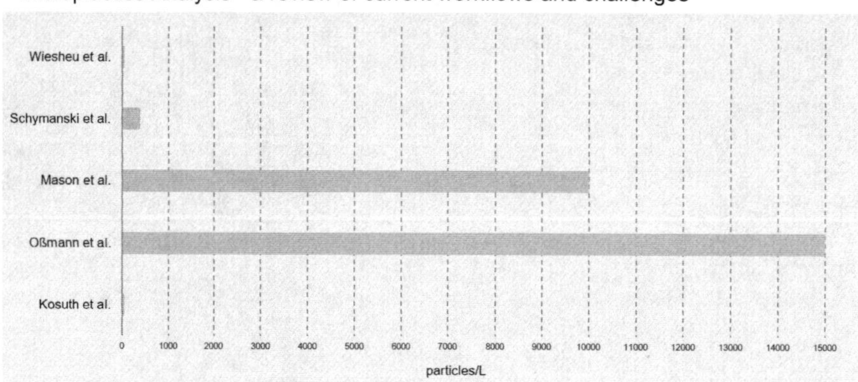

- Maximal particle concentrations reported in each of the studies. Please note that the studies are varying in the nature/origin of sample material as well as in the analytical procedure employed
- „The lower the minimal parcticle size studied, the higher the number of particles detected"

8

PALL Pall Corporation

Facts on Filtration – Three Typical Questions

- Can particles be found in filtered product?

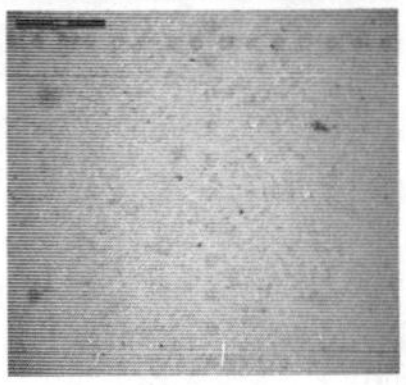

- Do filters add particles?

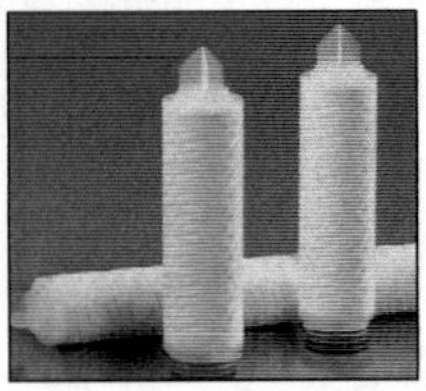

- Which filters remove particles?

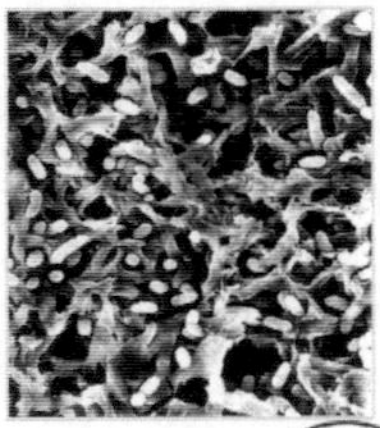

[11] Osumi et al., *PDA J. Pharma. Sci. & Technol.*, *v.* 50: pp. 30-34, (1996)

PALL Pall Corporation

Can particles be found in filtered product?

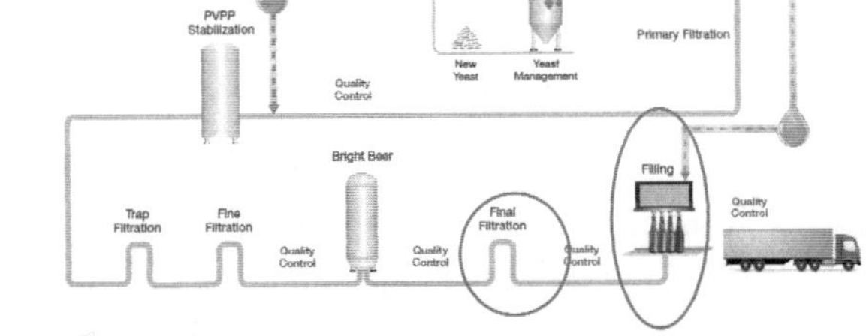

Several sources for particles

- Bottles, Kegs, containments
- Single use or re-usable
- Piping, valves
- Filling machine
- Caps, closures
- Labels
- Cleaning and rinsing solutions and their pre-treatment
- Service media used like water, CO_2, steam
- Surroundings, environment, air quality

PALL Pall Corporation

Do filters add particles?

- Every surface is to some extend releasing particles
- For years Pall has established routine QC to control particle release of filter cartridges
  - Particles are determined by membrane filtration and manual counting of all particles >5 µm by microscope.
  - New, unused filters are flushed by clean pre-filtered RO water and the total number of particles in the first 20 (!) L is measured.
  - The same is done after an interim flush of 20 L per 10“ element.
  - The LOD (Limit of Detection) of 500 particles per 20 L or 25 particles per L results from a factor of 5.79 used to calculate from the counted particles on a defined part of the total membrane surface to the complete area

20 L flush → analyze

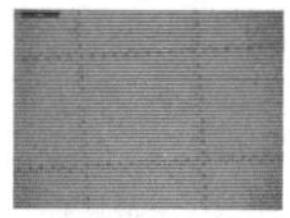

20 L flush → interim, discard

20 L flush → analyze

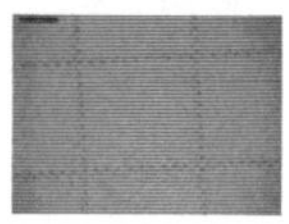

11

Pall Corporation

Do filters add particles?

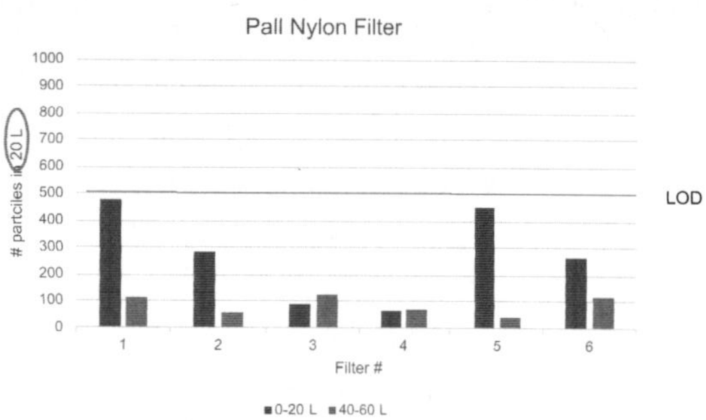

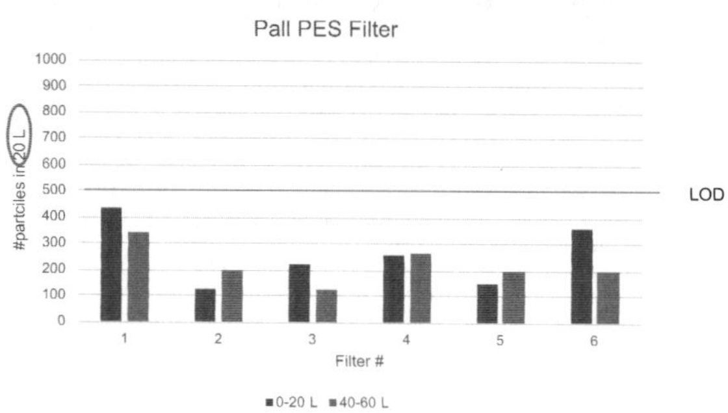

- Numbers of particles from new and unused material is controlled and extremely low
- Flushing reduces particles
- Materials of construction are food contact compliant

12

Pall Corporation

Counting of Particles in Different Matrices

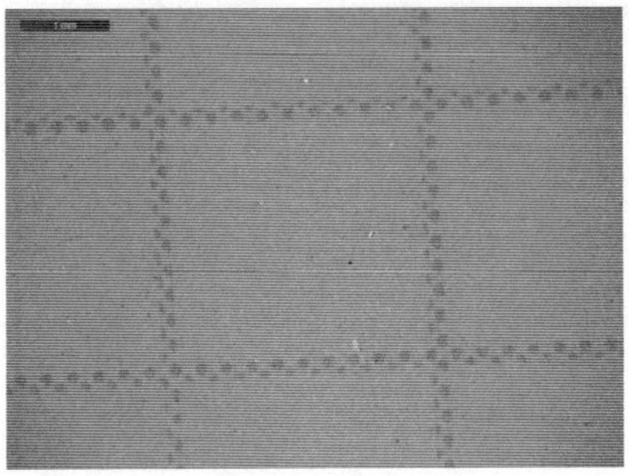

20 Liter pre-filtered RO water

0.8 µm membrane

few particles visible

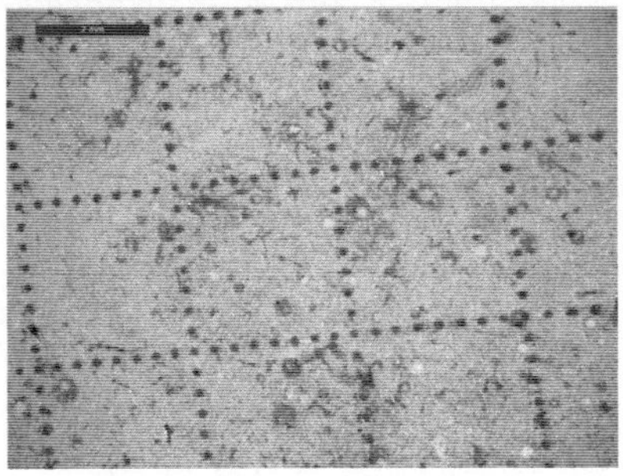

0.5 Liter alcohol free beer

0.8 µm membrane

lots of organic fibre like material

13

PALL Pall Corporation

Counting of Particles

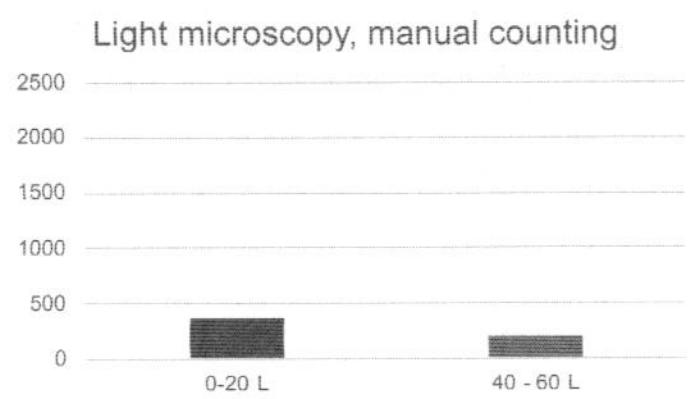

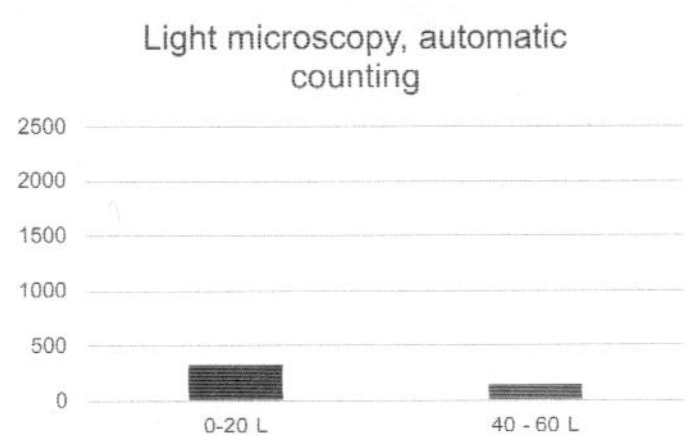

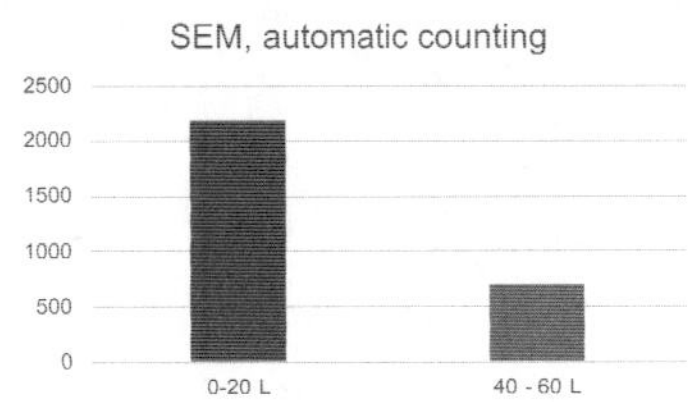

- Manual counting and automatic counting by light microscopy at 100 fold magnification lead to comparable results
- With a higher sensitivity by SEM, the absolute numbers are higher, the more sensitive the method – the more you see
- Particle reduction by flushing can be demonstrated by all methods

14

PALL Pall Corporation

Filtration to Remove Particles

- 1. Pre-filter
  - Relatively coarse matrix structure
  - Validated with test dust
  - Liquid challenge with solid particles
  - ß-value or removal efficiency [%]
  - Reduction circa 1000:1 at µm range
- 2. Final Filter
  - Relatively fine porous membrane structure
  - Validated with Microorganisms
  - Bacterial challenge tests in liquid
  - Titer reduction or Log Reduction Value
  - Reduction circa 100.000.000:1 at sub-µm range

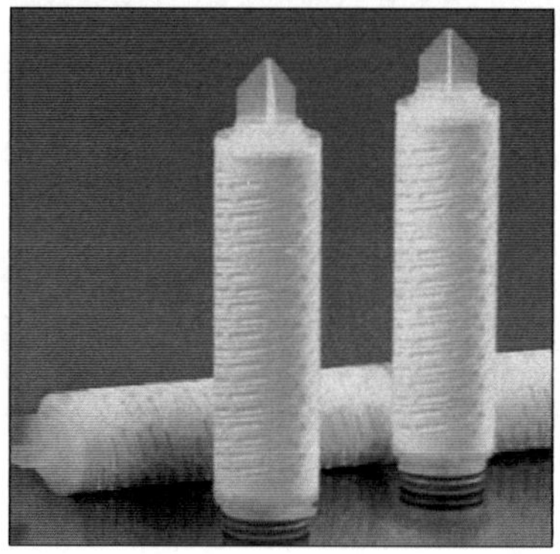

15

PALL Pall Corporation

Retention of Particles by Filter – Pre-filter

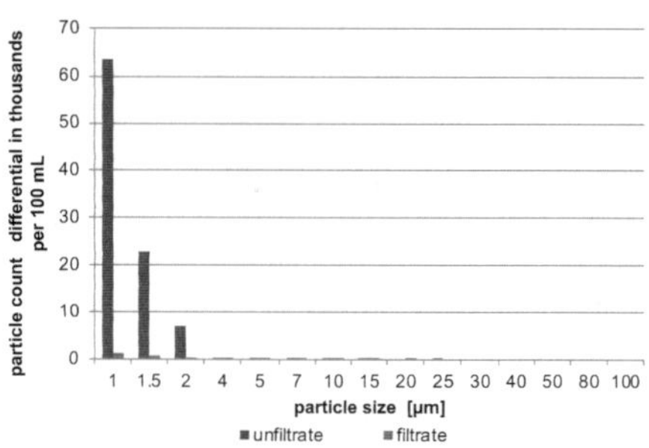

- Particle reduction by a pre-filter with a ß value of 10 for particles ≥ 1 µm or 90 % retention rate

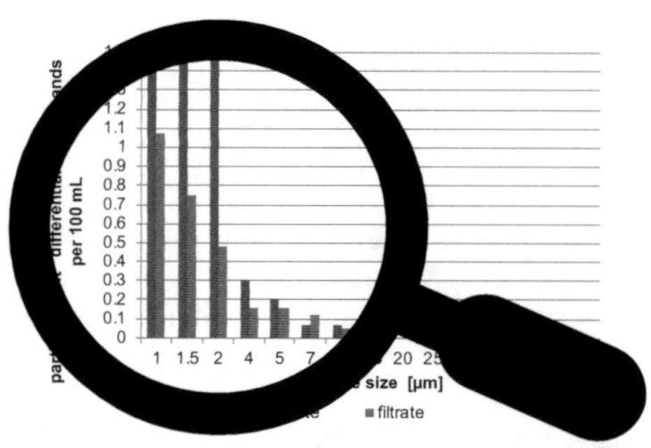

16

PALL Pall Corporation

Filter to Remove Particles – Some Mathematics

LRV	Titer Reduction	Removal Efficiency
1	10:1	90%
2	100:1	99%
3	1.000:1	99,9%
4	10.000:1	99,99%
5	100.000:1	99,999%
6	1.000.000:1	99,9999%
7	10.000.000:1	99,99999%
8	100.000.000:1	99,999999%
9	1.000.000.000:1	99,9999999%

17

Pall Corporation

Summary

- Plastic particles and residuals in food and beverages are a topic of high public interest
- No standard methods exist nor for sample preparation neither for counting and identification
- Toxicology of microplastics unclear, no limits exist so far
- Process steps after final filtration can contribute to the overall load of particles in the final product
- Amount and type of particles from polymeric filters are well controlled
- Flushing prior use reduces particle release during operation
- Methods to measure the reduction of microorganisms by filters can be used to conclude on micro particles (microorganisms = particles)
- Fine membrane filters reduce particles with high efficiency as they do with microorganisms
- The food and beverage industry would benefit from a fact driven shared position on microplastics

18

Pall Corporation

Literature

- 1. Plastik in Bier – das Reinheitsgebot war einmal, Daniel Beuschel, 24.08.2018, www.wir-essen-gesund.de
- 2. Mikroplastik in Mineralwasser und Bier, www.ndr.de/Ratgeber/Verbraucher, Sendedatum 02.02.2014,20:15
- 3. Mikroplastik in Bier und Mineralwässern gefunden, www.stern.de/gesundheit/ernährung vom 02.Juni 2014
- 4. Presence of microplastics and nanoplastics in food, with particular focus on seafood, EFSA Panel on Contaminants in the Food Chain (CONTAM), EFSA Journal, 2016. 14(6)
- 5. Mikroplastik in Lebensmitteln, Stellungnahme Nr. 013/2015 des BfR vom 30. April 2015
- 6. Presence of microplastics and nanoplastics in food, with particular focus on seafood, EFSA Panel on Contaminants in the Food Chain (CONTAM), EFSA Journal, 2016. 14(6)
- 7. Liebezeit, G. and E. Liebezeit, Synthetic particles as contaminations in German beers. Food additives & Contaminants: Part A,2014,31(9):p. 1574-8
- 8. Schymanski, Darena et al.,Analysis of microplastics in water by micro-Raman spectroscopy: release of plastic particles from different packaging into mineral water, Water Research 129 (2018) 154-162
- 9. Oßmann et al.,Small-sized microplastics and pigmented particles in bottled mineral water, Water Research 141 (2018) 307-316
- 10. Dr. Nils Rettberg, Microplastics Analysis- a review of current workflows and challenges, VLB Tagung 2019, Rust
- 11. Stellungnahme Deutscher Brauer-Bund vom 03. Juni 2014
- 12. Mikroplastik im Menschen nachgewiesen, www.kunststoffe.de/news vom 26.10.2018

19

PALL Pall Corporation

Thank you for your attention!
Questions?

20

PALL Pall Corporation

![Annex]

Annex

PALL Pall Corporation

6 BfR-Verbrauchermonitor | August 2018

Haben Sie von den folgenden Gesundheits- und Verbraucherthemen bereits gehört, oder haben Sie davon noch nicht gehört?

BfR-Verbrauchermonitor | August 2018 7

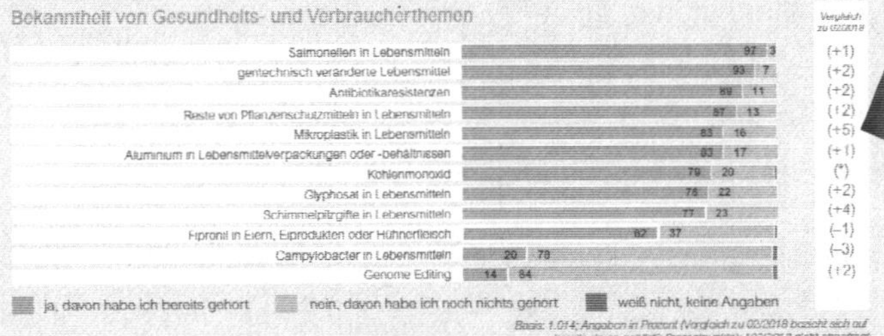

Basis: 1.014; Angaben in Prozent (Vergleich zu 02/2018 bezieht sich auf „bereits davon gehört": Prozentpunkte); *02/2018 nicht abgefragt

PALL Pall Corporation

Retention of Particles by Filter – Pre-filter

Beta Test Stand

Modified OSU F-2, ISO 4572, ANSI B.93.31-1973
(Single Pass)

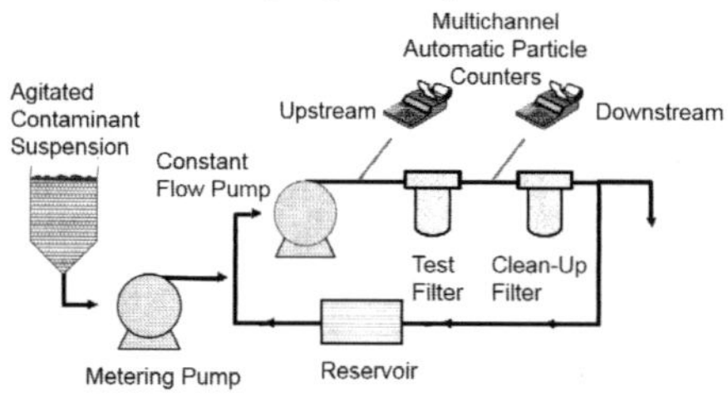

β is the ratio of the number of particles by count greater than a given size in the influent compared to those of the same size or larger in the effluent.

$$\beta_x = \frac{\text{Number of particles} \geq x \quad \text{IN}}{\text{Number of particles} \geq x \quad \text{OUT}}$$

$$\text{\% Removal Efficiency} = \frac{\beta_x - 1}{\beta_x} \times 100$$

23

PALL Pall Corporation

Retention of Particles/ Microorganisms – Final Filter

- Bacterial challenge test as a perfect test to determine very high reduction levels due to unique detection of a single microorganism in large volumes of a fluid

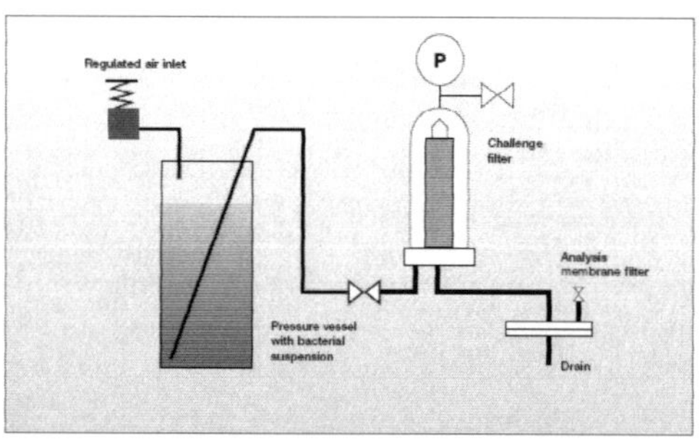

- The titer reduction for a filter is calculated as

$$\frac{\text{Number of organisms upstream}}{\text{Number of organisms downstream}}$$

- The LRV (Logarithmic Reduction Value) for each filter is defined as:

$$\log \frac{\text{Number of organisms upstream}}{\text{Number of organisms downstream}}$$

24

PALL Pall Corporation

L38 - FUNCTIONAL PREDICTION OF BEER CROSS-FLOW MEMBRANE FILTRATION EFFECIENCY FROM MALT QUALITY

Evan Evans[1]; Claudio Cornaggia[2]; Claudio Anna[2]; David Mangan[2]

1 - TBD Consulting and School of Natural Science, University of Tasmania; 2 - Megazyme, Ireland

Abstract

Cross-flow membrane technology (CFM) is increasingly being adopted by brewers for bright beer production, either replacing or being used in combination with more traditional methods, such as diatomaceous earth filtration (DE). A challenge with CFM is the need for improved control of malt quality for brewing. Current malt quality specifications do not fully inform brewers how a malt batch will perform during beer production. Prof Barry Axcell (SAB/Miller Chief Brewer) clearly identified the case for change, calling for malt quality specifications that describe the functional malt properties for beer quality and production. Brewers currently rely on conventional malt specifications including KI, wort viscosity, and ?-glucan level to predict CFM and DE filtration. However, many other malt-derived components such as arabinoxylans and proteins aren't effectively measured. One approach is to emulate the filtration process by adopting a cause agnostic approach as with the Small-scale Wort Filtration Test (SWIFT). This simple inexpensive test draws wort through a 0.45?m nylon filter, with the greater amount filtered indicating better filtration performance. Alternatively, enzymes assays can be conducted on the key enzymes expected to control filterability such as ?-glucanase, xylanase and cellulase using the Megazyme assays. In this trial consisting of 34 high quality European malts, viscosity was negatively correlated with both SWIFT (r=-0.756) and xylanase activity (r=-0.856). Multiple linear regression analysis (MLR) showed that 86.5% of variation in viscosity was predicted by xylanase, SWIFT, wort ?-glucan and friability. Conversely, MLR showed that 67.6% of variation in viscosity was predicted by viscosity, xylanase and KI.

Short biography

Evan Evans graduated with a B. Agr. Sc. (Hons) in 1986, followed by a Ph.D. in 1990, both at the University of Melbourne. In 1992, he joined the University of Adelaide where he developed his interest in malting barley and brewing. Between 2002 and 2013 he relocated to the University of Tasmania working towards improving malt quality to improve beer quality and the efficiency of the brewing process. Dr Evans is currently serving on the IBD Awards Committee and has authored more than 50 peer reviewd papers on brewing and malt quality issues. In 2005, Dr Evans was made a Fellow of the Institute of Brewing and Distilling.

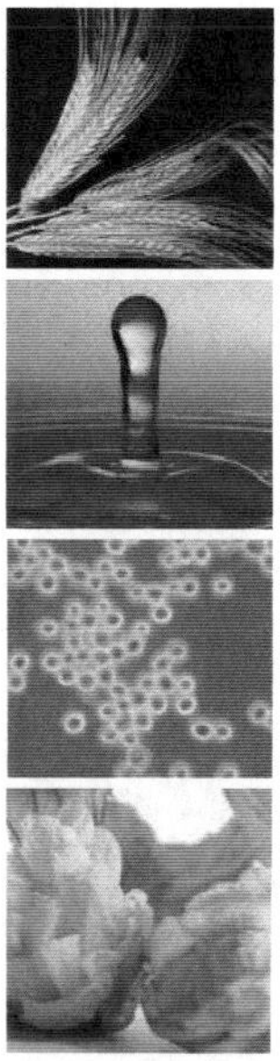

"New Insights into the cleanliness of washed returnable bottles"

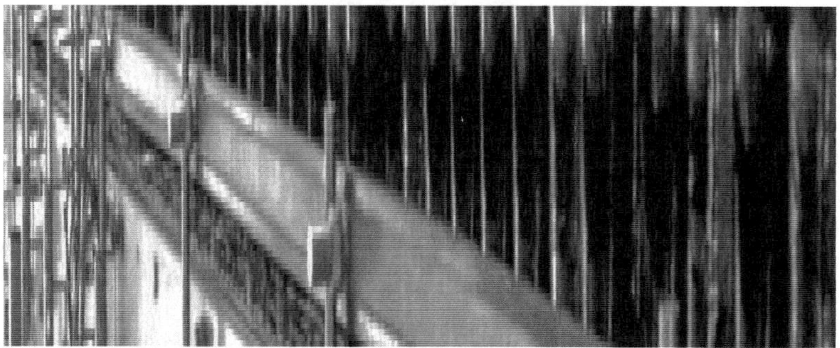

Versuchs- und Lehranstalt für Brauerei in Berlin (VLB) e.V.

Dr. R. Pahl

Dr. A. Ahrens and M. Holewa

Control of cleaning success

- Chemical methods
  - Residues of dirt and of cleaning agents
- Microbiological methods
  - No pathogens and product spoilers
- „Technological methods“
  - EBI
  - Customized control for particles
  - ...

Blank Slide

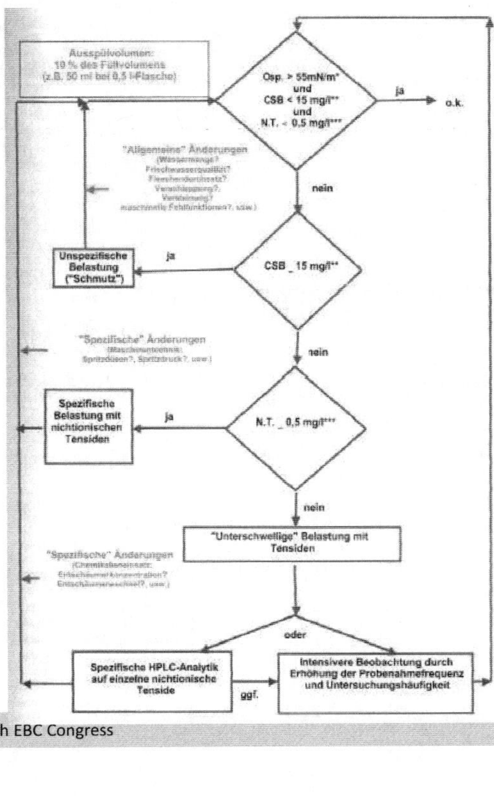
Ausspülvolumen:
10 % des Füllvolumens
(z.B. 50 ml bei 0,5 l-Flasche)
Osp. > 55mN/m* und CSB < 15 mg/l** und N.T. < 0,5 mg/l***
ja
o.k.
nein
"Allgemeine" Änderungen
Unspezifische Belastung ("Schmutz")
ja
CSB _ 15 mg/l**
nein
"Spezifische" Änderungen
Spezifische Belastung mit nichtionischen Tensiden
ja
N.T. _ 0,5 mg/l***
nein
"Unterschwellige" Belastung mit Tensiden
"Spezifische" Änderungen
oder
Spezifische HPLC-Analytik auf einzelne nichtionische Tenside
ggf.
Intensivere Beobachtung durch Erhöhung der Probenahmefrequenz und Untersuchungshäufigkeit

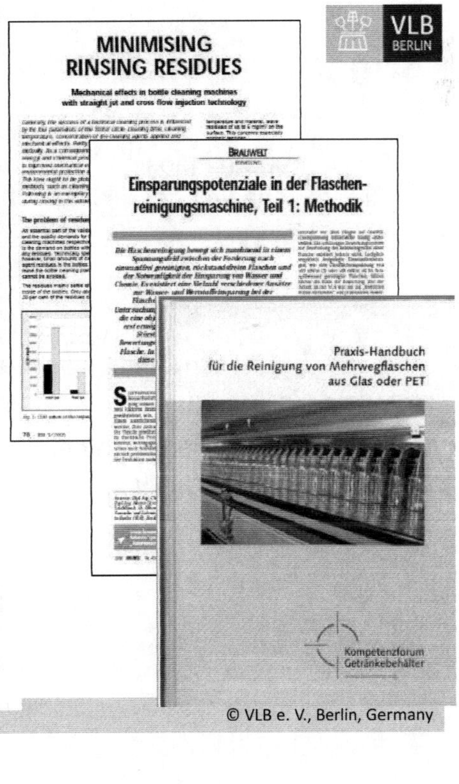
MINIMISING RINSING RESIDUES
Mechanical effects in bottle cleaning machines with straight jet and cross flow injection technology
BRAUWELT
Einsparungspotenziale in der Flaschenreinigungsmaschine, Teil 1: Methodik
Praxis-Handbuch für die Reinigung von Mehrwegflaschen aus Glas oder PET
Kompetenzforum Getränkebehälter

Three parameters

COD:	< 15 mg/l
Surface tension:	> 55 mN/m
Non-ionic tensides:	< 0,5 mg/l

o.k.

Spezifische HPLC-Analytik auf einzelne nichtionische Tenside

ggf.

Intensivere Beobachtung durch Erhöhung der Probenahmefrequenz und Untersuchungshäufigkeit

Trigger for renewed analyses

- Some modern machines have been sold with specifications of the surface tension after rinsing to be > 60 mN/m
- Question to be solved: is the threshold of > 55mN/m still up to date?
- Depends on many influencing factors
  - Age of machine
  - Technology of machine
  - Water consupmtion
  - etc

Methodology

+ Gauge: Krüss - K11 MK4, ring method.
+ Rinsing water: 10% of the bottle, rotate until good wetting of the complete surface.
    + A mapping of the bottles to positions on the conveyor belt of the bottle cleaning machine was not always possible.
+ Individual bottles were used as samples. Repeated (automated) measurement up to the equilibrium position, i.e. stop of the measurement when reaching a standard deviation < 0,1 mN/m
+ Maximum measurement repetition: 20 measurements, i.e. cancellation, also when no standard deviation < 0,1 mN/m from the last five measurements has been reached.
+ Test result: Average of the last five measurements.

Cleaning machines

+ Data collected

- SKUs (Type and volume)
- Year of construction
- One end / double end
- Output [bottles/h]
- Specific water consumption [ml/Flasche]
- Hardness of fresh water

Examples of tested washing machines

Probe	Zuordnung	Bottle type	FRM	Value mN/m	Sta.Dev.mN /m
Fl. 1-3	Anl.1 rechts	0,5 L-Steinie	Manufacturer 1, from 1994 * 40.000 b/h* doubleend * FW 5°GH * 120 ml/b * caustic1: 94 mS/cm * 70 °C	70,9	0,066
Fl. 4-6	Anl.1 links			71,0	0,048
Fl. 7-9	Anl.1 mitte			71,1	0,076
Fl. 10	Anl.1 mitte				
Fl. 11-13	Anl.3 rechts	0,5 L-Individual bottle	Manufacturer 2, from 1987 * 90.000 b/h * doubleend * FW 5°GH * 273 ml/b * caustic: 91 mS/cm * 69 °C	71,3	0,07
Fl. 14-16	Anl.3 links			71,3	0,041
Fl. 17-19	Anl.3 mitte			70,1	0,01
Fl. 20	Anl.3 mitte				

37th EBC Congress

Examples of tested washing machines

Probe	Zuordnung	Bottle	Machine	Value mN/m	Stand.dev. mN/m
Fl.1 = Zelle 1 *	1, Zelle 1, Abgabe rechts	1 L PET	?	57,3	0,138
Fl.2 = Zelle 5 *	1, Zelle 5 Abgabe rechts			64,6	0,116
Fl.3 = Zelle 10	1, Zelle 10, Abgabe rechts			58,6	0,099
Fl.4 = Zelle 15 *	1, Zelle 15, Abgabe rechts			65,3	0,129
Fl.5 = Zelle 20 *	5, Zelle 20, Abgabe mitte			51,9	0,585
Fl.6 = Zelle 25 *	5, Zelle 25, Abgabe mitte			55,6	0,144
Fl.7 = Zelle 28 *	Mischprobe Fl. 7-9: 5, Zelle 28, 32, 36, Abgabe mitte			56,7	0,143
Fl.8 = Zelle 32 *					
Fl.9 = Zelle 36 *					
Fl.10= Zelle 41 *	10, Zelle 41, Abgabe links			54,6	0,118
Fl.1 = Zelle 1 *	1, Zelle 1, Abgabe links	0,5 L Glass Individual bottle	?	57,8	0,176
Fl.2 = Zelle 5 *	1, Zelle 5, Abgabe links			51,9	0,117
Fl.3 = Zelle 10 *	1, Zelle 10, Abgabe links			54,3	0,161
Fl.4 = Zelle 15 *	1, Zelle 15, Abgabe links			57,0	0,173
Fl.5 = Zelle 25	5, Zelle 25, Abgabe mitte			56,4	0,093
Fl.6 = Zelle 30	5, Zelle 30, Abgabe mitte			55,4	0,099
Fl.7 = Zelle 34 *	Mischprobe Fl. 7-9: 5, Zelle 34, 38, 42, Abgabe mitte			59,3	0,154
Fl.8 = Zelle 38 *					
Fl.9 = Zelle 42 *					
Fl.10 = Zelle 48 *	10, Zelle 48, Abgabe rechts			61,1	0,176

37th EBC Congress

Evaluation 10 machines (mean values)

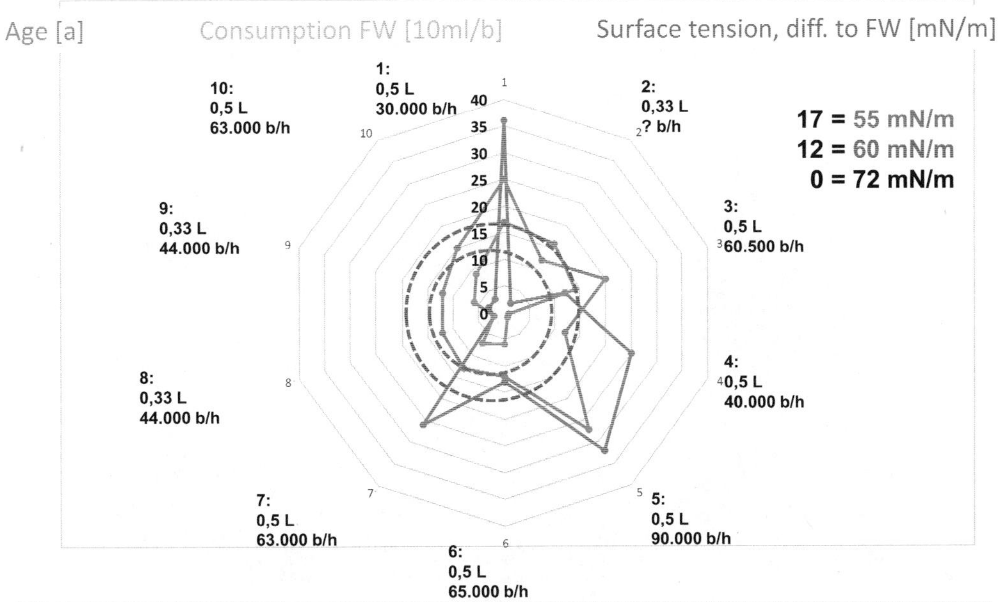

Investigation on particles in washed bottles

- Caustic in a solution is known to enrich with particles
  - Crown dust
  - Paper fibres
  - ...
- Particles that are not rinsed out of bottles can lead to problems
- Treatment of caustic becomes more complex today
  - Glue situation

➡target of investigation: particles >20 µm

Gravimetric determination

- Analoguos to MEBAK-Method for solids in lautered wort

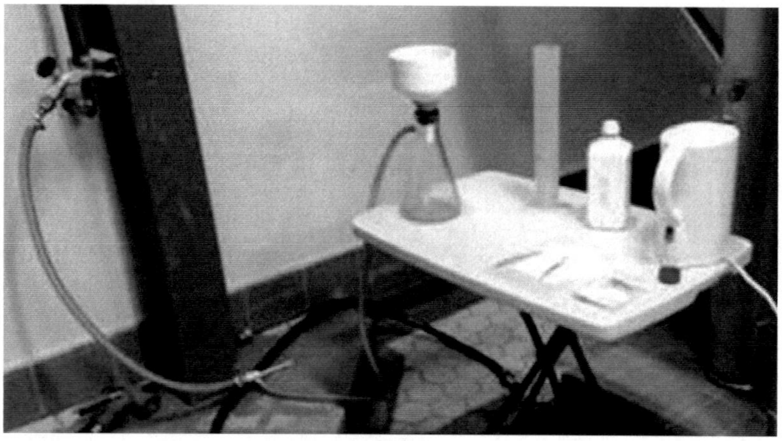

37th EBC Congress

Gravimetry

- Filters with a retention of Particles >20µm
- Single determination
- Clean bench and clean environment

➡Result: blank and main value = zero

➡solution: combining more bottles or more sensitive method

37th EBC Congress

Membrane-filtration + Microscopy

At least 12 bottles per sample

Filter of d=47mm and Poresize 0,2µm

Automated optical scan possible – digital microscope

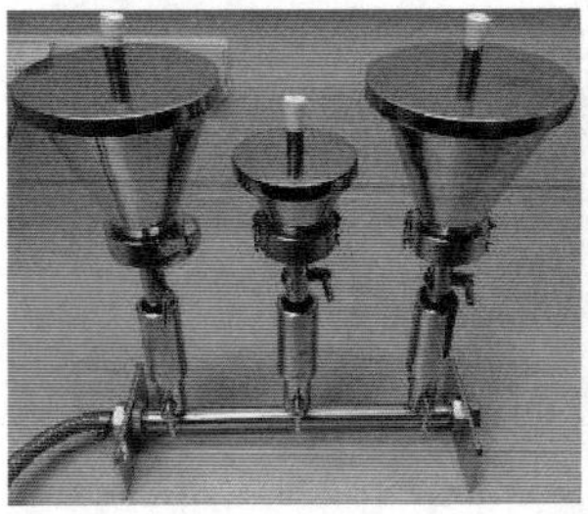

37th EBC Congress

Digital microscope

- High resolution function
  - 18 mega Pixels x 3CMOS
- Magnification > 1500x
- Rendering details on low coulour (contrast) graduation
- 3D display function
- Measuring included
- Shape recognition
- ...

Gefördert durch:

Bundesministerium für Wirtschaft und Energie

aufgrund eines Beschlusses des Deutschen Bundestages

37th EBC Congress

Membrane-filtration + Microscopy

- Example: Blank (left) vs. Sample (right)

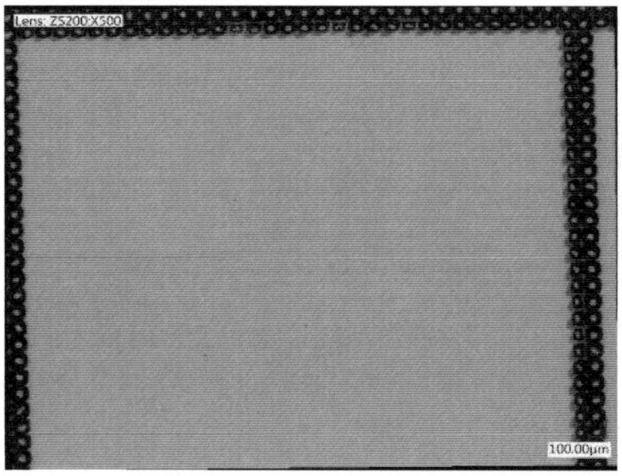

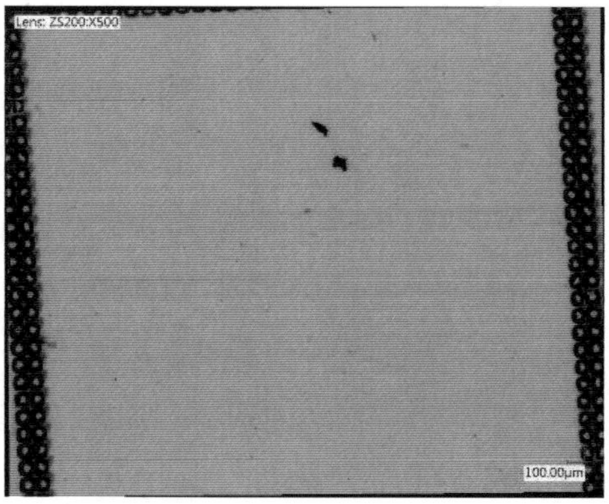

Membrane-filtration + Microscopy

- Particles from cleaned bottles

Membrane-filtration + Microscopy

- Particles on blank values

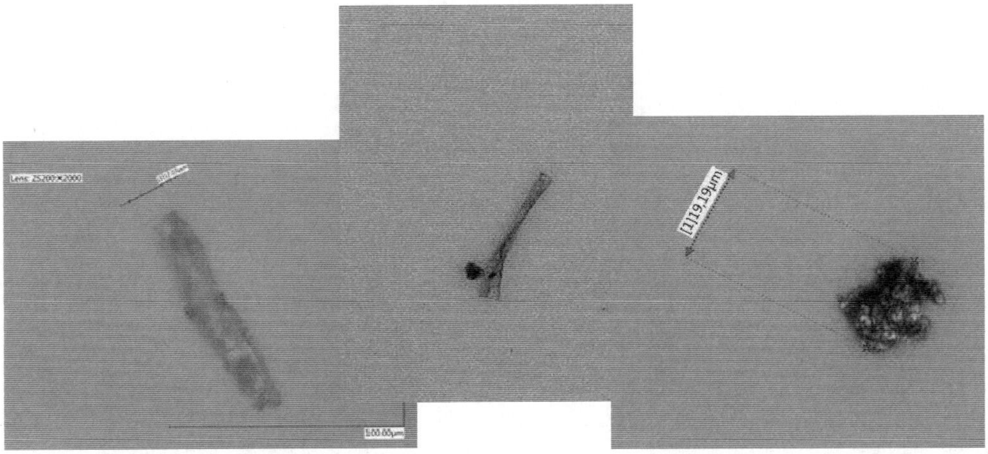

Membrane-filtration + Microscopy

- Automated control
- Diameter and contrast

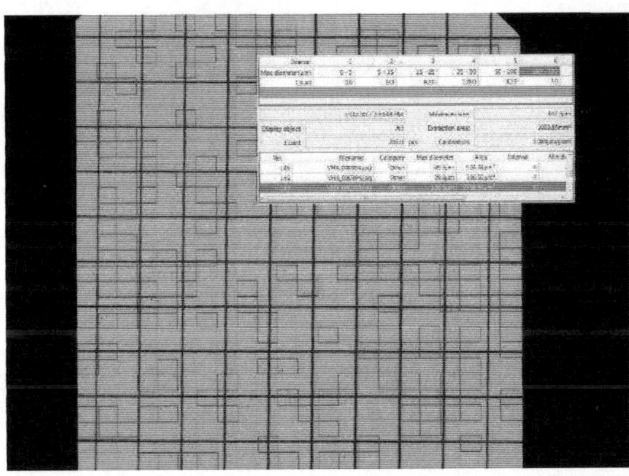

Untreated caustic

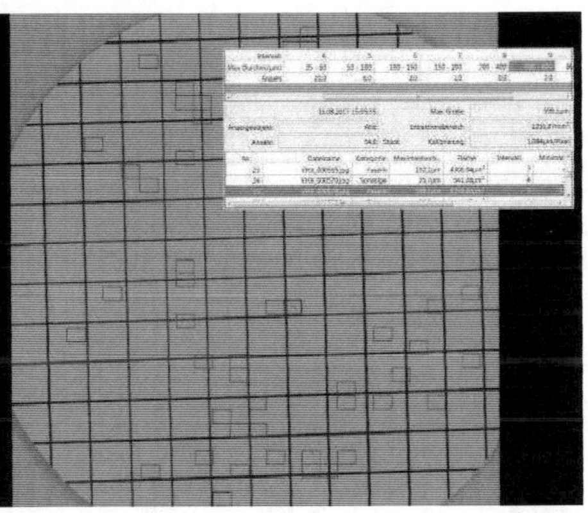

Treated caustic

Membrane-filtration + Microscopy

- Challenges in microscopic detection according to
  - Ambient light and membrane filter used
  - „pattern“ on membrane filter
  - transparent Particles difficult to get

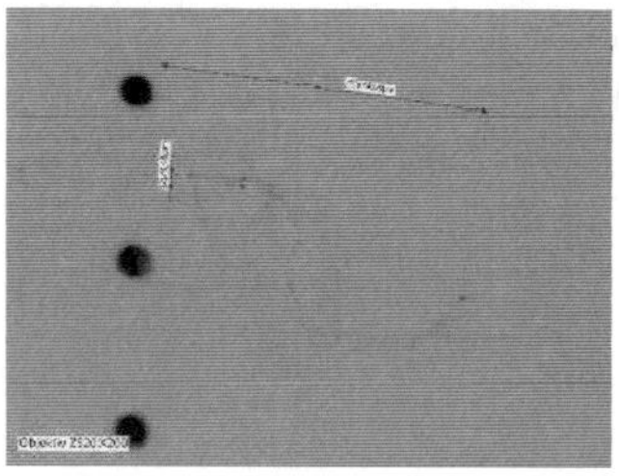

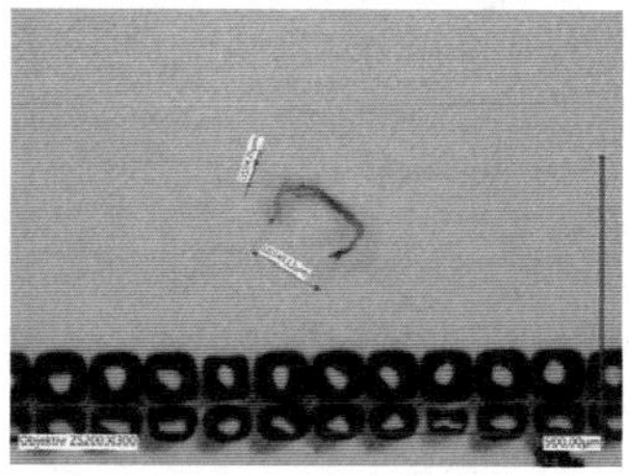

Membrane-filtration + Microscopy

Bachelor Thesis: Validation and optimization

- The „framework“
  - Rinsing water volume, syringe filter
  - Fume hood, ionized air
  - Filters: poresize, material, pattern print, roughness

Membrane-filtration + Microscopy

Bachelor Thesis: Validation and optimization

- Filters best without print-on pattern
- Cellulosenitrate smoother thanMicroglass
- Contrast needed to be adusted

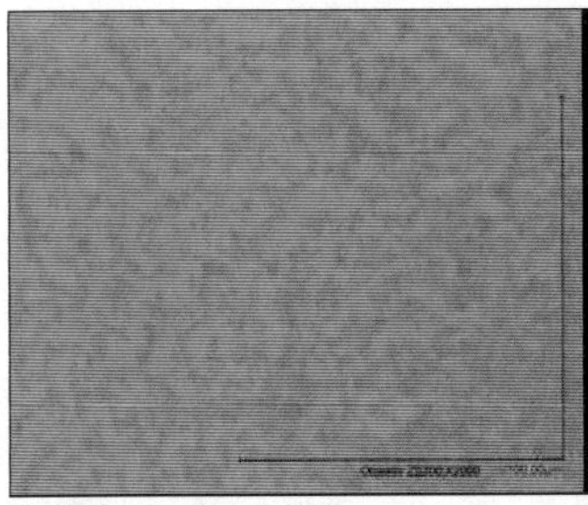

Cellulosenitrat 1,2μm

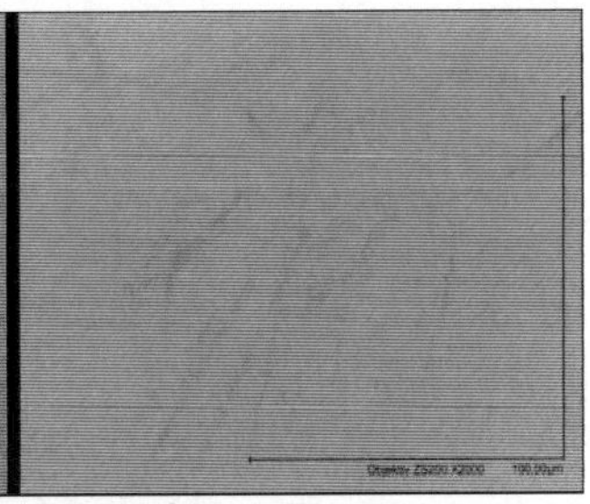

Microglass 1,6μm

37th EBC Congress

Membrane-filtration + Microscopy

Bachelor Thesis: Validation and optimization

- Cellulosenitrate with maximum contrast
- in Microglass particles from filter-material

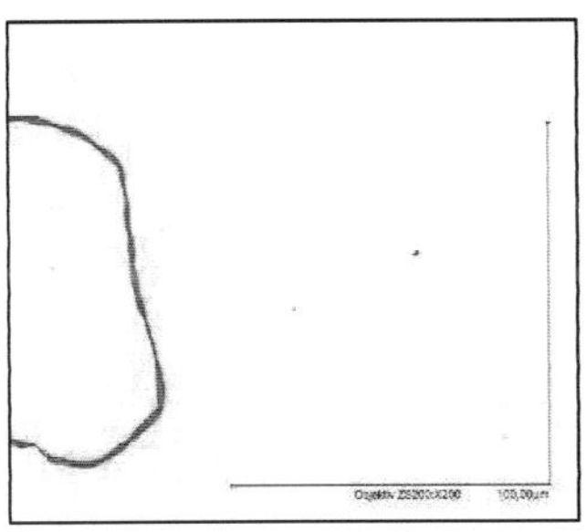

Cellulosenitrat 1,2μm

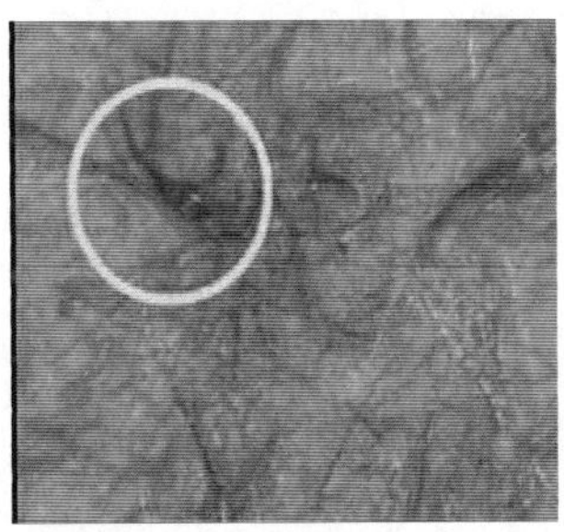

Microglass 1,6μm

37th EBC Congress

Membrane-filtration + Microscopy

Bachelor Thesis: Validation and optimization

- Brewery A:
  - 0,5 L Longneck
  - Sampling from bottle washer discharge
  - 4 samples

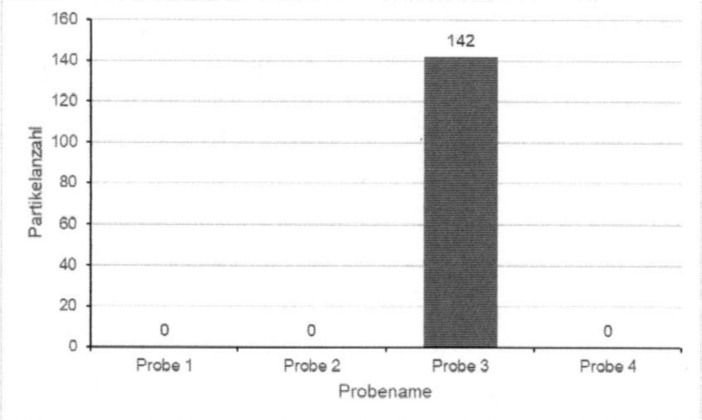

Membrane-filtration + Microscopy

Bachelor Thesis: Validation and optimization

- Brewery B:
  - 0,5 L NRW
  - Sampling from bottle washer discharge
  - triplicates
  - Aluminum foil particles

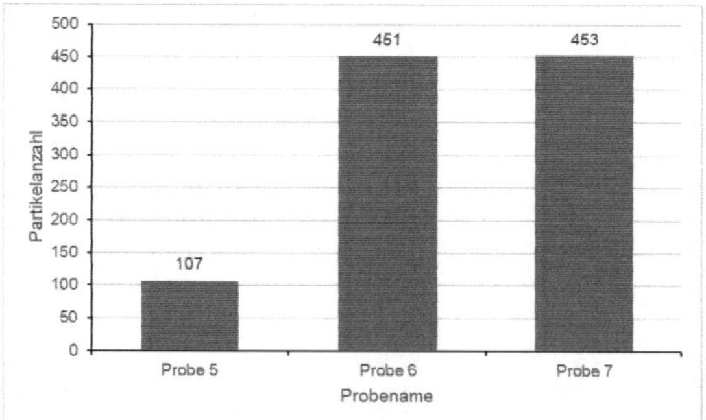

Membrane-filtration + Microscopy

Bachelor Thesis: Validation and optimization

- Brewery C:
  - 0,5 L Longneck
  - Sampling after EBI
  - triplicates

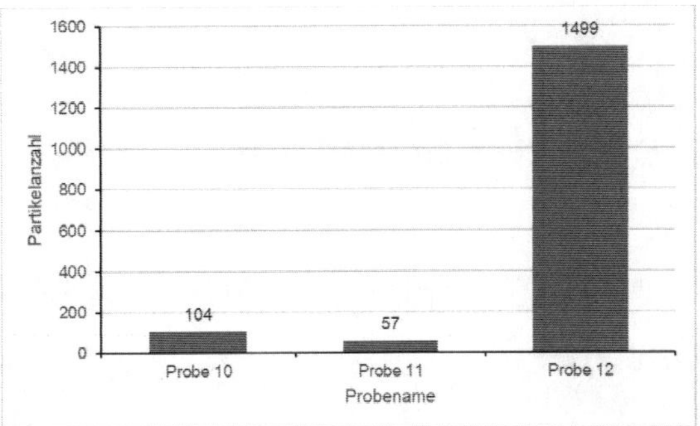

37th EBC Congress

Membrane-filtration + Microscopy

Bachelor Thesis: Validation and optimization

- Brewery C:
  - 0,5 L Longneck
  - Sampling after EBI
  - „problematic empties"

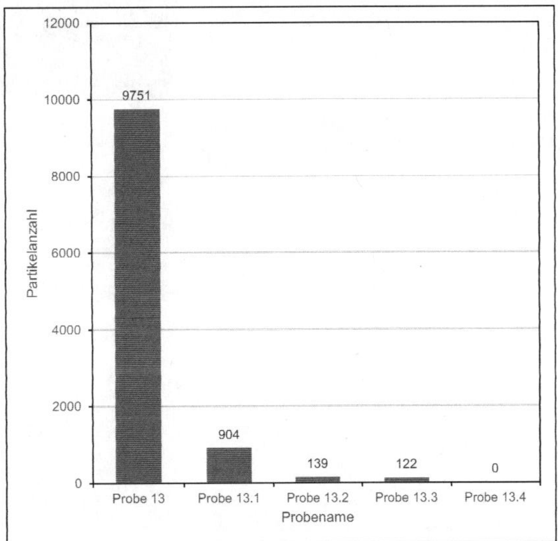

37th EBC Congress

Thanks for the attention!

VLB BERLIN

Acknowledgement for funding of digital microscope

Gefördert durch:

Bundesministerium für Wirtschaft und Energie

aufgrund eines Beschlusses des Deutschen Bundestages

INNO-KOM-Ost

INNO-KOM

IZ

Investitionszuschuss

VLB Berlin

Dr. R. Pahl / Dr. A. Ahrens / M. Holewa

pahl@vlb-berlin.org

www.vlb-berlin.org

37th EBC Congress

L40 - STATUS AND TRENDS IN BEER FILTRATION - A BREWERY MANAGERS' PERSPECTIVE

Stefan Lustig[1]

1 – DBMB (German Brew and Malt-Masters Association, Munich)

Abstract

A new generation of kieselgur-free filtration technologies has been gaining ground in the brewing industry during the last decade. This evolution triggers the neccessity for an overview on established kieselgur systems in comparison to recently developed kieselgur free concepts.

After an introductional view on biochemistry and technical process parameters of filtration, an assessment of parameters to influence filterability in the brewing process will be presented.

The specific process technology of modern filtration systems will be compared. Kieselgur technologies, concepts with alternative filter aids and membrane technologies will be discussed and evaluated in detail. This assessment will include key layout parameters of the systems, performance criteria and economic impact based on operational experience and supplier market information.

Methods to guarantee colloidal shelf life will be discussed in the framework of modern filtration systems as well as the required infrastructure of a state-of-the art filtration area.

An important focus is the role of filtration in the Supply Chain from a brewery managers' perspective of large scale as well as small and medium sized breweries. In today's competitive environment, aspects on health and safety, quality management, financial impact and LEAN/maintenance strategies are important criteria in the responsibility of leaders in the brewing department.

The overview on current filtration technologies will be completed by a quick glance into the role of filtration in craft breweries. An outlook on future developments will conclude this comprehensive review on filtration technology.

Short biography

Stefan Lustig is a brewing engineer graduated fromTU München-Weihenstephan. After completion of his PhD in research on flavor stability under guidance of Prof. Ludwig Narziss, he joined Beck's brewery as head of development & technology.

After several leadership roles in quality, production, packaging and plant management, he joined the board of Anheuser-Busch InBev Germany as Director of Brewery Operations.

Next he moved to Brau-Holding International, Germanys 6thlargest brewing company, a joint venture with Heineken involvement. As Chief Operating Officer and Managing Director he was responsible for Supply, covering a network of 12 breweries and the construction of a 3.5 Mhl greenfield brewery. Currently he is president of "Deutscher Braumeister und Malzmeister Bund", Germanys largest brewmasters association.

Stefan's experience combines know how on brewing science/technology with the operational world of a brewery manager. His expertise covers the supply chain of the brewing industry from raw materials to logistics.

L41 - SULPHITE CONTENT OF BEER : IMPACT OF RAW MATERIAL & FERMENTATION PARAMETERS

Sophie Schwebel[1]; Marc Schmitt[1]

1 – IFBM-Qualtech, Nancy, France

Abstract

Sulfur dioxide is a natural antioxydant. It can be added by the brewer or naturally produced by the yeast. It is identified as an allergen and thus must be labelled if its content is higher than 10 mg/l. In lager beer, the concentration in the final beer must be lower than 20 mg/l. However, beer SO_2 content can vary from one brew to another without any clear explanation for the brewer.

The purpose of this work is to evaluate the impact of raw material and process (fermentation) parameters on SO_2 content in lager beer.

In a first part, the transfer from malt to wort was investigated. In a second part, the influence of fermentation parameters were assessed.

Data collection was performed thanks a design of experiment built in order to study the influence of wort gravity, oxygenation and pitching rate with interactions. Several yeast generations were also tested. The influence of the parameters on fermentation duration and on sulphite production are discussed.

Short biography

She is a research engineer in the research and development department at IFBM.

She received a Higher National Diploma in Technology (DUT) in Biology and Biochemistry from Strasbourg University (2011), and a postgraduate degree in Biotechnology from ESBS (Ecole Supérieure de Biotechnologies de Strasbourg) in 2014.

Today, she participates to the projects management in the barley to beer chain at IFBM (technological issues solving, method development, participation to collaborative projects...).

EBC Congress 2019 5 June 2019

Yeast strain specific release of cystein-conjugated thiols like 4MMP by β-lyase activity

Korbinian Haslbeck

Forschungszentrum Weihenstephan
für Brau- und Lebensmittelqualität

TUM
Technische Universität München

Rising demand for special beers

High consumer interest in beers with special and intense flavors

Different strategies to differentiate/intensify beer flavor:

- new flavor hop breeds
- flavouring methods (dry hopping)
- engineered yeast for brewing?

Reference: Denby et al.: Industrial brewing yeast engineered for the production of primary flavor determinants in hopped beer. Nature Communications 9, Article number: 965 (2018)

- release of flavour precursor from **malt/hop** during **fermentation → monoterpene alcohols and thiols**

Using thiol potential - excursus in wine area

TUM Technische Universität München

4MMP varietal flavor of Sauvignon Blanc

4-<u>m</u>ercapto-4-<u>m</u>ethyl<u>p</u>entan-2-one

grape juice

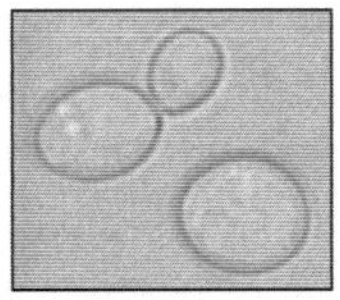

wine

no 4MMP detected | **no 4MMP detected** | **4MMP detected**

<u>2 possibilities</u>:
- *S. cerevisiae* forms 4MMP
- grapes contain 4MMP in bound form

  → **release by *S. cerevisiae* – β-lyase enzymes involved**

Free and bound thiols in hops

TUM Technische Universität München

Thiol	Structure	flavour descriptors	perception threshold in ng/l
3MH	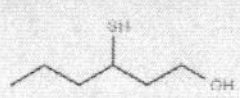	Rhubarb, grapefruit	55
3MHA	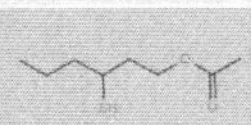	passion fruit, grape fruit	4
4MMP	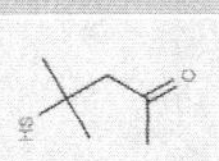	black currant, cassis, catty flavor	1.5
Precursor structure	**Structure**	**Concentration in hops**	
Cysteinylated precursors	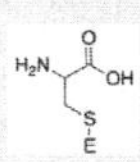	cysteine-3MH 129-1140 ng/g cysteine-4MMP 0-39 ng/g	
Glutathionylated precursors	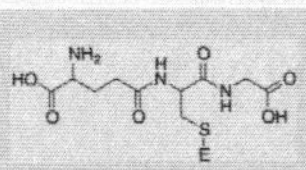	G-3MH 1283-19056 ng/g G-4MMP 0-3.5 ng/g	

<u>3</u>-<u>m</u>ercapto<u>h</u>exanol

<u>3</u>-<u>M</u>ercapto<u>h</u>exyl<u>a</u>cetate

4-<u>m</u>ercapto-4-<u>m</u>ethyl<u>p</u>entan-2-one

Accepted theory: β-lyase substrate in beer fermentation

Reference: C. Schönberger, Brauwelt Nr. 31-32. S 897-900 (2018)
A. Roland, S. Delpech and L. Dagan, BrewingScience, p. 170-175, 2017

Theory using hop thiol potential in beer fermentation

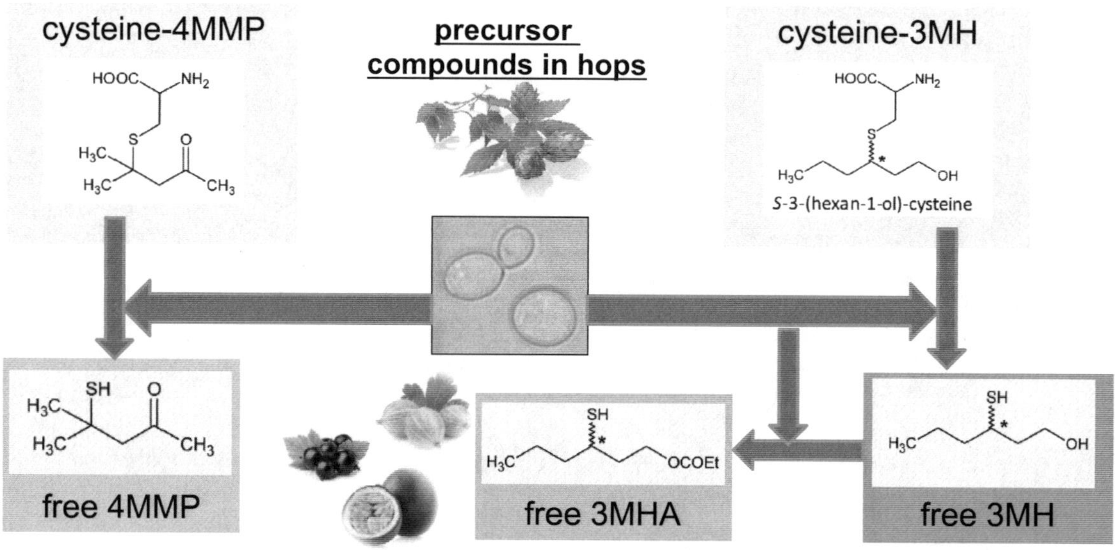

Reference: A. Roland et al. J. Agric. Food Chem. 58, 10684-10689, DOI: 10.1021/jf101996p (2010)

Improvement of release of bound thiols by selection of yeasts with increased β-lyase activity?

Reserch project

Improvement of release of bound thiols by selection of yeasts with increased β-lyase activity?

Working tasks:

1. Screening of yeasts having β-lyases-activity (*IRC7* gene)
2. Rating of β-lyase activity (selective media)
3. Beer fermentation using Mosaic hops

1. Screening *IRC7* gene in brewing yeasts

Theory

Potentially, several different genes responsible for β-lyase activity (→ release of bound thiols)
Genes: *BNA3*, *CYS*, *GLO1*, *IRC7*, *STR3*

<u>*IRC7* most important for release of bound 4MMP</u>

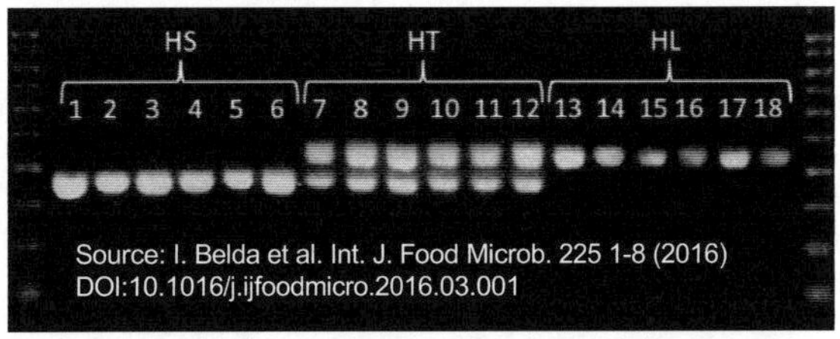

PCR products of the main three IRC7 genotypes
HS = short (181 bp), HT= short and long, HL = long (219 bp)

Test set-up for gene screening

<u>145 strains</u>:
82 top-fermenting
47 bottom-fermenting
15 *T. delbrueckii*

Primer
PF6 (5'-AGCTGGTCTGGAGAAAATGG-3')
PR7 (5'-TCTTCTGCGAGACGTTCAAA 3')
ITdF (5'-AGCTGGGCTCAAGGATATGC-3')
ITdR (5'-GTGACTCTTGAGACGCTCAA-3')

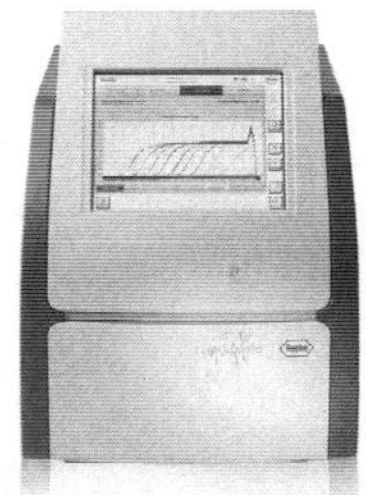

1. Screening *IRC7* gene in brewing yeasts

Results PCR

<u>Flokkulent bottom-fermenting strains</u>

long *IRC7*

short and long *IRC7*

short *IRC7*

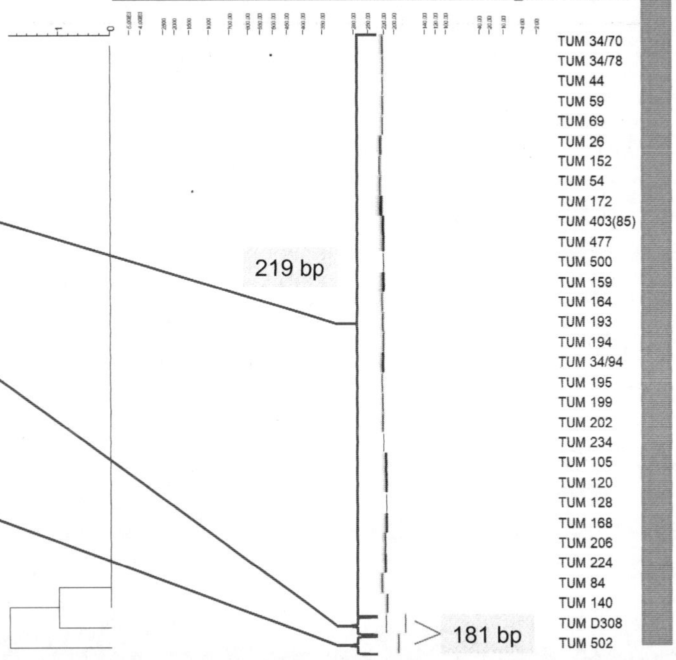

1. Screening IRC7 gene in brewing yeasts

TUM
Technische Universität München

Results

47 Bottom fermenting strains tested:
45 long *IRC7* gene
1 long and short *IRC7* gene
1 short *IRC7* gene

83 Top-fermenting strains tested:
62 long *IRC7* gene
21 long and short *IRC7* gene
- short *IRC7* gene

→
21 Wheat beer strains (90% long, 10 % short and long *IRC7*)
20 Ale (100% long *IRC7*)
23 Wine strains (50% long, 50 % short and long *IRC7*)
10 *S. diastaticus* (30% long, 70% short and long *IRC7*)
9 Others (90% long, 10 % short and long *IRC7*)

15 T. delbrueckii
12 long *IRC7* gene
- long and short *IRC7* gene
- short *IRC7* gene

145 strains tested: 82 % long, 15 % short and long, 1 % short, 2 % without *IRC7*
Saccharomyces* brewing yeasts: 95% long *IRC7

2. Differentiation of yeast strains by β-lyase activity

TUM
Technische Universität München

Selective medium – SMC-medium – based on Belda et al.

S-methyl-L-cysteine (SMC)

Medium with carbohydrate sources but no amino acids except cysteine in bound form

Hypothesis:
If cells grow in medium, yeasts are able to release thiols

Reference: I. Belda et al. Int. J. Food Microb. 225 1-8 (2016) DOI:10.1016/j.ijfoodmicro.2016.03.001

2. Differentiation of yeast strains by β-lyase activity

TUM
Technische Universität München

Set-up

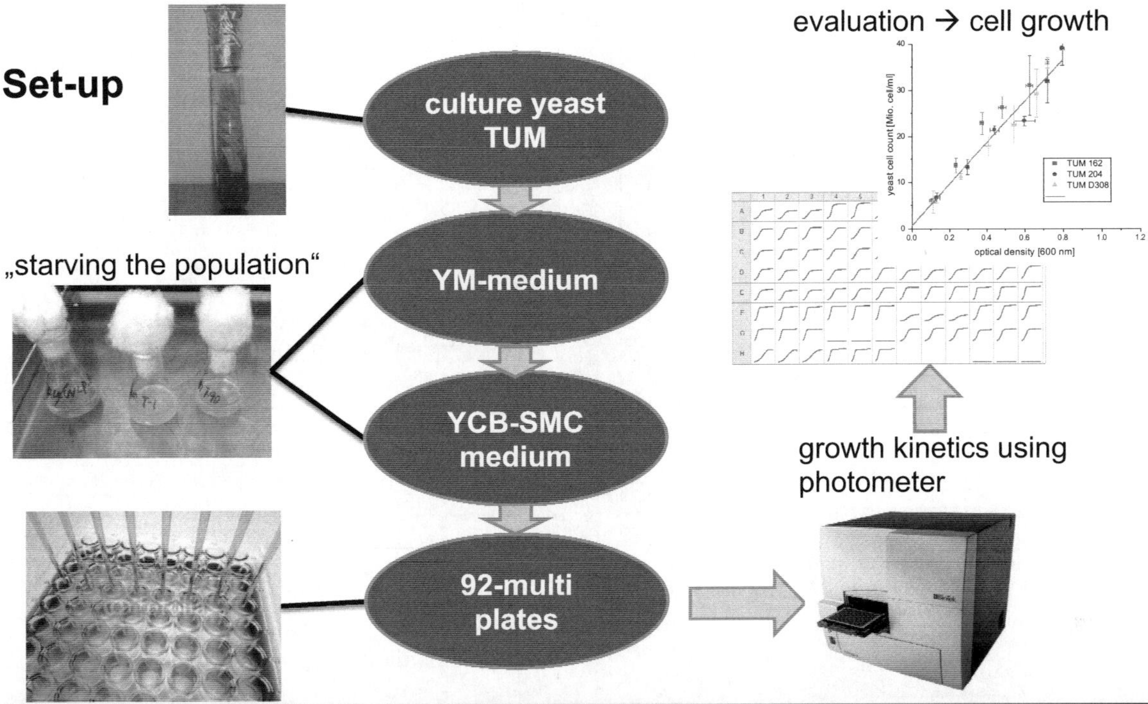

2. Differentiation of yeast strains by β-lyase activity

TUM
Technische Universität München

Results – growth on SMC-media

Diverse growth rates of the *Saccaromyces* yeasts:

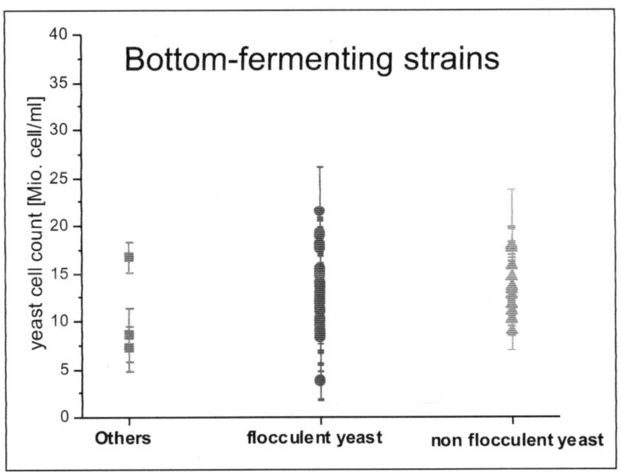

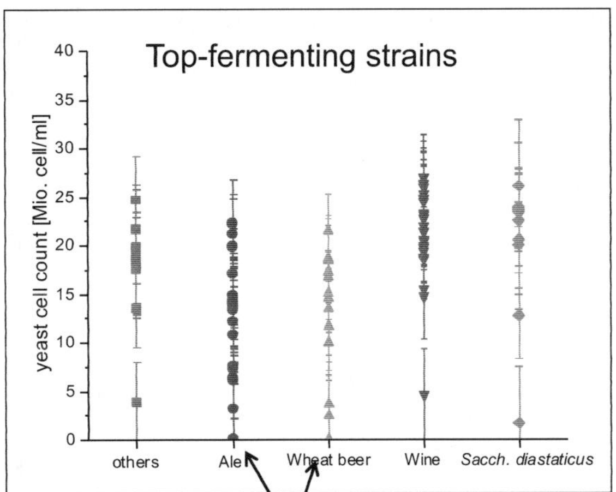

Top-fermenting strains in some cases no growth

Results – growth on SMC-media

Growth rates and yeast species

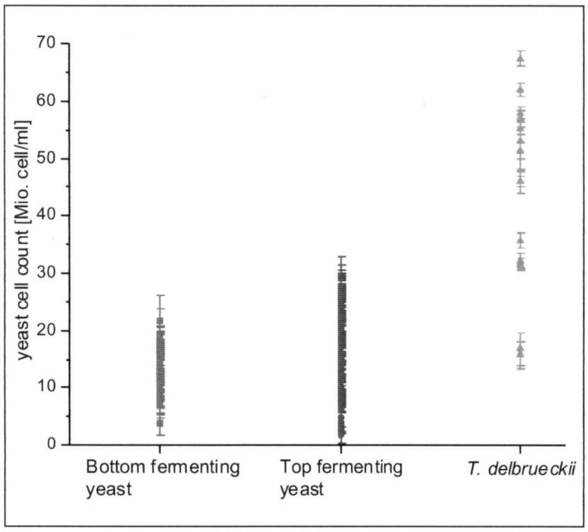

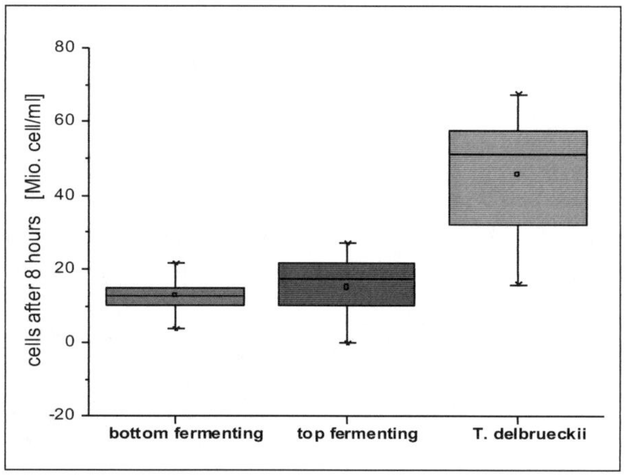

- Bottom-fermenting strains less variable
- Top-fermenting strains on average higher cell counts, however, in some cases no growth
- *T. delbrueckii* highest growth rates

Results and discussion

Combining results task 1 and 2

- Growth rate in the selective medium does not correlate with *IRC7* gene expression

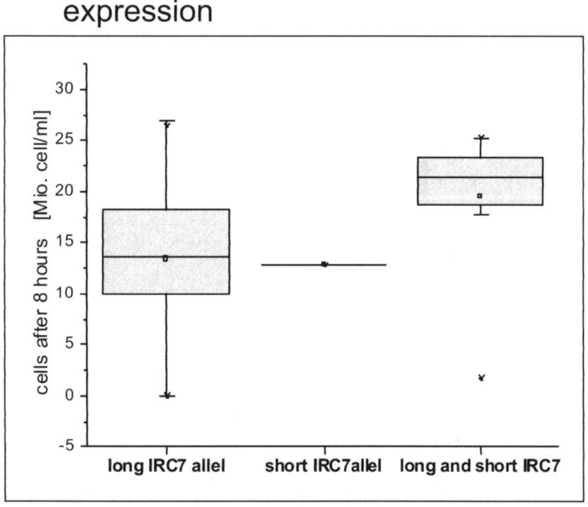

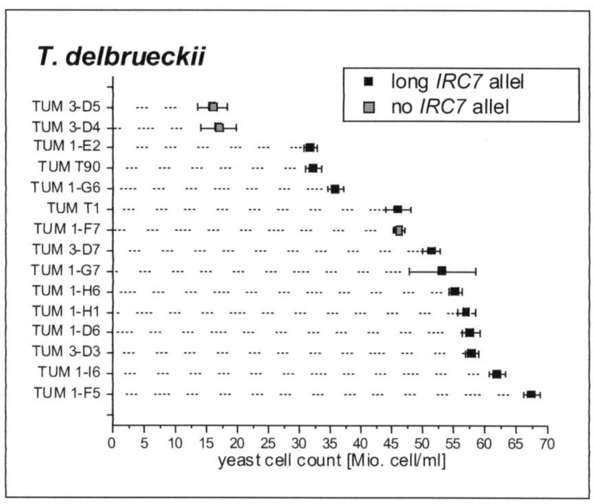

Growth of strains without *IRC7*

→ Further genes next to *IRC7* are involved in yeasts β-lyase activity; candidate: *STR3*

3. Beer fermentation using Mosaic hops

Resume task 1 and 2

Due to the screening results no clear statement is possible

Selection of different strains due to growth in the SMC-medium:

Selected yeasts for brewing trial

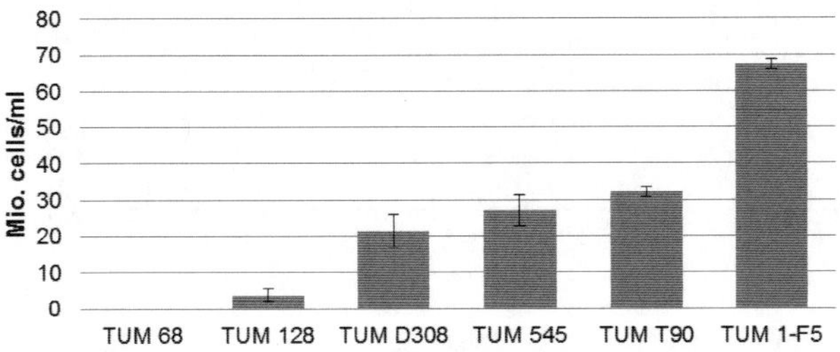

Top-fermenting	TUM 68
Bottom-fermenting	TUM 128
Bottom-fermenting	TUM D308
Top-fermenting	TUM 545
Special	TUM T90
Special	TUM 1-F5

Growth of the various strains in the selective medium for 8 h at 28℃

Hypotheses:

TUM 68 No growth in the selective medium, ie no release of bound thiols.
TUM 128 Low growth rate, moderate release of bound thiols.
TUM D308/545/T90 High growth rate, high release of bound thiols.
TUM T1-F5 Very high growth rate, very high release of bound thiols.

3. Beer fermentation laboratory scale set-up

US-Mosaic

Daughter of YCR 14 Simcoe®brand hop variety and a Nugget derived male

- Crop 2017, pellets type 90
- α-acids 12.1 %
- Total oil 2 ml/100 g

Pear
Gooseberry
Mandarin
Lime
Peach

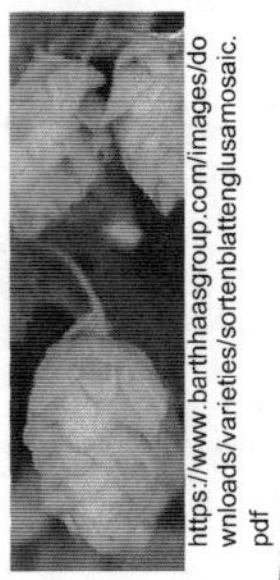

Yeast propagation using unhopped wort

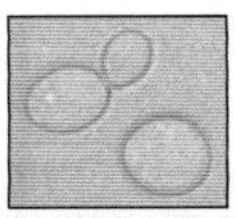

reference

Fermentation 2-l-scale of unhopped wort using 0.75 g/l mosaic, then cold storage 1 week

3. Standard beer analysis

Levels in a normal range
- TUM D308 lower final attenuation
- TUM 1-F5 maltotriose negative

		Alcohol % (v/v)	pH	Extract real % (w/w)
Bottom-fermenting	TUM D308	4.64	4.46	5.66
Bottom-fermenting	TUM 128	5.41	4.45	4.54
Top-fermenting	TUM 68	5.47	4.41	4.47
Top-fermenting	TUM 545	5.68	4.31	4.01
Special	TUM 1-F5	2.4	4.66	8.96
Special	TUM T90	5.07	4.26	3.27

<u>HS-GC-MS of beer samples:</u>

Linalool in samples: 69-76 µg/l
Geraniol in samples: 55-95 µg/l

<u>Several reactions known impact geraniol conc.</u>: Geraniol metabolism
Release of glycosidic bound geraniol
Transformation of geranyl acetate

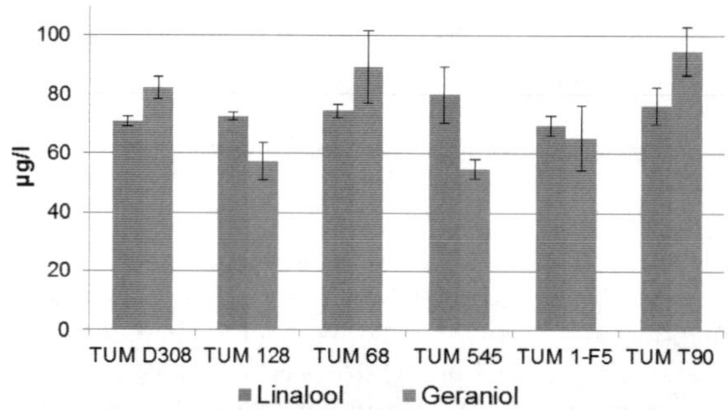

3. Beer analysis

Thiol analysis methods

External Laboratory
nanoLC-MS/MS
free 4MMP, 3MH in hops, wort and final beer
Precursor compounds

Laboratory BLQ TUM

free 4MMP, 3MH in wort and final beer

Agilent 6495B Triple Quadrupole LC-MS/MS

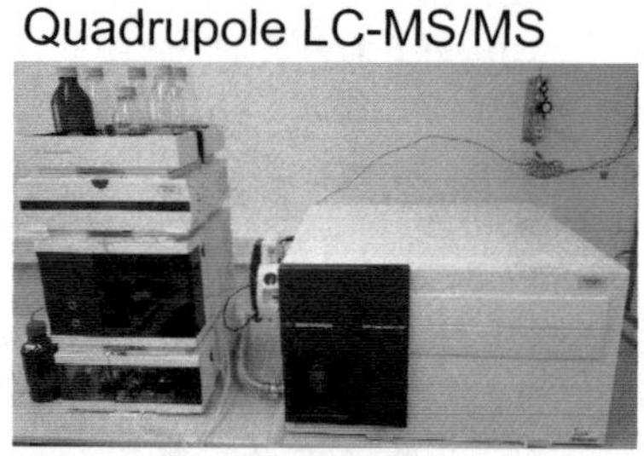

Method based on:
Roland et al.: First identification and quantification of glutathionylated and cysteinylated precursors of 3-mercaptohexan-1-ol and 4-methyl-4-mercaptopentan-2-one in hops (Humulus lupulus). Flavour Frag. J., 2016

3. Beer fermentation using Mosaic hops

Results 4MMP analysis of beer and hop samples

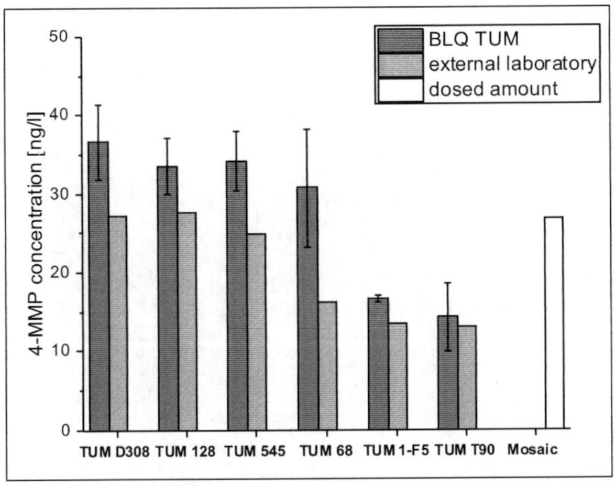

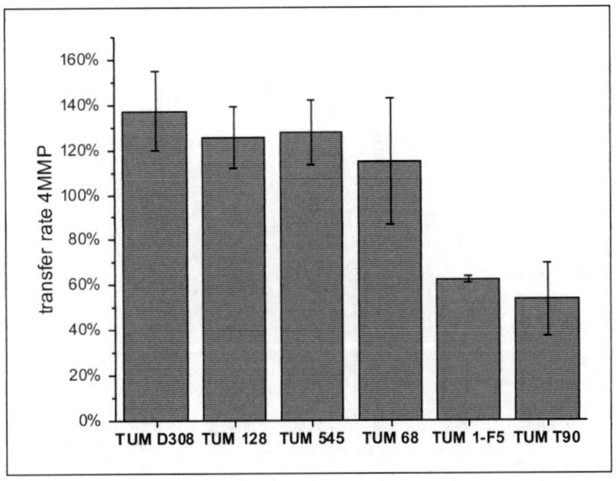

- No 4MMP in blanks
- Dosing 21 ng/l free 4MMP

→ Release of bound 4MMP in case of *Saccharomyces* yeasts assumed

19 Forschungszentrum Weihenstephan für Brau- und Lebensmittelqualität

3. Beer fermentation using Mosaic hops

Results 4MMP-precursor analysis

Tested compounds:

- cysteine-4MMP
- cysteinyl-glycine-4MMP
- γ-glutamyl-cysteine-4MMP
- gluthation-4MMP

No 4MMP-precursor detected in Mosaic hops
No 4MMP-precursor detected in beer samples

Concentration of 4MMP-precursor below detection limits:

- Detection limit precursor analysis: **~0.1 µg/g in hops**
  in beer max. 4MMP-release 16 ng/l (slide 19)
  → detection limit must be at **~12 ng/g in hops**
- Detection limit precursor analysis: **~0.1 µg/l in beer**
  28 ng/g free 4MMP in hops
  → free form most important source of 4MMP in hops*;
  detection limit must be **~ <10 ng/l**

*Roland et al.: First identification and quantification of glutathionylated and cysteinylated precursors of 3-mercaptohexan-1-ol and 4-methyl-4-mercaptopentan-2-one in hops (Humulus lupulus). Flavour Frag. J., 2016

20 Forschungszentrum Weihenstephan für Brau- und Lebensmittelqualität

3. Beer fermentation using Mosaic hops

Technische Universität München

Results 3MH analysis of beer and hop samples

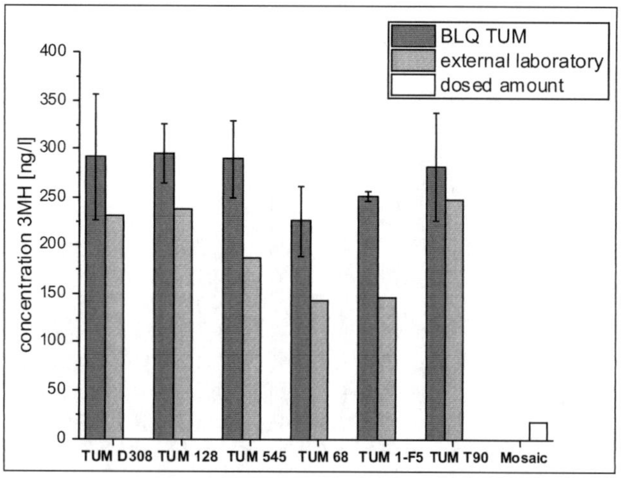

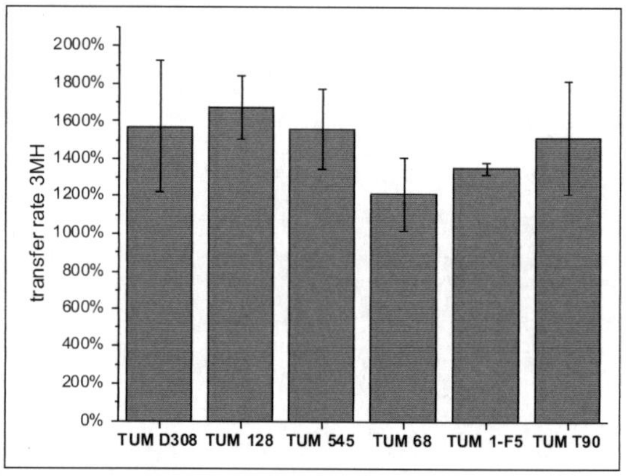

- No 3MH in blanks (exception TUM 128: 16,39 ng/l 3MH)
- Dosing 26 ng/l free 3MH

→ Release of bound 3MH

- No 3MHA detected

3. Results 3MH-precursor analysis

Technische Universität München

Dosed 3MH-precursors (hops):

cysteine-3MH	cysteinyl-glycine-3MH	γ-glutamyl-cysteine-3MH	gluthation-3MH
134 µg/l	**388 µg/l**	**nd**	**2580 µg/l**

Dosing 0.75 g hop pellets per liter

Sum 3MH-precursor: 3.1 mg/l

→0.41 % in hops

→Still potential for 3MH:

~ 1.2 % free 3MH (~270 ng/l) present in beer compared with bound 3MH (~22.2 µg/l)

~ 0.01 % of total dosed 3MH (free + bound) present in free form in beer

more than 99 % of 3MH occurre as precursors in hops; analysis of 10 hop varieties by Roland et al.

Roland et al.: First identification and quantification of glutathionylated and cysteinylated precursors of 3-mercaptohexan-1-ol and 4-methyl-4-mercaptopentan-2-one in hops (Humulus lupulus). Flavour Frag. J., 2016

3. Results 3MH-precursor analysis

Technische Universität München

3MH-precursor in beer samples:

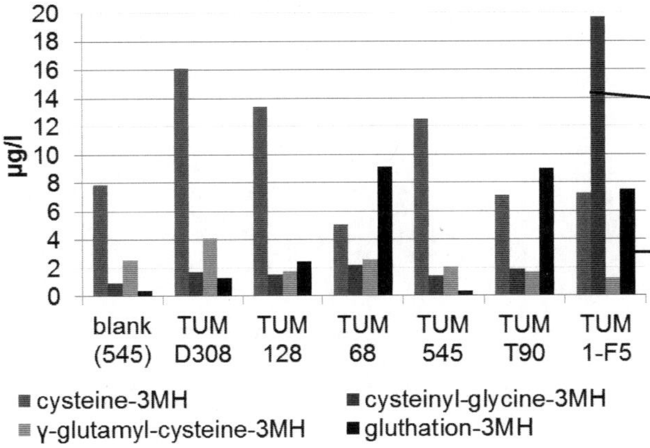

3MH-precursors in malt and hops

TUM 1-F5: Lowest ability to transform cysteinyl-glycine-3MH

TUM 68 and both ***T. delbrueckii*** low ability to transform gluthation-3MH

Conclusion

Technische Universität München

Hop yeast interaction leading to increase of thiol concentration during in beer fermentation

→ Positive influence of yeasts on hoppy beer flavor

→ Still potential for 3MH

~ 0.01 % of total dosed 3MH (free + bound) present in free form in beer

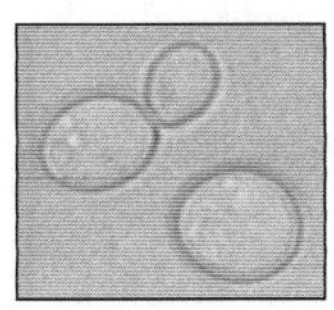

Outlook

- Extend gene screening (*STR3*)
- Influence of hopping timing on thiol release
- Reaction kinetics

Thank you for your attention …

TUM
Technische Universität München

and thanks to:

Dr. Yang Liu
Dr. Maximilian Michel
Dr. Mathias Hutzler
Dr. Martin Zarnkow
Prof. Mehmet Coelhan
Prof. Fritz Jacob
Dr. Christina Schönberger
Aofei Cheng

BARTH-HAAS GROUP
FOR YOUR SUCCESS

Korbinian Haslbeck
Alte Akademie 3
85354 Freising-Weihenstephan

Telefon: +49 (0) 8161 / 71-5744
Telefax: +49 (0) 8161 / 71-4181

E-Mail: korbinian.haslbeck@tum.de
www.blq-weihenstephan.de

Forschungszentrum Weihenstephan
für Brau- und Lebensmittelqualität

TECHNISCHE UNIVERSITÄT DRESDEN

Metabolization of glycated amino acids from wort by different *S. cerevisiae* strains

Michael Hellwig, Anna-Lena Kertsch, Thomas Henle

Chair of Food Chemistry, Technische Universität Dresden, Germany

DRESDEN concept
Exzellenz aus Wissenschaft und Kultur

37th EBC congress, Antwerp, 02-06 june 2019

Introduction – Maillard Reaction products (MRPs)

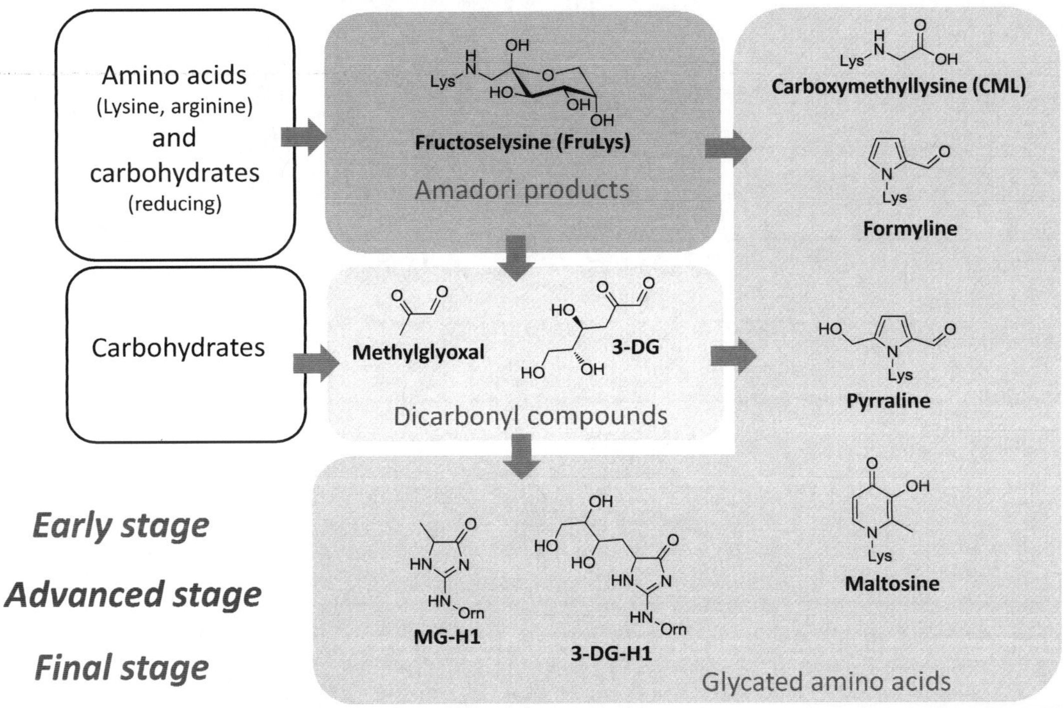

2

The Maillard reaction? – The Ling reaction!

First observation of non-enzymatic browning as dependent on amino acids and sugars

494

J. Inst. Brew., 1908 #1757

MEETING HELD AT THE CRITERION RESTAURANT, PICCADILLY, W., ON MONDAY, MAY 11TH, 1908.

Mr. ALFRED CHASTON CHAPMAN in the Chair.

The following paper was read and discussed :—

Malting.

By ARTHUR R. LING, F.I.C., Lecturer on Brewing and Malting at the Sir John Cass Institute.

The following observations will demonstrate the accuracy of my statement. If ordinary glucose be heated for a short time at about 130° with such a compound as asparagine, combination occurs, without however the production of colour. If the resulting product be then heated in a boiling water bath at 212°, much darkening occurs. No such production of colour is produced by heating the glucose alone. This is quite a parallel case to the production of colour in malt.

Ling (1908) *J. Inst. Brew.*

3

Glycated amino acids in malt

www.barcoo.com

Malt	EBC*	Fructose-lysine	Pyrraline	MG-H1
		[mg/kg]	[mg/kg]	[mg/kg]
Pilsner	6	317	10	3
Wheat	5	251	10	3
Melanoidin	79	978	174	6
Caramel	136	114	412	41
Chocolate	670	n.d.	1	8

* isothermal 65-°C mesh, n = 2.

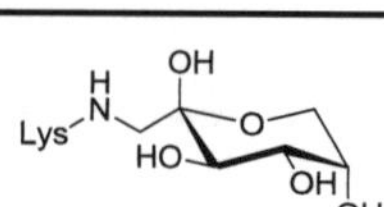

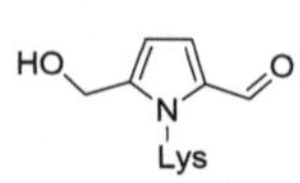

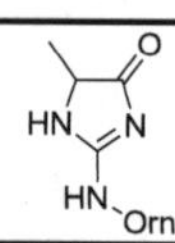

Hellwig, Henle (2019) in preparation

4

Fermentation of amino acids by *S. cerevisiae*

Ehrlich pathway („amino acid fermentation“)

branched-chain and aromatic amino acids

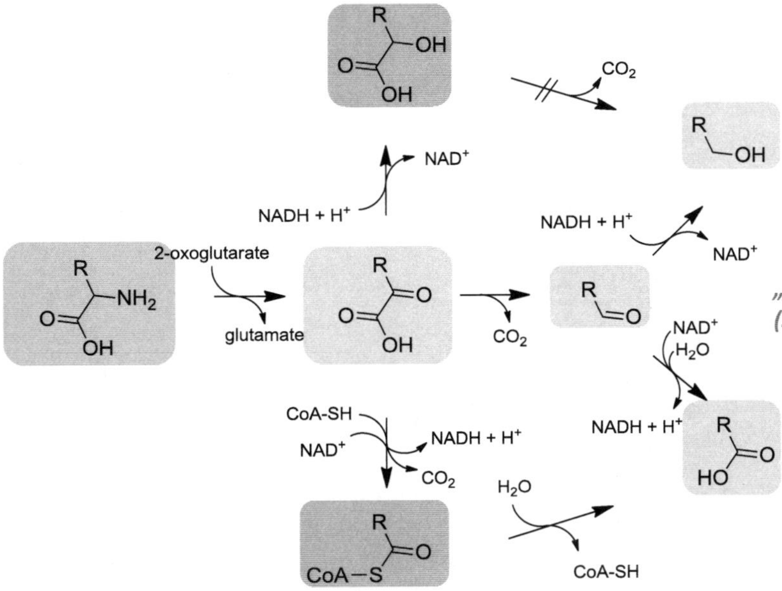

„fusel alcohols“

„fusel aldehydes“ (Strecker aldehydes)

„fusel acids“

Ehrlich, 1907; Hazelwood et al., 2008; Dickinson et al., 1998

5

Fermentation of amino acids by *S. cerevisiae*

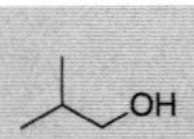

2-Methyl-1-propanol
15-23 mg/L

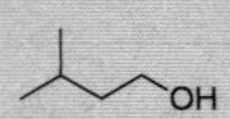

3-Methyl-1-butanol
6-60 mg/L

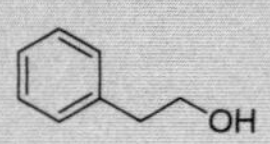

2-Phenylethanol
2-30 mg/L

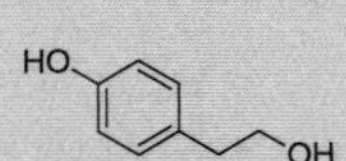

Tyrosol
5-22 mg/L

Ehrlich alcohols in beer

reaction products in a „reductive“ environment

key aroma compounds in fermented foods

no significant contribution to toxicity of alcoholic beverages

Can MRPs be metabolized by brewer‘s yeast?

Langos et al., 2013; Szlavko, 1973; Li et al., 2008; Dunkel et al., 2014

6

Research questions

Can MRPs be metabolized by brewer‘s yeast?

Which metabolites are formed?

Can these metabolites be quantified in beer?

7

Incubation of MRPs in the presence of *S. cerevisiae*

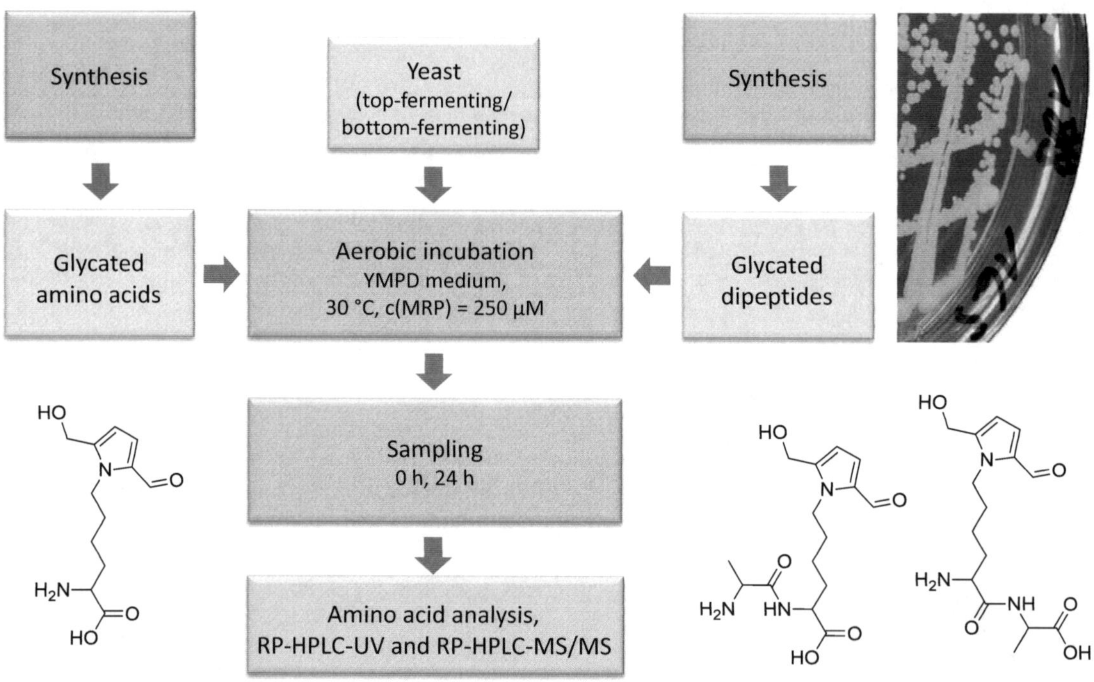

Hellwig, Börner, Beer, van Pée, Henle (2017) ChemBioChem, 18, 266-275.

8

Fermentation of glycated amino acids by *S. cerevisiae*

Identification of extracellular metabolites by HPLC-MS/MS

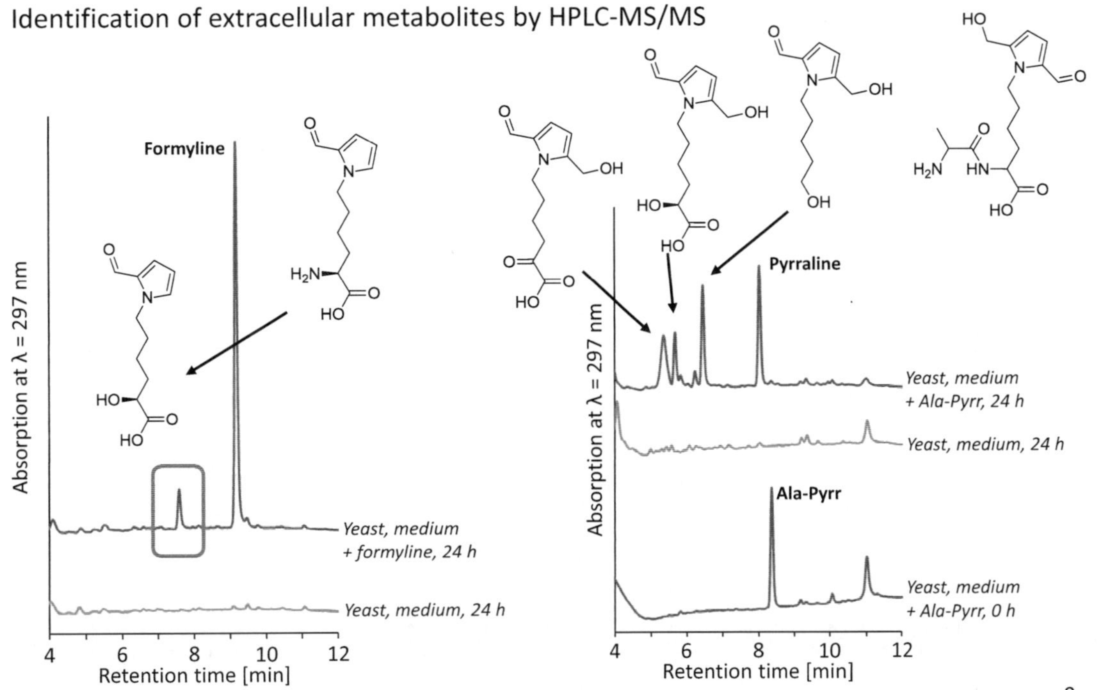

Hellwig, Börner, Beer, van Pée, Henle (2017) ChemBioChem 18, 266-275.

9

Fermentation of glycated amino acids by *S. cerevisiae*

Comparison of Ala-Pyrr metabolization between strains

DSM 1848 (bottom-fermenting) **DSM 70449 (top-fermenting)**

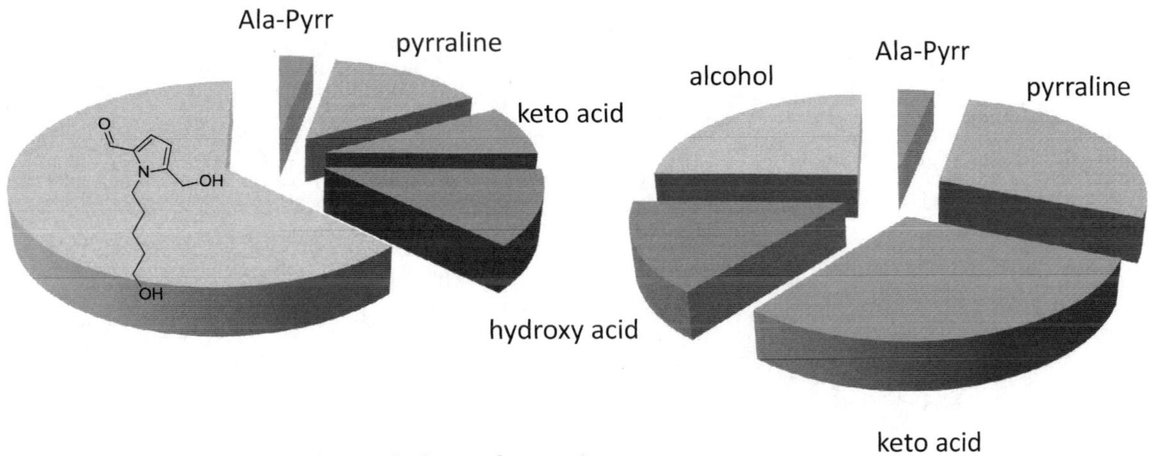

Percentage of individual metabolites after 24 h

Hellwig, Börner, Beer, van Pée, Henle (2017) ChemBioChem, 18, 266-275.

10

Interconversion of metabolites – pyrraline and formyline

Higher product yields in **minimal medium**

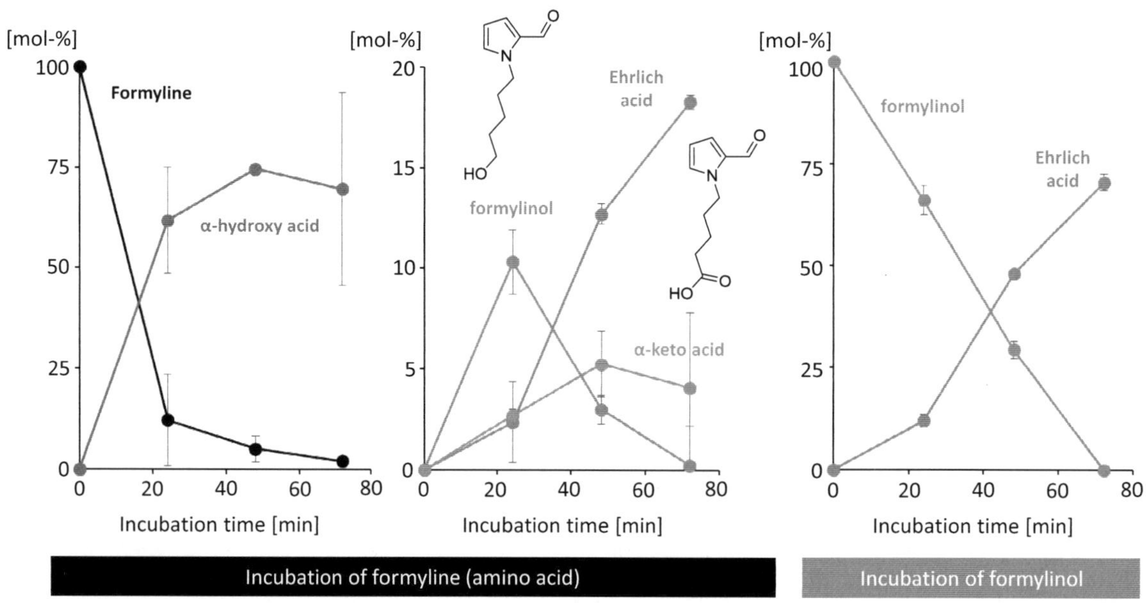

Transformation *vs.* transport

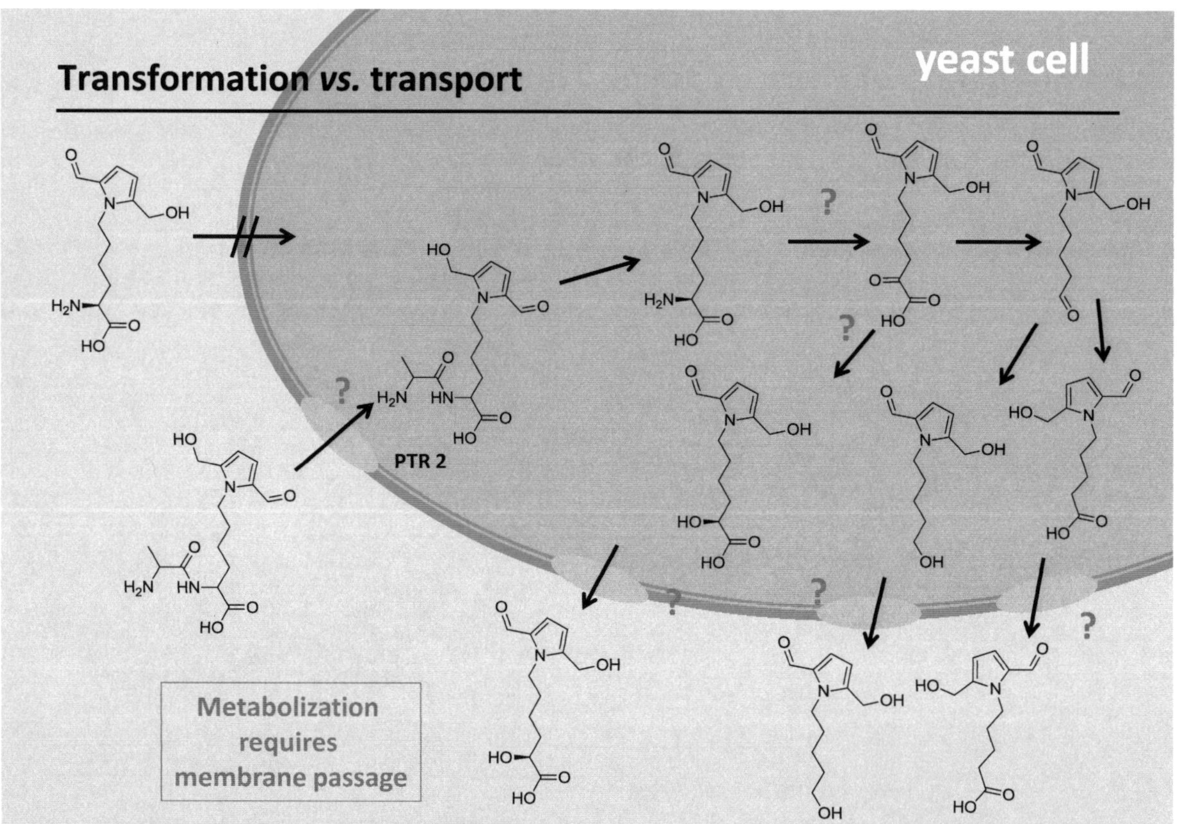

Hellwig, Börner, Beer, van Pée, Henle (2017) ChemBioChem, 18, 266-275.

Higher alcohols in beer

Quantification of Ehrlich alcohols by RP-HPLC-UV

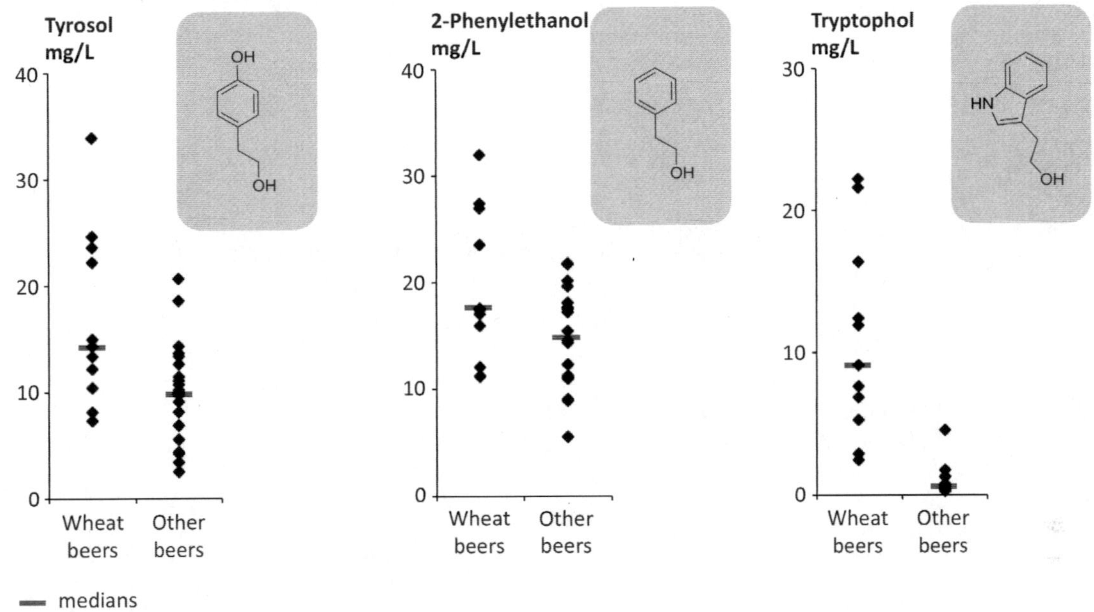

13

Hellwig, Beer, Witte, Henle (2018) J Agric Food Chem 66, 7451-7460.

MRP metabolites in beer

Analysis of MRP metabolites by RP-HPLC-MS/MS

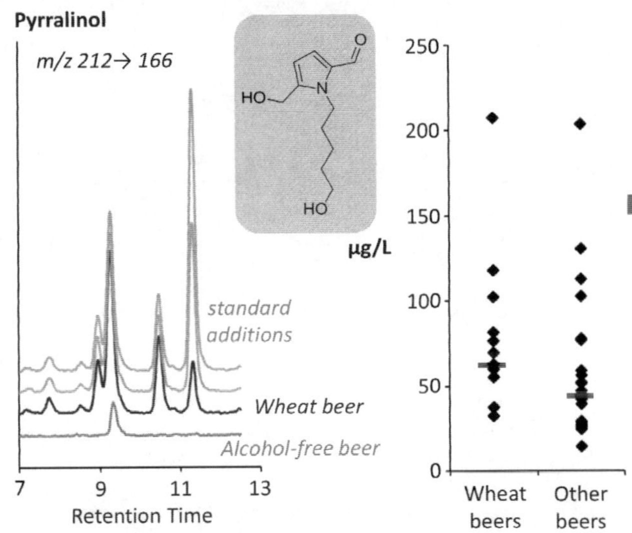

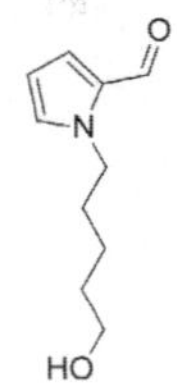

Pyrralinol up to 200 µg/L

Formylinol up to 50 µg/L

Maltosinol up to 7 µg/L

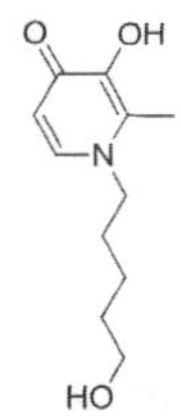

20-40% of the amount of parent MRPs in beer

14

Hellwig, Beer, Witte, Henle (2018) J Agric Food Chem 66, 7451-7460.

Summary and outlook

Can MRPs be metabolized by brewer's yeast? ✓

Which metabolites are formed?

new higher alcohols and α-hydroxy acids

Can these metabolites be quantified in beer?

✓

15

Thanks to...

Chair of Food Chemistry (Prof. Henle)

K. Schlosser *(Amino acid analysis)*

TECHNISCHE UNIVERSITÄT DRESDEN

Chair of Biochemistry (Profs. van Pée and Gulder)

C. Lüth, M. Wiegand *(Microbiology – Yeasts)*

TECHNISCHE UNIVERSITÄT DRESDEN

...financial support...

gefördert durch

Bundesministerium für Wirtschaft und Energie

aufgrund eines Beschlusses des Deutschen Bundestages (WiFö, 20624 BG)

DFG

16

L44 - A RAPID AND OBJECTIVE METHOD FOR IN SITU DETERMINATION OF A DEGREE OF SENSORY DETERIORATION OF BEER CAUSED BY AGING PROCESS

Jana Olšovská[1]; Tomáš Vrzal[1,2]; Karel Štěrba[1]

1 - Research Institute of Brewing and Malting; 2 - , Charles University, Faculty of Science

Abstract

A rapid, mobile and objective method for in situ determination of a degree of sensory deterioration of beer caused by the aging process was developed and validated. The method is based on current reflectometric determination of 5-(hydroxymethyl)furan-2-carbaldehyde (HMF), as according to our previous study, HMF is a very good indicator of a stale flavor in beer determined by a sensory analysis (correlation higher than 0.96). It requires a mobile reflectometer with test strips and therefore is available for everyone and everywhere outside the laboratory as a taste sensor in the meaning of "an artificial tongue". The reflectometric signal is transformed to the degree of the stale flavor using a specified algorithm, which was derived from this study and it is being patented. The method provides results comparable with the sensory analysis by a trained sensory panel for lager beers and also an objective reflection of beer deterioration for other beer styles. The stability of a QC sample was a crucial parameter for avoiding false positives/negatives, therefore, the assay of stabilization of HMF standard was simultaneously developed. The reproducibility of the method derived from a comparison of data sets from three individual reflectometers by t-test ($p<0.05$) and expressed as a relative deviation was 7.3%. Finally, monitoring of the stale flavor in real beer from the local market was proved. Nearly 56% of the tested samples were positive. Weak/medium deterioration was noted in 41% of the samples and 15% of samples were damaged significantly (medium/strong).

Short biography

Jana Olšovská has worked for nearly 15 years as an analytical chemist in a biomedical field. She is an expert in modern separation techniques such as HPLC with UV, FL and MS detection. She started to work in brewery research in 2011; since 2013 she has been an expert on the Analytical Committee of EBC. Presently, she is a manager of the Analytical Testing Laboratory EN ISO/IEC 17025:2005, a manager of the Sensory Laboratory and a member of a professional sensory panel of the Research Institute of Brewing and Malting, Prague. She is an author of more than 50 papers in international impacted journals and a new book "Sensory analysis of beer", Kvasny Prum, Czech Republic, 2017. She is also involved in pedagogical activities, she has already been a supervisor of many BSc., MSc., and Ph.D. students.

L45 - MORE SUSTAINABLE BREWING AND INCREASED FLAVOUR STABILITY DUE TO THE AB INBEV SIMMER & STRIP TECHNOLOGY

David De Schutter[1]; Mario Rottiers[2]; Mario Thiry[3]

1 - R&D Director Europe at AB InBev; 2 - Process Development Specialist; 3 - Global Process Development Specialist, Leuven, Belgium

Abstract

Wort boiling is the most energy-intensive step in the brewing process. A wide range of technologies for reducing energy usage or recovering energy have been developed in the past and are implemented in many breweries with different challenges and benefits.

The technology developed by AB InBev consists of a new disruptive way of conducting the 'boiling' step while the evaporation is replaced by a stripping of an inert gas. This gas is directly injected in the kettle and the wort is kept just below the boiling point. This technology aims to reduce both wort heat load and energy consumption.

In the regular boiling process, the energy is primarily used for the evaporation which is required for flavour stripping. DMS is traditionally considered as a marker for this purpose. To maintain highest quality standards, the monitoring of Strecker aldehydes with even lower volatility such as 3-methylbutanal and Maillard components like 2-furfural have been considered and measured in the brewhouse as well as in the final product. A third marker, 2-Ethyl-3,5-dimethylpyrazine, has been identified for its very low volatility and its extremely low sensory threshold (< 1ppb).

The efficiency of the stripping and the impact of the thermal load on flavour stability of the beer have been assessed and correlated with a standard forced aging test. The objective of this will present and discuss the impact of this new proprietary boiling technology on those markers during the process from boiling to final product.

Short biography

David De Schutter obtained his PhD degree at the Center for Malting and Brewing Science of the KULeuven under supervision of Prof. Delvaux and Prof. Derdelinckx. Before his career at AB InBev he made a passage outside the brewing industry at the R&D Center of Procter & Gamble in Brussels. He joined AB InBev as Product & Process Development Specialist where he was involved in the development of the first new generation non-alcoholic beer Then he became Global Director for Process Development and R&D Transformation in the Global Innovation & Technology Center in Leuven (GITEC). His current position is R&D Director for Europe at AB InBev. David is also representing AB InBev in the EBC Executive Committee, member of the BIR&D (Belgian Industrial Research and Development organization) and coordinator of the Brewing program of Fund Baillet Latour

L46 - INFLUENCE OF MASHING PARAMETERS ON FILTER CAKE STRUCTURE AND FILTERABILITY IN THE LAUTER TUN

Martin Hennemann[1], Martina Gastl[1], Thomas Becker[1], Florian Lehnhardt[1]

1 - TUM Munchen-Weihenstephan, Germany

Abstract

Lautering is the most time consuming step in wort production. It is an exceptional filtration process in food industry based on the combination of cake filtration and extraction of value-giving substances.

In the lauter tun, fine particles of mash contribute to the formation of a jelly-like layer on top of the filter cake. This blocking layer is known for filtration problems during lautering. Therefore, a reduction of the filtration limiting effects of the fine layer is desirable.

Attempts were made to reduce the size of this fine layer by incorporating fine particles homogeneously into the filter cake structure. In order to achieve this, varied mash parameters, for example pH-value or ion concentration, were tested. The influence on the homogeneity of the filter cake structure was analysed by particle size measurement. Wort filtration time was measured to determine filterability of mash with different parameters. To prevent the influence of the grist composition, finely milled malt was used for the filtration tests.

Short biography

Martin Hennemann is a research assistant at the Chair of Brewing and Beverage Technology of TU Munchen-Weihenstephan. Hennemann started his career with an apprenticeship as brewer and maltster. Afterwards, he attended the Doemens Academy in Grafelfing for advanced training as brewing technician in 2009. Hennemann received a Bachelor's and Master's degree in chemistry and biochemistry at the LMU Munchen and completed his Master's thesis in 2016 at the Birkbeck College in London. Since 2017, Hennemann deals with the lautering process as part of his PhD at the research group Raw Material based Brewing and Beverage Technology at TUM.

Development of the methodology for determination of the geographical origin of hops

- Iztok Jože Košir, Nives Ogrinc, Miha Ocvirk

INŠTITUT ZA HMELJARSTVO IN PIVOVARSTVO SLOVENIJE

Jožef Stefan
Institute
Ljubljana, Slovenija

EBC congress, Antwerpen 2019

Outline

- Importance of hop products authenticity
- Basics of stable isotope fractionation in nature
- Analytical methodology
- Results
- Conclusion

Importance of hop products authenticity

- Hop cones of the female plant have a huge value in brewing industry – key ingredient
- Resins and essential oils in lupulin glands
- Beer flavour and aroma are influenced on the hop variety used in the brewing process
- Hop quality depends on the varietal and geographical origin
- Usually hop cultivar gives the highest quality in the geographical area where it was breeded and differ regarding the geographical origin

Importance of hop products authenticity

- For this reasons brewers need to be sure about the geographical origin of hops
- Frauds in terms of false declarations of harvest data (geo. origin) could be a problem
- Rules on certification of hop products and trade conditions

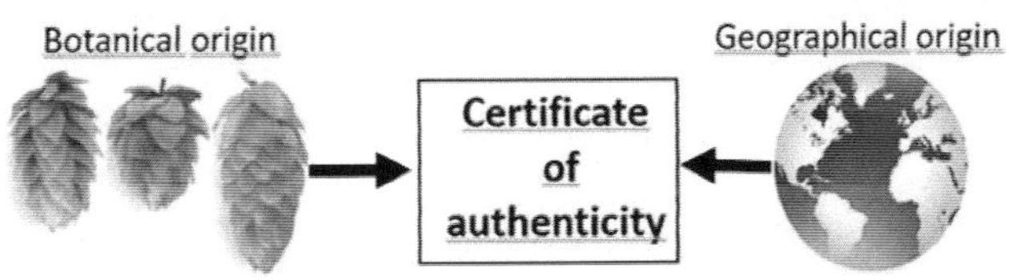

- Development of appropriate analytical methodologies that will enable its identification and validation is needed

- The most common methodology for authenticity studies is determination of metabolites (essential oil, resins composition)
- It's successful for determination of hop variety
- Not giving satisfactory results when the geographical origin is in question

- Solution could be the methodology not based on chemical composition of metabolites at the level of molecules but on elemental composition of the sample in the level of atoms.

Some basics of stable isotope fractionation and their occurance in the nature

Types of isotopic systems

92 naturally occuring elements

- 300 stable isotopes
  (H, C, N, O, S, Si, Cl, Mg, Ca, Fe, Hg, ...)

- 1200 radioactive isotopes
  (^{14}C, ^{40}K, ^{87}Rb, ^{235}U, ^{238}U, ^{210}Pb, ^{137}Cs, T, ...)

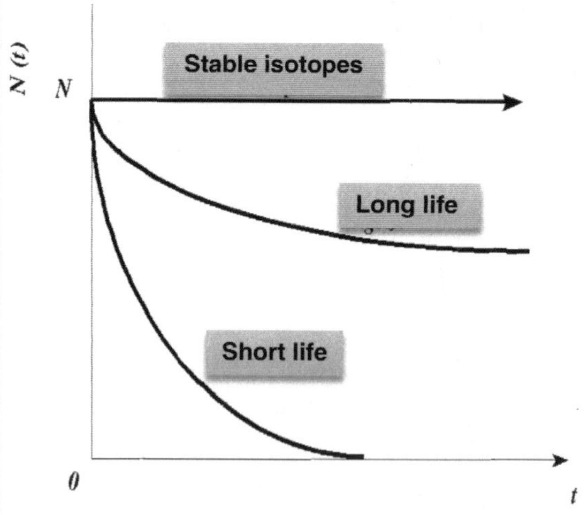

Stable isotopes

Organic matter basically consist of C, H, O and N.

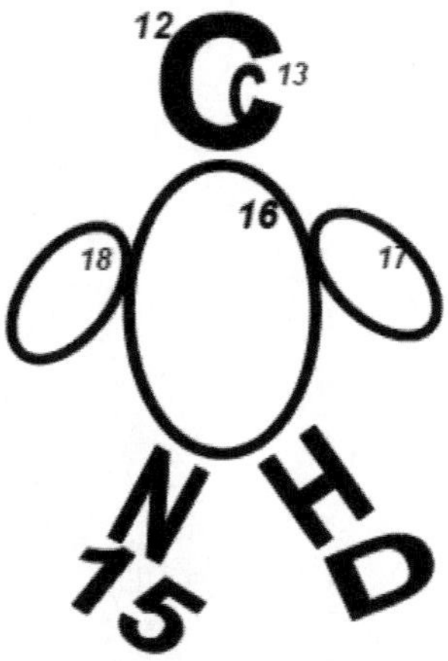

- All elements are present in different isotopes.
- Their content is influenced by natural processes.

Stable isotope fractionation

Definition: alteration of the distribution of stable isotopes as a result of a chemical or physical process

Examples:

Photosynthesis

$$CO_2 + H_2O \rightarrow C_6H_{12}O_6 + O_2 + H_2O$$

^{12}C preferentially incorporated into plant tissue (kinetic process)

Evaporation (thermodynamic process) ↑ ^{1}H ^{16}O

Lighter isotopes preferentially evaporated

$^{2}H/^{1}H$, $^{18}O/^{16}O$ increase

Two important types of fractionation:

→ Kinetic Processes (KIE) - depend primarily on differences in reaction rates of isotopic molecules

→ Thermodynamic Processes (EIE) - (equilibrium isotope distribution)

Distribution of deuterium on Earth

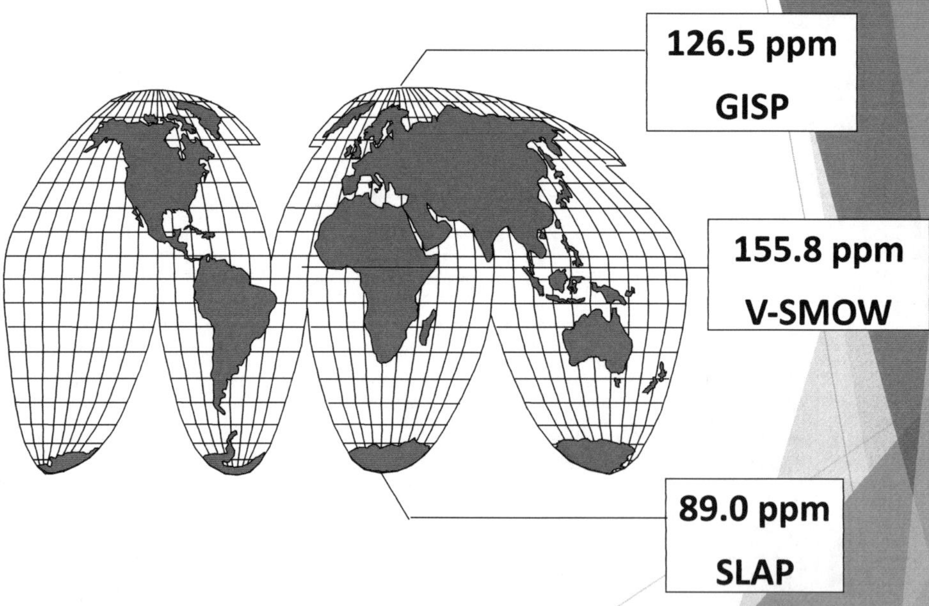

Origins of isotopes

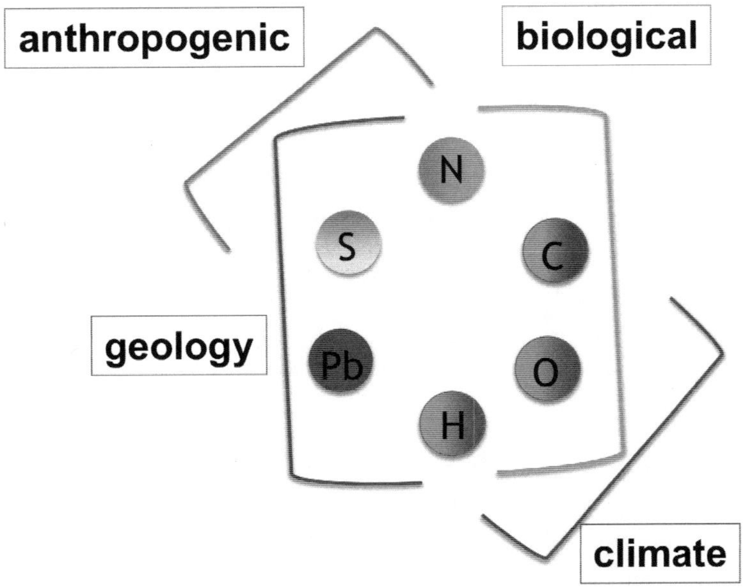

Analytical methodology – 5mg

- 1st set of samples
- Does the stable isotope ratios remain the same when hop is processed?
  - Different forms of the same variety and location
  - Samples of hop cones, pellets type 45 or 90, hop extract (in the case of cones – lupulin, green part, essential oil)

Hop cones

Hop pellets

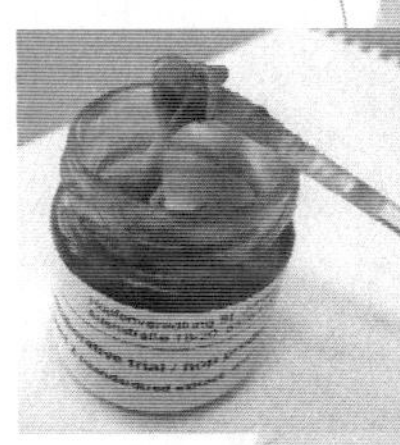

Hop extracts

Analytical methodology – 5mg

- 2nd set of samples
- Does the content of alpha-acids and essential oil affect the isotopic ratios?
  - Different varieties, from the same location
  - Samples of 5 different hop varieties from Slovenia

Analytical methodology – 5mg

- 3rd set of samples
  - Are we able to determine geographical origin?
  - 77 hop samples from the ten most important hop growing regions in the World

Determination of $\delta^{13}C$, $\delta^{15}N$, $\delta^{34}S$, δO^{18} , and D/H isotopic ratios by IRMS.

IRMS – Isotopic Ratio Mass Spectrometry

Determination of $\delta^{13}C$, $\delta^{15}N$, $\delta^{34}S$, δO^{18} , and D/H isotopis ratios.

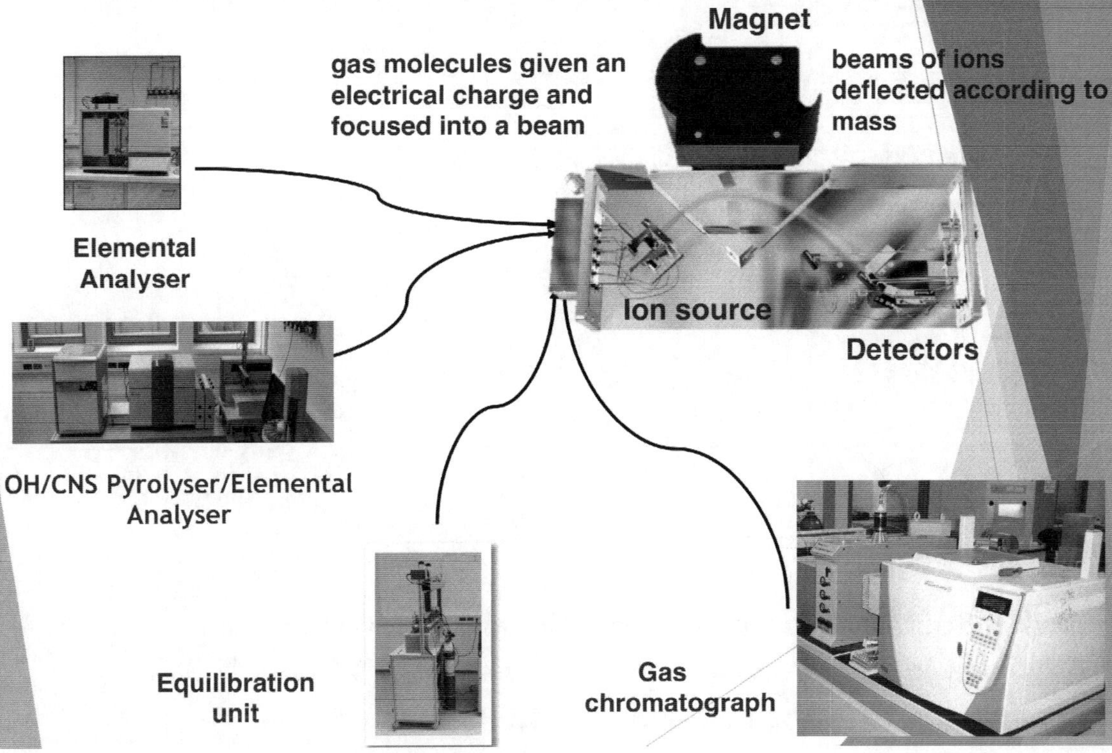

	n	Average ‰	SD	Min. ‰	Max. ‰
$\delta^{15}N$(‰)					
AUT	8	4,6	0,9	3,3	6,3
CZ	3	8,0	0,5	7,5	8,4
GER	23	5,9	0,6	4,8	7,2
SLO	16	4,7	0,6	3,7	5,6
USA_Y	4	3,9	0,5	3,4	4,7
USA_GL	6	3,0	0,8	2,2	4,6
GB	4	5,1	1,3	3,7	7,4
RSA	5	4,9	0,8	3,8	6,1
NZL	6	3,3	0,4	2,8	4,1
AUS	2	7,3	0,3	7,0	7,5
$\delta^{34}S$(‰)					
AUT	8	7,2	1,1	5,4	8,9
CZ	3	4,4	0,8	3,6	5,0
GER	23	5,3	0,7	4,1	6,8
SLO	16	4,0	0,7	2,7	4,8
USA_Y	4	4,0	0,4	3,5	4,4
USA_GL	6	2,1	1,1	0,7	3,8
GB	4	3,2	0,5	2,5	3,7
RSA	5	10,9	0,8	10,0	12,3
NZL	6	9,2	0,7	8,5	10,5
AUS	2	8,6	0,7	7,9	9,3

Results

In the Southern Hemisphere, the $\delta^{34}S$ values are significantly higher than in the samples in the Northern Hemisphere.

The $\delta^{15}N$ values of hops from the USA and NZL are the lowest.

The results of ANOVA comparison reveals that the best discrimination between samples from different regions was achieved using $\delta^{34}S$, followed by $\delta^{15}N$.

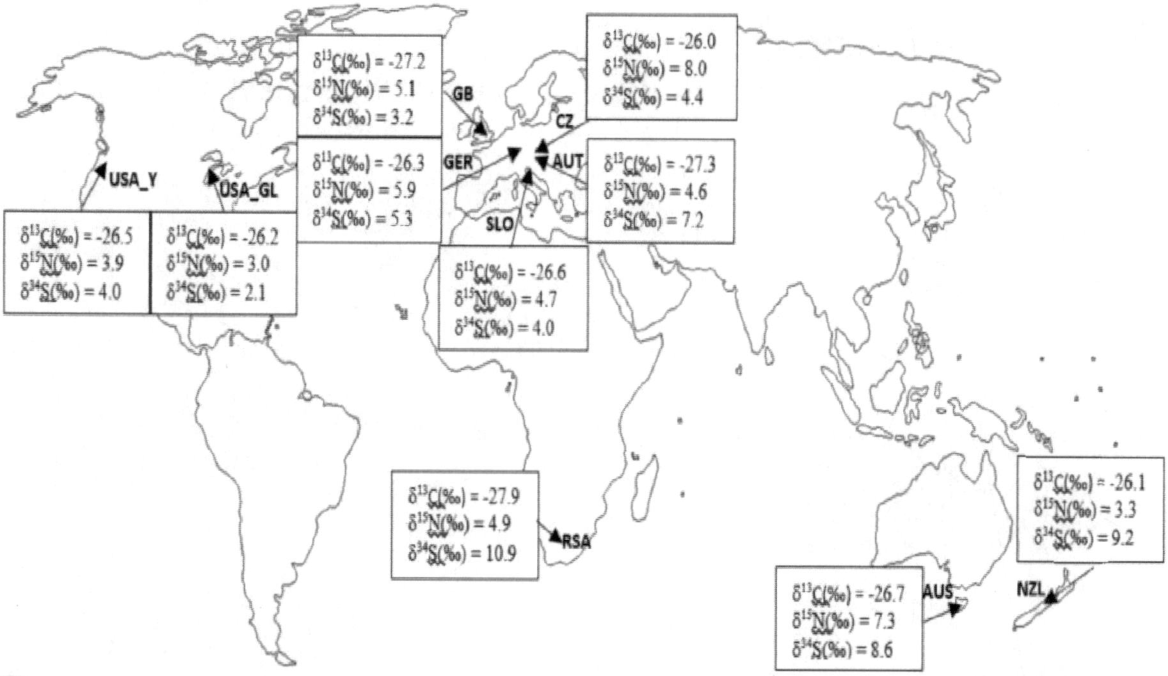

- **δ^{34}S in South hemisphere are significantly higher.**
- **δ^{15}N are lower in USA and NZL samples (3.0 ‰ to 3.9 ‰) in comparison to European samples (4.6 ‰ to 8,0 ‰).**

Results

Scatter plot of distribution of δ^{15}N / δ^{34}S

- NZL, RSA, AUS, USA_GL, AUT and CZ are forming their own separated groups
- Partly separated samples from GER, SLO, USA_Y in GB

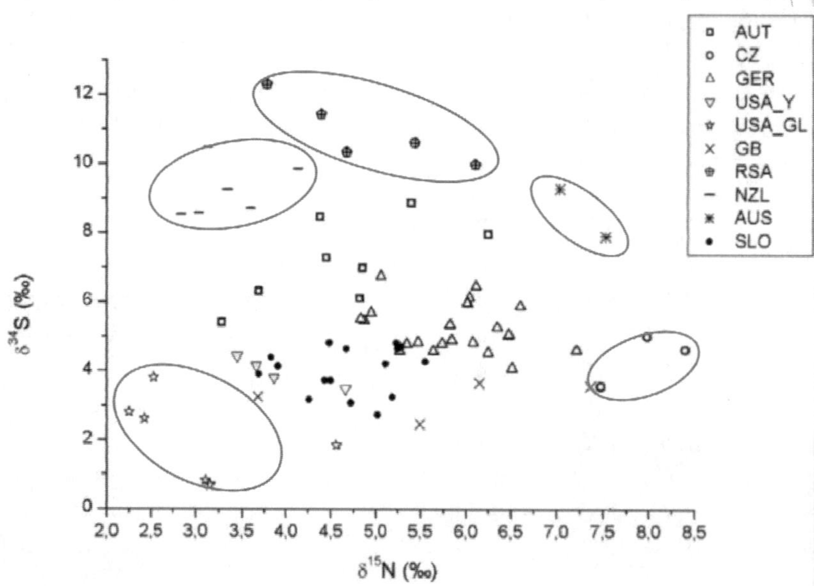

Results

Graphical distribution of European samples vs. others

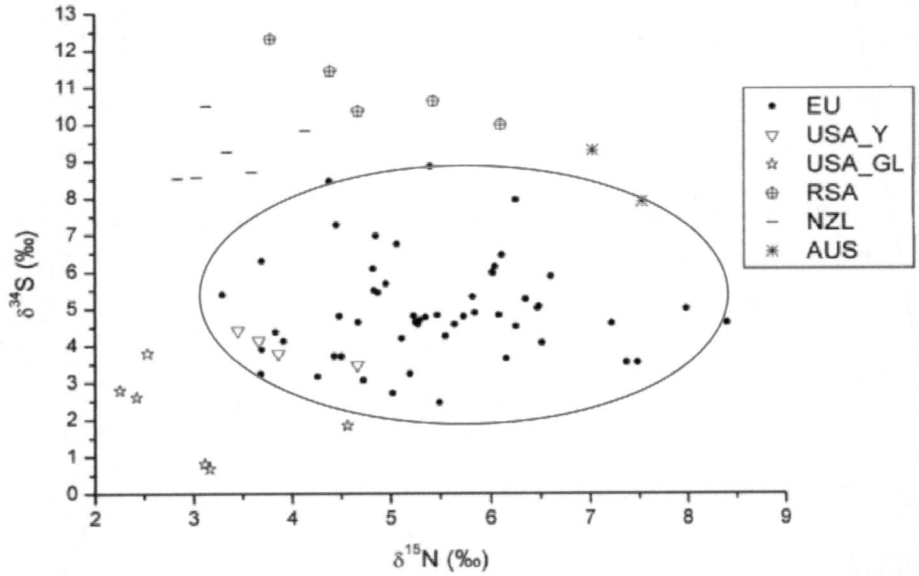

Ocvirk M., Ogrinc N., Košir I. J. Determination of the geographical and botanical origin of hops (*Humulus lupulus* L.) using stable isotopes of C, N and S., *J. Agric. Food Chem.* **2018** *66* (8), 2021-2026

Results European samples only

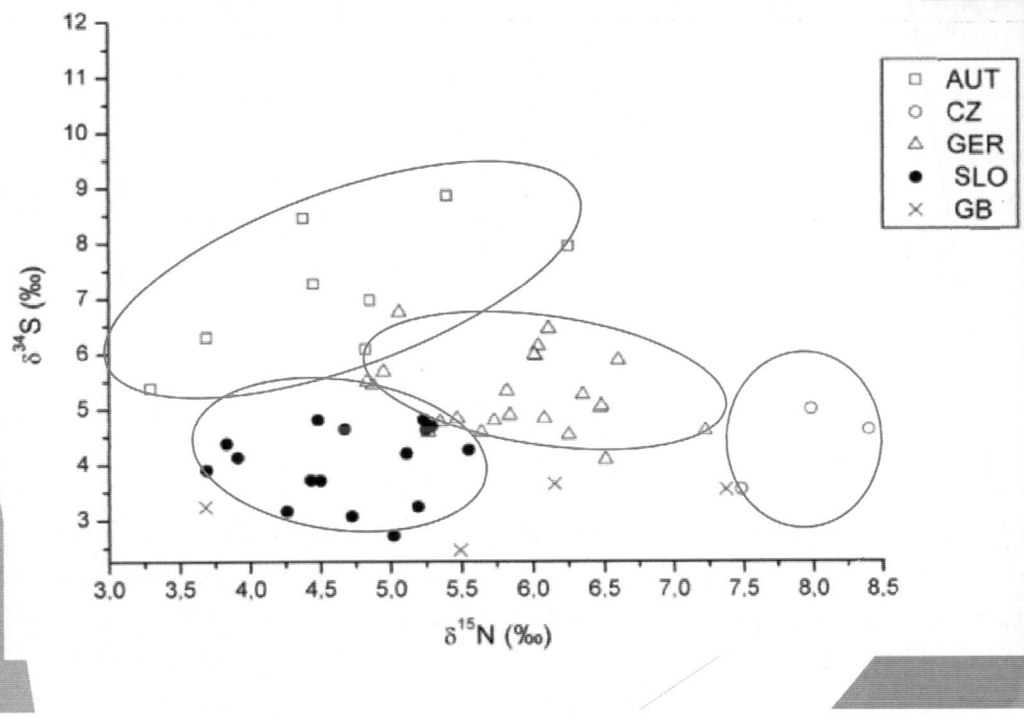

IN ADDITION
ELEMENTAL COMPOSITION OF HOPS

- The elemental content of the plant reflects the elemental composition of the soil
- 13 chemical elements: **Si, P, S, Cl, K, Ca, Ti, Mn, Fe, Zn, Br, Rb** in **Sr were measured by XRD**
- Exactly the same sets of hop samples, 0.5 g of pulverized samples, pressed into pellets

Classification of European samples

a. b.

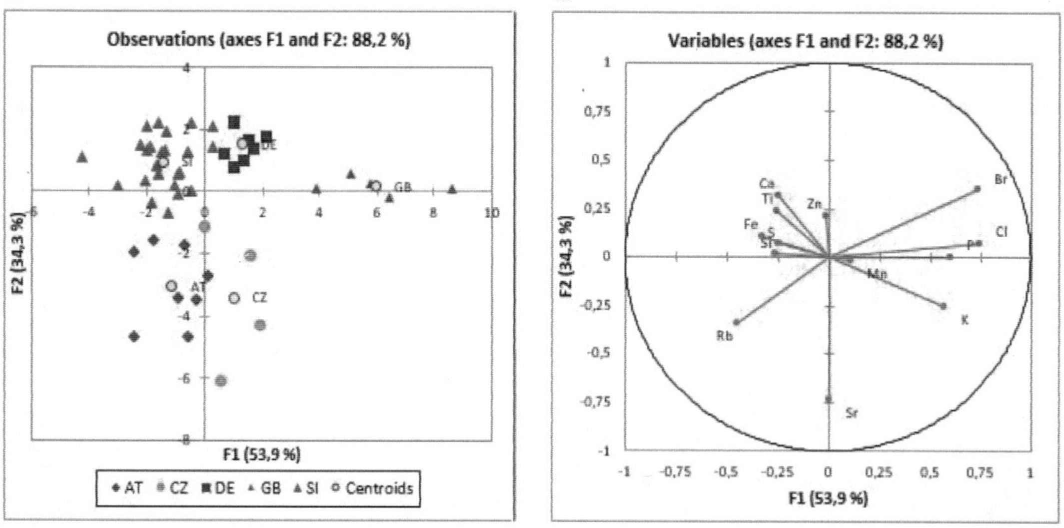

5 separated groups. The overall correct classification for this data set was 87%. The most influential parameters in the first discriminant functions (F1) were Ti (-0.82), Br (0.75) and Si (0.73). For F2, Sr (-1.39), Ca (1.02) and Fe (0.44).

Non European samples

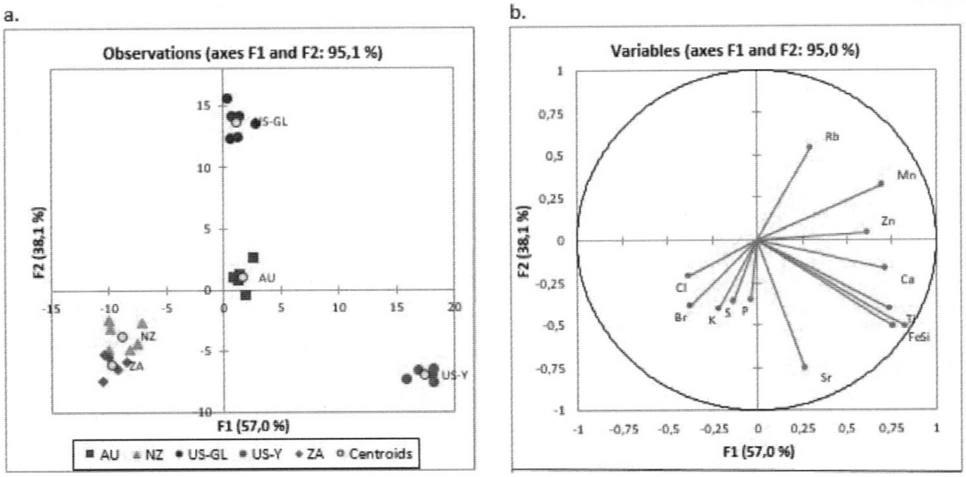

The overall correct classification for this data set was 93% based on the leave-one out cross-validation procedure. The most influential parameters in the first discriminant functions (F1) were S (-2.1), Sr (-1.9) and Si (1.7). For F2, Ca (3.2), Sr (-2.5) and Cl (-1.6).

Ocvirk M., NEČEMER m., Košir I. J. Determination of the geographical origins of hops (*Humulus lupulus* L.) by multi-elemental fingerprinting, *Food Chem.* **2019** *277*, 32-37.

Conclusions

- The use of stable isotopes ratios $\delta^{15}N$, $\delta^{34}S$, and $\delta^{13}C$ could be used as a new powerful tool for differentiating hops according to geographical origin
- could be of use to the brewing industry or to hop traders in resolving issues relating to authenticity.
- The methodology is straightforward, fast, and cheap, relative to standard GC or HPLC techniques and does not need any special sample pretreatment except for weighing the samples.
- Stable isotope ratios $\delta^{18}O$ and D/H does not give satisfactory results since the content of water in hop products is too low.
- Elemental composition could be usefull additional way for determination of the geographical origin.

Special thanks to

- Dr. Nives Ogrinc - Institute Josef Štefan, Ljubljana, Slovenia
- Colleagues from Germany, UK, Czech Republic, Austria, New Zealand, South African Republic, and USA for providing the samples from their regions

Thank you for your attention !

L48 - CHARACTERIZING THE DIVERSITY OF WILD HOPS (HUMULUS LUPULUS) FROM PORTUGAL

Julio Cesar Machado Junior[1]; Miguel A. Faria[1]; Zita E. Martins[1]; Armindo Melo[1]; Sara C. Cunha[1]; Florian Lehnhardt[2]; Hubert Kollmannsberger[2]; Martina Gastl[2]; Thomas Becker[2]; Isabel M. P. L. V. O. Ferreira[1]

1 - LAQV/ REQUIMTE, Departamento de Ciências Químicas, Faculdade de Farmácia, Universidade do Porto;
2 - School of Life Sciences Weihenstephan, Chair of Brewing and Beverage Technology, Technische Universität München

Abstract

Wild hops are an important source of new varieties and genetic variation for breeding. In this study, 109 accessions of wild Portuguese hops were characterized at the genetic (SNP), sensorial, and chemical levels (total oil, volatile profile, α/β-acids and xanthohumol); by HRMA, a semi-trained panel, and steam distillation, GC-MS, GC-MS-Olfactometry and HPLC, respectively. Additionally, 4 accessions were selected and their behavior in dry-hoping was evaluated by volatile compounds and sensorial analysis with a trained panel.

At the sensorial level Portuguese hops presented resinous, spicy, and floral aromas, whereas control commercial hops used (European and North-American varieties) were, in general, more citrus, fruity/ sweet, and floral. These differences were confirmed chromatographically by different volatile profiles presented. Relative area data showed that around 15% of compounds detected justify the big majority of volatiles mass, in Portuguese hops, whilst this value raise to 85% in commercial hop samples. Genetic analysis showed a great diversity amongst accessions and a clear cluster separation from most of the commercial references used, demonstrating the potential genetic richness of the germplasm. Concerning the chemical characterization total oil was, in average, 0.72 (± 0.35) mL/ 100g, ranging from 0.23 to 1.84 mL/ 100g. The α-acids contents were around or lower than 5%, β-acids ranged from 0.4% to 6.3% and xanthohumol contented up to 0.5% of weight. One particular sample should be underlined based on its α-acids content (10.6 ± 0.2%) and volatile profile showing high similarity with reputable varieties in the worldwide market, as Northern Brewer, Nugget and Citra®.

Short biography

PhD student in the Pharmaceutical Sciences program, specializing in Nutrition and Food Chemistry at the Faculty of Pharmacy of the University of Porto.

Master in Pharmaceutical Sciences from the Federal University of Paraná, graduated in

Pharmacy and Biochemistry from the Universidade do Vale do Itajaí.

Experience in clinical analysis, physical-chemical, and sensory analyzes of foods, with emphasis on beers.

i2 sys bio

VTT

Native yeasts from Ecuadorian chichas: insights into their potential application in beer production

Grijalva-Vallejos, N., Krogerus, K., Nikulin, J., Magalhães, F., Aranda, A., Matallana, E. & Gibson, B.

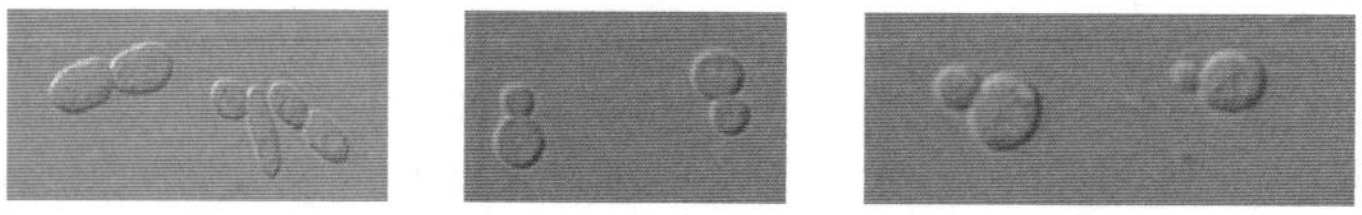

Secretaría de Educación Superior, Ciencia, Tecnología e Innovación

Alternative yeast for innovative brewing fermentations

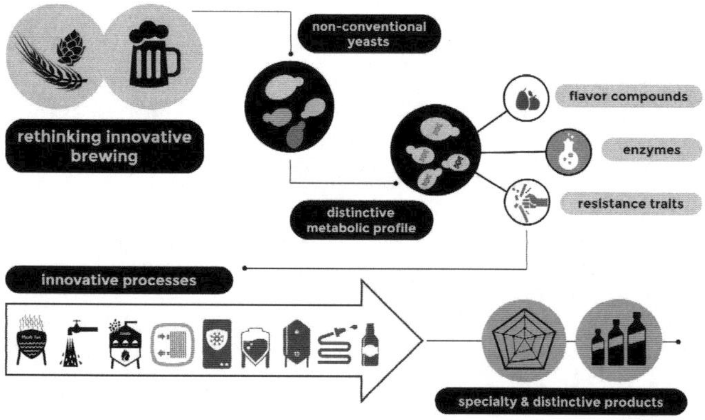

(Basso et al, 2016)

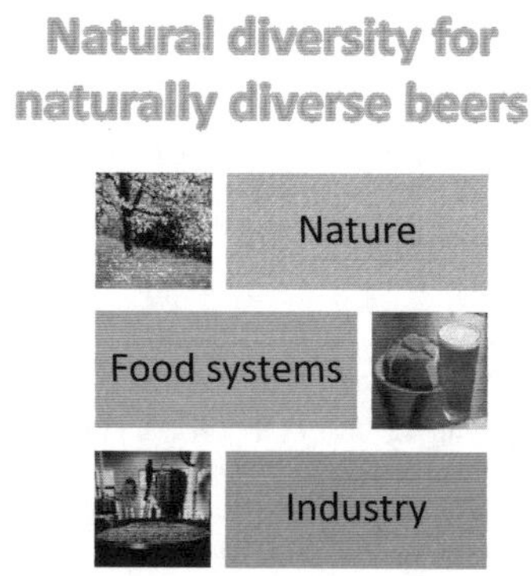

(Cubillos et al, 2019)

New yeasts in industrial beer production

Saccharomyces yeast

- New species, strains, natural hybrids
- Non domesticated
- Different genotypic and phenotypic profiles

Non-_Saccharomyces_ yeast

- Secondary metabolites, specific enzymes
- Sequential or co-inoculation
- *Dekkera/Brettanomyces* strains, *Saccharomycodes ludwigii*, *Wickerhamomyces anomalus* and *Torulaspora delbrueckii*

↑Flavouring
↓Stress
↓Undesirable compounds

Traditional beverages as a reservoir of diversity

Indigenous yeast strains, valuable tools

Native strains

Zygosaccharomyces bailii KBI 25.2

POF (-), moderately flocculent, maltotriose and maltose (-)

Final beer: ↑ fruitier , ↓worty character

(Bellut et al., 2018)

Native strains

Saccharomyces cerevisiae LBCM1045, LBCM1078

maltotriose and maltose (+), grow at low T°, low H_2S production

Final beer: good fermentative yields, ↓ diacetyl

(Araújo et al., 2018)

New hybrids

S. cerevisiae segregant diploid C-2-1d

Grow at low T°, highly flocculent

Final beer: ↑ acetate esters, distinct fruit traces

(Figueiredo et al., 2017)

Chichas

(Piló et al., 2018)

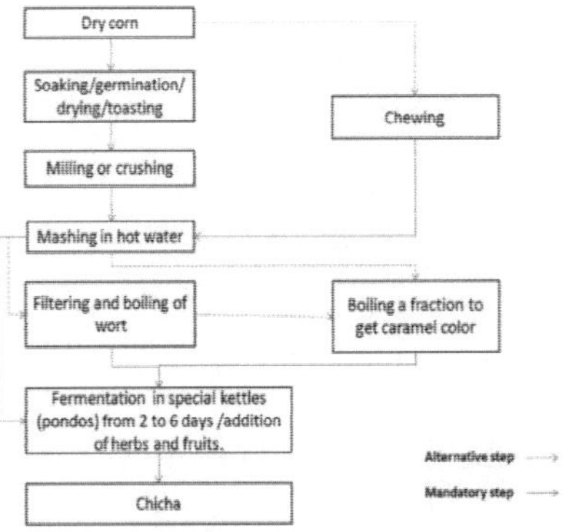

(Gomes et al., 2009)

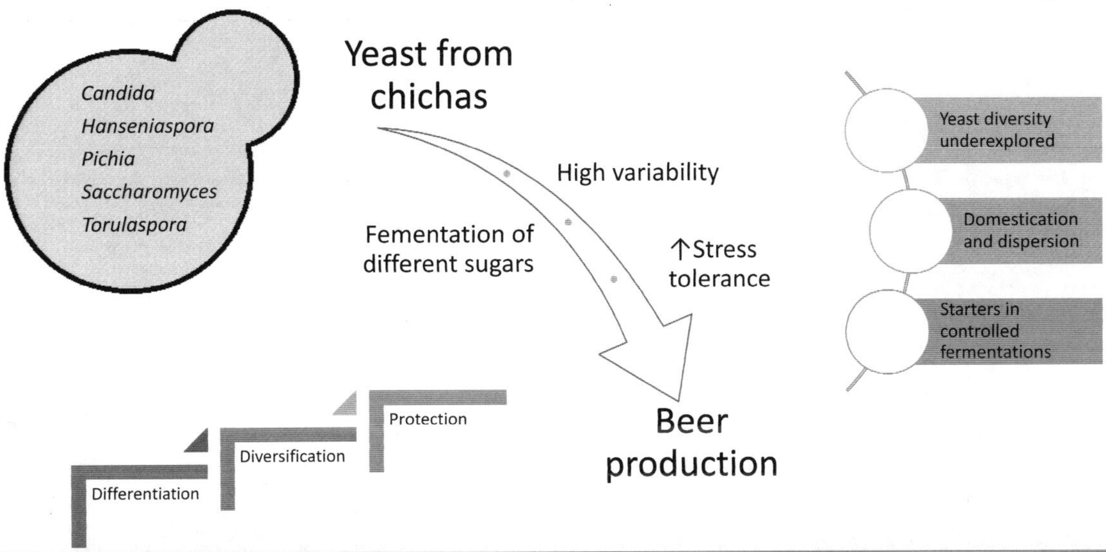

Objectives:

Investigate the application of novel yeast strains from traditional Ecuadorian chichas in the production of both beer and chicha.

Assess the ability of commercial yeast to ferment the traditional beverage

Materials and methods

Yeast Strains

Strain designation	Species	Origin
ERS1	*Saccharomyces cerevisiae*	Chicha, Ecuador
EYS4	*Saccharomyces cerevisiae*	Chicha, Ecuador
EYS5	*Saccharomyces cerevisiae*	Chicha, Ecuador
L2056	*Saccharomyces cerevisiae*	Lallemand, wine
A81062	*Saccharomyces cerevisiae*	VTT Culture Collection
A63015	*Saccharomyces pastorianus*	VTT Culture Collection
EGT2	*Torulaspora delbrueckii*	Chicha, Ecuador
D291	*Torulaspora delbrueckii*	Lallemand, wine

VTT

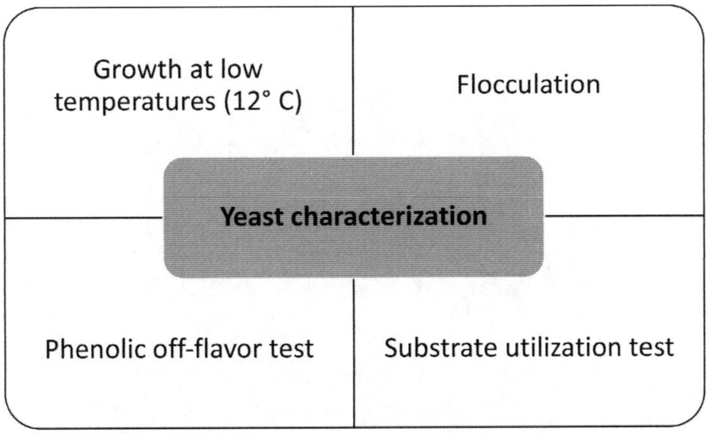
Growth at low temperatures (12° C)
Flocculation
Yeast characterization
Phenolic off-flavor test
Substrate utilization test

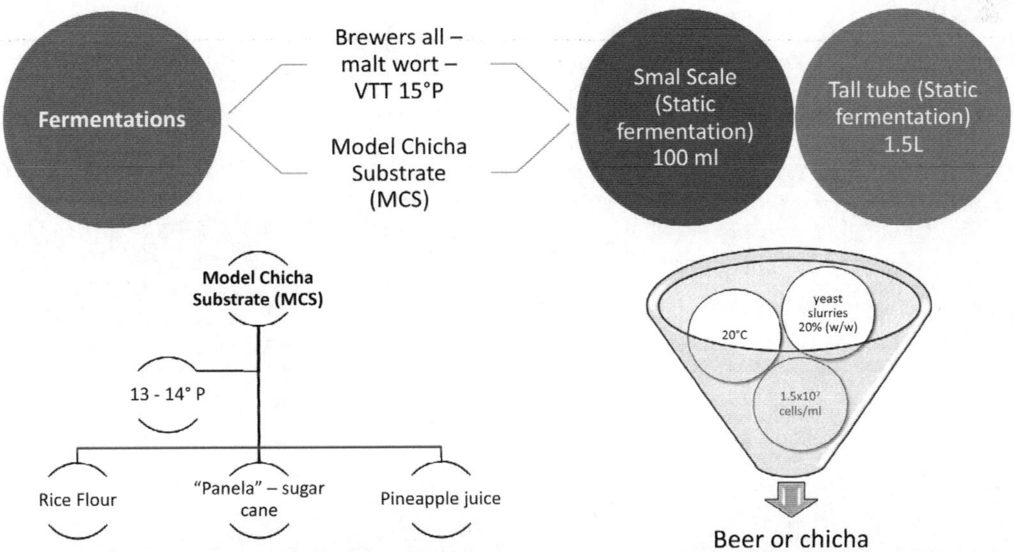
Fermentations
Brewers all – malt wort – VTT 15°P
Model Chicha Substrate (MCS)
Smal Scale (Static fermentation) 100 ml
Tall tube (Static fermentation) 1.5L
Model Chicha Substrate (MCS)
13 - 14° P
Rice Flour
"Panela" – sugar cane
Pineapple juice
20°C
yeast slurries 20% (w/w)
1.5x10⁷ cells/ml
Beer or chicha

Anton Paar
- Wort densities
- Alcohol content
- pH

HPLC
- Fermentable sugars
- Glycerol

HS-GC-FID Analysis
- Yeast-derived higher alcohols and esters

- Yeast viability
- Fresh and dry mass
- Mass loss (small-scale fermentation)

Results

Sugar utilization profile

Sugars	ERS1	EYS4	EYS5	L2056	A81062	A63015	EGT2	D291
	S. cerevisiae					*S. pastorianus*	*T. delbrueckii*	
Glucose	+	+	+	+	+	+	+	+
Fructose	+	+	+	+	+	+	+	+
Sucrose	+	+	+	+	+	+	+	+
Maltose	-	-	+	+	+	+	-	-
Maltotriose	-	-	+	+	+	+	-	-
Melibiose	-	-	-	-	-	-	-	-
Cellobiose	-	-	-	-	-	-	-	-
Raffinose	+	+	+	+	+	+	+	+

All Ecuadorian yeast strains, except EYS5, were incapable of fermenting maltose and maltotriose

Flocculation

Strain		Flocculation YPM %
S. cerevisiae	ERS1	18.46 ± 1.01 [b]
	EYS4	12.87 ± 2.19 [b]
	EYS5	3.95 ± 0.18 [a]
	L2056	5.02 ± 0.47 [a]
	A81062	**95.03 ± 1.51 [c]**
S. pastorianus	A63015	16.23 ± 2.71 [b]
T. delbrueckii	EGT2	3.63 ± 0.81 [a]
	D291	4.51 ± 0.17 [a]

None of the Ecuadorian strains was flocculent

POF production

Strain		POF
S. cerevisiae	ERS1	+
	EYS4	+
	EYS5	-
	L2056	+
	A81062	+
S. pastorianus	A63015	-
T. delbrueckii	EGT2	-
	D291	-

EYS5 was the only POF-negative strain

Growth al low temperatures (12°C)

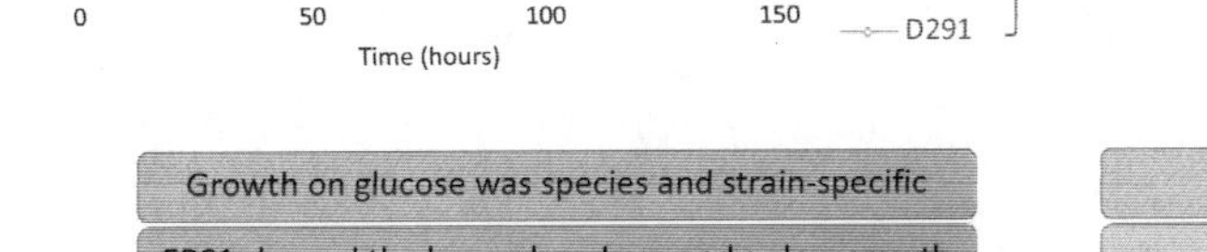

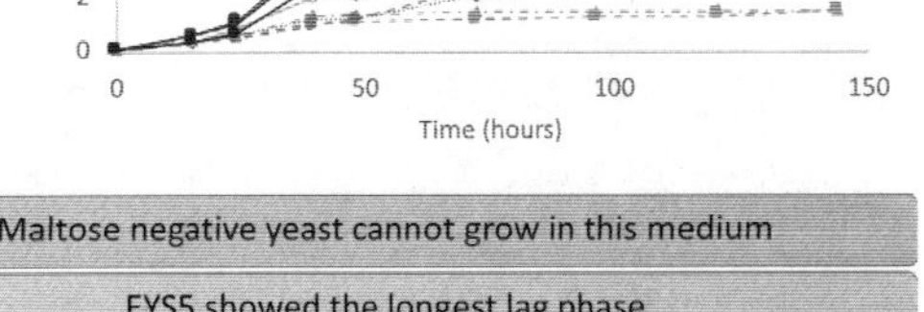

Growth on glucose was species and strain-specific

ERS1 showed the longer lag phase and a slow growth

EYS4 and EYS5 showed a similar growth

Maltose negative yeast cannot grow in this medium

EYS5 showed the longest lag phase

A63015 showed greatest ability to grow under these conditions.

1.5 L Beer wort Fermentation

15°P, 20°C

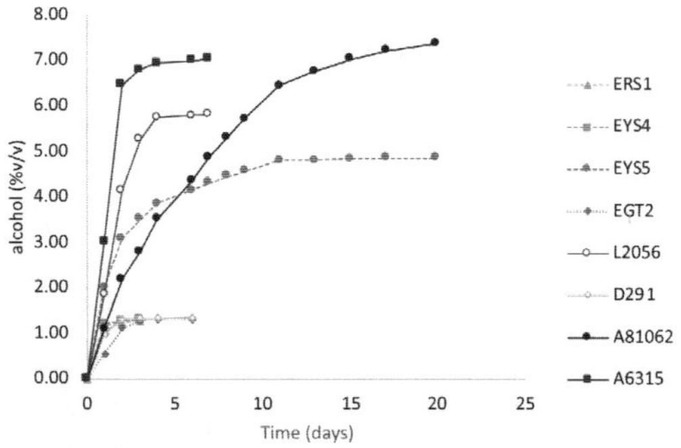

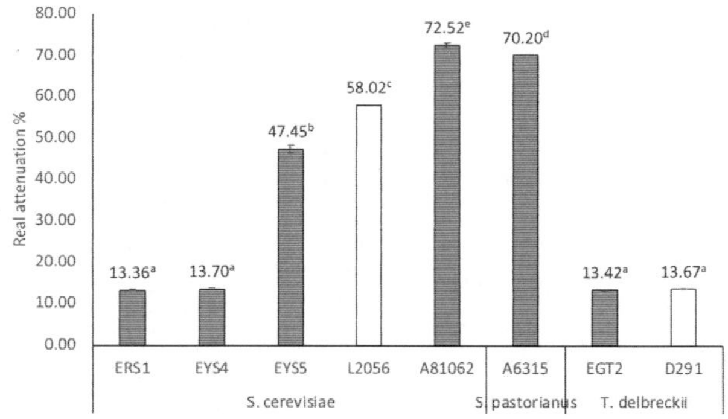

Fermentation ceased early for all maltose negative strains

A63015 showed the steepest decrease in extract

EYS5 and A81062 showed the slowest metabolism

Only beer strains reached a real attenuation ≥ 70%

HPLC analysis

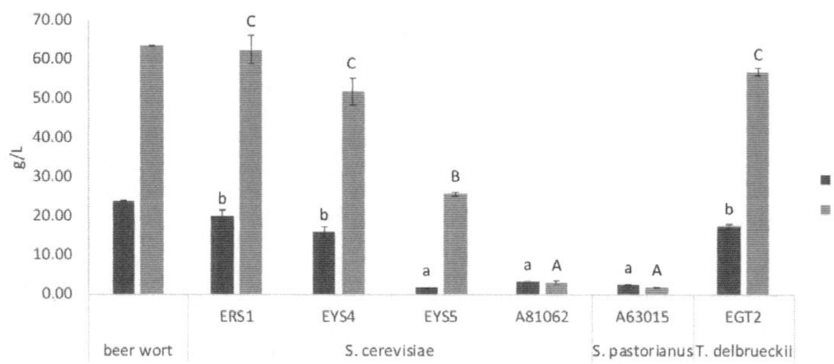

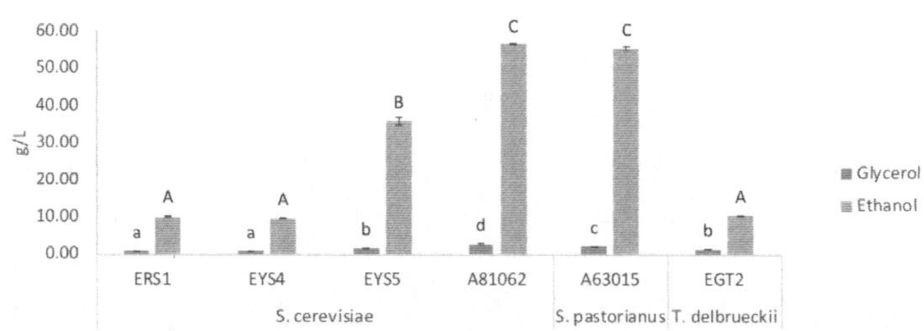

EYS5→ incomplete depletion of maltose

All strains: complete depletion of glucose and fructose

Maltose (-) strains showed similar results

Aroma

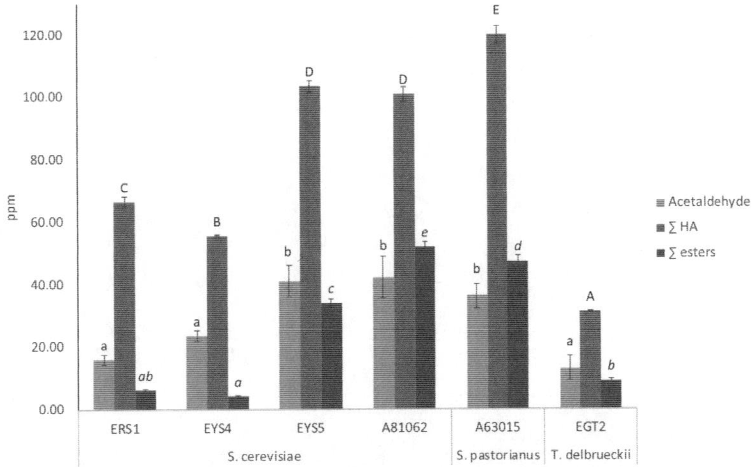

EYS5, A6315 and A81062: ↑ [ ] acetaldehyde

Higher alcohol concentrations varied depending on strain

EGT2: ↑ [ ] Ethylacetate

1.5 L Model Chicha Substrate (MCS) Fermentation

13 - 14°P, 20°C

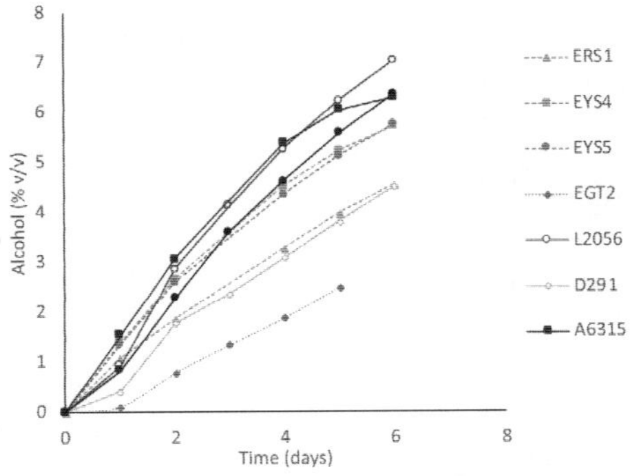

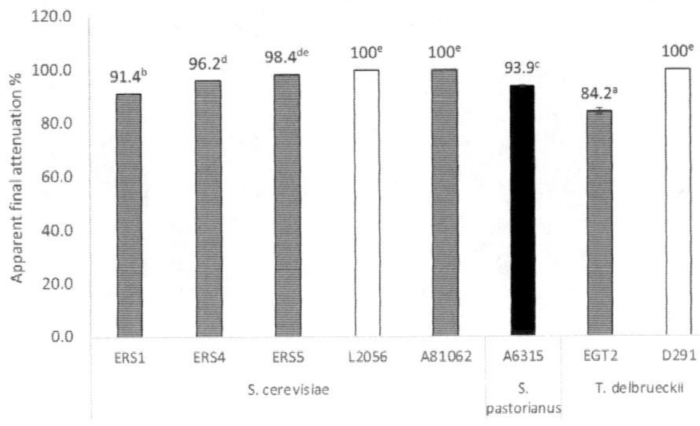

All strains were able to ferment the MCS

Fermentation performance of EYS4 and EYS1 show differences

L2056 showed the steepest decrease

All strains reached a real attenuation ≥ 84%

HPLC analysis

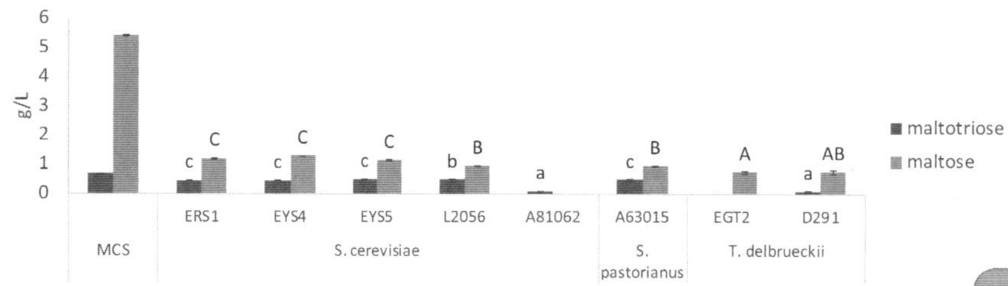

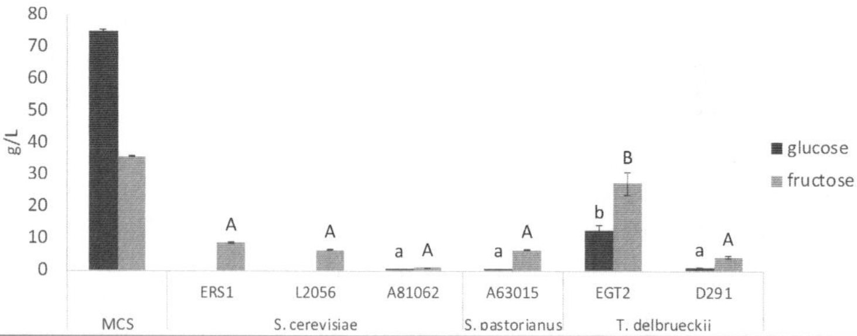

EYS4 and EYS5→ complete depletion of glucose and fructose

EGT2: incomplete depletion of glucose and fructose

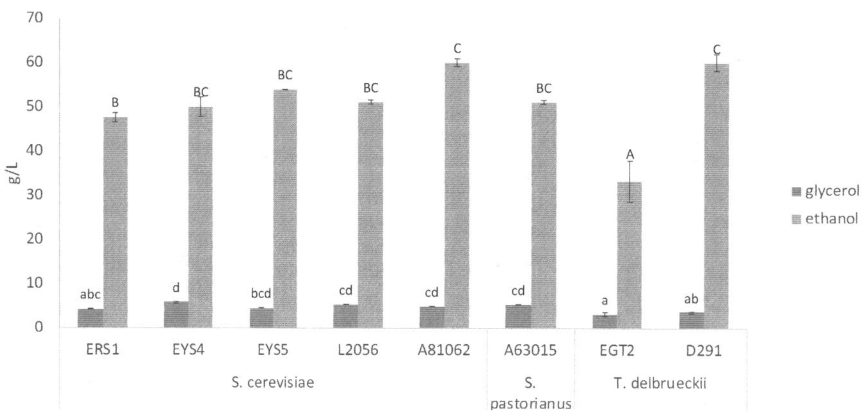

Beer and wine strains: ↑ [ ] Ethanol

EGT2 : lower [ ] Ethanol

Aroma

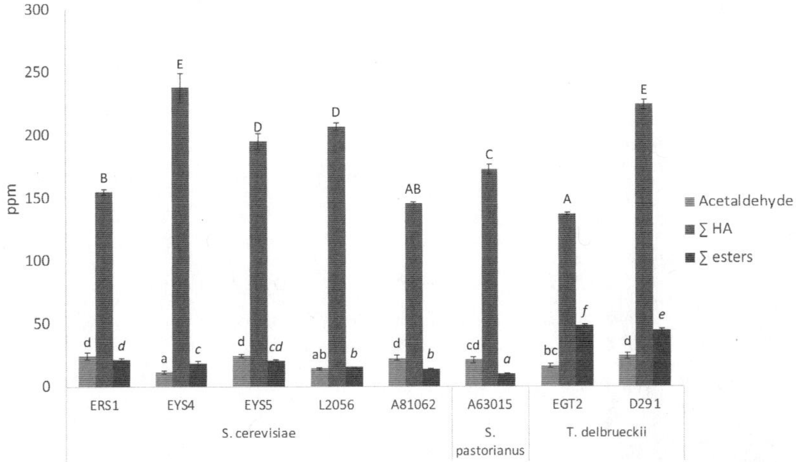

ERS1, EYS5 and EGT2: ↑ [ ] acetaldehyde

All chichas: ↑ [ ] higher alcohols > 135 g/L

T. delbrueckii strains showed the higher value of total esters followed by ERS1

CONCLUSIONS

- ❑Ecuadorian maltose negative yeast strains could be applied in low alcohol beer production. Final beer fermented with the *T. delbrueckii* strain EGT2, showed an interesting aroma balance.
- ❑The Ecuadorian *S. cerevisiae* yeast strain EYS5 showed a good potential to become a new alternative in the brewery industry.
- ❑Ecuadorian yeast strains have a good fermentative performance for chicha production, we found statistical differences between species and strains.
- ❑Wine and beer commercial yeast could be used for the production of Chicha

VTT

Secretaría de
Educación Superior,
Ciencia, Tecnología e Innovación

Dr. Brian Gibson

Dr. Frederico Magalhães
Kristoffer Krogerus
Jarkko Nikulin

i2sysbio

Dr. Emilia Matallana
Dr. Agustín Aranda

Bea Vallejo
Max Torrellas
Nubia Grijalva

L50 - CRITICAL ASSESSMENT OF CALIBRATION STRATEGIES FOR EFFECTIVE BEER FLAVOR ANALYSIS BY SOLID-PHASE MICROEXTRACTION (SPME)

Sarah Thörner[1]; Johanna Dennenlöhr[1]; Nils Rettberg[1]

1 - VLB Berlin - Research Institute for Beer and Beverage Analysis

Abstract

Solid phase microextraction (SPME) is the most commonly used automated sample preparation technique in beer flavor analysis. Interfaced to GC or GC-MS it is, amongst others, used in the analysis of hop aroma compounds, staling aldehydes, pyrazines, esters, and volatile fatty acids. The popularity of SPME relates to the availability of multiple SPME-fibers with different polarities. Those enable efficient extraction of volatiles from complex samples, whereas solvents and sample volumes are reduced to a minimum. Those features meet the demands of many analysts in breweries and academia.

Compared to other sample preparation techniques, SPME is rather straightforward on first view. However, the absolute recovery of some aroma compounds from beer or other brewing samples may only lie within a small percentage of their total amount (equilibrium extraction). Thus, matrix effects caused by either high or low alcohol content, pronounced hop aroma or fermentation by-product spectrum might strongly influence GC peak areas. Consequently some analysts complain about SPME performance being poor in terms of reproducibility and repeatability.

In the current paper we critically assess and compare different calibration strategies for SPME based flavor analysis. Focusing on hop aroma compounds and staling aldehydes we discuss the selection of suitable standards and matrices for matrix matched calibration. GC-MS data collected over the period of 24 month highlights the stability of properly designed SPME based methods. Also we outline a strategy to widen the method working range, which is important to the wealth of different beer styles currently available on the market.

Short biography

Sarah is a state approved food chemist who graduated from the University of Münster (WWU) in 2009. From 2010 to 2015 she performed her doctoral thesis on the "Characterization of enzymatic conversion of cis-[3,4-2H2]-3,4-epoxydecanoate in Saccharomyces strains" at TU Berlin. Simultaneously, Sarah was employed as a research associate at VLB (Research Institute for Special Analyses). Here, her field of activity included method development for various GC-MS applications, as well as the analytical assistance of versatile research projects. Since October 2015 Sarah is the Department Head for Special Analyses at the VLB Research Institute for Beer and Beverage Analysis. Her work is focused on the analysis of flavor and off-flavor in beer and beverages by GC, GC-MS, and GC-MS/MS.

L51 - INTERNATIONAL CALIBRATION STANDARDS FOR HOP AND BEER ANALYSIS – OVERVIEW AND NEWS

Biendl Martin[1]; Foster Robert, T.[2]; Maye John Paul[3]

1 - Hopsteiner HHV GmbH; 2 - Molson Coors; 3 - S.S. Steiner, Inc.

Abstract

More than 20 years ago, the first international cooperation was started with the purpose of production, release and maintenance of highly purified and stable HPLC calibration standards for use in the quantitative determinations of hop bitter components in hops and beer. Since 2001, the International Hop Standards Committee (IHSC), with experts from EBC, ASBC, BCOJ and IBD, has been responsible for this task. In addition to the International Calibration Extract (ICE), a carbon dioxide extract of assigned composition to be used for measurements of alpha- and beta-acids in hops and hop products, dicyclohexylamine complexes of various bitter acids present in beer have also been made available. These include iso-alpha-acids, rho-iso-alpha acids and hexahydro-iso-alpha-acids. Only the tetrahydro-iso-alpha-acids hop standards are released in pure form. In addition to iso-alpha-acids that are formed during wort boiling, considerable amounts of humulinones and non-isomerized alpha-acids can be transferred into beer via dry-hopping, where they contribute both to the analytical IBU and sensory bitterness values. In order to quantitate these contributory hop bitters in HPLC standard analysis (i.e. Analytica-EBC 9.47 method), new high purity standards needed to be synthesized. The IHSC recently succeeded in preparing both humulinones and alpha-acids as stable dicyclohexylamine complexes and approved each as new international calibration standards (ICS). Additionally, in consideration of ongoing worldwide research on its health benefits and new dry-hopping techniques in-use, in 2018, a pure xanthohumol has been made available as ICS. Finally, after purity validation and approval, the IHSC supplies the packaged hop standards to Labor Veritas (Switzerland) and Scientific Societies (USA) who sell them worldwide under the EBC and ASBC labels. To show all the steps and cooperation involved behind releasing a new standard, an example of how the production and purity assignments of dicyclohexylamine-humulinones is described.

Short biography

Martin Biendl received a Ph.D. degree in organic chemistry from Regensburg University in 1990. He is head of the R&D/Analytical Department at the German branch of the Hopsteiner group, one of the largest international hop growing, trading and processing firms. His research experience is in the field of hop-related needs for the brewing industry and beyond. He is the representative of the International Hop Industry Cooperation in the EBC Analysis Committee and, since 2001, chairman of the Hops Subcommittee. As EBC representative he is also co-chairman of the International Hop Standards Committee. In 2007 he was elected to the board of the Association of the German Hop Trade

L52 - THE INFLUENCE OF DRY-HOPPING ON SELECTED CHEMICAL-PHYSICAL CHARACTERISTICS OF BEER

Martin Zarnkow[1]; A. Stallforth[1]; K. Haselbeck[1]; S. Cocuzza[2]; F. Jacob[1]

1 - TU München Research Center for Brewing and Food Quality; 2 - Hopsteiner, Simon H. Steiner, Hopfen, GmbH

Abstract

In the course of this work it was investigated how a linearly increasing hopping rate affects the selected chemical-physical characteristics of beer. The aim was to analyse the modifications of the following characteristics: bitter units, α and iso α acid, humulinone, total polyphenols, flavanoids, iso- and xanthohumol¬, real extract, alcohol by volume (ABV), pH value, foam stability, a-pinen, b-pinen, b-myrcen, d-limonen, caryophyllen, humulene, a-terpeniol, damascenon, geraniol, sensory evaluation and turbidity due to dry hopping. It was also necessary to check whether the plant material added by dry hopping showed significant effects on the examined characteristics. For this purpose dry hopping trials using spent hop pellets from a CO_2-extraction were performed.

First a base beer was brewed and fermented. This young beer was the basis for the subsequent dry hopping experiments with a hopping rate ranging from 0 to 1500 g/hl. In addition, a test series has been dry hopped with spent hop pellets (1000 g/hl).

With regard to the bitter units it can be stated that dry hopping with both Cascade and spent hop pellets has no significant influence. On the other hand, the iso-α-acid decreased by 1,4 mg/l per 100 g/hl when dry hopped with Cascade pellets. Due to the high α-acid content in the reference beer, no clear increase can be observed. The humulinone content increases by 1 mg/l per 100 g/hl when dry hopped with Cascade pellets.

In summary, dry hopping has a significant effect on some characteristics of beer. Clearly modifications show the decrease in iso-α-acid and foam stability, as well as the increase in humulinone, pH value and total polyphenols. In contrast, no definitely conclusion can be made to the influence of dry hopping on the flavanoid, xanthohumol and alcohol content as well as the turbidity and the bitter units.

Short biography

Apprenticed as a brewer and maltster from 1989-1991 at a small brewery in Frankonia. Finishing a Diplom-Ingenieur (FH) graduation, option brewing technology, 1996 at the TU München Weihenstephan. Worked as a brew master for one year in a medium-sized brewery in Germany. Since 1997 at TU München. Head of research and development at research institute Weihenstephan for brewing and food quality. Finished 2010 his external PhD research at the University College of Cork, Ireland on the subject: "Proso Millet (Panicum miliaceum L.) a Sustainable Raw Material for the Malting and Brewing Process".

GENETIC AND BIOCHEMICAL APPROACH TO STUDY THE IMPACT OF TERROIR ON THE BREWING VALUE OF HOPS

Ann Van Holle, Hilde Muylle, Tom Ruttinck, Anita Van Landschoot, Geert Haesaert, Dirk Naudts, Denis De Keukeleire, and Isabel Roldán-Ruiz

Introduction

Preliminary study of terroir significance

Hop terroir study

Genetic fingerprinting

Biochemical fingerprinting

Case studies

cv. Amarillo

cv. Cascade

Conclusions

PRELIMINARY STUDY OF TERROIR SIGNIFICANCE

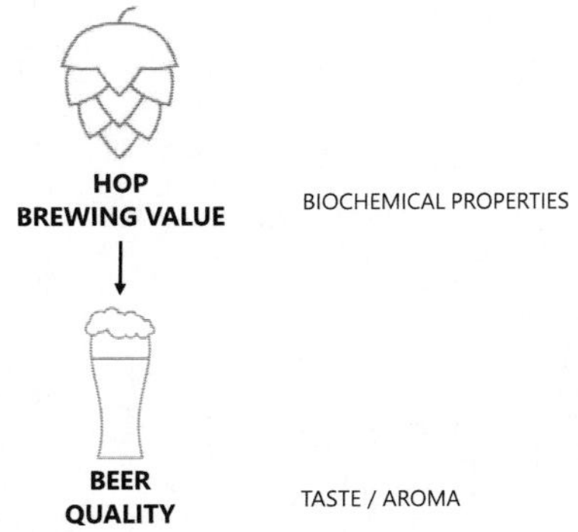

UNIVERSITEIT GENT

PROEF BROUWERIJ

PRELIMINARY STUDY OF TERROIR SIGNIFICANCE

Specific case: Amarillo single-hop beer

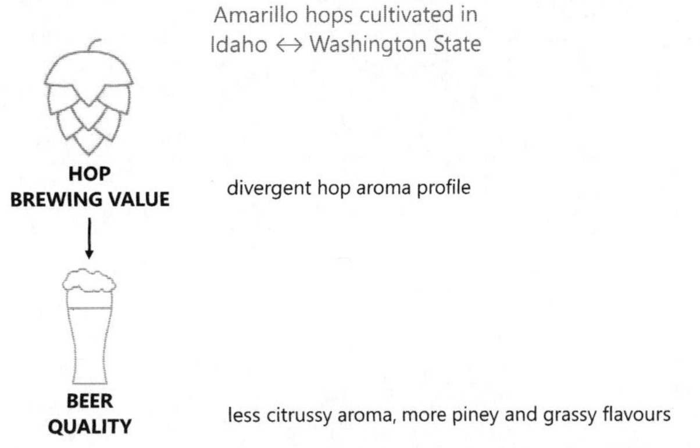

UNIVERSITEIT GENT

PROEF BROUWERIJ

Van Holle A., Van Landschoot A., Roldán-Ruiz I., Naudts D., and De Keukeleire D.
The brewing value of Amarillo hops (*Humulus lupulus L.*) grown in northwestern USA: A preliminary study of terroir significance.
J. Inst. Brew., 2017, 123(3): 312-318.

Preliminary study of terroir significance

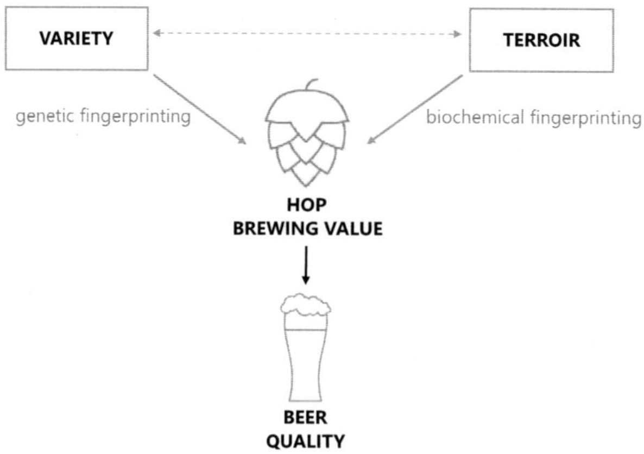

UNIVERSITEIT GENT

PROEF BROUWERIJ

5

Introduction

Preliminary study of terroir significance

Hop terroir study

Genetic fingerprinting

Biochemical fingerprinting

Case studies

cv. Amarillo

cv. Centennial

Conclusions

HOP TERROIR STUDY

Genetic and biochemical fingerprinting HOPS

Biochemical and sensory profiling single hop BEERS

UNIVERSITEIT GENT

PROEF BROUWERIJ

7

HOP TERROIR STUDY

http://blog.mikkeller.dk/mikkeller-launches-terroir-series

RECIPE

NEIPA style

Original gravity 16°P
Alcohol 7% ABV

Hop dosage (pellets T90)
- Late hopping: 250 g/hL
- Dry hopping: 1000 g/hL

UNIVERSITEIT GENT

PROEF BROUWERIJ

8

HOP TERROIR STUDY

Genetic and biochemical fingerprinting HOPS

- Genetic fingerprinting
  - SNP markers (GBS genotyping-by-sequencing)

  → Authenticity control of hop batches toward varietal origin

UNIVERSITEIT GENT

PROEF BROUWERIJ

Van Holle A., Muylle H., Ruttink T., Van Landschoot A., Haesaert G., Naudts D., De Keukeleire D., and Roldán-Ruiz I. Single Nucleotide Polymorphisms and biochemical markers as complementary tools to characterize hops (Humulus lupulus L.) in brewing practice. *J. Agric. Food Chem.*, 2019, 67: 3761-3771.

Phylogenetic relationships among 56 hop varieties

Cluster analysis (Nei, UPGMA)

based on 1,830 polymorphic SNP markers

- 48 unique genetic fingerprints
- 3 groups of somaclonal variants (identical genetic fingerprints)

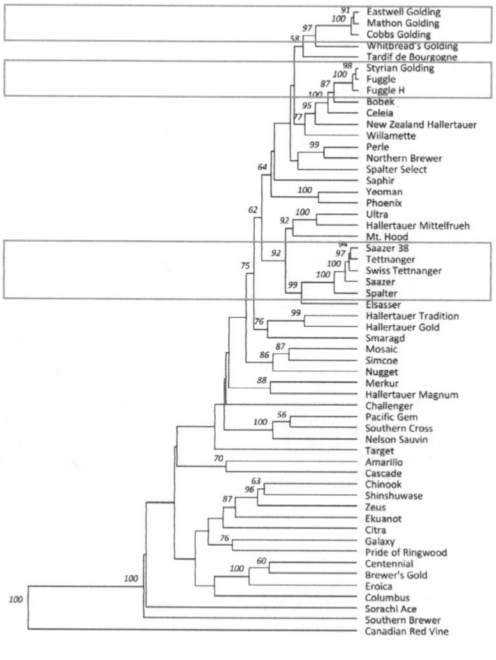

UNIVERSITEIT GENT

PROEF BROUWERIJ

HOP TERROIR STUDY

Genetic and biochemical fingerprinting HOPS

- Genetic fingerprinting
- Biochemical fingerprinting
  - Hop acids (ASBC Hops-6A) and HSI (ASBC Hops-12)
  - Hop oil content (EBC 7.10)
  - Hop aroma profiling (in-house HS-SPME-GC-MS method)

  → Classification of hops according to growth location

UNIVERSITEIT GENT

PROEF BROUWERIJ

Van Holle A., Muylle H., Ruttink T., Van Landschoot A., Haesaert G., Naudts D., De Keukeleire D., and Roldán-Ruiz I. Single Nucleotide Polymorphisms and biochemical markers as complementary tools to characterize hops (Humulus lupulus L.) in brewing practice. *J. Agric. Food Chem.*, 2019, 67: 3761-3771.

HOP TERROIR STUDY

Biochemical and sensory profiling single hop BEERS

- Aroma profiling (HS-SPME-GC-MS)
- Sensory evaluation
  - Triangle tests
  - Descriptive analysis

ODOUR main / side impression(s)

AROMA main / side impression(s)

TASTE

- Bitter intensity
- Bitterness quality
- After-bitterness quality
- Astringency

GLOBAL APPRECIATION

UNIVERSITEIT GENT

PROEF BROUWERIJ

12

CASE STUDY – AMARILLO

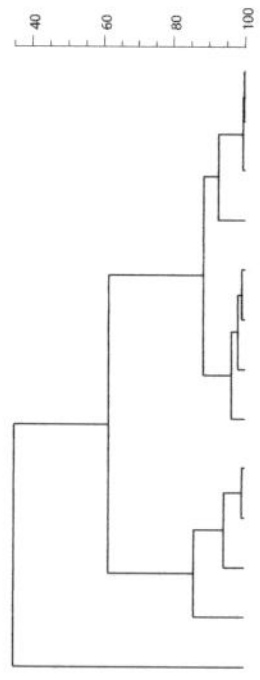

α-acids (%)	α/β	oil (mL/100g)			
11.0	1.8	2.4		AMA_WA(1)_2017	Washington, USA
9.8	1.7	2.0		AMA_WA(2)_2017	Washington, USA
9.9	1.4	2.5		AMA_WA(1)_2015	Washington, USA
7.9	1.4	2.0		AMA_WA(2)_2015	Washington, USA
8.8	1.3	2.5		AMA_WA(1)_2016	Washington, USA
7.5	1.3	1.6		AMA_WA(2)_2016	Washington, USA
7.5	1.3	1.8		AMA_WA(3)_2016	Washington, USA
7.5	1.3	2.2		AMA_WA(3)_2015	Washington, USA
6.7	1.5	0.9		AMA_ID(1)_2015	Idaho, USA
8.9	1.6	1.1		AMA_ID(1)_2016	Idaho, USA
7.0	1.4	1.4		AMA_ID(1)_2017	Idaho, USA
6.1	1.3	0.8		AMA_DE(1)_2017	Germany
8.8	2.4	2.5		CEN_WA(1)_2016	Washington, USA

Cluster analysis (Pearson, UPGMA) based on hop aroma profile

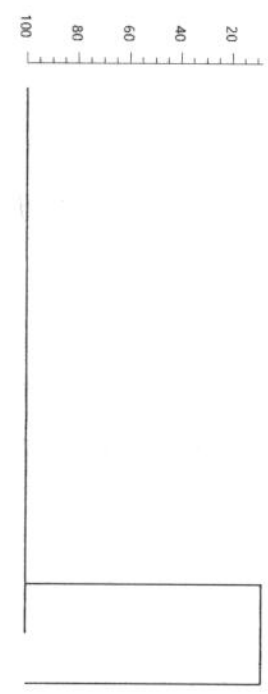

Cluster analysis (Pearson, UPGMA) based on SNP genotyping data

CASE STUDY – AMARILLO

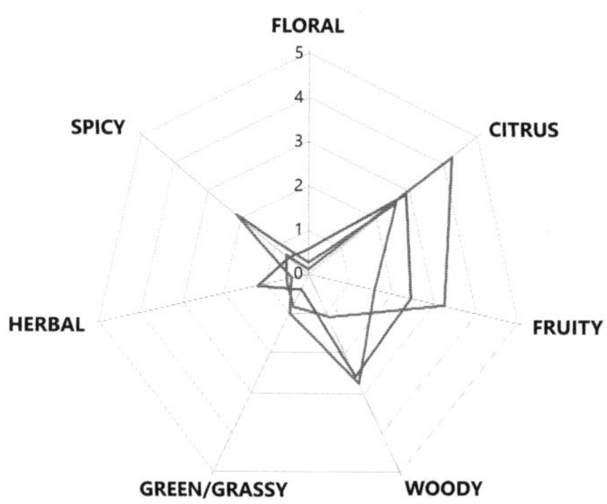

	WA, USA	ID, USA	Germany
CITRUS	grapefruit, orange, lemon	grapefruit	orange, tangerine
FRUITY	tropical fruits, lychee, apricot		green apple, tropical fruits
WOODY		resin	resin
SPICY		pepper	

UNIVERSITEIT GENT

PROEF BROUWERIJ

CASE STUDY – AMARILLO

	WA, USA	ID, USA	Germany
Bitter intensity			
(score 0-8)	5	4	4
Bitterness quality			
pleasant	18	11	14
neutral	1	5	5
unpleasant	1	5	0
After-bitterness			
pleasant	14	10	11
neutral	5	6	7
unpleasant	1	5	1
not perceivable	0	0	0
Astringency			
(score 0-8)	3	4	4
Global appreciation			
(score 0-10)	6.4	5.3	5.7

TRIANGLE TESTS

WA, USA vs ID, USA: 20/25 correct

WA, USA vs Germany: 13/24 correct

UNIVERSITEIT GENT

PROEF BROUWERIJ

CASE STUDY – CASCADE

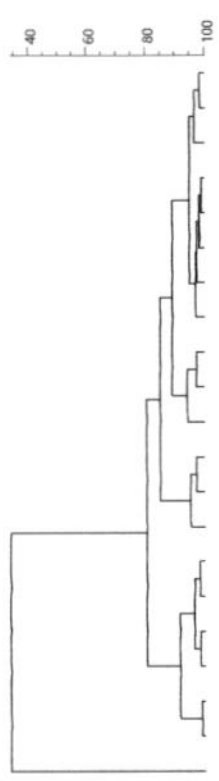

α-acids (%)	α/β	oil (mL/100g)		
7.2	1.0	1.9	**CAS_OR(1)_2015**	Oregon, USA
6.1	1.0	0.9	**CAS_WA(1)_2016**	Washington, USA
8.6	1.2	1.3	**CAS_OR(1)_2016**	Oregon, USA
6.8	1.3	1.1	**CAS_OR(1)_2017**	Oregon, USA
6.4	1.0	0.9	**CAS_WA(2)_2016**	Washington, USA
4.8	0.7	1.0	**CAS_ID(1)_2015**	Idaho, USA
6.1	1.0	0.9	**CAS_ID(1)_2016**	Idaho, USA
4.8	0.9	0.9	**CAS_ID(1)_2017**	Idaho, USA
6.2	1.2	0.8	**CAS_WA(1)_2015**	Washington, USA
6.8	1.1	0.8	**CAS_WA(2)_2015**	Washington, USA
6.4	1.1	1.4	**CAS_WA(1)_2017**	Washington, USA
7.9	1.3	1.1	**CAS_AU(1)_2016**	Australia
6.4	1.3	1.1	**CAS_AU(1)_2018**	Australia
7.6	1.4	0.8	**CAS_AU(1)_2017**	Australia
6.9	1.0	1.6	**CAS_DE(1)_2015**	Germany
6.5	1.1	1.3	**CAS_BE(1)_2016**	Belgium
8.4	1.2	0.9	**CAS_DE(1)_2016**	Germany
6.3	0.9	1.0	**CAS_DE(1)_2017**	Germany
5.1	1.0	0.8	**CAS_BE(1)_2015**	Belgium
4.7	1.3	0.4	**CAS_BE(1)_2017**	Belgium
8.8	2.4	2.5	**CEN_WA(1)_2016**	Washington, USA

Cluster analysis (Pearson, UPGMA) based on hop aroma profile

Cluster analysis (Pearson, UPGMA) based on SNP genotyping data

CASE STUDY – CASCADE

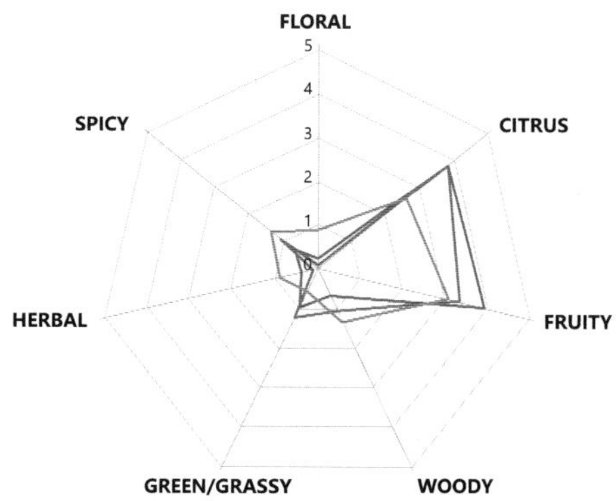

	Germany	Australia	WA, USA
CITRUS	grapefruit, tangerine	grapefruit	grapefruit
FRUITY	lychee, apple	tropical fruits	tropical fruits
WOODY		resin	
GREEN			grassy
HERBAL		green tea	
SPICY		pepper	

UNIVERSITEIT GENT

PROEF BROUWERIJ

17

CASE STUDY – CASCADE

	Germany	Australia	WA, USA
Bitter intensity			
(score 0-8)	3	5	4
Bitterness quality			
pleasant	16	14	18
neutral	5	5	2
unpleasant	0	1	0
After-bitterness			
pleasant	11	13	18
neutral	8	2	2
unpleasant	0	5	0
not perceivable	2	0	0
Astringency			
(score 0-8)	3	4	3
Global appreciation			
(score 0-10)	5.8	6.3	6.7

TRIANGLE TESTS

Germany vs Australia: 20/24 correct

Australia vs WA, USA: 21/24 correct

Germany vs WA, USA: 17/26 correct

UNIVERSITEIT GENT

PROEF BROUWERIJ

18

Introduction

Preliminary study of terroir significance

Hop terroir study

- Genetic fingerprinting
- Biochemical fingerprinting
- Case studies
  - cv. Amarillo
  - cv. Cascade

Conclusions

Conclusions and Future Perspectives

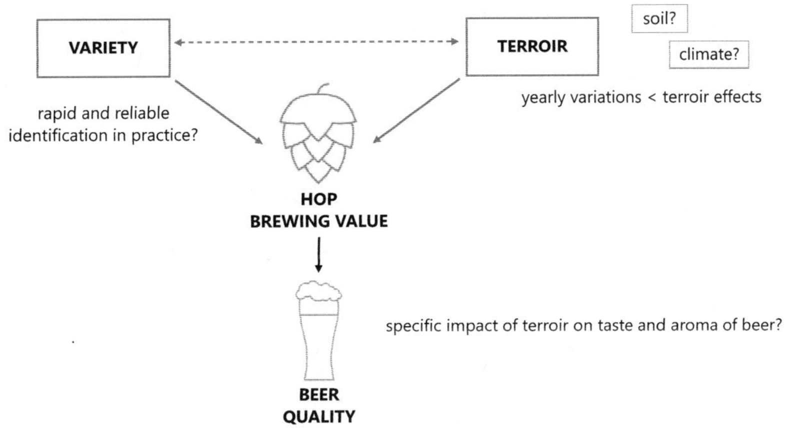

UNIVERSITEIT GENT

PROEF BROUWERIJ

20

THANKS TO

- De Proefbrouwerij team
- Hopgrowers/-distributors
- Research partners

  prof. Geert Haesaert
  prof. dr. Isabel Roldán-Ruiz
  ereprof. dr. Anita Van Landschoot
  ereprof. dr. Denis De Keukeleire
- YOU FOR YOUR ATTENTION

水と生きる SUNTORY

37TH CONGRESS EBC ANTWERP EUROPEAN BREWERY CONVENTION 02–06 June 2019

Effect of harvest time on polyphenol profiles of Saaz hops and beer taste

Takeshi Kumagai, Takako Inui, Takamasa Hasegawa, Taichi Maruhashi, Akira Kogin and Seisuke Takaoka

Suntory Beer Ltd., Beer Development Department, Japan

Outline

- Background
- Effect of hop harvest time
- Effect of hop addition timing
- Conclusion and future work

Hops are the spirit of beer

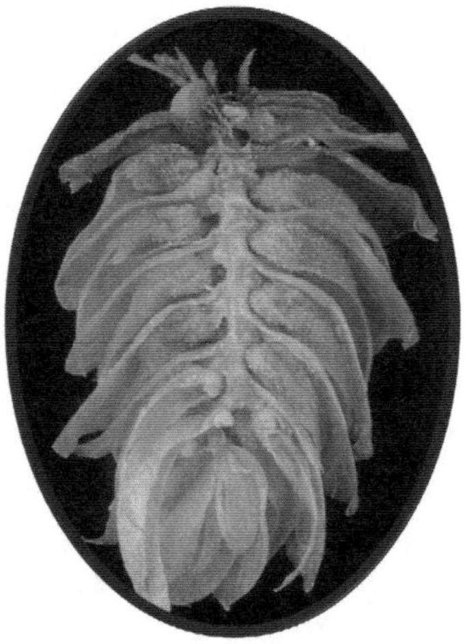

Aroma
Terpenes etc.

Taste
Polyphenols etc.

Bitterness
Alpha acid etc.

Foam
Alpha acid etc.

What are the influential factors that control the quality of hops in beer during the brewing process ?

Effect of harvest time on hops flavor profiles

Phases of Saaz hops cultivation over the year

Aroma profiles with respect to harvest time

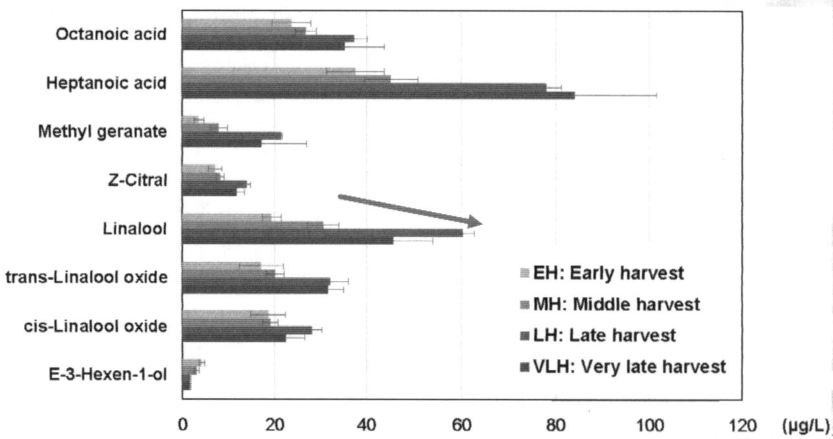

Inui et al. (2016), WBC.

Effect of harvest time on thiols in hop cones

5/31

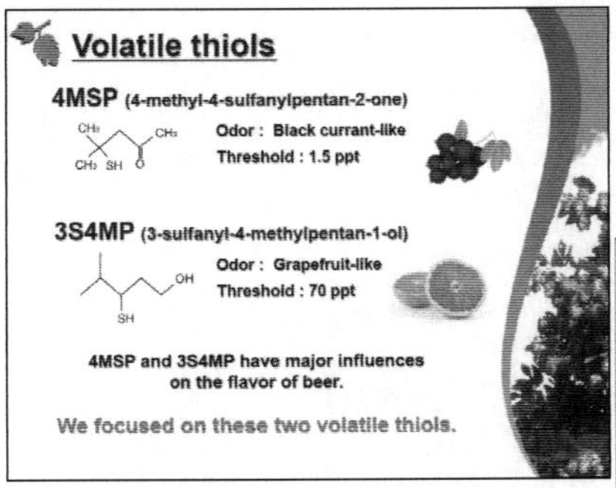

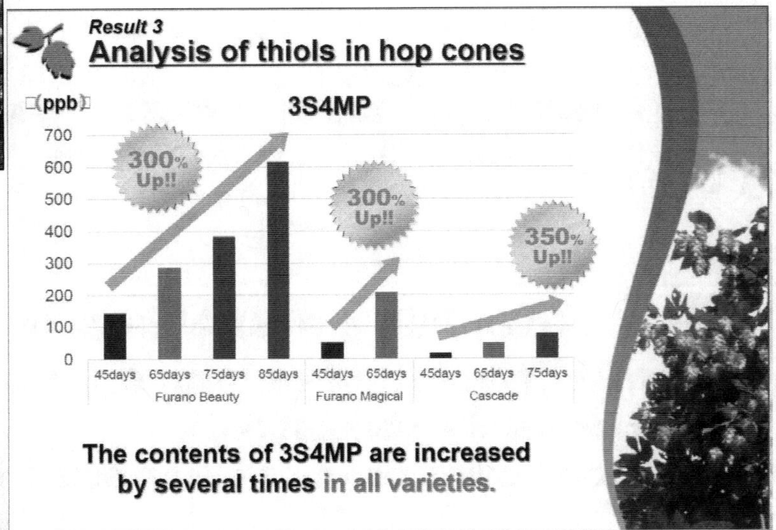

Uemoto et al. (2017), ASBC Annual Meeting.

Our focus

6/31

What about the effect of hop harvest time on beer taste?

Chemical composition of hops

Chemical composition	Weight (%)	Influence on beer flavor	
Moisture	8～12		
Alpha-acid	2～12	+++	▶Bitterness
Beta-acid	1～10	+	
Oil	0.5～5.0	++	▶Aroma
Polyphenol	**2～5**	++	▶Taste
Protein	1		
Amino acids	0		
Cellulose	40～50		
Monosaccharide	2		
Pectin	2		
Lipid and Fatty acid	～25		
Mineral	10		

Brewery Convention of Japan (1999)

Our goal

7/31

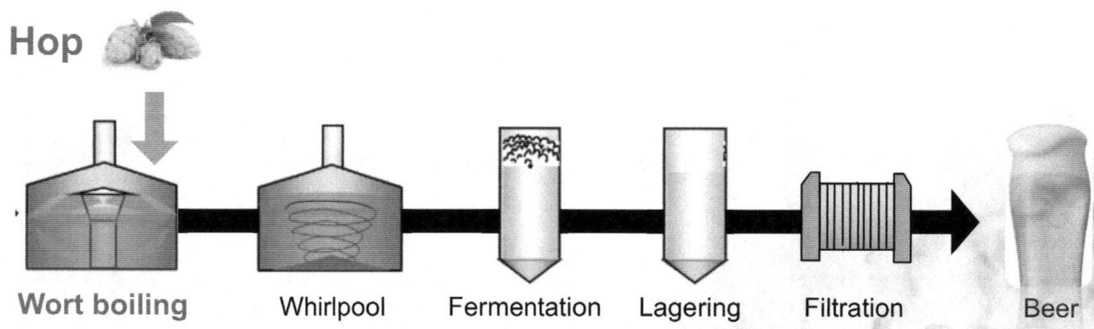

To obtain a desirable beer taste by controlling hop polyphenol profiles

Hop itself – Harvest time

Wort boiling process – Hop addition timing

Experimental procedure in this study

8/31

Trial 1. What is the effect of the hop harvest time on beer taste?

Trial 2. What are ideal hop addition conditions based on hop polyphenol profiles?

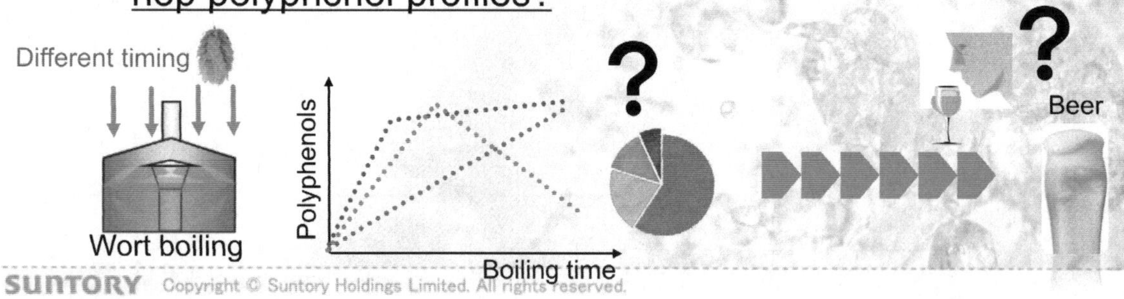

Outline

- Background
- **Effect of hop harvest time**
- Effect of hop addition timing
- Conclusion and future work

Brewing trial Ⅰ (materials and methods)

Hops

- 2017 crop Saaz variety with three different harvest times

Sample	Harvest time*	α-acid	β-acid	Oil mL/100g
EH (Early harvest)	37 days	2.5%	5.0%	0.47
MH (Middle harvest)	47 days	2.8%	5.1%	0.64
LH (Late harvest)	65 days	2.9%	6.8%	0.90

*Harvest time is defined as the number of days from flowering to harvest.

- Other conditions:
  - same growing area, same clone, root age (18), pruning date (mid-April)

Brewing

- Brewing Size: 100 L
- Hop addition: 129 g pulverized hops were added at the beginning of wort boiling
- Wort boiling time: 90 min
- Wort: All malt beer, 12.5% original gravity
- Fermentation: With lager yeast
- Fermentation temp.: 12.0 °C

Sensory evaluation of final beer

	EH (Early harvest)	MH (Middle harvest)	LH (Late harvest)
Bitterness (BUs)	10.9	13.0	14.0
Total Polyphenol (ppm)	173	158	169

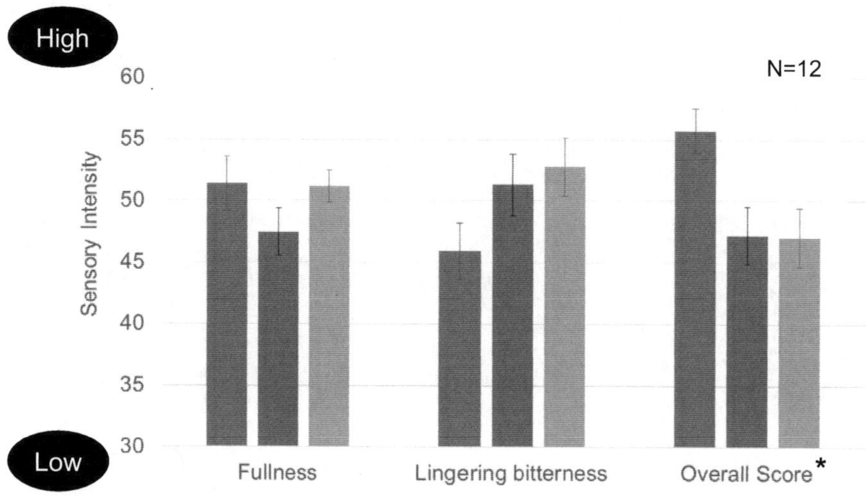

*Overall score was evaluated, not including the aspect of aroma

Identification of key polyphenol components

Fractionation of hop polyphenols

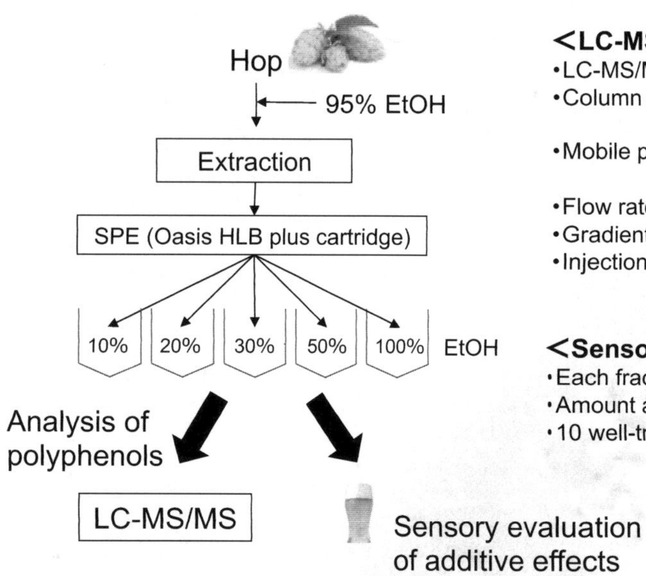

<LC-MS/MS conditions>

- LC-MS/MS : SHIMADZU LCMS-8040
- Column : Luna® 5 μm C18(2) 100 Å, LC Column 150 × 4.6 mm
- Mobile phase A: water containing 0.1% formic acid
  B: acetonitrile containing 0.1% formic acid
- Flow rate : 0.7 mL/min
- Gradient (min/%B) : 0/5, 5/5, 35/98,40/98,40/5,50/5
- Injection : 10 μL

<Sensory evaluation>

- Each fraction was added to un-hopped beer
- Amount added: A typical amount contained in beer
- 10 well-trained panelists

Asano et al. (2018), Brewing summit

Additive effects of polyphenols on beer taste

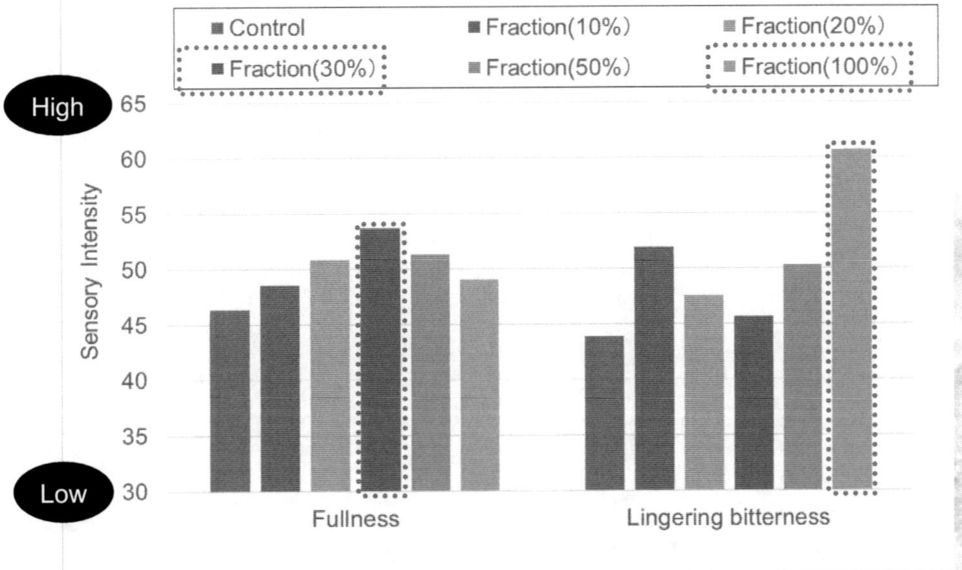

- Fraction of 30% ethanol contributed to fullness
- Fraction of 100% ethanol contributed to lingering bitterness

Hop polyphenol composition in each fraction

Fraction (30 % EtOH)

Fraction (100 % EtOH)

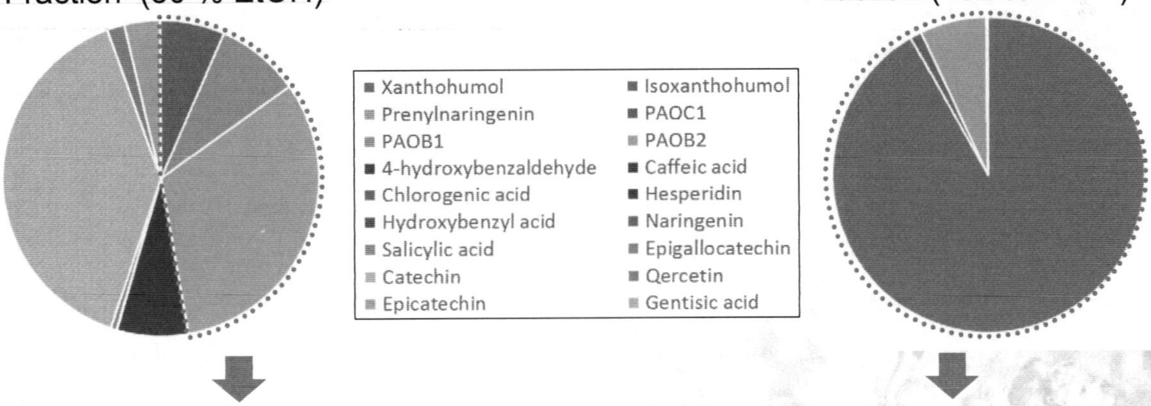

Proanthocyanidin oligomers (PAO2,3)

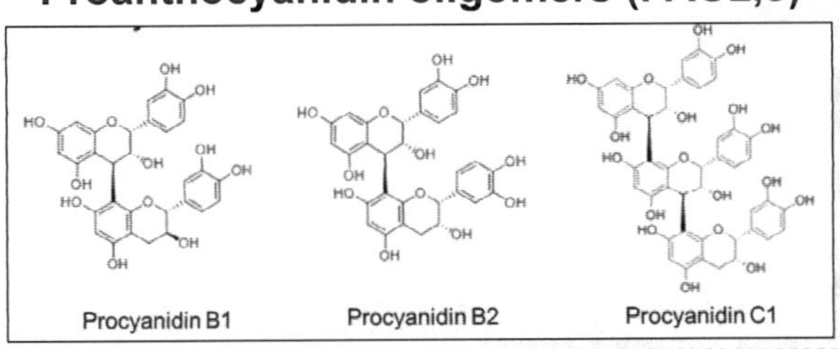

Prenylflavonoids

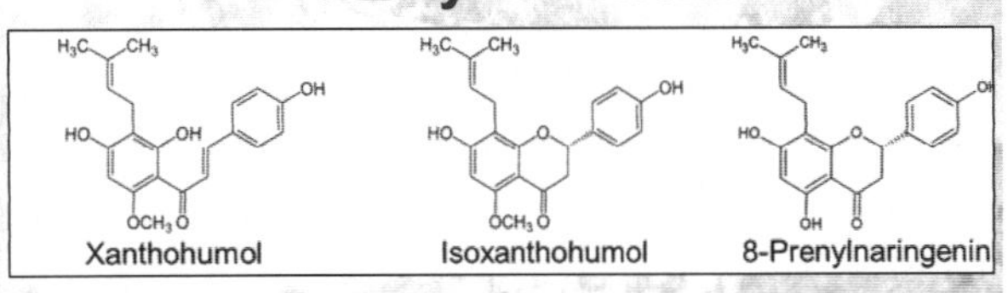

Effect of PAO2,3 on beer taste

EtOH 30% Fraction

↓

Colum (Sephadex LH-20)

by EtOH

Fra-1 | Fra-2 | Fra-3 | Fra-4 | Fra-5

SPE (Oasis HLB plus cartridge)

PAO2,3

Sensory additive test

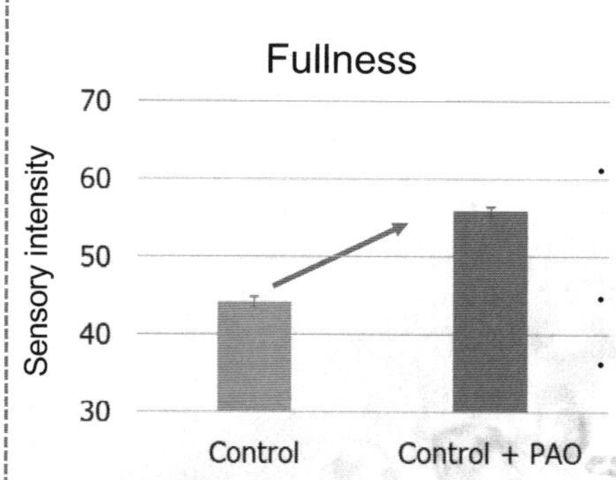

- Fraction was added to all malt beer.
- Amount added: 0.25 ppm
- Ten well-trained panelists

PAO2,3 (B1/B2 and C1) contributed to beer **fullness**

Effect of prenylflavonoids on beer taste

Control

CO2Ex. hop

↓

Boiling

↓

Freeze dry

α-Acid

Test

Hop

↓

Purification of Lupulin

↓

Boiling

↓

Freeze dry

Prenylflavonoids + α-Acid

Sensory additive test

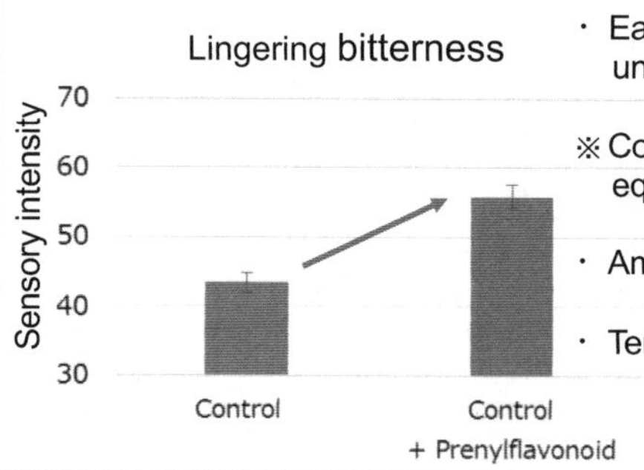

- Each fraction was added to un-hopped beer

※Control contained α-acid equal to test.

- Amount added: 1.2 ppm
- Ten well-trained panelists

Prenylflavonoids contributed to beer **lingering bitterness**

Polyphenol profiles of final beer

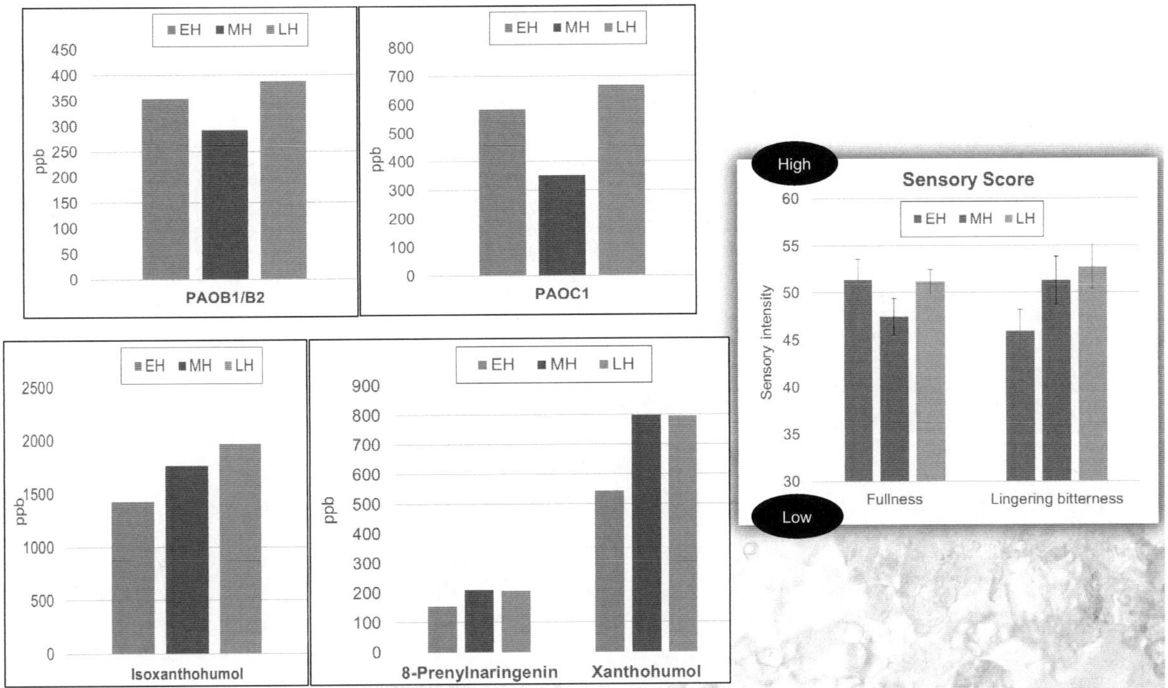

Long-term evaluation of hops

We evaluated hop polyphenol profiles over four consecutive years

Design of experiment

Harvest year (2011–2014) **X** **Harvest time** (8/15–9/21)

Harvest year	2011				2012				2013				2014			
Farm	Rybnany				Rybnany				Steknik				Steknik			
Pruning date	April 15				April 16				April 20				April 15			
Harvest time(days)	22	29	50	57	23	32	51	60	26	37	53	61	29	40	57	64
Sample	E	M	L	VL	E	M	L	VL	E	M	L	VL	E	M	L	VL

Other conditions:

- Virus free
- Same clone
- Same type of soil (brown)
- Root age (10–20 years old)

- E: Early harvest
- M: Middle harvest
- L: Late harvest
- VL: Very late harvest

 Inui et al. (2017), EBC Congress

Polyphenol profiles of hops over 4 years

19/31

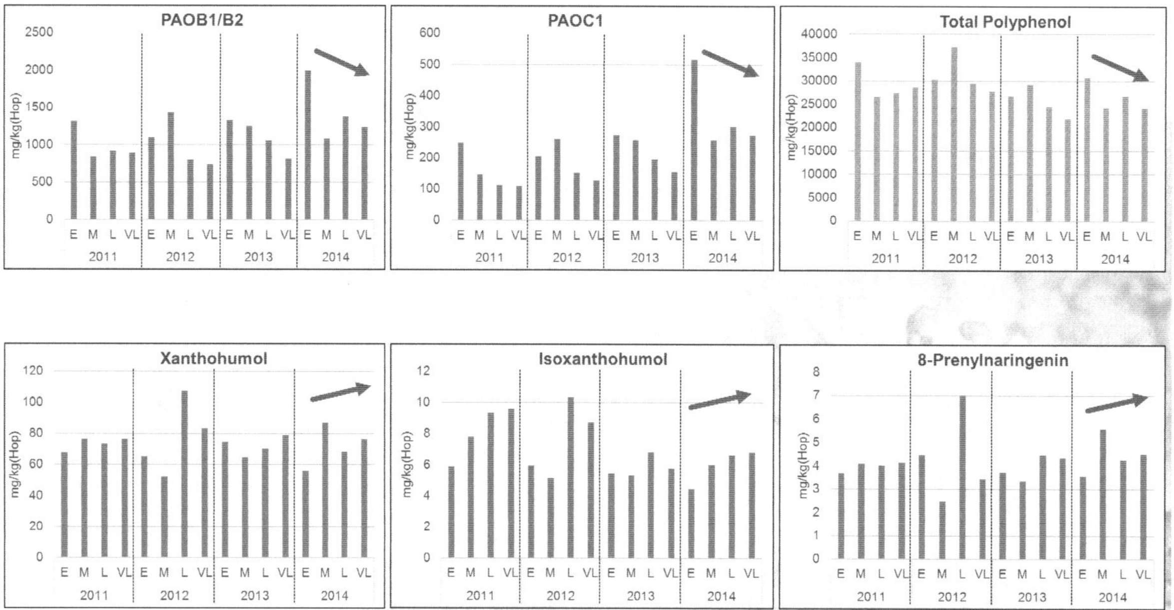

Inui et al. (2017), EBC Congress

Summary (1)

20/31

There were significant sensory differences between beers brewed with hops that were harvested for different times

Particular hop polyphenols affected the taste of beer
- PAO2,3 contributed to fullness in beer
- Prenylflavonoids contributed to lingering bitterness in beer

Polyphenol profiles in hops were significantly changed by their harvest time
- the contents of PAO2,3 are higher at earlier harvest times
- the contents of prenylflavonoids are higher at later harvest times

Outline

- Background
- Effect of hop harvest time
- Effect of hop addition timing
- Conclusion and future work

Evaluation of the behavior of polyphenols during wort boiling

Preliminary experiment - wort boiling condition

- Scale: 100 L pilot brewery
- Hop: 2017 crop Saaz variety
- Hop addition:
  150 g pulverized hops were added at the beginning of wort boiling
- Boiling time: 90 min
- Wort: All malt beer, 12.5% original gravity
- Sampling: every few minutes
- Polyphenols analysis: LC-MS/MS

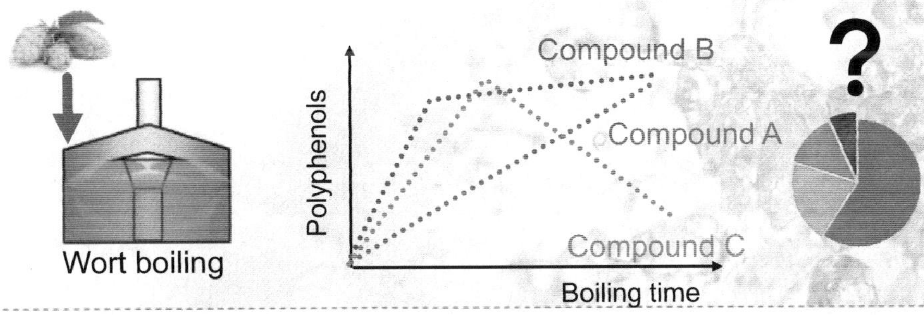

The behavior of polyphenols during wort boiling

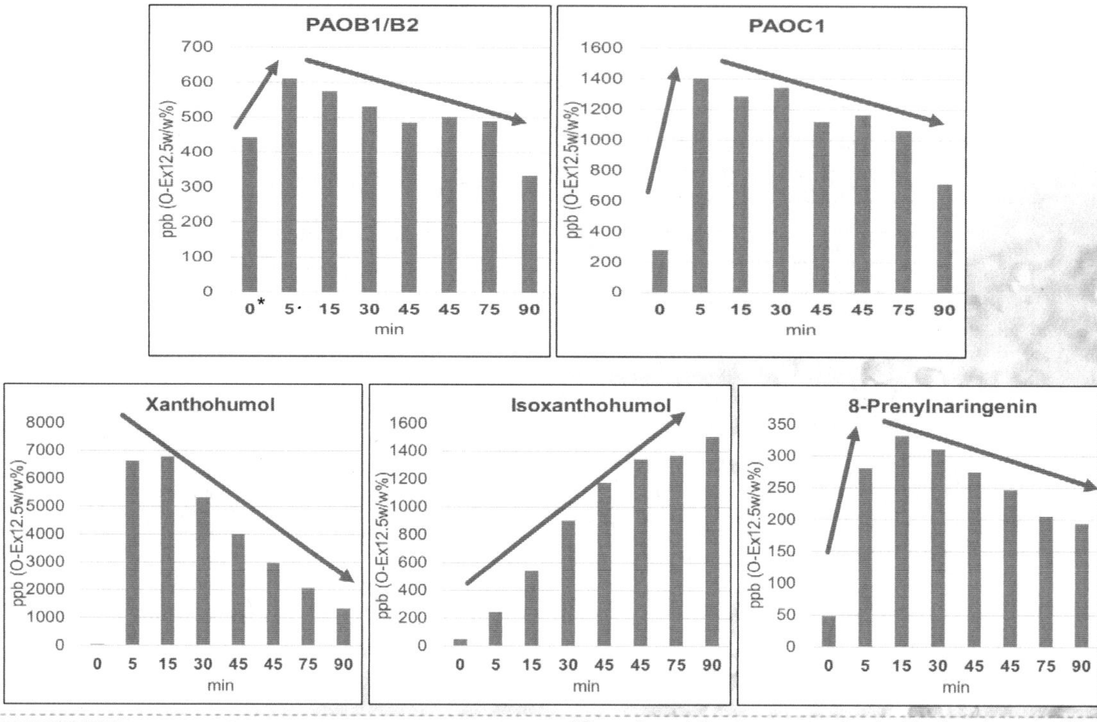

*Samples at 0 min were collected before hop addition.

Brewing trial Ⅱ (materials and methods)

How can we obtain a desirable taste, based on hop polyphenols?
~Trade-off between the positive and negative character in beer taste~

Hops

- 2017 crop Saaz variety with early harvest time

Sample	Harvest time	α-acid	β-acid	Oil mL/100g
EH (Early harvest)	37 days	2.5%	5.0%	0.47

Brewing (100 L pilot brewery)

- Hop addition

Sample	Hop addition timing (from the start of boiling)	Amount
E0	0 min	129 g/HL
E45	45 min	129 g/HL
E90	90 min	129 g/HL
E0/45/90	0, 45, and 90 min	43 g/HL per each timing

- Other conditions: same as brewing trial 1

25/31

Sensory evaluation of final beer

	E0	E45	E90	E0/45/90
Bitterness (BUs)	10.9	11.4	7.6	11.3
Total Polyphenol (ppm)	173	165	159	165

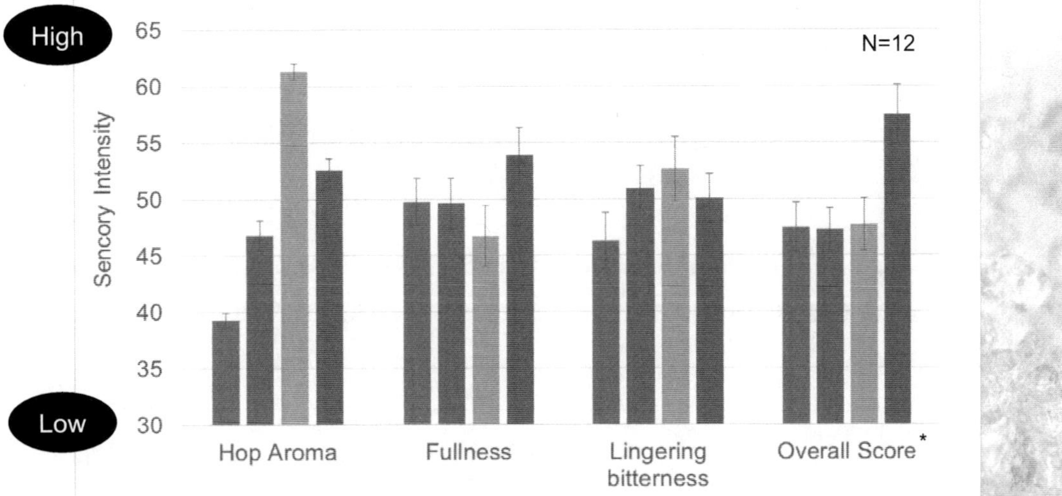

*Overall score was evaluated, not including the aspect of aroma

26/31

Polyphenol profiles of final beer

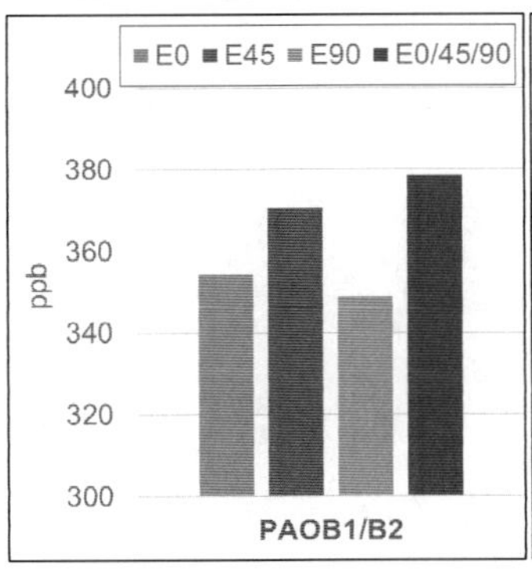

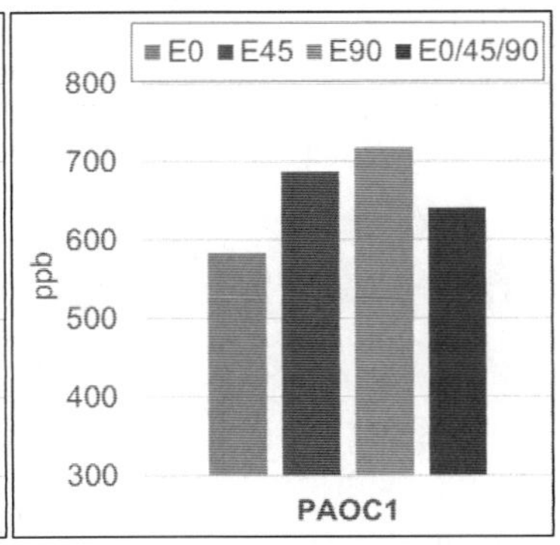

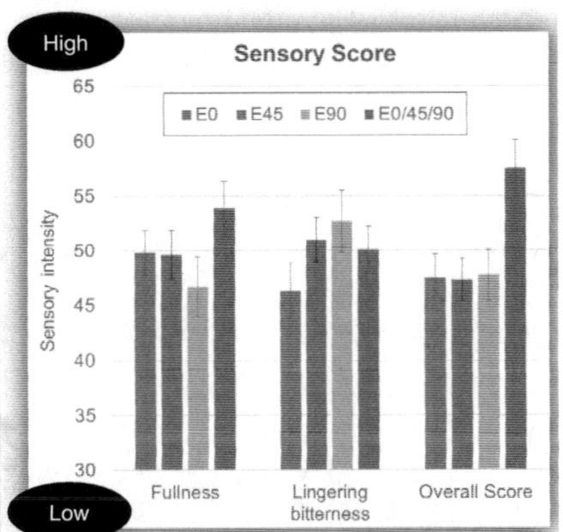

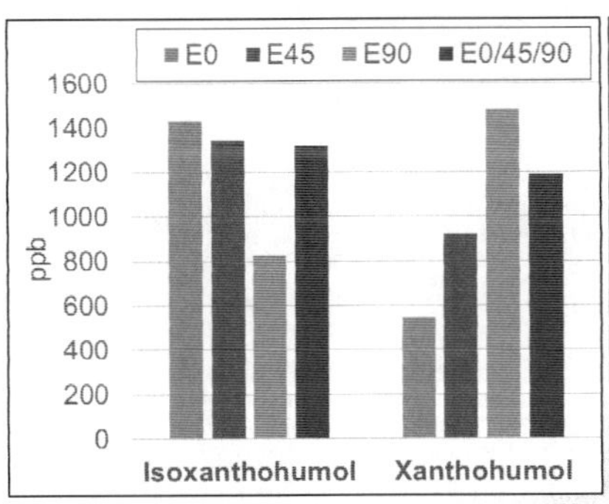

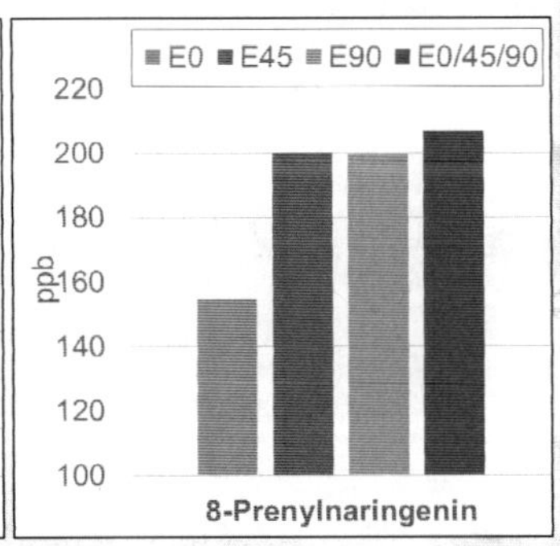

Summary (2)

During wort boiling, hop polyphenol profiles changed
- Rapid increase and gradual decrease of PAO2,3 and 8-prenylnaringenin (extraction, degradation, polymerization, precipitation, etc.)
- Continuous decrease of xanthohumol and increase of isoxanthohumol (isomerization)

Polyphenol profiles of beer and its taste were greatly affected by controlling hop addition timing

The addition of hops multiple times could be one of the solutions to obtaining greater fullness and lower lingering bitterness

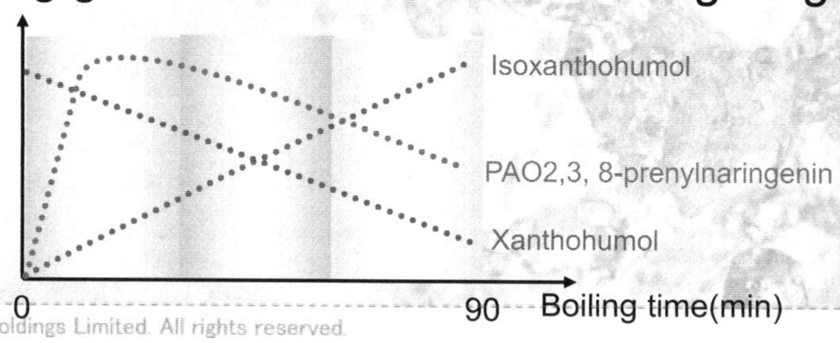

Outline

- Background
- Effect of hop harvest time
- Effect of hop addition timing
- Conclusion and future work

Conclusions

Hop polyphenols significantly affected the beer taste
Polyphenol profiles of beer can be controlled by altering the harvest time and also the timing of hops addition

Materials	×	How to use

Choice depending on the purpose

- Early harvested hops for more fullness and less lingering bitterness
- Late harvested hops for aroma

×

- Early addition for less lingering bitterness
- Late additinon for aroma
- Combination of timing and times

Effective and optimum use of hops provide good opportunities to create desirable character in beer

Future work

The following studies are required:

Comprehensive effects of other ways and factors for controlling hops and beer polyphenol profiles through the entire brewing process
(e.g., hop processing method, various hop addition conditions)

Contribution of other hops polyphenols to beer taste
(e.g., monophenol, flavonol, flavanol, flavanone)

Acknowledgments

- Hop Research Institute Co., Ltd. Saaz, Czech Republic

- Zatec Hop Company Ltd. Saaz, Czech Republic

Thank you for your kind attention

SUNTORY

L55 - INSIGHTS ON THE LAGER BEER VOLATILE TERPENIC COMPOUNDS THROUGH AN ADVANCED CHROMATOGRAPHIC TOOL

Cátia Martins[1]; Tiago Brandão[2]; Adelaide Almeida[3]; Sílvia M. Rocha[1]

1 - Departamento de Química & QOPNA, Universidade de Aveiro, Campus Universitário Santiago, 3810-193 Aveiro, Portugal; 2 - Super Bock Group, Via Norte, 4465-764 Leça do Balio, Portugal; 3 - Departamento de Biologia & CESAM, Universidade de Aveiro, Campus Universitário Santiago, 3810-193 Aveiro, Portugal

Abstract

Beer volatile terpenic compounds are originated from a variables' network (plant raw-materials metabolism, yeast metabolism [1], etc), and may undergo modifications during brewing. Therefore, for increasing their molecular understanding, this work aims to achieve the comprehensive characterization of volatile terpenic compounds on lager beer. It was used an advanced multidimensional chromatographic methodology previously optimized [2,3]. Several lager beers produced at different countries and breweries were analyzed.

This targeted analysis allowed the putative identification of 94 mono and sesquiterpenic compounds, from 6 chemical families: alcohols, aldehydes, esters, ketones, hydrocarbons, and oxide. Their volatile profiles allowed to achieve the lager beer terpen-typing, i.e. the samples' category clustering depending on 2 main classes: macro and micro-brewer beers. As far as we know, this is the most detailed screening of lager beer terpenic compounds in beer, which may be further applied in large-scale studies in different contexts, e.g. to comprehend distinctive beer styles or for beer typing.

Acknowledgements: FCT/MEC, for the financial support to QOPNA (FCT UID/QUI/00062/2013) and CESAM (FCT UID/AMB/50017/2013, project PEst-C/MAR/LA0017/2013). FCT for the financial support of C. Martins PhD grant (SFRH/BD/77988/2011), through the program POPH/FSE.

References:

[1] Martins C., Brandão T., Almeida A., Rocha S. M. J. Sep. Science, **2017**, *40*, 2228–2237.

[2] Martins, C.; Brandão, T.; Almeida, A.; Rocha, S.M. J. Sep. Science **2015**, *38*, 2140-2148.

[3] Martins, C.; Brandão, T.; Almeida, A.; Rocha, S.M. Food Res Int. **2018**, *114*, 199-207.

Short biography

Cátia Martins is MSc in Food Biotechnology (University of Aveiro). In 2018, she obtained the degree of Doctor in Biochemistry from University of Aveiro. The main objective of her PhD thesis was to advance in lager beer aroma comprehension through yeasts and beers volatile metabolomic profiling. Currently, she has a Pos-Doc scholarship at University of Aveiro.

L56 - ARABINOXYLANS IN THE BREWING PROCESS - AN UNDERESTIMATED SUBSTANCE CLASS?

Michael Kupetz[1]; Martina Gastl[1]; Thomas Becker[1]

1 - Lehrstuhl für Brau- und Getränketechnologie, TU München, Germany

Abstract

Problem-free processing of raw materials with a short processing time and optimal yield is a basis in modern brewing. According to the literature, cytolytic substances like β-glucans and arabinoxylans in malt are particularly important due to their impact on lautering and filterability. For this reason, a method for the classification of the raw materials with respect to arabinoxylan and β-glucans is important. According to EBC-analytica, β-glucans can be analysed using microtiter scale, optimal malt extraction methods are known and standard values can be found. However, comparable information is not available for arabinoxylans. Thus, aim of this study was to investigate various influencing factors like the impact of oxygen due to cross-linking and extraction temperatures on total arabinoxylans content. After identification of an optimal mashing procedure, standard values for malt were recorded. In addition, molecular weight distribution of arabinoxylans in malts with different modifications was investigated. Total arabinoxylans ranged between 400 and 1000 mg/l in barley malt, whereby a higher total concentration could be measured in high-modified varieties. In addition, these varieties had a higher proportion of high-molecular-weight fractions (> 50 kDa) of up to 35%. Besides, activity of arabinoxylan-degrading enzymes such as xylanase and arabinofuranosidase are important for characterization of raw materials. It could be shown that medium-modified varieties had a higher xylanase activity. Characterizing the substance class of arabinoxylans and their degrading enzymes with the help of the described methods can result in a better monitoring of brewing processes, which not only affects process times, but the beer quality.

Short biography

Michael Kupetz graduated in 2011 as a graduate engineer for brewing and beverage technology at the Technical University of Munich. In July 2017 he received his doctorate at the Chair of Brewing and Beverage Technology in Weihenstephan on the topic "Comparative identification of filtration-inhibiting substances in the membrane and kieselguhr filtration of beer". In addition, he is head of the Analytical Laboratory for malt, wort and beer quality since July 2016. His research topics are in the fields of method development with a focus on polysaccharides and method automation as well as filterability of beer.

P002 - NON-DESTRUCTIVE DETERMINATION OF HOP CONSTITUENTS BY NEAR INFRARED SPECTROSCOPY (NIRS)

Christian Temme[1]; Florian Schüll[2]; Kirsten Kramer[3]

1 - RMI Analytics; 2 - HVG; 3 - Eurofins QTA

Abstract

NIRS is a rapid, non-destructive and cost effective analytical technique which is gaining wide acceptance in a variety of industries. For over 30 years, there have been efforts to establish NIRS analysis for hops and hop products but the accuracy has been a concern for commercial applications. The raw data from NIRS, a multi-point spectrum, contains broad absorbance features of several overlapped constituents, requiring chemometrics and multivariate regression for the analysis. To implement NIRS successfully, two key elements are critical: the quality of the data and careful construction of the calibration. Data from a highest cost, high quality instrument is useless if the calibration is not constructed properly and conversely, if the spectral quality is low or lab results are of high error, then even an expert chemometrician cannot create a useful calibration. In a new approach, Fourier Transform (FT) instrumentation combined with robust chemometric modeling has allowed NIRS to be implemented with success in accuracy, reliability and ruggedness for determining hop constituents. We will present the method developments and quality indicators for the NIR calibrations for the prediction of moisture (EBC 7.2), lead conductance value (EBC 7.4 and 7.5), alpha-acid (EBC 7.7), and total oils (EBC 7.10) in pellets and powder. Based on these results we reasonably expect that further hop constituents can be calibrated in the near future and that this valid and rapid method is applicable to both, screening and routine purposes within the whole processing chain.

Short biography

Christian Temme is an analytical chemist with a long history in research and development of analytical methods and their application in the food and beverage industry.

He is the CEO of RMI Analytics GmbH in Hamburg, one of only a few independent advisory firms in the world specialized in brewing raw materials. He holds a PhD in Analytical Chemistry from the University of Jena and was the leading author of several international publications for the development of new analytical methods in selected environmental and food matrices.

P006 - NON-DISCRIMINANT ANALYSIS OF BEER AGEING CHEMISTRY

Laura Knoke[1]; Henning Schiebenhöfer[1]; Sarah Thörner[1]; Nils Rettberg[1]

1 - VLB Berlin - Research Institute for Beer and Beverage Analysis

Abstract

Economic production of high quality and stable beer is a major concern of brewers. Their brewing operations aim to maximize shelf life by reducing e.g. thermal stress and oxygen uptake. Multiple dedicated quality control systems, using either targeted instrumental and/or sensory analysis are used in industry. Whereas common instrumental methods target a specified list of analytes (e.g. aldehydes), sensory provides a comprehensive picture of beer flavor. Flavor (in)stability depends on beer style and obviously relates to interactions between several thousand molecules. Its prediction and evaluation based on analytical markers frequently fails or at least does not necessarily correlate well with sensory.

In order to tackle the complexity of events during beer staling, it seems reasonable to use analytical techniques that enable non-discriminative data collection of which untargeted metabolomic profiling (UMP) is a promising approach. UMP aims to map all metabolites (molecules < 1200 Da) of a particular sample and thereby generates a metabolic profile that is used for comparison within a sample set. UMP commonly uses chromatography interfaced with (high-resolution) mass spectrometry and bioinformatics to analyze the complex datasets.

The current work shows results of an UPLC-ESI-Q-ToF based UMP study focusing on beer ageing. Statistical analysis enabled rapid estimation of the similarities and differences within the sample set (different beer styles), which is compared with results from targeted instrumental analysis and sensory evaluation. Further, identification of metabolites that evidently distinguished beers within the sample set is described.

Short biography

Laura Knoke graduated from Technical University Berlin (Germany) with Diplom-Ingenieur (graduate engineer) in biotechnology in 2016. She is now a research associate at the Department for Special Analysis at the Research and Teaching Institute for Brewing in Berlin (VLB). Laura's scientific work is focused on the development of an untargeted metabolomics workflow and its application to brewing related topics.

P007 - SENSORIAL AND ANALYTICAL CHARACTERIZATION OF THE NEW HÜLLER "SPECIAL FLAVOUR HOPS

Klaus Kammhuber[1]

1 - Bavarian Hop Research Center

Abstract

Craft Brewers want hop varieties with unusual flavors, such as fruity, citrusy and floral aroma notes. The hop research center in Hüll has responded to this trend and has approved six new hop varieties in the recent years, which correspond to these properties.

These are Mandarina Bavaria, Hallertau Blanc, Huell Melon, Polaris, Callista and Ariana. In the presentation, the sensorial properties and a detailed presentation of the analysis data will be presented. In addition to the routine determination of the bitter substances by HPLC, Wöllmer analyzes were also carried out to determine the unspecific soft and hard resins for getting statements about the bitter quality.

The aroma substances of the new Hüller varieties are compared with the variety Cascade, as this variety is a standard for the Craftbrewers. Sensory-important sulfur compounds such as 4-mercapto-4-methyl-pentanone, 3-mercaptohexanol and 3-mercaptohexyl acetate are also detectable in the new Hüller cultivars. Extensive biogenesis studies on the oils have also been made, and it has been found that the total oil content and the composition are highly depending on the harvest date. Sulfur compounds increase strongly only at late harvest dates. Hot and dry years even increase the oil content.

Finally, the total polyphenol content was determined according to EBC 7.14 and individual polyphenols such as quercetin and kaempferiol by HPLC. This extensive analytical presentation is certainly very helpful for the brewers to orient themselves in the selection of their hop varieties.

Short biography

born 26.3.1962 in Bad Reichenhall (Bavaria)

1983 - 1990 study of chemistry at the University of Regensburg

1990 Ph.D. in organic chemistry

1991 starting at the Bavarian hop Research Center in Hüll

since 1998 head of the department hop chemistry at the Institute in Hüll

Special interests in aroma compounds and polyphenols

P012 - EFFECT OF HOP VARIETIES ON BEER AROMA DURING DRY HOPPING

Nathalie Brignier[1]; Damien Steyer[1]; Céline Clayeux[1]; Christophe Marcic[2]; Francis Heitz[3]; Antoine Wuchner[3]; Bernadette Laugel[3]

1 - TWISTAROMA; 2 - Faculté de Pharmacie de Strasbourg; 3 - Comptoir Agricole

Abstract

In recent years, the dry hopping technique was developed to produce unique beers by revealing all the aromatic potential of hops. The aim of this study was to investigate the impact of exposure time during dry hopping (4, 8 and 12 days) and hop varieties (Aramis, Barbe Rouge and Mistral) on the sensory properties and volatile compounds of dry-hopped lagers. Volatile compounds were followed from the raw material (by SPME) to the final beer by SBSE-TD-GC-MS for 4, 8 and 12 days exposure time, then beers were tasted by sensorial analysis. All the data were statistically analysed and showed that exposure time has a minor influence whereas hop varieties modify the concentration and the chemical diversity of the volatile compounds in beer. Sensory ranking tests of beers revealed that these 3 hops are different and have unique aromatic signatures.

Short biography

Damien Steyer, engineer in Biotechnology (ESBS, Strasbourg, France), PhD in Biochemistry, CEO of TWISTAROMA. To complete his doctoral work at INRA (Colmar, France) on the influence of yeast strains on wine aroma, he developed a technique that assess flavour profiles of wine and beer. This technique combined with his expertise and knowledges of flavors, allowed him to create TWISTAROMA in 2011. This company offers innovative techniques for compounds analysis (antioxidants, volatiles, vitamins...) for food industry.

P014 - EVALUATION OF WATER AND ETHANOL EXTRACTS OF LEMON VERBENA, SAGE, CINNAMON AND HIBISCUS IN THE AROMATIC AND ORGANOLEPTIC PROFILE OF BEER

Avgoustinos Evripiotis[1]; Panagiotis Tataridis[1]

1 - Department of Wine, Vine & Beverage Sciences, University of West Attica, Ag. Spyridonos str., 12243 Egaleo, Greece

Abstract

The expansion of the craft beer market has led the brewers to continuously try to create new types of beers with special organoleptic characteristics. Herbs and spices were always used in the brewing process to enhance the flavors and the aromas of the final product. In this study, four different kinds of herbs and spices and their water and alcohol extracts were analyzed. These are: Aloysia citrodora, Salvia officinalis, Hibiscus sabdariffa and Cinnamomum cassia. Using a GC-MS SPME method, the aromatic compounds of the previous samples were identified and semi-quantified. The results were assorted into basic aromatic groups and compared to show the differences between the two extraction methods. A tasting panel graded and evaluated the previous samples based on their aroma and flavor intensity of each aromatic group as well as their addition to commercial beer at five different concentrations for each extract. The results from the sensory evaluation show the preferences of the tasting panel for each extraction method and the preferred concentration in the final product. For Aloysia citrodora, beer with alcohol extract with a concentration of 2g/L of herb in the final beer was rated higher. For Salvia officinalis, it was beer with water extract of 3.5g/L. For Cinnamomum cassia, both methods had similar ratings with the preferred concentrations of water extraction of 3.5g/L and alcohol extraction of 2g/L. For Hibiscus sabdariffa, both methods had similar ratings and the chosen concentration was 3.5g/L. This methodology is a useful tool for evaluating the addition of botanicals in beer.

Short biography

Avgoustinos Evripiotis graduated in 2015 with a B.sc. in Physics from National and Kapodistrian University of Athens. In 2017, he obtained his M. Sc. degree in Brewing Science from the department of Wine, Vine & Beverage Sciences of the University of West Attica. During his master degree, he did an Erasmus internship in KU Leuven in the Faculty of Engineering Technology under the supervision of Prof. Filip Van Opstaele in "Aromatic compound analysis of different distillation methods with Juniper berries". He is a passionate homebrewer since 2009 and a founding member of the Hellenic Homebrewer Association. In 2018 he is opening his own craft brewery in Athens, Greece.

P020 - FACTORS AFFECTING HUMULINONE FORMATION IN HOPS

Philip Wietstock[1]; Keil Franziska[1]; Methner Frank-Jürgen[1]

1 - Technische Universität Berlin

Abstract

Humulinones are a constituent of hops which have a bitter value of 0.66 as compared to iso-α-acids. They are more soluble than α-acids, and are thus readily dissolving almost completely in beer during dry-hopping. The level of humulinones in hops and hop products therefore directly affects the bitterness perception of the beer produced thereof. Objective of this study was to study factors affecting humulinone formation in hops during hop processing and storage.

Eight hop varieties (HHA, HHE, HHT, HHS, HPE, HTU, HHM, HHS, crop 2017) from which two of them were from two different farms from the Hallertau region were used in this study. Hop acids, hop volatiles, and metal ion concentrations were quantified in unbaled raw hops by LC-UV/VIS, SAFE-GC/MS, or ICP-OES, respectively. Hops were ground in liquid nitrogen, then pressed into tablets, were vacuum-sealed, and stored for 4 weeks at 20 °C in the dark. All data were ultimately correlated using XLSTAT software.

Humulinone concentration ranged from 0.03 % to 0.16 % in unbaled raw hops and 0.14-0.46 % in pressed and stored hop samples. After pressing and 4-week storage, humulinone concentration increased by 182-507 %. While total humulinone concentration in raw and stored hops correlated positively with α-acid concentration, the % humulinone per % α-acid formed correlated negatively with α-acid concentration. Consequently, it appears that humulinones per % α-acid formed are higher the less α-acid is contained in the hops. Further results will also be presented, and a correlation of analytical parameters with humulinone formation will be discussed. This study features new insights into factors affecting humulinone formation in hops.

Short biography

Philip Wietstock is a post-doc at the Technische Universität Berlin, Germany. After graduating his biotechnology studies with qualification as a Dipl.-Ing. at the Technische Universität Berlin (2009), he was working for one year as a researcher at the Department of Food Science and Technology at the Oregon State University, Corvallis, USA. In 2011, he transferred to the Technische Universität Berlin where he finished his dissertation thesis in 02/2017.

P025 - NORWEGIAN PRODUCTION OF MALTING BARLEY AND HOPS

Mette Goul Thomsen[1]; Mauritz Aassveen[1]; Ragnar Eltun[1]

1 - NIBIO

Abstract

In Norway, there has been a rapid development in production of beer with special qualities, and the demand for Norwegian raw material – malt, hops and herbs – are increasing. In the present project, we found that we could produce malt and hops with good quality in Norway. However, two-row barley varieties can only be grown in South-eastern -and Mid-Norway. In areas with shorter growing season one depend on six-row varieties with grain size and size distribution, less favourable for malting. In old varieties, special challenges tied to long and weak straw may lead to lodging and reduced grain quality. Our study also showed large yearly variation in yield and malt quality under Norwegian weather conditions.

Also for hops we found, large differences between both varieties as well as cropping methods and years in yield and quality. In plastic tunnels yield was two - three times higher than in traditional cropping systems. In a multiple analyses, we found that the factors length of longest primary shoot, clone and cropping method together explained 82% of the variation in cone yield. Increased yield in plastic tunnels may partly be due to increased degree-days as well as measures like irrigation from nozzles in the tunnel roof reducing infestation by powdery mildew. Content of α-acid, essential oils and aroma profile were analysed/described for the Norwegian hops in the project.

Short biography

Dr. Mette Goul Thomsen is a Researcher at Norwegian Institute of Bioeconomy Research. She has a PhD in plant biology and weed management in organic farming systems. For the last four years she has completed a number of studies on cultivation and quality of Norwegian hops. She has also worked with quality in malt and adaptation of the malting process to increase malt yield from barley produced in the Northern region.

P026 - ION MOBILITY SPECTROMETRY BASED HEADSPACE DETECTION OF VOLATILE ORGANIC COMPOUNDS PRODUCED BY MOLD FUNGI

Alexander Erler[1]; Daniel Riebe[1]; Toralf Beitz[1]; Hans-Gerd Löhmannsröben[1]; Thomas Kunz[2]; Daniela Grothusheitkamp[2]; Frank-Jürgen Methner[2]

1 - University of Potsdam; 2 - Technical University Berlin

Abstract

Mold fungi such as Fusarium, Alternaria, Aspergillus and Penicillium on malting barley cause major economic losses in malting and brewery facilities. Possible proxies for the detection are their (semi)volatile metabolites. For a confident detection in varying surroundings, characteristic marker compounds or patterns have to be found. The search for these substances was performed by headspace-gas chromatography (GC)-mass spectrometry (MS) investigations using two different ionization sources, a conventional electron ionization (EI) source and a new atmospheric pressure chemical ionization (APCI) source based on soft X-radiation. These results where than transferred to ion mobility (IM) spectrometry using the same APCI source. The aim of this work was the development of a mobile detection method for the on-site detection of mold fungi in grain stores.

Volatile metabolites of four mold fungi on different agars and barley grain were investigated. In total, 79 metabolites were found by the combination of EI/APCI-MS, 72 metabolites were detected by APCI-MS and 66 metabolites were found in APCI-IM spectrometry. These metabolites belong to different substance classes, such as substituted aromatics, terpenes and sesquiterpenes. In addition to metabolites unspecific to the individual fungus species, characteristic patterns allowing the confident discrimination of different fungi genera and even different species of one fungus genus were found. Despite the lower resolution of IM spectrometry in comparison to MS, most of the specific metabolites could be separated and detected in GC-APCI-IM spectrometry. Thus, APCI-IM spectrometry can potentially be used as a mobile monitoring method for the on-site detection and classification of mold fungi.

P039 - TR15245: A NEW TWO-ROW MALTING BARLEY COMBINES DESIRABLE AGRONOMICS, DISEASE RESISTANCE AND QUALITY INCLUDING REDUCED PHYTATE

Ana Badea[1]; William Legge[1]; James Tucker[1]; Adam Carter[1]; Marta Izydorczyk[2]

1 - Agriculture and Agri-Food Canada, Brandon Research and Development Centre; 2 - Canadian Grain Commission, Grain Research Laboratory

Abstract

TR15245, a new two-row hulled malting barley line with reduced phytate (carries the phytate reducing *lpa1-1* barley mutation), was recently developed at Agriculture and Agri-Food Canada, Brandon Research and Development Centre. This line combines superior agronomic performance (13% higher yield than AC Metcalfe check, 8% higher than CDC Copeland check, similar to AAC Synergy check) and physical grain quality (high plumpness, low thins similar to AC Metcalfe check, fair test weight and high kernel weight) with moderate resistance to surface borne smuts, stem rust (carries the *Rpg1* gene) and spot-form net blotch, and intermediate resistance to Fusarium Head Blight (FHB) and spot blotch. It has an acceptable malting quality profile with fine extract and viscosity similar to the checks and alpha amylase slightly higher than CDC Copeland. It has lower peeled and broken grains, diastatic power, soluble protein, Kolbach index, and FAN levels than the checks; low beta-glucan levels and higher friability. The reduced phytate barley malt from TR15245 may improve yeast health and fermentation efficiency. Grains and brewer's spent grains with low phytate may also help reduce environmental pollution.

Short biography

Dr. Ana Badea is a research scientist – barley breeding and genetics with Agriculture and Agri-Food Canada Brandon Research Centre (AAFC-BRDC) based in Brandon, Manitoba, Canada since 2012. As leader of a multi-disciplinary program, she integrates principles of plant breeding, agronomy, pathology, physiology, cereal chemistry, molecular genetics, and genomics to develop cultivars for two spring barley classes, two-row hulless for food and two-row covered malting, with improved agronomic performance, disease resistance and quality profiles to serve traditional and emerging markets. She is also an adjunct professor within the Department of Plant Science, Faculty of Agricultural and Food Sciences at University of Manitoba in Winnipeg, Manitoba, Canada and an Associate Editor with the Canadian Journal of Plant Science (CJPS).

P040 - EFFECTS OF PLANT GROWTH REGULATOR APPLICATION ON THE MALTING QUALITY OF BARLEY

Tricia Mcmillan[1]; John O'donnovan[2]; Breanne Tidemann[3]; Marta Izydorczyk[4]

1 - Grain Research Laboratory, Canadian Grain Commission; 2 - AAFC Lacombe AB; 3 - AAFC, Lacombe AB; 4 - Grain Research Laboratory, Canadian Grain Commission, Winnipeg MB Canada

Abstract

Barley can be cultivated in a wide range of environments; however, adverse climatic conditions and abiotic stresses may negatively affect the yield and quality. Heavy rainfalls and strong winds are known to cause lodging in barley, which can negatively affect kernel weight and plumpness, result in a higher susceptibility of grain to fungal attack, and lower its potential for being selected for malting. Plant growth regulators (PGRs) are synthetic compounds that can modify plant growth and reduce lodging. The objective of this study was to assess the effect of PGR application on the malting quality of barley grown in western Canada. The effect of three PGRs (Ethephon, Chlormequat, and Trinexapac) on CDC Copeland was assessed using a factorial randomized complete block design. The results showed that application of any of the three investigated PGRs reduced the kernel weight of barley. The kernel weight in PGR-treated barley was reduced by 1.7 to 6.5% compared to the non-treated grain with ethephon instigating the greatest effect. Application of PGRs did not negatively affect other quality parameters of barley grain. The smaller kernels of specifically ethephon- and trinexapac-treated barley showed good hydration during steeping, good germination vigour, and good grain modification during malting as indicated by high levels of starch converting enzymes, high Kolbach indices, and low levels of wort β-glucans. The chlormequat-treated barley, however, produced malt with somewhat reduced levels of enzymes and poorer endosperm modification indicating possibly a different mode of action of chlormequat chloride compared to other two PGRs tested in this study.

Short biography

Marta S. Izydorczyk is currently a research scientist and program manager of Milling and Malting and Research on Barley and other Grains at the Grain Research Laboratory, Canadian Grain Commission, and an adjunct professor at the Department of Food Science, University of Manitoba in Winnipeg, Canada. Her research program focuses on identifying and characterizing constituents and molecular mechanisms, which are responsible for functionality, quality, and performance of barley and other grains in food products. She is responsible for assessment of malting quality of Canadian barley. Her research program aims at developing effective strategies for improving the quality of barley and for better adapting it to current and future needs. Marta also has responsibilities in representing the Canadian Grain Commission in domestic and international forums, establishing liaisons and technology transfer activities with producers, industry, and marketers, providing advice and responding to queries from clients and public.

P044 - INFLUENCE OF POST-FERMENTATION PROCESSES ON HOP AROMA IN BEER

Ray Marriott[1]; Liridona Ferizi[1]

1 - Totally Natural Solutions Ltd

Abstract

Maximising hop aroma and flavour in beer has been key to the development of the craft beer sector and there has been many developments to improve the efficiency of hop aroma transfer from hops to beer. However there has been little attention paid to the preservation of the hop aroma in post fermentation processes and this study looks at the influence of centrifugation, stabilisation, filtration and pasteurisation on the retention of key aroma molecules in beer.

We have investigated the changes in hop aroma molecules using headspace SPME-GC-MS with model beer systems in the laboratory and also in commercial breweries where individual batches have been tracked from post-fermentation to final pack. Bright beer and cloudy beer styles such as New England IPA have both been investigated as the "cloud" particles have a particular influence on hop aroma stability.

This work has shown that filter aids and filter materials selectively absorb certain groups of hop molecules resulting in significant changes in the hop aroma and flavour profile but centrifugation and stabilisation has much less impact.

This work has led to a better understanding of hop aroma stability post fermentation and the development of new products that exhibit better shelf-life stability.

Short biography

Ray is Innovation Director of Totally Natural Solutions Ltd a company specialising in the development of hop aroma and flavour products. Ray Marriott spent 35 years in industry working on the extraction and application of natural products and the development of hop products before moving to academia in 2008. Ray then spent time at York University and Bangor University developing "green" processing techniques before returning to the hop industry. Ray has a biochemistry degree from Cambridge and a PhD in terpene chemistry from Bath University.

P046 - THE CONTRIBUTION OF STALING ALDEHYDES TO THE FLAVOR (IN)STABILITY OF TOP FERMENTED, HOPPY BEERS

Christian Schubert[1]; Sarah Thörner[1]; Jörg Maxminer[1]; Nils Rettberg[1]

1 - VLB Berlin - Research Institute for Beer and Beverage

Abstract

Many consumers highly appreciate hoppy beer styles such as Pale Ales, India Pale Ales (IPA), or Double IPAs. These beers are now distributed and consumed country- or even worldwide, making consistent quality and flavor stability a growing concern for their producers. To date, various studies have investigated flavor stability in Lager beer, whereas very little research has been carried out in order to understand flavor changes in hoppy top fermented beer.

In order to investigate the flavor chemistry of hoppy beer, eleven German craft beers were selected by a Design of Experiment (DoE) approach using basic beer parameters (ABV, IBU, and Color) as input variables. The beers were subjected to a six month storage trial at 4 °C and 20 °C. Chemical changes were tracked by application of multiple targeted and non-targeted instrumental assays on eight separate occasions and complemented with sensory evaluation.

The current paper discusses the contribution of staling aldehydes to flavor changes in hoppy top fermented beer. It is commonly known that trace amounts of staling aldehydes negatively contribute to the overall flavor impression of aged lagers and our work reveals that some of the fresh tested beers already contained high aldehyde levels. The aldehyde distribution differed from those typically observed in lagers and these demonstrated a remarkable increase throughout storage. However, these findings did not necessarily correlate well with sensory, suggesting the high importance of known or unknown yeast and/or hop derived volatiles in flavor (in)stability of some top fermented, hoppy beers.

Short biography

Christian Schubert is a trained brewer and maltster, graduated from Technical University Berlin (Germany) with a Master of Science (brewing and beverage technology) in 2016. He is now a research associate at the Department for Special Analysis at the Research and Teaching Institute for Brewing in Berlin (VLB). Christian's scientific work is focused on the ageing stability of beer especially top fermented, hoppy beers.

P055 - ASSESSMENT OF QUALITY CONTROL METRICES FOR MEASURING BEER FRESHNESS IN THE LAB USING THE MICROESR TECHNOLOGY

Mangethe Zwane[1]; Dr Boris Gadzov[1]; Evelyne Canterranne[1]; James Jugg[1]

1 - FlavorActiV

Abstract

Beer freshness is mainly measured using three metrices in the laboratory; Lagtime, T150 and Area. This study will assess the stability of the QC method and propose an improved alternative to these metrices namely Freshness index, Oxidation index and Oxidation capacity. A better-quality method for measuring beer freshness will also be presented and seeks to improve sensitivity and robustness of the assessment of beer freshness in a quality control laboratory. Furthermore, all the quality control metrices will be assessed on their correlation to beer oxidation levels when tested by a sensory panel.

Short biography

FlavorActiV produces Pharmaceutical Quality GMP Flavour Standards for training and maintaining the abilities of professional tasters, combined with FlavorActiV's proprietary taster proficiency software, in use by 50,000+ tasters globally, a trained, monitored panel brings validation to analytical instruments. The combination of both human and instrument sensory is where FlavorActiV adds value and mitigates risk for brand owners.

P056 - INTRODUCTION OF A STANDARD BEER FOR QUALITY CONTROL ASSESSMENT OF BEER FRESHNESS IN THE LABORATORY USING THE MICROESR TECHNOLOGY

Mangethe Zwane[1]; Dr Boris Gadzov[1]; Evelyne Canterranne[1]; James Jugg[1]

1 - FlavorActiV

Abstract

The study will introduce a reference beer standard that will be able to assess quality control laboratories that measure beer freshness using the microESR. The reference beer is a stable reference standard which can be used to compare laboratory performance. The results will show the assessment of the stability of the reference beer over a period of three months. Also, a comparative study will be presented to assess the robustness of the quality control metrices across 4 laboratories using the reference beer standard.

Short biography

FlavorActiV produces flavour standards to Pharmaceutical Quality GMP Standards for training and maintaining the abilities of professional tasters. FlavorActiV's multi-lingual trainers deliver sensory training globally and monitor tasters with FlavorActiV's proprietary taster proficiency software, in use by 50,000+ tasters globally. Trained, monitored panels bring validation to analytical outputs of sensory instruments. The combination of both human and instrument sensory is where FlavorActiV adds value and mitigates risk.

P060 - CAN BEER CONSUMPTION CONTRIBUTE TO THE DAILY INTAKE OF ESSENTIAL MINERALS ELEMENTS?

Edgar Pinto[1]; Isabel Ferreira[1]; Agostinho Almeida[1]

1 - University of Porto -Faculty of Pharmacy

Abstract

Minerals deficiencies are a major public health problem in many developed and developing countries. The micronutrient deficiencies of major public health concern are the Fe and Zn deficiencies, which cause several health problems such as impairment in cognitive performance, lowered immunity to infections, poor learning capacity, among others. However, Se and I deficiencies are also of great relevance due to their high prevalence in developed countries nowadays [1].

Beer is an extremely rich product, containing a wide variety of compounds. Beer usually has a well-balanced amount of minerals [2]. However, it is known that the mineral composition of beer can greatly varies depending on factors such as the quality of raw products (including water), brewing process, type of beer, among others. Nowadays, much is known about the concentration of major minerals, like Ca, Mg, Na and K, in beer. Yet, little information exists on the concentration of microminerals such as Fe, Zn, Se and I. As beer can be an important source of these minerals for the human diet, more information is needed about the general mineral composition of beer.

This study aimed to determine the concentration of minerals in beers from the European market. Using the powerful analytical technique inductively coupled plasma – mass spectrometry (ICP-MS), the concentration of several macro and micro minerals is being determined in several hundred beer samples. The results of this comprehensive study will be presented in the present communication.

[1] WHO, 2014. Vitamin and Mineral Deficiency, A Global Progress Report - Unicef.

[2] Powl, P, 2009. Metals in beer. In: *Beer in Health and Disease Prevention*. ISBN: 978-0-12-373891-2

Short biography

Edgar Pinto (EP) has a PhD degree in Sustainable Chemistry from University of Porto (2014). He has a vast experience in the fields of analytical, food and environmental chemistry, namely in the analysis of both organic and inorganic compounds in food and environmental matrices using liquid chromatography tandem mass spectrometry (LC-MS/MS) and inductively coupled plasma-mass spectrometry (ICP-MS). His scientific path has been focused on the study of minerals transfer in the water-soil-plant system under normal and stress conditions. In last few years, he has been focused on the subjects of mineral malnutrition, biofortification, and food composition analysis.

P064- WHITE BEER HAZE STABILITY

Olivier Caille[1]; Dominique Gallo[2]; Laurence Van Nedervelde[2]; Stéphane Meulemans[1]; Philippe Janssens[1]; Yves Gosselin[1]

1 - FERMENTIS Lesaffre for beverage; 2 - Meurice Institute

Abstract

Yeast derivatives are yeast cellular fractions obtained by inactivation or autolyzation of living yeast cells, and subsequently fractionated and purified the components thereof. Multiple applications of yeast derivatives have been explored in food and wine. In white beer, sensitive proteins and polyphenols form complexes responsible of haze. Could yeast provide sensitive proteins as a clean label solution for stable haze ?

Fermentis is dedicated to the development of fermentation and characterization solutions for beverages. It has developed a yeast derivative rich in protein (Yeast Protein Extraction) able to provide a stable homogeneous haze in white beers.

The haze has been investigated systematically for a range of YPE dosing rates, in various types of beer matrixes chosen based on specificities of the brewing process (stabilization protocol, filtration/centrifugation system, bottle conditioning) and the beer composition (cereals, proteins and polyphenols). In parallel, YPE has been fractionated according to the protein molecular weight to characterize haze active fraction.

The YPE product increase the haze in White beer and create a haze in Blond beers. Surprisingly, a stable haze could be created in PVPP-treated beers. Haze induced by YPE is strictly related to the total polyphenol. Alcohol and yeast content show no influence. Beers made with a specific enzyme treatment generates unstable haze over time. This YPE should be added during the bottling process.

The work presented reveals that 100% yeast derived protein extract as a clean label product is an innovative and sustainable solution for maintaining or creating stable haze in white beer.

Short biography

I get two different degrees: An Industrial engineer degree in brewing and fermentation technologies at Meurice Institute (Belgium) and a Master at Université Libre de Bruxelles, Belgium (1996-2002).
My first professional experience concerned a work of eight years on antibiotic resistance mechanisms in pathogenic bacteria (PhD thesis in Geneva 2002-2007 and three years post-doctoral fellowship at Harvard Medical School in Boston 2007-2010).
Afterwards, as a second post-doctoral, I developed molecular tools for improving bacterial bioethanol production at CNRS (Centre National de la Recherche Scientifique) in collaboration with Deinove.
My last professional experience before I join Fermentis pointed out the biotransformation, by the gut microbiota, of ellagitanins contained in pomegranate peel into polyphenol molecules active against inflammatory mechanisms and muscular mass loss, (2013-2016).
I joined Fermentis' team in September 2016 as the technical manager of yeast derivatives uses in fermented beverages.

P067- RENEWABLE ENERGIES IN BREWERY MICRO GRIDS

Holger Schmidt[1]

1 - ABB Automation Products GmBH

Abstract

Figures have revealed that nearly 90 per cent of wasted food is lost before products even reach supermarket shelves. In many developing countries, power supplied by the public grid can be unstable, often resulting in blackouts and power surges, which can be devastating to food manufacturers.

Manufacturers are now looking to Industrial Internet of Things (IIoT) technologies, including microgrids, to solve the problem.

Microgrids, or small-scale power networks, which often incorporate on-site renewable energy sources like wind and solar, as well as customized battery storage systems, are providing plants with sustainable energy that is not only more efficient, but also has a lower carbon footprint.

In this presentation XXX will explore how microgrids can ensure a stable and reliable power supply for breweries and how this can increase automation efficiency and extend the lifetime of electronic components.

As the world grows more aware of the environmental impact of fossil fuels, manufacturers are increasingly turning to renewable energy as a power supply. We will discuss how renewable energy can be integrated into existing plant infrastructure with minimal disruption for maximum benefit.

Short biography

Mr. Schmidt completed his training as a Brewer and Malter at the Beck Brewery in 1989. Graduating as a Brewmaster in Weihenstephan, Mr. Shmidt began working as a sales and project engineer for KHS. He held similar positions with APV (now SPX) and Huppmann (now GEA). He was responsible for the global food and beverage marketing at Endress+Hauser and became involved with the European Hygienic Engineering and Design Group (EHEDG), and currently leads the Subgroupo 37 "Hygienic Sensors" and has been part of the advisory board since 2015. Mr. Schmidt joined ABB in 2017 as a Global Key Account Manager working with German machine builders.

P068 - UNLOCKING PRODUCTION EFFICIENCY

Holger Schmidt[1]

1 - ABB Automation Products GmBH

Abstract

Overall equipment effectiveness (OEE) is a powerful tool for measuring manufacturing productivity and improving the performance of machines and industrial assets. While measuring OEE is best practice, where do you begin when collating and analyzing data?

This presentation will explore the concept of holistic plant assessments (HPA) and the vital role they play in establishing a starting point for understanding and unlocking production efficiency as well as improving sustainability and overall plant efficiency.

Using the example of an international brewery, we will explore how a HPA is conducted, focusing on six key areas; power, heat, compressed air, refrigeration, motor controls and automation. We will show how plant managers can develop and implement their own action plans and how this can affect OEE by improving plant efficiency, productivity and profit margins.

Short biography

Mr. Schmidt completed his training as a Brewer and Malter at the Beck Brewery in 1989. Graduating as a Brewmaster in Weihenstephan, Mr. Shmidt began working as a sales and project engineer for KHS. He held similar positions with APV (now SPX) and Huppmann (now GEA). He was responsible for the global food and beverage marketing at Endress+Hauser and became involved with the European Hygienic Engineering and Design Group (EHEDG), and currently leads the Subgroupo 37 "Hygienic Sensors" and has been part of the advisory board since 2015. Mr. Schmidt joined ABB in 2017 as a Global Key Account Manager working with German machine builders.

P069 - THE RE-USE OF PVPP-REGENERATION-CAUSTIC FOR CIP-CLEANING

Deniz Bilge[1]

1 - VLB Berlin

Abstract

PVPP-caustic is used in order to regenerate PVPP for beer stabilization. By elevating the pH with caustic polyphenols are removed from the stabilizer. After regeneration the caustic looks dark and wasted wherefore it is usally dumped into the drain. Its dark appearance makes it look unusable but that is not representing its real cleaning power, since the main components carried in this caustic are polyphenols and their concentration is rather low. Furthermore it is free of microorganisms since it is processed hot. So why not use it in CIP operations? The idea came up to re-use this caustic for cleaning purposes. Doubts against this idea were possible dangers of discolorations, deposits or foaming and therefore a harm to the cleaning result or the brewing equipment. A research project was carried out in order to evaluate the properties and the processability of this former waste product as cleaning medium in CIP-operations in the cold block and Brewhouse. The results were positively surprising. The PVPP-caustic showed the same cleaning properties like new caustic solutions. Stains were easily removable from plastic material and stainless steel. Tiles just needed a simple detergent treatment when stained with this caustic. Even breweries working with lost cleaning media could cover their caustic needs with PVPP-caustic.

Short biography

After an apprenticeship at Beck Co. with subsequent work as crafter in the filtration cellar a two year occupation followed in a craft brewery in Montreal/Canada. After returning to Germany and work in the fermentation cellar at Beck Co. again studies in fermentation technology followed at the Technical University in Berlin. In a five year contract at the Technical University´s chair of brewing Deniz Bilge worked as assistent to the professor. In this time he worked as head of the malt and beer laboratory and later as chief of the pilot plant. He furthermore accomplished a PhD-thesis on ultrafiltration in this time. Jobs in the technical support of breweries at Sopura and as plant manager in industrial brewery operations in Greece followed before returning to the VLB in Berlin.

P075 - FIRST EXPERIENCE OF THE 'BREWERY 4.0' – THE INTERNET OF THINGS CONCEPT APPLIED IN AN OPERATIONAL BEER FILTRATION UNIT

Dr. Roland Folz[1]; Mark Mol[1]

1 – Pentair, Netherlands

Abstract

First experience of the 'Brewery 4.0' – The Internet of Things concept applied in an operational beer filtration unit

Process manufacturers, like the brewer of today, face constant pressure to improve operational performance. Boosting efficiency and productivity is commercially interesting to do thereby outperforming market rivals, while manufacturing processes are complex so this objective is not easily accomplished. "Internet of Things" technology enables connected operations, giving process manufacturers the ability to collect data from connected equipment, sensors and devices that can be combined with data and intelligence from business systems, leading to benefits in maintenance, operational costs and product quality.

The use of IoT technology:

1. enables maintenance alerts when the equipment deviates from its prescribed parameters which will conserve energy, reduce costs and eliminate machine downtime.
2. enables the monitoring of production lines starting from the malting process down to the packaging of final products leading to a better control of operational costs.
3. aggregates product- and quality data from various stages of a product cycle. All of these inputs can be analyzed to identify and correct quality issues.

In this presentation it will be shown how IoT works in the typical manufacturing process of a brewery. Questions and challenges are popping up which must be answered and tackled, like which data are needed from the process, is it possible to measure these data on a real time basis and how to present these data to involved personal like operators, supervisors and managers. The Beer Membrane Filter system is shown as a worked out example and experiences from first commercial systems are evaluated.

Short biography

Dr. Roland Folz is the Director of Technology & Innovation for Pentair's Food & Beverage division. Since February 1, 2014, Folz is responsible for innovation, product design, solutions development, and the global R&D function.

Folz has 17 years of experience in the brewing and beverage industries. Prior to joining Pentair, he headed the VLB department Brewing & Beverage Science and Applications (BBSA), located in Berlin, Germany.

Folz holds a Ph.D. in Engineering Technology and a Diploma for Brewing Technology from the Technical University Berlin, Germany.

P079 - HOP OIL PROFILING TO MITIGATE CLIMATE CHANGE IMPACT ON TRADITIONAL HOP VARIETIES

Ray Marriott[1]; Colin Wilson[1]

1 - Totally Natural Solutions Ltd

Abstract

In recent years many traditional hop varieties such as Saaz, Hallertau Hersbrucker, Cascade and Styrian Golding have exhibited declining oil content which has resulted in decreased aroma impact and lower yields of hop oils. It is not clear that this is entirely a result of climatic change [1,2], although these growing areas have seen hotter and drier summers, or whether it is a combination of hop age, climate and other factors.

This decrease in oil content has an impact on both end product quality and economics and to mitigate this the essential oil of each affected hop variety has been profiled to identify key components and to help create similar aroma and flavour profiles from higher yielding hop varieties from areas less exposed to climate change.

This work has been carried out using both analytical and sensory assessment and oil profiles recreated using hop oils, fractions and individual impact molecules.

Short biography

Ray is Innovation Director of Totally Natural Solutions Ltd a company specialising in the development of hop aroma and flavour products. Ray Marriott spent 35 years in industry working on the extraction and application of natural products and the development of hop products before moving to academia in 2008. Ray then spent time at York University and Bangor University developing “green” processing techniques before returning to the hop industry. Ray has a biochemistry degree from Cambridge and a PhD in terpene chemistry from Bath University.

P082 - IMPACT OF KILN TEMPERATURES ON THE AROMA AND ENZYMATIC POTENTIAL OF HOPS

Scott Lafontaine[1]; Lindsey Rubottom[1]; Thomas Shellhammer[1]

1 - Oregon State University, Department of Food Science and Technology

Abstract

This study evaluated the impact of kiln temperature on dry-hop aroma potential of Amarillo® and Simcoe® hops across 2 different harvest years. Hops were dried over a range of temperatures (47°C, 60°C, 77°C and 82°C) on pilot scale and commercial scale kilns. The hop storage index increased with kiln temperature, however the concentration of humulones were not significantly impacted by kiln temperature. Furthermore, kiln temperature had only a minimal impact on total oil content of dried hops. In terms of aroma potential in beer, hops dried at high and low temperatures attributed similar aroma intensities and qualities which were more intense compared to intermediate temperatures. Drying higher temperatures significantly reduced the dextrin degrading enzymatic activity of hops relative to lower temperature treatments. These results suggest that higher drying temperatures and shorting drying times can produce hops with similar aroma potential and lower enzymatic activity compared to hops kilned at lower temperatures.

Short biography

Scott Lafontaine is a Ph.D. candidate in Dr. Tom Shellhammer's Laboratory in the Department of Food Science at Oregon State University. His research focuses on understanding the molecular constituents that drive hop flavor and aroma in dry-hopped beer. He assists in the teaching of residential and continuing brewing analytical and quality education courses at OSU. Prior to joining the Shellhammer Laboratory at OSU, he received his Master of Science in Chemistry at OSU in 2015, during which his studies focused on analytical environmental chemistry.

P085 - WHAT CAN CRYO HOPSR AND AMERICAN NOBLE HOPS™ DO FOR YOUR BEER?

Cynthia Almaguer1, Patrick Jensen1, Missy Raver1, Emily Breneman1, Jacqueline Hite1, Brenda Padilla1, Aidee Padilla1, Nick Ziegler1

1 - Yakima Chief, Louvain-la-Neuve

Abstract

The Cryo HopsR process is an innovative cryogenic hop processing technology, proprietary of Yakima Chief Hops, used to separate whole cone hops into two different hop products for brewing purposes. Cryo HopsR and American Noble Hops™ are produced simultaneously when hop cones are separated into purified lupulin and bract fractions, respectively. The separation process occurs at very low temperatures and in a nitrogen-rich atmosphere avoiding the rupture of the lupulin glands. If these glands are damaged, this would consequently cause the oxidation of hop resins and oils. It is in the lupulin glands that the main brewing principles of hops, resins and essential oils, are synthesized and accumulated. It is the purpose of this work to characterize and profile these innovative brewing hop products. Further, the functionality of Cryo HopsR and American Noble Hops™ as brewing products is examined. To achieve this, brewing trials were conducted using Cryo HopsR, American Noble Hops™, and traditional hop products (e.g. whole cone hops) to dry hop beers. These beers were chemically analyzed and compared. The hop oil profiles of these beers were determined by stir bar sorptive extraction coupled with GC-MS. These experiments showed that adding the same amount of Cryo HopsR will yield almost twice the amount of hop constituents and lower vegetal notes compared to traditional hop products. In contrast, American Noble Hops™ could be used as an alternative to traditional hop varieties used in mildly hopped beer styles. Both products offer the brewer new options to deliver desirable aromas and flavors.

P086 - CHARACTERIZATION OF MALTS BY SENSORY ANALYSIS AND ANALYSIS OF VOLATILES

Michael Féchir[1,2]; Veronika Mall[2]; Klaas Reglitz[2]; Jens Voigt[1]; Martin Steinhaus[2]

1 - Trier University of Applied Sciences; 2 - Leibniz-Institute for Food Systems Biology at Technical University of Munich

Abstract

Besides water, malt is quantitatively the most important raw material for beer production and the main source of fermentable extract and colour. Malt may also have an impact on the overall olfactory and retronasal profile of the final product beer. A trained sensory panel was used to characterize the aroma of trial malt types produced in a pilot plant model system by varying technological malting parameters. In parallel, malt volatiles were analyzed by applying a semiquantitative HS-trap-GC-MS method. Results were compared to industrially produced reference malts. Volatiles that varied most significantly between malt types were selected as marker substances to be exactly quantified. Varied malting parameters led to significant differences in the sensory properties as well as in the volatile composition according to statistical evaluation by ANOVA and Bonferroni Multiple Comparison. PLSR modeling with sensory and analytical malt data showed, that several quantified marker substances may serve as promising indicators for malt aroma properties, although it was not assessed whether these compounds causally contributed to the aroma or not. Ongoing research will apply the approach also to beer and brewing intermediates to assess and predict the influence of malt type and malting parameters on the overall aroma properties of beer.

Short biography

Michael Féchir is a researcher in Food Technology at Trier University of Applied Sciences and PhD candidate at the Technical University of Munich. His research focusses on the characterization of malts by sensory analysis and analysis of their volatiles. He is a certified expert for sensory evaluation of beer and mixed beer drinks (DLG) and successfully completed his M. Eng. in Food Economy (graduation as 2nd best of year 2015, Master Thesis at Mich. Weyermann Specialty Malts, Bamberg) and his B. Eng. in Food Technology (2014, Bachelor Thesis at Merck, Darmstadt) at Trier University of Applied Sciences.

P088 - UNDERSTANDING THE COMPONENTS OF SPECIFIC WEIGHT IN MALTING BARLEY

Aaron Hoyle[1,2]; Maree Brennan[3]; Gail Jackson[2]; Steve Hoad[1]

1 - Scotland's Rural College; 2 - University of Edinburgh; 3 - Faculté des Sciences et Technologies

Abstract

Specific weight is one of the longest standing measures of grain quality for cereals because it is a measure of the weight of grain per unit volume, and is therefore of economic importance for transport and batch processes such as malting and brewing. A high specific weight is thought to be indicative of high starch content and benefit maltsters. Accordingly, specific weight influences the value of barley grain and high specific weight is an important breeding target. However, the reliability of specific weight as a malt quality indicator is disputed. Understanding components of specific weight is an essential step towards linking this measure to efficiency/productivity of the malting process. Initial work has shown that specific weight is the product of grain packing efficiency and grain density. The determinants of these two components are being investigated to determine whether they are beneficial for malt production. The grain density component was further dissected by stratifying samples from five spring barley malting cultivars into different density classes. Elemental analyses, enzymatic starch assays and starch granule analyses were carried out to determine if composition correlates with density. Nitrogen content (r^2=0.34, p<0.01) and volume of starch B granules (r^2=0.28, p<0.01) have significant positive correlations with grain density. However, grain starch content does not correlate with density. This could have wide implications for the use of specific weight as a quality measure in the malting supply chain, especially since density (a component of specific weight) carries detrimental characteristics like nitrogen content and increased B granule volume.

Short biography

Aaron Hoyle is a PhD student entering his 3rd year at Scotland's Rural College (SRUC) and the University of Edinburgh (UoE). Aaron's project is on the grain quality measure specific weight, an important quality measure for malting barley.

P090 - EVALUATION OF GERMINATION PARAMETERS DURING MALTING REGARDING PHENOLIC ACID RELEASE DURING WORT PRODUCTION USING RESPONSE SURFACE METHODOLOGY

Torsten Seewald[1]; Max Biermann[1]; Frank-Jürgen Methner[1]

1 - Technische Universität Berlin, Institute of Food Technology and Food Chemistry, Chair of Brewing Science

Abstract

The release of phenolic acids during wort production has a considerable influence on the formation of key aroma compounds during the production of beers, which are fermented with POF^{+} yeasts. The most common phenolic acids are ferulic acid and p-coumaric acid, as precursors of 4-vinylguaiacol and 4-vinylphenol, respectively. However, in parallel to these desired precursors, undesired precursors such as cinnamic acid, which can be enzymatically decarboxylated to vinylbenzene (styrene), are also released. Since vinylbenzene has a carcinogenic potential, it is essential to minimize the release of cinnamic acid. The aim of the study was to evaluate the influence of the germination process during malting on the release of desired and undesired phenolic acids.

Therefore, small scale malting trials with two barley varieties (*Solist* and *Quench)* and one wheat variety (*Dacanto*) were performed using response surface methodology. Within the scope of this experimental design the germination temperature and the aeration rate were varied. Congress worts were produced and malt quality was evaluated according to MEBAK®. Ferulic acid, p-coumaric acid and cinnamic acid were quantified using HPLC-DAD.

The results indicate that the varied germination parameters lead to a significant change of the phenolic acid release. Especially, the modification of the malts from the barley varieties *Solist* and *Quench* contribute to an optimized ratio between the desired and undesired phenolic acids.These results can contribute to achieve a targeted malt modification for an increased release of the desired ferulic acid and p-coumaric acid, while simultaneously minimizing the cinnamic acid.

Short biography

Torsten Seewald is a scientific assistant at the Technische Universität Berlin, Germany, where he studied food technology and graduated as Diplom-Ingenieur. During his studies, he was working on several research projects at the Chair of Brewing Science and completed industrial placements at WILD Dairy Ingredients GmbH and Herbstreith & Fox KG. Currently, he is working in the field of malting and oxidative processes in wort and beer.

Keyword's : Malting, Germination, Phenolic Acid Release

P098 - USE OF WOOD CHIPS IN BREWING : IMPACT OF PROCESS AND MATERIALS

Brabant Pierre[1]; Schmit Marc[1]

1 - IFBM

Abstract

More and more beers are made with wood chips. La Tonnellerie Radoux (French for Radoux copperage) and IFBM collaborated together to understand in beer what is already known with wine and wood.

(The aim of this study is to understand the impact of process, dosage and wood species on the final result.)

Different timings were also tested from Brewhouse to ageing tanks. Two types of wood were tested, French oak and American oak with different dosages.

Analysis part was approach from the perspective of beer, performing standard analysis but also looking for molecules of interest brought from wood, molecules which are already well-known in wine.

The goal was to control the taste from the process control monitored with analytical data.

Short biography

Pierre Brabant is the pilot plan manager at IFBM. He studied brewing and malting at ENSAIA engineer school in Nancy, France. After working in production of soft drinks he joined IFBM in 2016. He is in charge of the malting plant, the brewery plant and the beverages studies. He is a trainer in the IFBM craft brewing training program.

P100 - HIGH PROLYL-ENDOPEPTIDASE ENZYME DOSAGES FOR THE PRODUCTION OF GLUTEN-FREE WHEAT BEER: A LAB-SCALE CASE STUDY

Hellen Watson[1]; Anneleen Decloedt[2]; Dana Vanderputten[3]; Jessika De Clippeleer[1,3]; Anita Van Landschoot[1], Filip Van Opstaele[4]

1 - Ghent University, Faculty of Bioscience Engineering, Department of Biotechnology, Brewing Science and Technology Research group, Ghent, Belgium; 2 - Ghent University, Faculty of Veterinary Medicine, Laboratory of Chemical Analysis, Ghent, Belgium; 3 - University College Ghent, Faculty of Science and Technology, Department of Biosciences and Food Sciences, Brewing Science and Technology Research group, Ghent, Belgium; 4 - KU Leuven, Department of Microbial and Molecular Systems (M2S), Cluster for Bioengineering Technology (CBeT), Laboratory of Enzyme, Fermentation and Brewing Technology (EFBT), Technology Campus Ghent, Gebroeders De Smetstraat 1, B-9000 Ghent, Belgium

Abstract

Gluten-free is one of the fastest growing markets. Also in the beer-market today's consumers have a thirst for gluten-free beer. The enzyme prolyl-endopeptidase from *Aspergillus niger* (AN-PEP) has showed to be effective in the reduction of gluten and gluten peptide levels in barley malt beers to below the gluten-free (< 20 ppm) threshold. AN-PEP enables the production of gluten-free malt beers in a controlled and reproducible manner [1]. This creates new opportunities for breweries in the gluten-free beer-market and it also drives the need for a range of different gluten-free beer styles.

In this study the effect of the AN-PEP enzyme (1- to 16-times the recommended standard dose) on an industrial wheat beer wort (13°P) with high gluten concentrations (i.e. > 2700 ppm) was studied. The addition of a high AN-PEP dosage successfully lowered the gluten and gluten peptide concentration to < 20 ppm and made the final wheat beer gluten-free. Characteristic for wheat beers are a pleasant fruity aroma, a whitish-yellow colour, intense beer foam and high turbidity due to the lack of a beer clarifying step. Therefore, the impact of elevated AN-PEP enzyme dosages on latter quality and sensory attributes of the final wheat beer was evaluated in comparison to the untreated reference.

[1] Watson H.G., Vanderputten D., Van Landschoot A., Decloedt A.I. (2019) Applicability of different brewhouse technologies and gluten-minimization treatments for the production of gluten-free (barley) malt beers: pilot- to industrial-scale, *J. Food Eng.* 245, 33-42

Short biography

Hellen Watson graduated in 2014 as M. Sc. Biochemical Engineering, Ghent University, Belgium. After graduating she started to work at the Brewing Science & Technology Research Group of Ghent University as a PhD student. Her doctoral study focusses on the occurrence and influence of processing on gluten and toxic gluten peptides in fermented beverages. The technological aspect of her work has successfully led to the implementation of gluten-minimization treatments in industrial brewhouses for the production of gluten-free malt beers. Her work also gives more insight into the gluten proteome of non-gluten-free and gluten-free malt beers.

P102 - BREWING WITH 10/20 % LENTIL MALT ADDITION IN A MICROBREW SCALE

Jonas Trummer[1]; Hellen Watson[2]; Jessika de Clippeleer[2]; Aleksander Poreda[1]

1 - University of Agriculture in Krakow, Faculty of Food Technology, Department of Fermentation Technology and Technical Microbiology, Krakow, Poland; 2 - Ghent University, Faculty of Bioscience Engineering, Department of Biotechnology, Laboratory of Brewing Science and Technology, Ghent, Belgium

Abstract

Lentils are a common source of food throughout the whole world. It has been proven, that commercial green lentils can be germinated and kilned in order to get a malt which is suitable for wort production. In this research, the addtion of 10 and 20 % lentil malt, in order to produce a beer, was done in a microbrew scale of 80 l. Analysis that have been performed were: Chemical-technical analysis; sugar composition; gluten content; amino acid composition and sensory analysis. The analysis showed sufficient results compared to the reference brews (100 % barley malt). Overall, the use of lentil malt in beer production has been determined as possible, for usage of 10 and 20 % addition.

Short biography

Jonas Trummer is a Phd student at the University of Agriculture in Krakow within the European Joint Doctorate in Food Science (PhD School in malting and brewing).

Graduated from the TU Munich-Weihenstephan where he accomplished the B.Sc. and the M.Sc. in Brewery Science and Beverage Technology. He did exchange programs at Corvinus University Budapest and University College Cork, where he wrote his master thesis about the influence of LAB on malt quality. Before starting the PhD in 2017, he was working for two years for a medium sized brewery in Germany in the position of a quality controller.

P103 - HOW RESEARCH ON ALTERNATIVE GRAINS CAN BOOST THE BREWING INDUSTRY

Jessika De Clippeleer[1,2] Julie Demeulenaere[1], Dana Vanderputten[2]
1 - Ghent University, Faculty of Bioscience Engineering, Department of Biotechnology, Laboratory for Brewing and Fermentation Science & Technology, Valentin Vaerwyckweg 1, B-9000 Ghent, Belgium; 2 - University College Ghent, Faculty of Science and Technology, Department of Biosciences and Food Sciences, Laboratory for Brewing and Fermentation Science & Technology, Valentin Vaerwyckweg 1, B-9000 Ghent, Belgium

Abstract

Worldwide, beer lovers increasingly appreciate the creativity and authenticity of the brewer through sustainable innovations. The use of raw alternative grains in the brewing process offers a possible answer to this increasing market demand for natural, authentic and innovative products. Mixing of unmalted grains such as barley, wheat, rice and corn is already common practice. In contrast to these traditional crops, little attention is paid to the use of unprocessed alternative grains such as kamut, unicorn, bucket, amaranth, teff, etc. It is therefore not self-evident for a brewer to integrate these raw materials into the brewing process when not knowing which technological problems will arise and what influence these grains have on the visual characteristics and organoleptic quality of the final beer. The aim of our research is to provide the brewer with hands-on knowledge for the processing of raw alternative grains into beer, related to their availability and brewing technological relevant characteristics, through analytical characterization of different selected alternative grains and their corresponding worts and beers.

Short biography

Starting from October 1st, 2017, Jessika De Clippeleer was appointed a full-time position as Professor (tenure track) in brewing and technology of fermentation processes at the Biotechnology Department of Ghent University, Belgium. She is Head of the Research Group Brewing Science and Technology of Ghent University and University College Ghent, involved in research and industrial services. She is Lecturer in biochemistry, brewing technology, industrial microbiology, and biomolecules' analysis and separation.

Jessika graduated in Biochemical Engineering Technology (KAHO Sint-Lieven, Ghent) in 2000, and received her MSc degree in Food Technology specialisation Food Chemistry (Wageningen University) in 2003. Subsequently, she was working at the Enzyme, Fermentation and Brewing Technology (EFBT) lab of KU Leuven, Technology Campus Ghent, as lecturer and researcher. In 2013, she received her PhD in Bioscience Engineering from KU Leuven on Flavour stability of pale lager beer - Chemical-analytical characterisation of critical factors related to wort production and hopping.

P116 - HYDROGEN PEROXIDE FORMATION AND -INACTIVATION BY MALT-DERIVED THIOLS DURING MASHING

Matthias Baldus[1]; Sarah Majetschak[1]; Frank-Jürgen Methner[1]

1 - Technische Universität Berlin, Institute of Food Technology and Food Chemistry, Chair of Brewing Science

Abstract

Nowadays, malt-derived thiols are well established antioxidants in beer. In mash, thiols may inactivate hydrogen peroxide (H_2O_2) and thereby diminish oxidative reactions responsible for beer quality deterioration. However, in the presence of transition metal ions and molecular oxygen, thiols may also form hydrogen peroxide via reductive chelation. The present work aimed to assess the involvement of thiols and the transition metal copper (Cu^{2+}) on oxygen consumption and hydrogen peroxide formation in mash. In model solutions, the addition of 250 µM of thiols (cysteine (Cys), glutathione (GSH)) together with Cu^{2+} was followed by significant O_2 consumption within 15 minutes at 60 °C. At this stage H_2O_2 formation was observed, which correlated to the respective oxygen consumption. O_2 consumption and H_2O_2 formation was significantly higher in the Cys treatments as compared to GSH. In the further course of incubation the H_2O_2 concentration was degraded to ~70% in the Cys treatment while for the GSH treatment the H_2O_2 concentration remained constant. During mashing (Pilsner malt, grist-water ratio: 1:4) O_2 was completely degraded within 5 minutes. The addition of maleimide (thiol-blocking agent) significantly decreased O_2 consumption demonstrating the potential involvement of thiols in O_2 reduction and H_2O_2 formation in mash. As O_2 consumption could not be diminished completely it is proposed that further malt-derived substances are involved in these mechanisms. The data of this work contribute to the understanding of the behavior of thiols in oxidative processes during mashing and can be used for further approaches to increase wort and beer quality.

Short biography

Prof. Dr. Frank-Jürgen Methner was graduating with a Diploma in Engineering at Technical University of Berlin in 1881. From 1982 to 1986 he was PhD-Student at TU Berlin graduating 1987. From 1987 til 2004 he was working as a Director at Bitburger Brauerei, Bitburg, Germany, with responsibilities in Brewing Technology, Development and Quality Management. Since winter-semester 2004/2005 he is leading the Department of Brewing Science at the TU Berlin.

P119 - A NEW METHOD FOR SCREENING INDUSTRIAL YEASTS WITH FERULIC ACID DECARBOXYLATION ACTIVITY BASED ON SNP MARKERS

Chen Lu[1]; Hua Yin[1]; Jianjun Dong[1]

1 - State Key Laboratory of Biological Fermentation Engineering of Beer

Abstract
The compound 4-vinylguaiacol, which is produced by the decarboxylation of ferulic acid in certain yeasts, is an important flavour compound in the brewing industry. This decarboxylation ability varies significantly in different beer yeasts and most strains possessing this ability are classified as ale yeasts. In this study, both lager and ale strains were used for fermentation tests and the decarboxylation ability was quantified. Two lager and three ale strains were identified with high ferulic acid decarboxylation activity. Single nucleotide polymorphism(SNP) analysis of PAD1 and FDC1 was conducted by gene sequencing and alignment. In addition, transcription analysis of the two genes was also performed with a GeXP genetic analysis system. A positive relationship between SNPs and ferulic acid decarboxylation activity was determined, and the premature termination codon introduced by base substitution and insertion resulted in the loss of this ability in beer yeasts, whereas other SNPs had a relatively limited influence on it. The GeXP results revealed that genes containing nonsense mutations exhibited low transcription levels, which could be explained by nonsense-mediated mRNA decay. These findings will further the understanding of the relationship between SNPs and biological traits, and will hopefully provide a promising strain selection method for industrial yeasts based on SNP markers.

Short biography

Lu Chen studied Food Science at Ocean University of China. From 2008-2013 she completed her PhD thesis on the influencing mechanism and detection method of PYF of barley malt. In 2010 she began employment as a senior project manager at Tsingtao Brewery Corporation and from 2012 she was responsible for the organisation and implement of research direction in yeast and fermentation at the state key State Key Laboratory of Biological Fermentation Engineering of Beer and focused on the research on yeast metabolism, strain screening, brewing technology and microbiological control.

P123 - REDUCTION OF DIACETYL FORMATION DURING BREWING USING CRISPR/CAS9 GENETICALLY MODIFIED YEAST

Konstantina Giannakou[1]; Ida Kallehauge Nielsen[1]; Christopher Workman[1]; Tomas Strucko[1]; Mikael A. Petersen[2]; Timothy Hobley[1]; Jakob B. Hoof[1]; José L. Martinez[1]

1 - Technical University of Denmark; 2 - University of Copenhagen

Abstract

The formation of diacetyl in beer causes a buttery off-taste. Silencing the genes ILV2 and ILV6, which are responsible for the synthesis of the diacetyl precursor, α-acetolacetate, has previously been shown to have great potential in reducing the diacetyl content. Breeding or mutating yeast with silenced genes is a rather time consuming process. In this work, we have tested the gene editing potential of the CRISPR/Cas9 system for this purpose. Single and double knock-out mutants were created. Results show that two base pairs were successfully removed, which also caused a frameshift mutation. No extra DNA residues remained after the manipulations resulting in a strain, which appears to have suffered a "natural" mutation. . The time frame for this was approximately 3 months, which is much reduced compared to classical strain improvement. The yeast strains were tested for beer fermentation in small scale conditions. GC-MS analysis indicated an up to 50 % reduction in diacetyl in beer produced with mutant strains compared to wild type. This study provides a novel way of constructing yeast strains with reduced off-flavor production during fermentation.

AUTHOR INDEX

AUTHOR INDEX

AUTHOR INDEX